Conference collection

AF771052

Nuclear Weapons and Related Security Issues

Washington, DC, USA
21-22 April 2017

Editors

Pierce Corden
Emeritus State Department, Bethesda, MD, USA

Tony Fainberg
Emeritus Homeland Security Department, Bethesda, MD, USA

David Hafemeister
Emeritus California Polytechnic St. Univ., San Luis Obispo, CA, USA

Allison Macfarlane
George Washington University, Washington, DC, USA

All papers have been peer reviewed.

AIP Publishing

Melville, New York, 2017
AIP Conference Proceedings

Volume 1898

To learn more about AIP Conference Proceedings visit **http://proceedings.aip.org**

Editors

Pierce Corden
Emeritus State Department
4908 Glen Cove Parkway
Bethesda, MD 20816
USA

Email: Pierce.corden@yahoo.com

Tony Fainberg
Emeritus Homeland Security Department
4500 Wetherill Road
Bethesda, MD 20816
USA

Email: Fainberg666@verizon.net

David Hafemeister
Emeritus California Polytechnic State University
Physics Department
553 Serrano Drive
San Luis Obispo, CA 93405
USA

Email: dhafemei@calpoly.edu

Allison Macfarlane
George Washington University
Elliott School of International Affairs
Suite 403
1957 E St., NW
Washington, DC 20052
USA

Email: amacfarlane@gwu.edu

Cover image data courtesy of Edward Ifft.

ISBN 978-0-7354-1586-7
ISSN 0094-243X
Printed in the United States of America

AIP Conference Proceedings, Volume 1898

Nuclear Weapons and Related Security Issues

Table of Contents

PREFACE

Nuclear Weapons and Related Security Issues

At the end of the Cold War, in the early 1990's, it was generally hoped that the excessive dangers of nuclear weapons were being addressed with the significant reductions of nuclear weapons by Russia and the U.S. The considerable cooperation between nuclear laboratories of the two states greatly constrained the "loose nukes" problem. On campuses, this optimism reduced the interest in the nuclear weapon topics, as it was naively moved beyond these issues to those of energy and environment. Unfortunately, we now see a world that is more threatened by the proliferation of nuclear weapons, the rise of terrorism, and the increased separation of Russia and its actions.

The two-tiered Nuclear Non-Proliferation Treaty (NPT) commits the five nuclear-weapon parties (the permanent members of the UN Security Council: US, UK, Russia, China and France – the P5) not to assist any state in acquiring nuclear weapons, while non-nuclear-weapon parties commit not to acquire nuclear weapons and to implement IAEA safeguards to prevent "diversion of nuclear energy for peaceful uses to nuclear weapons…." India, Pakistan, Israel and North Korea are not NPT parties. This requires dealing in parallel with the capabilities of these four non-NPT states, the P5 states and the 185 non-nuclear NPT parties.

Many now believe that a nuclear weapons confrontation is more likely to happen, but at a lower level of confrontational intensity. South Asia is often mentioned as a possible venue for nuclear weapon use. We don't know if this is true, but there is some evidence for this conclusion. The global nuclear nonproliferation regime is challenged today, as it has been over the past several decades. It has had to deal with events in India, Iran, Iraq, Israel, Libya, North Korea, Pakistan, South Africa, and Syria.

This is the fifth short course *proceedings* on nuclear weapons offered by the *Forum on Physics and Society* over the past thirty–five years. A list of the courses follows.

Short Course on the ARMS RACE, San Francisco Hilton Hotel (*El Corale Press*, Cal Poly U, 1982, 179 p.)

Physics, Technology and the Nuclear Arms Race, Baltimore Convention Center (*AIPCP* 104, 1983, 372 p.)

Nuclear Arms Race Technologies in the 1990s, George Washington University (*AIPCP* 178, 1988, 476 p.)

Nuclear Weapon Issues in the 21st Century, George Washington University (*AIPCP* 1596, 2014, 272 p.)

Nuclear Weapons and Related Security Issues, George Washington University (*AIPCP* 1898, 2017, 299 p.)

The 5th short course was held on April 21–22, 2017 at the Elliott School of International Affairs of The George Washington University in Washington, DC. It was cosponsored by the APS *Forum on Physics and Society*, the GWU Elliot School of International Affairs, GWU Nuclear Science and Security Consortium, American Association of Physics Teachers and Federation of American Scientists. It was organized by Pierce Corden (former State/ACDA, AAAS), Tony Fainberg (former Department of Homeland Security and Institute for Defense Analyses), David Hafemeister (former SFRC–Senate, State/ACDA, NAS, Cal Poly) and Allison Macfarlane (GWU Elliott School, former Chair, Nuclear Regulatory Commission). The 2017 event covered nuclear weapon issues in four sessions. The 2017 short course power point presentations can be downloaded from the GWU Elliott School, located at https://blogs.gwu.edu/nuclear-policy-talks, and click on Short Course on Nuclear Weapons and Related Security Issues and then click here. The 2013 power points are at http://elliott.gwu.edu/nuclear-weapon-issues-21st-century.

The short course examined the following issues:

Strategic Nuclear Weapons: US policy towards the nuclear forces of Russia and China, alert status of nuclear weapons, nuclear modernization trends, global–strike hypersonic weapons, the monitoring of nuclear warheads and materials, Richard Garwin's contributions, the history of the Soviet H–bomb, Sakharov and Gorbachev nuclear reductions, the history of the Soviet H–bomb, cooperation between US and Russian nuclear weapon laboratories.

Nuclear Weapons and Related Security Issues
AIP Conf. Proc. 1898, 010001-1–010001-2; https://doi.org/10.1063/1.5009205
Published by AIP Publishing. 978-0-7354-1586-7/$30.00

Comprehensive Nuclear-Test-Ban Treaty: Fundamentals of the CTBT, seismic monitoring, radioxenon and radionuclide monitoring, CTBT on-site inspections, stockpile stewardship, the connection between the proliferation of nuclear weapons and the global ban on nuclear testing.

Multilateral Arms Control: The control of conventional arms, space weapon technology, ballistic missile defense effectiveness, chemical and biological weapon controls.

Nuclear Proliferation: Evolution of the non-proliferation regime, the nuclear programs of North Korea and Iran, the future of the NPT regime, the role of safeguards to ensure compliance, the Argentina-Brazil four party approach, proliferation from nuclear power infrastructure.

Terrorism: Probabilistic projections of terrorism, the threat of nuclear terrorism, technologies to counter nuclear terrorism, technologies to counter aviation security threats, drone warfare, summary of countering terrorist threats.

Acknowledgement

The participants of the short course on nuclear weapons issues appreciate the kind and gracious hospitality shown by the GWU Elliott School of International Affairs. We are particularly thankful for the help of Samantha D'Introno and Zlata Gogoleva. The bulk of this book was written by our faculty, with bio-briefs in Appendix B. In addition, we thank those who contributed their *Physics Today* reprinted papers: Matthias Auer, Edward Geist, Robert Hanrahan, Kirk Levedahl, Mark Prior, Victor Reis, Frank von Hippel, Alex Wallerstein. In addition Sig Hecker summarized his recent book, *Doomed to Cooperate*, on the US/Russian nuclear cooperation.

Pierce Corden
Former State Dept.
Bethesda, MD

Tony Fainberg
Former Homeland Security
Bethesda, MD

David Hafemeister
Cal Poly Univ.
San Luis Obispo, CA

Allison Macfarlane
Elliott School, GWU
Washington, DC

Left to Right: 120 attendees. Co-organizers: Tony Fainberg, David Hafemeister Allison Macfarlane, Pierce Corden.

The Future of U.S.-Russia Nuclear Arms Control

Steven Pifer

Senior Fellow
The Brookings Institution
Washington, DC 20036, USA

Abstract: Nuclear arms control has long made contributions to U.S.-Soviet and U.S.-Russian security, but the current regime is at risk. The 1987 Intermediate-range Nuclear Forces Treaty may be headed for collapse. Both the United States and Russia are modernizing their strategic forces, and the fate of the 2010 New Strategic Arms Reduction Treaty is unclear. In the unlikely case that the sides are prepared to go beyond New START, there are ways to address further reductions and related issues. A collapse of the arms control regime, on the other hand, would mean the end of constraints on U.S. and Russian nuclear forces, a significant loss of transparency, and potential costs to U.S. security.

INTRODUCTION

Nuclear arms control has long made significant contributions to security and stability between the United States and Soviet Union, and later between the United States and Russia. Those contributions date back to the 1960s, when Washington and Moscow launched the first Strategic Arms Limitation Talks (SALT). The specific treaties and agreements resulting from the negotiations were valuable, as was the dialogue itself, which helped the sides come to a common understanding of terms such as strategic stability (a situation in which the sides do not have incentives to use nuclear weapons first in a crisis).

Unfortunately, the bilateral nuclear arms control is now at risk. After briefly recounting the history of nuclear arms control negotiations between Washington and Moscow, this paper examines the challenges to the regime, describes steps to preserve it, outlines measures that could build on the regime to achieve further reductions in nuclear weapons, and notes the potential costs to U.S. security should the regime collapse.

A SHORT HISTORY OF U.S.-RUSSIA NUCLEAR ARMS CONTROL

The SALT negotiations, launched in the late 1960s, produced two agreements in 1972. The Anti-Ballistic Missile (ABM) Treaty banned nationwide missile defenses and limited the United States and Soviet Union each to two ABM deployment sites, each with no more than 100 missile interceptor launchers (a 1974 protocol limited each side to a single deployment site of no more than 100 launchers). The second agreement, the Interim Offensive Agreement, limited each side to the number of intercontinental ballistic missile (ICBM) silos existing as of July 1, 1972 and limited each side to the number of submarine-launched ballistic missile (SLBM) launch tubes on ballistic missile submarines existing or under construction as of May 26, 1972.

As a result of the Interim Offensive Agreement, the Soviet Union was permitted a larger number of ICBM and SLBM launchers. This was offset by the fact that the United States had begun the process of putting multiple-independently targetable reentry vehicles (MIRVs) on its missiles, so that the Minuteman III ICBM, for example, could carry three warheads instead of one. The Interim Offensive Agreement also did not constrain strategic bombers, an area of American advantage.

Negotiations began soon thereafter on a SALT II Treaty. They concluded in 1979, producing a treaty that limited each side to no more than 2,250 strategic nuclear delivery vehicles, defined as ICBM launchers, SLBM launchers and strategic bombers. Sublimits constrained the number of MIRVed ballistic missile launchers.

The Interim Offensive Agreement and the SALT II Treaty limited the numbers of launchers (with SALT II including strategic bombers), but neither agreement constrained warheads. Over the course of the 1970s and 1980s,

Nuclear Weapons and Related Security Issues
AIP Conf. Proc. 1898, 020001-1–020001-11; https://doi.org/10.1063/1.5009206
Published by AIP Publishing. 978-0-7354-1586-7/$30.00

both sides significantly increased the numbers of their strategic nuclear weapons by placing MIRVs on their ICBMs and SLBMs and placing air-launched cruise missiles (ALCMs) on their strategic bombers.

TABLE 1: Operational U.S. and Soviet Strategic Offensive Forces, 1974[1]

Type of forces	U.S.	USSR
ICBM launchers	1,054	1,575
SLBM launchers	656	660
Intercontinental bombers	496	140

TABLE 2: U.S. and Soviet Strategic Offensive Launchers, 1979[2]

Type of forces	U.S.	USSR
ICBM launchers	1,054	1,398
MIRVed ICBM launchers	550	608
SLBM launchers	656	950
MIRVed SLBM launchers	496	144
Heavy bombers	573	156
Heavy bombers equipped to carry ALCMs	3	0

SALT II never entered into force. The Carter administration withdrew it from consideration after the discovery of a Soviet combat brigade in Cuba and the Soviet invasion of Afghanistan led to a downturn in U.S.-Soviet relations. President Reagan took office in 1981 having called SALT II "fatally flawed," but his administration pursued a policy of not undercutting the SALT II limits until 1986.

The period 1987-1992 saw major breakthroughs on nuclear arms reductions. Following the U.S. deployment in Europe of Pershing II ballistic missiles and ground-launched cruise missiles (GLCMs) to counter the Soviet SS-20 MIRVed ballistic missile, the United States and Soviet Union concluded the Intermediate-range Nuclear Forces (INF) Treaty in 1987. That treaty banned all U.S. and Soviet ground-launched ballistic and cruise missiles with ranges between 500 and 5,500 kilometers, including the Pershing IIs, GLCMs and SS-20s. By 1991, the two countries had destroyed some 2,700 missiles pursuant to the treaty.

Washington and Moscow concluded the Strategic Arms Reduction Treaty (START, later referred to as START I) in 1991. That treaty limited each side to no more than 6,000 accountable warheads on no more than 1,600 ICBM launchers, SLBM launchers and strategic bombers. START I required both sides to reduce both strategic nuclear delivery vehicles and total accountable warheads (each side was estimated to have about 10,000 strategic warheads at the end of the 1980s). Following the collapse of the Soviet Union at the end of 1991, Russia, Belarus, Kazakhstan and Ukraine agreed in the May 1992 Lisbon Protocol to take on the Soviet START I obligations. The latter three also agreed to eliminate all of the nuclear weapons on their territory, leaving Russia as the only nuclear weapons state in the post-Soviet space.

TABLE 3: U.S. and Soviet Strategic Offensive Forces, 1990[3]

Type of forces	U.S.	USSR
Deployed ICBM launchers	1,000	1,398
Warheads attributed to ICBM launchers	2,450	6,612
Deployed SLBM launchers	672	940
Warheads attributed to SLBM launchers	5,760	2,804
Deployed heavy bombers	574	162
Warheads attributed to heavy bombers	2,353	855
Total launchers and bombers	2,246	2,500
Total warheads attributed to launchers and bombers	10,563	10,271

In addition to formal agreements, the United States, Soviet Union and, in 1992, Russia undertook unilateral measures—referred to as the presidential nuclear initiatives—to reduce nuclear arms. Those initiatives reduced thousands of nuclear weapons, particularly non-strategic nuclear weapons not covered by START I or the INF Treaty.

The United States and Russia concluded the START II Treaty in early 1993. That agreement limited each side to no more than 3,000-3,500 accountable strategic warheads and banned heavy ICBMs and MIRVed ICBMs. The treaty, however, became entangled in U.S.-Russian differences over issues such as missile defense and NATO military action against Serbia in 1999. It never entered into force.

The George W. Bush administration took office in 2001 with a different approach to arms control. Wanting to deploy a national missile defense in a manner that would not be consistent with the ABM Treaty, the administration withdrew from the treaty in 2002. Although the Bush administration had little interest in formal limitations on strategic offensive forces, the president acceded to a request from Russian President Putin, and, following a short negotiation, the United States and Russia concluded the Strategic Offensive Reductions Treaty (SORT) in May 2002. The treaty limited each side to no more than 1,700-2,200 operationally deployed strategic warheads, but it contained no agreed definitions, counting rules or verification measures; as a result, it was unclear whether the two sides even counted the same weapons under the limit. (The START I Treaty, however, was still in effect until December 2009.)

TABLE 4: U.S. and Russian Strategic Offensive Forces, July 2009[4]

Type of forces	U.S.	Russia
Deployed ICBM launchers	550	465
Warheads attributed to ICBM launchers	1,600	2.001
Deployed SLBM launchers	432	268
Warheads attributed to SLBM launchers	3,264	1,288
Deployed heavy bombers	206	76
Warheads attributed to heavy bombers	1,052	608
Total launchers and bombers	1,188	796
Total warheads attributed to launchers and bombers	5,916	3,897

President Obama sought to accelerate nuclear arms reductions and, in 2009, laid out a vision for a world without nuclear weapons. He also he made clear that much had to happen first and that, as long as nuclear weapons existed, the United States would maintain a safe, secure and effective nuclear deterrent. In May 2010, the United States and Russia concluded the New START Treaty (sometimes referred to as START III in Russia). That treaty limits the United States and Russia each to no more than 1,550 deployed strategic warheads on no more than 700 deployed strategic delivery vehicles. A deployed warhead is a warhead on a deployed ICBM or SLBM; each deployed strategic bomber is accounted as one deployed warhead. (Deployed ICBMs and SLBMs are ballistic missiles in silos or submarine launch tubes or on mobile ICBM launchers.) The treaty also constrains each side to no more than 800 deployed and non-deployed ICBM and SLBM launchers and strategic bombers (non-deployed ICBM and SLBM launchers are launchers that contain no missiles). New START entered into force in February 2011. Its limits take full effect in February 2018, and it has a term of ten years—that is, until February 2021. The treaty can be extended by up to five years by agreement by the sides.

TABLE 5: U.S. and Russian Strategic Forces under New START, March 2017[5]

Type of forces	U.S.	Russia
Deployed ICBMs, deployed SLBMs and deployed heavy bombers	673	523
Warheads on deployed ICBMs and SLBMs, and warheads counted for deployed heavy bombers	1,411	1,765
Deployed and non-deployed launchers of ICBMs and SLBMs, and deployed and non-deployed heavy bombers	820	816

TABLE 6: Planned U.S. Strategic Forces, 2018

Type of system	Deployed Strategic Delivery Vehicles	Deployed & Non-Deployed Launchers & Bombers
Minuteman III ICBMs	400	454
Trident II D-5 SLBMs	240	280
B-2/B-52H bombers	60	66

The two agreements currently governing U.S. and Russian nuclear weapons are the INF Treaty and New START. The Obama administration sought to go beyond New START and achieve further reductions, proposing a new round of U.S.-Russian negotiations that would include not just deployed strategic warheads but also non-strategic nuclear weapons and reserve strategic warheads, opening a possibility that the two countries might discuss their entire nuclear stockpiles. Russia, however, declined to negotiate on further reductions, raising instead its concerns about issues such as U.S. missile defense (particularly in Europe), advanced conventional strike weapons and third-country nuclear forces. Whether these issues are genuine concerns or a Russian pretext not to engage in further nuclear arms cuts is unclear; it is probably a mixture of both.

THE INF TREATY IN PERIL

By 2017, the overall U.S.-Russian relationship had fallen to its lowest point since the end of the Cold War. The countries differed sharply over issues such as Russia's aggression against Ukraine and the conflict in Syria. Moscow has adopted a more threatening rhetoric, including nuclear saber-rattling, and the number of encounters between NATO and Russian military aircraft and warships has increased dramatically. Moreover, many quarters in the United States remain unhappy about Russian interference in the 2016 presidential election. Arms control is possible when political relations are strained, but it is difficult.

An urgent question is the fate of the INF Treaty. Russia assumed the Soviet INF Treaty obligations following the collapse of the USSR in 1991 (as did Belarus, Kazakhstan and Ukraine). In 2007, senior Russian officials began to express concern about the treaty, noting that it banned the United States and Russia from having intermediate-range missiles, while third countries were beginning to acquire them in significant numbers. Those countries included China, North Korea, India, Pakistan, Iran and Israel—all closer to Russia than the United States.

Given the large number of Russian strategic forces and other nuclear weapons not constrained by the INF Treaty, it was unclear why Moscow needed intermediate-range missiles to counter third countries. In late 2007, the Russians proposed to multilateralize the treaty, an idea briefly endorsed by the United States, but the idea gained no traction with third countries.

In 2014, the Obama administration charged Russia with violating the treaty by developing a prohibited ground-launched cruise missile (press reports indicated that the Russians had tested the missile). The administration sought to bring Moscow back into full compliance with the treaty but reported no progress. In early 2017, press reports alleged that Russia had begun deploying the intermediate-range GLCM. In March, the vice chairman of the Joint Chiefs of Staff confirmed that.

The charge apparently concerns an extended range version of the SSC-7 (Iskander K) cruise missile. (The Iskander K has a range of less than 500 kilometers and thus is unconstrained by treaty.) The extended range missile is reportedly called the SSC-8.

Moscow consistently denies any violation of the INF Treaty. The Russians claim that U.S. officials have not provided specific information to back up the American charge. U.S. officials state that the Russians have received enough information so that they could understand the charge. (Apparently due to concerns about protecting sources and methods, Washington has not released all of the information that it has collected regarding the violation.)

For its part, Moscow levels three charges regarding U.S. violations of the INF Treaty, some of which date back many years. First, the Russians assert that ballistic missiles used as targets in U.S. missile defense tests are intermediate-range ballistic missiles prohibited by the treaty. Second, they claim that armed, unmanned aerial vehicles (drones) are in effect GLCMs banned by the treaty. Third, the Russians charge that the Mk-41 vertical launch system used for the "Aegis Ashore" SM-3 missile interceptor site in Romania (and to be made operational in Poland in 2018) could launch intermediate-range cruise missiles as well as SM-3 interceptor missiles.

Obama administration officials rejected the Russian charges. They allowed, however, that, if Moscow were to address the Russian violation in a serious manner, the U.S. government would be prepared to consider measures to assuage Russian concerns, such as transparency visits to the SM-3 interceptor site.

The Trump administration is reviewing the INF Treaty issues. One question for it to consider is whether to continue to seek to bring Russia back into compliance, a more difficult challenge now that Russia has moved from testing to deploying its GLCM. That may not be possible, but maintaining the treaty would appear to be in the U.S. interest, particularly as the Pentagon has no formal requirement for an intermediate-range missile and no plans to build one. In any case, it would be unwise for the United States to withdraw at this point. Given the limited information available on the Russian violation, Washington could end up getting the blame for the treaty's demise, while Moscow would be free to deploy intermediate-range missiles, perhaps adding a ballistic missile in the 500-5,500 kilometer range band.

Bringing Russia back into compliance requires that the United States develop some leverage with Moscow. This could consist of military and/or diplomatic steps.

Certain military steps make sense in any case. The Pentagon needs to explore what measures can be taken to defend against cruise missiles. Even if Russia reverses its deployment of the intermediate-range GLCM, it is beginning to replicate U.S. deployments of conventionally-armed air- and sea-launched cruise missiles. Third countries are also beginning to develop cruise missile capabilities.

Some—including members of Congress—have suggested that the United States develop its own intermediate-range missiles as a counter to the Russian GLCM. The prospect of a U.S. Pershing III or new GLCM in Europe could give the Kremlin pause and remind the Russians how much they did not like American intermediate-range missiles in Europe in the 1980s.

Before proceeding down this path, however, the administration and Congress should consider two questions. First, can the Pentagon budget, already facing many competing demands, afford new intermediate-range missiles? Second, if the missiles were built, would NATO agree to deploy them? NATO fell into crisis over deploying U.S. missiles in 1983, and there is no reason to think that today's alliance would be any more unified on the question. Intermediate-range missiles in the United States pose no threat to Russia.

A second option would be to position in the European area conventionally-armed weapons systems that are not prohibited by the INF Treaty. For example, the U.S. Air Force could deploy conventional B-52 or B-1 bombers periodically to Fairford Air Base in Britain and stockpile Joint Air-to-Surface Standoff Missiles (JASSMs) there for potential use by the bombers. The U.S. Navy could increase the cruises of surface ships and submarines carrying sea-launched cruise missiles in the northern European area. Calls by the USS Florida or USS Georgia—converted Trident submarines each capable of carrying up to 154 sea-launched cruise missiles—in ports around the North Sea would get the Kremlin's attention. The U.S. Navy might even consider home-porting warships with sea-launched cruise missiles in Europe.

These measures, since they do not contravene the INF Treaty, would likely be more acceptable in Europe. They would be cheaper and could be taken more quickly than building new intermediate-range missiles. The measures could be reversed if Russia ceased its violation of the treaty.

Washington should also consider diplomatic steps. The Russian violation is a treaty issue between Washington and Moscow, but Russian intermediate-range GLCMs cannot reach the United States (Alaska might be at some risk, but only if the Russian military deployed missiles where it has not done so in the past). As with the SS-20 in the 1970s and 1980s, the direct threat is to the countries of Europe and Asia.

The U.S. government should engage its allies in NATO and in the Far East, friends such as Sweden and Finland, and other countries such as China, inform them about the details of the Russian violation, and point out that they are potential targets of the new missile. The goal should be to multilateralize the problem—that is, to increase pressure on Moscow from third countries who should not want to see a new threat to their security. Would this suffice to change the Kremlin's calculation? That is an unknown. But having countries other than just the United States complain to Russia about the violation—preferably at senior levels—would increase the prospects that Moscow might reconsider.

If there were political decisions in Washington and Moscow to address the other side's concerns in a serious way and preserve the INF Treaty, solutions could be found to the violations that each has charged, restoring full compliance with the treaty.

The Special Verification Commission (SVC) was established by the treaty to handle, among other things, concerns about compliance. That is the forum for technical-level talks that could work out solutions to the compliance concerns. Although the SVC includes Belarus, Kazakhstan and Ukraine (also successor states to the Soviet Union as regards the INF Treaty), their presence should not hinder the main work between U.S. and Russian experts.

A starting point for addressing the U.S. charge that Russia has deployed an intermediate-range ground-launched cruise missile would be for the SVC to work out the procedures for an exhibition of the SSC-8 to U.S. experts and a briefing regarding the missile's characteristics and capabilities, especially its range. A particular issue would be the capacity of the missile's fuel tanks. One possibility is that the U.S. experts would be satisfied that the missile was not a treaty violation. (This has an antecedent: in the 1980s, the U.S. government charged the Soviet Union with violating the Threshold Test Ban Treaty by conducting nuclear tests with yields in excess of 150 kilotons. The sides conducted a joint verification experiment in 1988, which entailed on-site monitoring of a Soviet nuclear test. That led to a revision of U.S. algorithms and the conclusion that earlier Soviet tests had not exceeded the 150-kiloton limit.)

If the exhibition did not resolve the question, the issue could be returned to the SVC. U.S. and Russian experts there could consider further steps. Of course, if the exhibition (or subsequent agreed measures) showed that the

SSC-8 had a range of between 500 kilometers and 5,500 kilometers, Russia would have to destroy all missiles and their associated launchers. The INF Treaty spelled out measures to accomplish this, though the SVC might have to review and consider whether those measures needed adaptation.

The SVC likewise would work out procedures for addressing Russian concerns about U.S. compliance. The INF Treaty has a provision that, in effect, allows ballistic missiles that might be of intermediate range to be used as targets in missile defense tests. Presumably, this would be of interest to the Russian military, which may want to test its missile defenses against targets that might otherwise be regarded as prohibited intermediate-range ballistic missiles. Experts in the SVC could craft language that would make clear the distinction between permitted target missiles and prohibited intermediate-range missiles. Additional measures might limit the number of target missiles each side had at any one time and/or constrain them to declared sites associated with missile defense tests.

The difference between a cruise missile, which flies from point A to point B and detonates on the target, and an armed unmanned aerial vehicle, which flies from point A to point B, where it releases a weapon before returning to point A or flying on to point C, seems obvious. However, experts in the SVC could develop language to clarify the difference between prohibited ground-launched cruise missiles and permitted armed unmanned aerial vehicles.

Of the three Russian charges of U.S. violations, the charge concerning the Aegis Ashore appears to have the most substance. Aegis Ashore utilizes the Mk-41 vertical launch system to launch SM-3 missile interceptors. The Russians argue that Mk-41 vertical launch systems on U.S. Navy warships can launch other weapons, including sea-launched cruise missiles and the SM-6, an air defense missile that is being given a secondary surface-to-surface mission.

One way to address the Russian concern would be to modify the Mk-41 vertical launch systems on land so that they were observably different from those on warships; ideally, this would be a functionally-related observable difference rather than a cosmetic one. A second way to address the concern would be for the SVC to work out procedures—which would then require the approval of Romania and Poland—under which Russian experts could periodically visit the Aegis Ashore sites and randomly specify one or two of the 24 Mk-41 launch tubes to be opened to confirm that they held SM-3 missile interceptors, not missiles of some other type.

Working out the procedures and language to resolve the sides' compliance concerns would not be easy and would require political guidance from capitals to SVC experts to find workable solutions. That said, these solutions offer a way to preserve the treaty—provide there are political decisions to do so in Moscow and Washington.

STRATEGIC MODERNIZATION PROGRAMS AND NEW START

Another potential challenge to the U.S.-Russian nuclear arms control regime is that both countries are conducting or on the verge of conducting major programs to modernize all three legs of their respective strategic triads.

Russia is well along in its program. Among other things, it is building the new Borey-class ballistic missile submarine, the Bulava SLBM and the SS-27 ICBM. The Russians are developing a new heavy ICBM—the Sarmat—as a successor the SS-18 and plan to reopen the production line for the Tu-160 Blackjack bomber.

While the Russian strategic modernization program is substantial, much of what the Russians are doing is replacing old weapons with new weapons. Due to the economic contraction that followed the collapse of the Soviet Union in 1991, Russian strategic forces received only limited funding in the 1990s and early 2000s. Had funds been available, Russia likely would have replaced some of its older systems earlier. The current Russian modernization program, moreover, appears sized to fit within the limits of the New START Treaty.

The more worrisome aspects of Russian nuclear policy have to do with the modernization of non-strategic nuclear arms and nuclear doctrine. Russia still maintains a panoply of ground-, air- and sea-based non-strategic nuclear weapons, while the U.S. non-strategic nuclear arsenal contains only the B61 gravity bomb. The modernization of its non-strategic nuclear arsenal, doctrinal questions and loose talk about nuclear weapons by Russian officials have raised concern in the West that Russia may be lowering its threshold for nuclear use.

The U.S. strategic modernization program will accelerate in the 2020s and, like the Russian program, largely consists of replacing old weapons with new. In the next decade, the Pentagon plans to be building the Columbia-class ballistic missile submarine, a new ICBM (referred to as the ground-based strategic deterrent), the B-21 strategic bomber and the long-range standoff option (LRSO—a new nuclear-armed air-launched cruise missile). In addition, the U.S. Navy will continue its life extension program for the Trident II SLBM.

The U.S. program appears sized to fit within the limits established by New START. The B61 gravity bomb is going through a life extension and modernization program, which will result in a single variant for use by strategic bombers and dual-capable aircraft such as the F-35 (dual-capable aircraft can carry nuclear weapons or conventional ordnance).

Some have questioned the size of the planned U.S. program, asking whether the United States needs and can afford to build 12 Columbia-class ballistic missile submarines and a force of 400 deployed ICBMs. The U.S. military plans to have a force of 700 deployed SLBMs, ICBMs and bombers under New START, where a lower number could—at some cost in flexibility—still carry 1,550 deployed strategic warheads and result in significant cost savings.

Particular questions have arisen about the LRSO. The U.S. Air Force developed nuclear-armed air-launched cruise missiles in the 1970s because of concern that the B-52 could not penetrate advanced Soviet air defenses. The B-21, however, supposedly will have advanced stealth and electronic warfare capabilities that should allow it to defeat sophisticated air defenses, which would appear to weaken the case for the LRSO.

The Trump administration's view on these issues is not yet known. Secretary of Defense Mattis has expressed support for new ballistic missile submarines and ICBMs as well as for the B-21 bomber, but he has said that he wants to review the requirement for the LRSO. The Trump administration has launched a nuclear posture review and a ballistic missile defense review, which may be completed by the end of 2017. The likelihood that the administration will choose to reduce the planned modernization program appears to be low absent a new arms reduction agreement or major budget problems for the Department of Defense. While some non-governmental experts have called for eliminating the ICBM leg of the triad, the nuclear posture review almost certainly will reaffirm that the three legs should be maintained.

Each leg of the triad has its advantages. SLBMs on board ballistic missile submarines at sea are the most survivable leg, and they carry the largest portion of U.S. deployed strategic warheads. ICBMs provide a hedge against an anti-submarine warfare breakthrough and confront a potential adversary with the fact that attacking the ICBMs would require pouring hundreds of nuclear warheads into the American heartland, guaranteeing a U.S. nuclear response. Bombers can be used in conventional operations and can be used for signaling purposes, e.g., the overflight of strategic bombers over South Korea sends a message to Pyongyang.

President Trump's late January phone call with President Putin raised concern about the status of New START; Mr. Trump reportedly did not know what the treaty was and then dismissed it as a bad deal. Just a few weeks later, however, the commander of Strategic Command and vice chairman of the Joint Chiefs of Staff made clear that the U.S. military supports and wants to preserve New START. (In addition to the caps on Russian strategic forces, the U.S. military appreciates the information that New START provides in semi-annual data exchanges, notifications and the right to conduct up to 18 inspections per year in Russia.)

A White House official in June said the United States is on track to meet the New START limits by the February 2018 deadline and assessed that Russia likewise will be within the limits by that date. As for the question of extending New START by up to five years after its 2021 expiration date, the U.S. government likely will consider that question after completion of the nuclear posture review.

The U.S. military likely will favor extension, and Moscow traditionally prefers to have some cap on U.S. strategic nuclear forces. The extension question thus may turn on views elsewhere in the Trump administration. Congress may also have a view, though Congressional approval is not required for extension. Should there continue to be questions regarding Russian compliance with the INF Treaty, or if that treaty has collapsed, that could decrease the likelihood of New START extension.

GOING BEYOND NEW START?

While it may appear unlikely that the Trump administration would seek to negotiate reductions going below New START levels, the possibility should not be totally discounted. Further nuclear reductions, properly structured, could have advantages for the United States.

Reductions would reduce nuclear risk and the number of nuclear weapons that might be targeted at the United States. A well-designed agreement could result in force structures that were more stabilizing. Reductions could result in substantial cost savings. They would demonstrate U.S. commitment to disarmament and the Non-Proliferation Treaty—an important point for non-nuclear weapons states.

It should be noted, however, that resolution of compliance concerns related to the INF Treaty would almost certainly be a precondition for any negotiation of further reductions.

Secretary of State Tillerson and Russian Foreign Minister Lavrov have agreed in principle on resuming strategic stability talks, though the details remain to be worked out. Ideally, those talks will be wide-ranging. They may have value even if they do not produce negotiations aimed at specific agreements, in that that they can help the sides better understand the other's concerns and motivations.

If the U.S. government were interested in seeking further nuclear reductions, including non-strategic nuclear weapons and reserve strategic warheads, one approach would seek to negotiate a single aggregate limit that would cover all U.S. and Russian nuclear weapons (with the possible exception of those in the elimination queue, which could be covered by a different regime).

For example, the sides might agree that each could have no more than 2,200 total nuclear weapons—approximately half of what each side is believed to have at present. Within that aggregate limit of 2,200, there could be a sublimit of no more than 1,000 deployed strategic warheads. Above the 1,000 sublimit, each side would be free to choose its mix of non-strategic nuclear weapons and reserve strategic warheads within the overall limit of 2,200.

As for limits on delivery vehicles, the sides could seek to reduce the limit on deployed strategic delivery vehicles from New START's 700 to no more than 500. Given that most delivery vehicles for non-strategic nuclear arms are dual-capable, it would be best not to try to constrain such systems.

Such a treaty would require more intrusive verification measures than New START, in particular for monitoring the number of nuclear weapons not on deployed strategic delivery vehicles, e.g., at storage sites. Negotiating such measures would prove difficult but perhaps not insurmountable, if all non-deployed nuclear weapons were restricted to declared storage sites.

Such a regime might hold some interest for the Russians. First, it does not appear that they will deploy 700 strategic delivery vehicles as allowed by New START; many projections suggest they will be at 500-550. Reducing the limit on deployed strategic delivery vehicles down to 500 thus would eliminate a U.S. numerical advantage. Second, the lower number of permitted deployed strategic delivery vehicles would reduce the current U.S. advantage in "upload" potential. Most U.S. SLBMs and ICBMs have been "downloaded" so that they carry fewer warheads than they are capable of carrying. Since the United States maintains a large number of reserve strategic warheads, if New START broke down or if there was a major crisis, the U.S. military could upload warheads on to its SLBMs and ICBMs, significantly increasing the total number of deployed strategic warheads.

However, these points alone might not be sufficient for the Russian side. If—and it very much remains an "if "question—the Trump administration desired to seek further nuclear arms reductions, it would likely need to address at least some Russian concerns on issues such as missile defense and precision-guided conventional strike.

Missile defense has long been a problematic issue on the U.S.-Russia agenda. Curiously, Moscow seems to be more concerned about the deployment in Europe of U.S. SM-3 missile interceptors, which lack the velocity and are ill placed to defend against Russian strategic ballistic missile warheads, than about the ground-based mid-course defense, which deploys ground-based interceptors in Alaska and California designed to defend against strategic ballistic missile warheads.

In any event, were the Trump administration to want to pursue a negotiation with Russia on additional nuclear reductions beyond those required by New START, it likely would have to take some steps in the area of missile defense.

One step might be to reiterate the U.S. offer, made in 2013, of an executive agreement on transparency regarding missile defense. The Obama administration proposed that the United States and Russia annually exchange data that, for key elements of missile defenses (e.g., missiles, launchers and radars) would specify the current number and the number projected for each year for the succeeding ten years. The idea underlying this proposal was to share sufficient information so that each side could understand whether the other's missile defenses posed a threat to its strategic offensive forces. If it saw such a threat emerging, it would have sufficient time to take compensatory measures.

A second step might entail a cap on the number of SM-3 interceptors to be deployed in Europe. Right now, the European phased adaptive approach is open-ended as to numbers, though the current plan is for 24 interceptors each in Romania and Poland, in addition to those on U.S. warships operating in European waters. The U.S. government could consider, after consultation with NATO, articulating to the Russians the maximum number of SM-3 interceptors it would deploy in the European area. It could, of course, add a caveat: the level might have to change if there were a dramatic change in the missile threat to Europe.

Another step to consider is altering the European phased adaptive approach. The Joint Comprehensive Plan of Action limits Iranian nuclear capabilities, not its ballistic missile program. However, an Iranian ballistic missile with a conventional warhead poses a significantly lesser threat than one with a nuclear warhead. The U.S. government could consider, after consultation with NATO and, in particular, with Poland, completing work on the SM-3 site in Poland but not deploying SM-3 interceptors there, again provided that there was no dramatic change in the missile threat to Europe. (The United States in that case should consider deploying some other military capability in Poland with at least the same number of personnel.)

These three steps could move in the direction of addressing Russia's expressed concern about missile defense in Europe. That said, it is difficult to see the Trump administration embracing such ideas and, if it did, Congressional Republicans might express concern.

If there were a serious discussion of missile defense, Moscow would have to give up its demand for legally-binding limits—that is, a treaty. The current political reality in Washington is that no treaty that limited U.S. missile defenses—even if the limits did not affect U.S. programs in any meaningful way—would have a chance of receiving consent to ratification in the Senate.

A second area that would likely have to be addressed in the context of additional reductions of nuclear arms would be precision-guided conventional strike weapons. The Russians have expressed concern about such systems, with some analysts fearing that the United States might be tempted in a crisis to strike Russian strategic targets with conventional weapons, hoping that it could do so without risking nuclear retaliation.

Were the United States (or Russia, for that matter) to decide to place conventional warheads on ICBMs or SLBMs, those warheads would be captured by New START's limit of 1,550 deployed strategic warheads. That limit makes no distinction between nuclear and conventional warheads.

Both the United States and Russia are working on hypersonic glide vehicles (HGVs). These are launched by a ballistic missile but do not fly a ballistic trajectory and hence are not captured by New START's definitions and limits. HGVs instead dive back down and "glide" at high speeds along the upper reaches of the atmosphere. If successfully developed, HGVs would prove to be expensive means for delivering conventional weapons to long-range targets; indeed, the Obama administration described such weapons as filling a "niche" need. Small numbers might make long-range HGVs susceptible to limitation in the context of a new agreement reducing nuclear arms. U.S. and Russian officials have a chance to discuss possible constraints on these systems before they are fully developed or deployed.

The tougher near-term challenge will be conventionally-armed air- and sea-launched cruise missiles. The United States deploys these in large numbers, and Russia now deploys comparable systems. Both militaries likely regard these weapons as key elements of their power projection capabilities, and they would be very difficult to limit. However, as a starting point, U.S. and Russian officials might begin a dialogue on conventionally-armed air- and sea-launched cruise missiles and their impact on the nuclear balance between the two countries.

Russian officials have also called for the inclusion of third-country nuclear forces in any future nuclear arms reduction agreement. This would prove difficult. Just starting with Britain, France and China, the disparity in nuclear weapons numbers between those countries and the United States and Russia is so large as to make negotiation of equal limits an impossible task. That may be why Russian officials, who call for a multilateral negotiation, have never offered a proposal for how such a multilateral nuclear arms control agreement would be structured.

A more plausible way to address this issue would be for the United States and Russia, in the context of a new bilateral nuclear arms reduction agreement, one that ideally would include all their nuclear weapons, to ask Britain, France and China to undertake unilateral, politically-binding commitments. In those commitments, each of the three countries would undertake not to increase the total number of its nuclear weapons so long as the United States and Russia were implementing their agreement on nuclear reductions. Such an approach would begin to bring Britain, France and China into the nuclear arms limitation process while pushing off the day of a difficult multilateral negotiating attempt. The U.S.-Russian agreement might also contain a provision allowing an opt-out of further reductions were a third-country to dramatically increase its nuclear weapons numbers in a manner that endangered U.S. or Russian security.

These kinds of ideas on missile defense, precision-guided conventional strike weapons and third-country nuclear forces could offer a means to bridge the differences between stated U.S. and Russian positions if Washington were to seek to negotiate further nuclear reductions. The main question, of course, is whether these ideas would prove of interest to the Trump administration or to the Kremlin.

COMPREHENSIVE TEST BAN TREATY

A related area that the U.S. government could move on is ratification of the 1996 Comprehensive Test Ban Treaty (CTBT), which bans all nuclear testing. The Senate failed to give consent to ratification in 1999, due to two major concerns: the U.S. ability to maintain a reliable nuclear stockpile absent nuclear testing, and questions about whether the United States could be sure that other countries were not cheating.

There are very good answers to these questions today. The Stockpile Stewardship Program, launched in the mid-1990s, has developed techniques that have allowed the directors of the national nuclear labs and the commander of Strategic Command to certify that the stockpile is reliable. Testing is not needed. In fact, using some of the new technologies, the labs have learned things about how nuclear weapons work that they did not understand when conducting a nuclear testing program.

As for verification, there have been improvements in U.S. national technical means, and the CTBT Organization has established a global monitoring system, comprising some 300 stations with sensors to detect nuclear tests in the atmosphere, in the ocean or underground. It is believed that that system can detect underground tests down to one kiloton and, in some geologies, down to .1-.2 kilotons. All of the North Korean nuclear tests were detected and quickly reported to the CTBT Organization by multiple stations.

Other reasons argue for the United States to ratify the CTBT. First, for political reasons, it is very difficult to see a resumption of nuclear testing at the Nevada National Security Site (formerly the Nevada Test Site). Nevada fought strongly against the storage of nuclear waste at the site, and the city of Las Vegas—located about 60 miles south—has almost three times the population now as it had in 1992, when the last nuclear test was conducted.

Second, the United States has conducted more nuclear tests than the rest of the world combined and gained a great deal of knowledge from those tests. It now has the most sophisticated system for ensuring the reliability of its nuclear weapons without having to test. In these circumstances, freezing the ability of others to test, to gain new knowledge about nuclear weapons, and perhaps to build more advanced nuclear weapons would lock in a U.S. advantage.

All that said, Senate Republicans have shown no enthusiasm for the CTBT, and the chances that the Trump administration would pursue ratification appear slim.

COSTS OF A COLLAPSE OF THE NUCLEAR ARMS CONTROL REGIME

As noted above, steps could be taken to improve upon the nuclear arms control regime. However, prospects for significant new progress do not appear high. Preservation of the existing arms control regime—primarily the INF and New START treaties—may be the best that can be hoped for until there is a substantial change in the political atmospherics between Washington and Moscow. Just maintaining the existing regime, particularly the INF Treaty, may not be easy.

As it thinks through how it plans to handle nuclear arms control and related issues, the U.S. government should consider what would happen were the nuclear arms control regime to break down, that is, if the INF Treaty collapsed and New START not extended beyond 2021. That would mean that, for the first time in nearly 50 years, there would be no negotiated agreements governing the nuclear relationship between the United States and Soviet Union/Russia. That could raise significant costs and risks for U.S. security.

The end of New START would mean the end of the transparency measures contained in that treaty—the semi-annual data exchanges, notifications and inspection opportunities. Replacing the information gained through those measures would cost many billions of dollars, and it might not be possible to gain some kinds of information—such as the number of warheads on individual Russian ICBMs and SLBMs—through U.S. national technical means. The U.S. government would have to go back to worst-case assumptions regarding Russian nuclear forces, which invariably would lead to more expensive decisions about how to equip and operate U.S. nuclear forces.

Moreover, the United States would lose the caps that constrain the overall level of Russian strategic nuclear forces. That might not automatically trigger a new arms race; cost considerations, for example, could be a factor for both. However, the end of New START would mean that the sides' militaries no longer were bound by the treaty's limits. Likewise, the end of the INF Treaty would allow the Russians to deploy unlimited numbers of intermediate-range GLCMs and ballistic missiles against American allies and friends in Europe and Asia and against U.S. forces in those areas. A new arms race could be possible.

A new arms race could raise several questions. As numbers grew, there would also be greater uncertainty. Would U.S. allies support a U.S. build-up, and, for that matter, would Democrats in Congress support funding for a larger nuclear arsenal?

The Russians would not face these kinds of questions and likely could add nuclear weapons more cheaply than could the United States. Indeed, engaging in a nuclear arms race might mean competing in an area where Russia has a comparative advantage, as opposed to other areas, such as advanced conventional strike weapons, where the comparative advantage lies with the United States.

The end of the nuclear arms control regime would also have negative consequences for the Non-Proliferation Treaty, where non-nuclear weapons states would take it as one more indicator that the nuclear weapons states are not serious about their NPT commitment to disarm. That in turn could lead to greater global support for the nuclear ban treaty.

Finally, third countries could react in ways harmful to U.S. security. China, which has modernized its strategic nuclear forces at a modest rate, could change its policy and accelerate its nuclear build-up.

In sum, the end of the nuclear arms control regime would have significant downsides for the United States and its allies. It makes more sense for U.S. policy to strive to maintain and, if possible, improve that regime.

REFERENCES

1 Thomas H. Wolfe, *The SALT Experience,* Cambridge, MA: Ballinger Publishing Company, 1979, p. 97.

2 U.S. and Soviet "Statement on Data on the Numbers of Strategic Offensive Forces as of the Date of the Signature of the Treaty," Treaty between the United States of America and the Union of Soviet Socialist Republics on the Limitation of Strategic Offensive Arms, June 18, 1979.

3 Data from "Memorandum of Understanding on the Establishment of the Data Base Relating to the Treaty between the United States of America and the Union of Soviet Socialist Republics on the Reduction and Limitation of Strategic Offensive Arms."

4 Department of State Fact Sheet, "START Aggregate Numbers of Strategic Offensive Forces (as of July 1, 2009, as compiled from individual data submissions of the Parties." Note that the START I counting rules significantly overstated the total number of U.S. and Russian strategic forces, for three reasons. First, a number of the ICBM silos and SLBM tubes were empty but, because the silos and tubes had not been eliminated according to START I procedures, they still counted as carrying deployed ICBMs and SLBMs, with their attributed warheads. Second, four U.S. Trident submarines had been converted to carry conventional sea-launched cruise missiles instead of Trident SLBMs, and a number of U.S. B-1 bombers had been converted to carry conventional weapons only, but these systems continued to count under START I rules. Third, most U.S. ICBMs and SLBMs had been downloaded to carry fewer warheads than the number with which they were attributed.

5 Department of State, Bureau of Arms Control, Verification and Compliance Fact Sheet, "New START Treaty Aggregate Numbers of Strategic Offensive Arms," April 1, 2017.

The United States Should Forego a Damage-Limitation Capability Against China[1]

Charles L. Glaser

Institute for Security and Conflict Studies
Elliott School of International Affairs
George Washington University
Washington, DC 20052
USA

Abstract. Bottom Lines

- THE KEY STRATEGIC NUCLEAR CHOICE.
  Whether to attempt to preserve its damage-limitation capability against China is the key strategic nuclear choice facing the United States. The answer is much less clear-cut than when the United States faced the Soviet Union during the Cold War.

- FEASIBILITY OF DAMAGE LIMITATION.
  Although technology has advanced significantly over the past three decades, future military competition between the U.S. and Chinese forces will favor large-scale nuclear retaliation over significant damage limitation.

- BENEFITS AND RISKS OF A DAMAGE-LIMITATION CAPABILITY.
  The benefits provided by a modest damage-limitation capability would be small, because the United States can meet its most important regional deterrent requirements without one. In comparison, the risks, which include an increased probability of accidental and unauthorized Chinese attacks, as well as strained U.S.–China relations, would be large.

- FOREGO DAMAGE LIMITATION.
  These twin findings—the poor prospects for prevailing in the military competition, and the small benefits and likely overall decrease in U.S. security—call for a U.S. policy that foregoes efforts to preserve or enhance its damage-limitation capability.

The Key Strategic Nuclear Choice

The key strategic nuclear choice facing the United States is whether to attempt to preserve its "damage-limitation" capability—that is, the ability to reduce significantly the damage China can inflict against the U.S.

[1] This policy brief summarizes the key arguments in Charles L. Glaser and Steve Fetter, "Should the United States Reject MAD? Damage Limitation and U.S. Nuclear Strategy toward China," *International Security*, Vol. 41, No. 1 (Summer 2016), pp. 49-98.

Nuclear Weapons and Related Security Issues
AIP Conf. Proc. 1898, 020002-1–020002-8; https://doi.org/10.1063/1.5009207
Published by AIP Publishing. 978-0-7354-1586-7/$30.00

homeland in an all-out nuclear retaliatory attack. The United States currently has the ability to destroy much of China's strategic nuclear force (missile silos, ballistic missile submarines, command and control, etc.) in a first strike, which would somewhat reduce the impact of a Chinese retaliatory second strike. U.S. decisions about whether to attempt to preserve this modest damage-limitation capability could greatly affect U.S. force structure, U.S. and Chinese incentives in a crisis, and the U.S. political relationship with China as well as potentially with its regional allies.

Attempting to preserve this damage-limitation capability will require the United States to make large investments in ISR (intelligence, surveillance, and reconnaissance) and possibly in missile defenses designed to defeat China's strategic nuclear forces. In contrast, if the United States decides to forego a damage-limitation capability, its current forces could still perform the full array of required strategic deterrent missions, including limited nuclear attacks designed to deter conventional war and unlimited retaliation designed to deter all-out nuclear war. A damage-limitation strategy requires competition with China's nuclear forces; foregoing damage limitation will enable the United States to avoid this competition.

Feasibility of Damage Limitation

What counts as significant damage limitation? During the Cold War, the United States understood damage limitation as the ability to deny the Soviet Union an "assured destruction capability" against the United States, which Secretary of Defense Robert McNamara defined as the ability to destroy 20 to 25 percent of a country's population and 50 percent of its industrial base in retaliation, after suffering an all-out attack against its nuclear forces. It was estimated that the 200 1-megaton warheads, or the equivalent, could inflict this level of destruction.

More recent estimates that include a fuller set of nuclear effects, fire in particular, show that 40 megaton weapons could inflict this level of damage on the United States. In addition, these calculations do not include the impact of attacks on energy systems, information and communication systems, and major ports and other transportation nodes. If a highly integrated modern industrial economy is fragile, the destruction could be much greater. To acknowledge the uncertainties and their potential implications, we believe a damage-limitation range is more appropriate than a single threshold—from 10 medium-yield warheads to 40 equivalent megatons on cities in a retaliatory attack against the United States. Analysts can reasonably disagree about where within this range reducing the size of a Chinese attack would begin to produce a significantly better outcome for the United States.

China currently deploys about 20 silo-based inter-continental ballistic missiles (ICBMs) and 25 mobile ICBMs capable of delivering warheads against cities in the United States. China is modernizing and expanding its ICBM force, which is predicted to reach 100 mobile ICBMs by 2030, and also may soon deploy up to 60 submarine-launched ballistic missiles (SLBMs) on its new generation ballistic missile submarine. China's silo-based missiles are highly vulnerable to attacks from extremely accurate U.S. nuclear missiles; we also believe that China's current-generation ballistic missile submarine is highly vulnerable to U.S. anti-submarine warfare capabilities. Consequently, for at least the next decade, China's ability to launch retaliatory strikes will depend on the survivability of its mobile ICBMs.

During peacetime, China's mobile missiles are stored separately from their warheads in garrisons. Their ability to survive a U.S. attack therefore depends on mating warheads with missiles and moving them well out of garrison before the United States attacks in a severe crisis or conventional war. China must also be able to launch its missiles from unprepared positions because the United States is likely to have identified most prepared launch sites.

To preserve its damage-limitation capability, United States would need the ability to find and track Chinese mobile missiles, or to intercept Chinese warheads in flight with a national missile defense (NMD) system, or both. To track mobile missiles, the United States would need a large constellation of space-based radars or a much larger, but overall much less expensive, constellation of small imaging satellites. We expect, however, that China could defeat these radars with a combination of hiding and decoys, as well as active attacks against U.S. satellites. To intercept warheads in flight, the United States would need a larger and more capable NMD. Here again, we believe that China could defeat this with a combination of countermeasures and attacks that defeat or destroy U.S. tracking radars. In short, if China operates its missile force effectively and pursues technologically feasible reactions, the United States has poor prospects for maintaining a significant damage-limitation capability.

Benefits and Risks of a Damage-Limitation Capability

If critical to ensuring U.S. security, then the United States' best option might be to pursue a damage-limitation capability, even though the probability of success is low. We find, however, that in fact the benefits are small and overall pursuit of a damage-limitation capability could reduce U.S. security.

The United States will be able to continue to meet its extended deterrence commitments to its key East Asian allies without a damage-limitation capability. Most importantly, the combined capabilities of the United States and these allies should be sufficient to prevent China from winning a major conventional war and thereby sufficient to deter large Chinese conventional attacks. In addition, even without a U.S. damage-limitation capability, the possibility that a large conventional war could escalate to a nuclear war should contribute to deterring a large Chinese conventional attack. The United States can enhance this aspect of its nuclear deterrent by planning limited nuclear options (LNOs); a U.S. doctrine that called only for all-out retaliation could fuel Chinese doubts about the credibility of U.S. threats. LNOs should be small and not planned against Chinese nuclear and command-and-control targets. Given the deterrent value of the other components of the U.S. deterrent, the additional deterrent value of a damage-limitation capability is small and will decline further as China modernizes its forces.

Compared to these limited benefits, the potential costs and risks of striving to preserve the U.S. damage-limitation capability are large. Holding vulnerable China's ability to retaliate could generate a variety of escalatory dangers including accidental, unauthorized, and inadvertent nuclear attacks, and early intentional limited nuclear escalation driven by concerns about U.S. preemptive attacks. In addition, U.S. efforts to preserve a damage-limitation capability would fuel strategic nuclear competition that would almost certainly add to strains in U.S.–China relations, which would in turn increase the probability of conflict. Finally, the systems that the United States would likely deploy in pursuit of an effective damage-limitation capability—including an extensive national ballistic missile defense (BMD) designed to protect against Chinese ICBMs and space-based sensors designed to track mobile missiles—would be very expensive.

Forego Damage Limitation

The United States should not try to enhance or preserve whatever damage-limitation capability it now possesses. In part this reflects a technical judgment: China's continuing deployment of mobile missiles has the potential to fully erode U.S. damage-limitation capabilities, although in the mid-term this will depend upon how China operates its forces and in the longer term will depend upon whether the United States develops and deploys systems that can reliably find and destroy mobile targets. More important, however, is that the value of a damage-limitation capability is small. If a damage-limitation capability were necessary for protecting vital U.S. interests, then a case could be made for investing great sums in the pursuit of exotic and unproven technologies to acquire one, even if the probability of limited success were low. The United States, however, faces a much less daunting security environment. The United States can meet its extended deterrence commitments to its key East Asian allies without a damage-limitation capability. Given the military and political risks of adopting the competitive military policies, the overall result of pursuing a damage-limitation capability would be increased U.S. insecurity.

U.S. NUCLEAR STRATEGY TOWARD CHINA

The following outline provides some background for the preceding analysis of a damage-limitation strategy and sketches the central points in that analysis.

I. Basic purposes of US nuclear strategy

- Deter nuclear attacks against the US homeland
- Deter attacks against US allies—that is, extend deterrence
- Reassure US allies about the effectiveness of its deterrent, thereby helping preserve the alliance
- Reduce the damage/costs the US would suffer in a nuclear war, either by:
  - Averting escalation to all-out nuclear war
  - Reducing the adversary's ability to damage the United States

II. Basic strategy questions about US nuclear policy toward China

- Should the United States pursue a "damage-limitation" capability?
- Should the US threaten first use of nuclear weapons to deter conventional attacks against its allies?
  - The questions are related: a damage-limitation capability might enhance extended deterrence
- Most of this outline addresses the first question; it also provides a short sketch of an analysis of the second question

III. Roadmap of analysis damage-limitation question

- Is damage limitation feasible?
  - What counts as damage limitation?
  - Vulnerability of Chinese forces: fixed silos, mobile missiles, SSBNs, C2
  - Ballistic missile defense
- How large would the benefits be?
  - Reducing costs of all out war
  - Deterrence—homeland and extended
  - Reassuring allies
- How large would the costs be?
  - Accidental and unauthorized attacks; tactical windows
  - Political costs of military competition

IV. Background

- Why the question of US-China nuclear strategy is becoming more important
  - Tensions between the U.S. and China are growing and China's goals in East Asia are uncertain
  - China's conventional capabilities have grown considerably
  - The combination of tensions and growing conventional capabilities raises the possibility of a conventional war, which could escalate to a nuclear war
- Why damage limitation is the key question
  - A strategy designed for damage limitation requires a force dedicated to destroying Chinese forces and command and control; it necessarily competes with Chinese forces and is likely to drive an arms race

- In contrast, a strategy that accepts full U.S. vulnerability to Chinese nuclear retaliation requires small and less sophisticated U.S. nuclear forces, less advanced intelligence and reconnaissance assets, and smaller ballistic missile defense. This approach is much less competitive and would help to avoid a nuclear arms race with China

- The policy choice is less straight forward than during Cold War
  - The Soviet Union deployed over 10,000 warheads on thousands of long-range missiles
  - China now deploys about 20 silo-based ICBMs and 25 mobile ICBMs; it may soon deploy up to 60 SLBMs
  - Given current projections, China is likely to have around 100 warheads on mobile missiles by 2030
  - Technological advances have increased the ability to find forces compared to hide them; the combination of China's relatively small strategic nuclear force and the greater difficulty of deploying survivable forces against the United States raises the possibility of that a dedicated U.S. policy could produce some damage-limitation capability

V. What counts as damage limitation?

- Cold War levels:
  - McNamara: the United States need to be able to destroy 20-25% of Soviet population and 50% of its industry following a full Soviet attack against U.S. forces
    - At the time, this capability was found to require 200 equivalent megatons (EMT) of retaliatory capability
    - More sophisticated calculations done in the 1980s, which included fire effects as well as the blast effects in the early calculations, found that 40 EMT could inflict this level of retaliation
- How should we define damage limitation?
  - More than one definition is reasonable; they capture different features of massive destruction.
    - There was little analysis of this question during the Cold War, because the levels of retaliation were so large
  - One possibility is to define damage limitation at the level at which the costs of additional damage are insignificant: a traditional flattening of the damage curve argument
  - Another possibility is to define the level at which a modern state cannot recover; this approach requires a more holistic analysis of the full range of effects of nuclear damage inflicted on a modern economy
- Need for comprehensive technical analysis:
  - An adequate analysis of whether a full-scale damage-limitation attack would provide a "better" outcome for the United States must take into account the full range of economic, environmental and social/political costs of nuclear retaliation against the United States
  - Considering only the blast, or the blast and fire, effects underestimates the damage and thereby provide unwarranted support for a damage-limitation strategy

VI. Survivability of ICBMs

- All fixed silos are highly vulnerable to U.S. nuclear attack

- They could also be vulnerable to attacks by conventional weapons, if the United States chooses to deploy conventional ballistic missiles in sufficiently large numbers

- Mobile missiles: survivability depends on:
  - How China operates its forces during a crisis—does it put its forces on alert early in a crisis, does it effectively mate warheads and missiles, etc.
  - Whether China can launch from unprepared positions
    - This is necessary because the United States likely has identified all prepared positions and can destroy them
  - US ability to locate missiles:
    - US space-based surveillance, including space-based radars; and possibly other types of surveillance. The emerging revolution in small satellites is increasing the possibilities for near real-time tracking of mobile missiles
  - Chinese will be able to pursue a variety of countermeasures to defeat US surveillance; possibilities include:
    - Deploying decoy missiles
    - Hiding mobile missiles
    - Attacking or otherwise defeating US satellites
  - The prospects for damage limitation will depend on the outcome of this measures-countermeasures competition

VII. Survivability of Ballistic Missile Submarines (SSBN) and Command and Control (C2)

- China is deploying a second-generation ballistic missile submarine:
  - BUT, it will almost certainly be vulnerable to US anti-submarine warfare (ASW) capabilities for the foreseeable future
- China's nuclear command and control is vulnerable; "solutions" are available but none of them will be easy for China to implement, nor are they risk free:
  - Ensuring the survivability of links between political leaders and launch commanders via mobility, diversity, and redundancy
  - Predelegation of launch orders to military commands or even missile launch officers
  - Launch on warning:

VIII. Effectiveness of US Ballistic Missile Defense (BMD)

- Current US BMD system is not oriented against China
- China will be able to defeat an expanded version of the current US system:
  - Extreme difficulty of midcourse discrimination
  - Attacks against BMD radars can cripple the entire missile defense system
- Nevertheless, it is important for US analysts to appreciate that China sees US BMD as a serious threat to Chinese retaliatory capabilities
  - Even if current US capabilities are too small and not sufficiently advanced, China worries about future deployments that could be much large and more advanced

IX. Benefits of a Damage-Limitation capability

- Reducing costs of all out war:

    - US ability is currently relatively small and declining
- Homeland Deterrence: generally quite easy—retaliatory threats are highly credible
- Extended Deterrence: Benefits are small because deterrence is already highly effective
    - US-Japan alliance conventional forces can defend Japan against a large-scale conventional attack; other scenarios appear unlikely to depend significantly on the details of U.S. nuclear capabilities
    - The possibility of nuclear war will cast a deterrent shadow over major crises, even if the United States can not limit damage
        - Limited nuclear options can add credibility to US nuclear threats
    - Taiwan is the exception:
        - The U.S. ability to protect Taiwan from conventional coercion via a blockade or eventually an invasion is decreasing
        - In the case of Taiwan, China cares much more about the outcome than does the United States, which would confer a bargaining advantage in a nuclear confrontation over Taiwan
- Reassuring allies: benefits are small, because US commitments can be highly credible without a damage-limitation capability

X. Cost of a damage-limitation capability

- In broad terms, a U.S. damage-limitation strategy creates pressures for both the United States and China to use nuclear weapons earlier in a crisis than they might prefer, because the United States might launch a massive counterforce attack and China's anticipatory moves to undermine this attack could create pressure for the U.S. to attack before it loses the option
- Escalatory pressures and risks
    - Accidental Chinese launch via pressure to launch quickly
    - Unauthorized Chinese launch via predelegation of launch authority to military commanders
    - Incentives for the United States to attack early, before Chinese forces are fully alerted and dispersed and therefore more survivable
- Political costs:
    - China will view U.S. efforts to maintain a significant damage-limitation capability as reflecting malign U.S. motives
        - China views its retaliatory capability as necessary for protecting its security
        - U.S. plans to defeat China's retaliatory capability are viewed as a threat to China's security and the United States is viewed as determined to prevent China's rise
    - Nuclear competition will strain the US-China relations, increasing the probability of dangerous crises

XI. Conclusions re damage limitation

- Prospects for significant damage limitation over the medium to long term appear to be poor
    - Does depend, however, on where the threshold for damage limitation is set
- Benefits of feasible damage-limitation capabilities are small:
    - The United States already possess effective deterrent capabilities

- Costs are significant:
  - A damage-limitation posture generates an array of escalatory pressures and strains political relations
- Bottom-line: US should forego efforts to preserve its damage-limitation capability
  - However, other analysts disagree along all of these dimensions

XII. No First Use (FU)?

- The US has extended deterrence commitments to Japan and South Korea
- A policy dedicated to deterring conventional attacks requires a policy of first use:
  - This was NATO strategy for deterring the Soviet Union/Warsaw Pact during the Cold War
- The marginal deterrent value of FU threats depends on:
  - Whether conventional forces are sufficiently effective to deter Chinese conventional attack
  - Whether nuclear weapons, essentially independent of US doctrine, contribute to deterrence of conventional war when the US lacks a damage limitation capability
- Reassurance: do US allies believe FU is necessary?
- Risks of FU: early use, unauthorized use

ALERT STATUS OF NUCLEAR WEAPONS

Hans M. Kristensen

Director, Nuclear Information Project
Federation of American Scientists
1725 DeSales St., NW (6th floor)
Washington, DC 20036

Abstract. Nuclear Alert Forces. Four nuclear-armed states deploy nuclear warheads on alert, ready to be used on relatively short notice: United States, Russia, France and Britain. Combined, the four countries deploy an estimated 1,869 nuclear alert warheads. Russia and the United States deploy 1,749 alert warheads combined, or 94% of all alert warheads.

Despite some debate about possible need to increase readiness of nuclear forces (China, Pakistan), the five other nuclear-armed states (China, Pakistan, India, Israel and North Korea) are thought to store their warheads separate from launchers under normal circumstances. The overall number of alert warheads has remained relatively stable during the past five years.

TABLE I: Country Stockpiled Warheads/Alert Warheads.
(*a*) It is possible that Russia has more warheads on alert depending on the readiness level of road-mobile ICBMs and non-deployed submarines.
(b) French and British alert forces are thought to have lower readiness than U.S. and Russian alert forces and take longer to launch.
(c) In addition to stockpiled warheads, Russia and the United States have large numbers of retired, but still relatively intact, warheads awaiting dismantlement. For a global overview of nuclear forces, visit: https://fas.org/issues/nuclear-weapons/status-world-nuclear-forces/

United States 4,000/852
Russia 4,300/897*a*
France 300/80*b*
Britain 215/40*b*
China 270/0
Pakistan 140/0
India 120/0
Israel 80/0
North Korea (10-20)/(0)

Total 9,425*c*/1,869

Nuclear Weapons and Related Security Issues
AIP Conf. Proc. 1898, 020003-1–020003-6; https://doi.org/10.1063/1.5009208
Published by AIP Publishing. 978-0-7354-1586-7/$30.00

US NUCLEAR ALERT FORCES

An estimated 852 warheads with 228 MT on prompt alert (ready to launch <15 min). Alert warheads roughly 21% of total stockpile:

ICBMs: 392 warheads (125 MT)
Surge: 300 warheads (upload)

SSBNs: 460 warheads (103 MT)
Surge: 1,460 warheads (deployed + uploads)

Bombers: 0 warheads
Surge: 816 warheads (deployed + uploads)

Non-Strategic: 0 warheads
Surge: 300 warheads (bombs)

Alert Rate Assumptions:

ICBMs: Minuteman III (98%)
SSBNs: 5 on hard alert (each with 20 tubes)
Bombers: None
Non-strategic: None

US Nuclear Alert Forces: Trends

ICBM alert trends

- Overall force size declining slightly to 400 missiles
- Warhead download completed in 2014 (upload capability retained for Mk12A/W78 equipped missiles)
- Next ICBM (GBSD) will be MIRV-capable

SSBN alert trends

- Near term: SSBN tube reduction from 24 to 20 (no overall warhead reduction)
- Long term: Next SSBN class tube reduction from 20 to 16
- SSBN fleet reduction from 14 to 12 with next SSBN (possibly sooner after last Ohio refuelings)

RUSSIAN NUCLEAR ALERT FORCES

An estimated 897 warheads with 516 MT on prompt alert (ready to launch <15 min). Alert warheads roughly 21% of total stockpile:

ICBMs: 686 warheads (492 MT)
Silo: 65-95% readiness
Mobile: 15-20% readiness (It is possible that the mobile ICBM alert level is higher, in which case the surge capacity would be less.)
Surge: 391 warheads (mainly mobile)

SSBNs: 211 warheads (24 MT)
Surge: 429 warheads

Bombers: 0 warheads
Surge: 616 warheads (at bases + reserve)

Non-Strategic: 0 warheads
Surge: ~1,480 warheads (all except ABM/air-defense)

Alert Rate Assumptions:

ICBMs (silo): SS-18 (90-95%); SS-19 (65%); SS-27 Mod 1 (90%); SS-27 Mod 2 (90%)
ICBMs (mobile): SS-25 (15%); SS-27 Mod 1 (20%); SS-27 Mod 2 (20%)
SSBNs: 3 at sea; 3 at pierside (~30%)
Bombers: None
Non-strategic: None

Russian Nuclear Alert Forces: Trends

ICBM alert trends

- Overall force size declining but leveling out around 310+ launchers
- Increased emphasis on MIRV to keep warhead parity with US
- 10 years ago no mobile ICBMs had MIRV; by 2022 ~all will have MIRV
- Greater share of warheads on mobile (lower alert level but harder to find)
- Mobiles increase duration of combat patrol tours

SSBN Alert Trends

- SSBN operations increasing some (despite what you hear, not Cold War level and not as much as attack submarine operations), probably 2-4 on alert at sea plus 2-4 at pier side
- Warhead loading increasing (6 MIRV on SS-N-32; 4 on SSN-23; 3 on SS-N-18); hedge emerging?

FRENCH NUCLEAR ALERT FORCES

An estimated 80 warheads with 8 MT on alert. Alert warheads roughly 27% of total stockpile:

SSBNs: 80 warheads (8 MT)
Surge: 80 warheads

Bombers: 0 warheads
Surge: 40 warheads

Non-Strategic: 0 warheads
Surge: 14 warheads

Alert Rate Assumptions:

SSBNs: 1 at sea at all times; second boat on standby
Bombers: None
Non-strategic: None

French Nuclear Alert Forces: Trends

SSBN alert trends

- Warhead loading on some SLBMs has been reduced to allow more tailored targeting.
- M45 SLBM has recently been completely replaced by M51 SLBM with "significantly greater range and payload capacity, as well as greater accuracy."

• 100 kt TN75 warhead is being replaced with 150 kt TNO warhead.

BRITISH NUCLEAR ALERT FORCES

An estimated 40 warheads with 4 MT on alert. Alert warheads roughly 19% of total stockpile:

SSBNs: 40 warheads (4 MT)
Surge: 80 warheads

Alert Rate Assumptions:

SSBNs: 1 at sea at all times; second boat on standby
Bombers: None
Non-strategic: None

BRITISH NUCLEAR ALERT FORCES: TRENDS

SSBN alert trends

- SLBM loading has been reduced from 16 to 8
- Warhead loading has been reduced from 48 to 40
- Operationally available warheads has been reduced from 180 to 120
- Next generation SSBN will have 12 instead of 16 tubes

PARTIAL DEALERTING STEPS

Since end of the Cold War, nuclear alert rates have been reduced in several ways:

- Strategic bombers were taken off alert in 1991.
- ICBM warhead loading reduced (United States); New START treaty included ICBM MIRV ban but was abandoned in favor of ballistic missile defenses against rogue states.
- SLBM warhead loading reduced (United States).
- De-targeting initiatives: ICBMs/SLBMs targeted at open ocean areas during peacetime (note: de-targeting is *not* dealerting).
- Non-strategic forces taken off alert. Most warheads destroyed but remaining placed in central storage. Some still deployed on bases near launchers (US bombs in Europe; French cruise missiles).

Promises To Dealert

US presidential candidates have repeatedly promised to take nuclear weapons off alert. President George W. Bush lamented that Cold War differences “were expressed in a dangerous military confrontation that resulted in thousands of nuclear weapons pointed at each other on hair-trigger alert.” Barack Obama promised to “work with Russia to take U.S. and Russian ballistic missiles off hair-trigger alert.” Yet, neither Bush nor Obama did so but instead retained status quo of nuclear alert postures.

Russia, France and Britain do not seem interested.

Reduced Alert Status Seen As A Good Thing

At the end of the Cold War Russia had “12,000 [strategic] nuclear weapons, most of those on alert …[By 1999], under START I, the Russians have about 2,000 nuclear weapons on alert [out of a total of 6,100 strategic warheads that year]. Under START II [out of 3,000–3,500 deployed strategic warheads], they’ll be down to about 1,000 nuclear weapons on alert. Under START III, if all goes [as planned, that number will] be around 700 nuclear weapons”. *Gen. Eugene Habiger, former STRATCOM Commander.*

Yes – No: The Dealerting Debate

Debate is very polarized. Proponents of reducing or even dealerting nuclear forces argue that doing so would reduce the competitive dynamic and worst-case scenario attitude in nuclear force planning and operations, as well as reducing or eliminating the risks of accidental launch of nuclear forces.

Opponents of reducing – certainly dealerting – nuclear forces argue that alert nuclear forces help maintain crisis stability and that dealerting would create significant risks of a re-alerting race in a crisis.

While there are important crisis stability and verification issues in both alerted and de-alerted postures, we find that even if all US and Russian nuclear forces were de-alerted and one side secretly re-alerted, the aggressor could not be confident in carrying out a disarming first strike because a sufficient number of highly capable forces would survive to provide a devastating retaliation. Despite concern about realerting effect, both US and Russian escalation strategies rely on significant realerting to signal and boost nuclear postures in a crisis.

US warhead "hedge" of non-deployed warheads explicitly provides "the option for [the] leadership to increase the number of operationally deployed forces in proportion to the severity of an evolving crisis."

Realerting In Existing Alert Postures

Despite concern about re-alerting effect, both US and Russian escalation strategies rely on significant re-alerting to signal and boost nuclear postures in a crisis.

US warhead "hedge" of non-deployed warheads explicitly provides "the option for [the] leadership to increase the number of operationally deployed forces in proportion to the severity of an evolving crisis."

The US has more warheads in non-deployed hedge than in deployed force (56% vs 44%). Russian ratio similar but hedge is mainly (77%) non-strategic.

Realerting of bombers is part of bomber exercises. During the Constant Vigilance alert exercise at Minot AFB in May 2015, a dozen B-52 bombers were loaded with unarmed nuclear air-launched cruise missiles.

New nuclear cruise missile mission includes rapid uploading: "**The LRSO will provide a rapid and flexible hedge against changes in the strategic environment and limitations of the other two legs of the Triad**. Under the New START Treaty, each strategic bomber counts as one launcher and one warhead, regardless of the number of nuclear cruise missiles and bombs in our inventory. This provides a **rapid upload capability** to hedge against geopolitical or technical surprise." [*Brian McKeon, OSD, Testimony before Congress, February 26, 2015*]

Disparity In US And Russian Alert Postures

Significant differences in military postures can have significant implications:

- Russian ICBM force has more than twice the number of warheads of US ICBM force.
- US has 150 deployed strategic launchers more than Russia; drives Russian emphasis on MIRV.
- Future Russian ICBM force will be entirely MIRVed; US ICBM force is downloaded.
- Larger US ICBM force with fewer warheads can threaten smaller Russian ICBM force carrying more warheads (400 warheads on 400 ICBMs threatening 1,000 warheads on 316 ICBMs).
- Larger US SLBM force with more warheads has significant counterforce capability in addition to secure retaliation mission.

Shaping Nuclear Alert Futures

While intended to ensure crisis stability, alert forces can also have other negative effects:

- Nuclear alert postures drive worst-case and short-fused strategy, posture planning and decision making. Alert forces require adversaries to maintain forces on alert: cyclic alert relationship.
- US-Russian nuclear alert postures might gradually inspire smaller nuclear-armed states to also increase the readiness of their nuclear forces (China, Pakistan).
- When combined with conventional offensive and defensive capabilities, nuclear alert postures might stimulate smaller nuclear-armed states to increase nuclear force loadings: MIRV (China and Pakistan).

- While there are risks with both alerted and dealerted postures, a realerting race that takes three months under a dealerted posture seems preferable to an alert race that takes less than three hours under current highly alerted posture.

In our 2012 UNIDIR report, we discuss alert postures and issues more. Available online at: http://www.unidir.org/files/publica;ons/pdfs/reducing-alertrates-of-nuclearweapons-400.pdf

Crises Shaping Alert Postures

Recent East-West crisis shows signs of affecting nuclear readiness postures and strategies:

- Russian vocal nuclear threats and putting nuclear forces on alert.
- Escalate-deescalate strategies entertaining limited and early use.
- Talk of increasing escalation planning and readiness of dual-capable aircraft.
- OPLAN in effect for EUCOM with increased nuclear bomber support.

  Signaling with alert forces: bomber strike exercises and forward deployments and SSBN port visits.

CONCLUSIONS

-- Four countries (United States, Russia, France, Britain) have nearly 1,900 nuclear warheads (94% are Russian/US warheads) on alert on ballistic missiles, capable of launching in relatively short time.

-- Some smaller nuclear-armed states (China and Pakistan) are debating whether to increase the readiness of their nuclear forces.

-- Overall alert level has been relatively stable the past five years but some force structures are changing (Russian mobile ICBM MIRV; US warhead kill effectiveness).

-- Signaling is increasing with nuclear alert forces and forces that would be re-alerted in a crisis.

-- US presidential candidates promised try to take nuclear weapons off "hair-trigger" alert but have not done so.

-- Alert-countries oppose reducing alert levels, certainly de-alerting, arguing alert forces create stability and dealerting would create risks for a re-alerting race in a crisis.

-- Opponents of nuclear alert argue the posture is dangerous, that considerable partial de-alerting has been done without incidents, and that fear of re-alerting race is exaggerated.

Nuclear Weapons Modernization: Plans, Programs, and Issues for Congress*

Amy F. Woolf

Specialist in Nuclear Weapons Policy,
Congressional Research Service
Library of Congress
Washington, DC 20540
USA

(The views expressed are those of the author and do not reflect on either the Congressional Research Service or the Library of Congress.)

Abstract. The United States is currently recapitalizing each delivery system in its "nuclear triad" and refurbishing many of the warheads carried by those systems. The plans for these modernization programs have raised a number of questions, both within Congress and among analysts in the nuclear weapons and arms control communities, about the costs associated with the programs and the need to recapitalize each leg of the triad at the same time. This paper covers four distinct issues. It begins with a brief review of the planned modernization programs, then addresses questions about why the United States is pursuing all of these modernization programs at this time. It then reviews the debate about how much these modernization programs are likely to cost in the next decade and considers possible changes that might reduce the cost. It concludes with some comments about congressional views on the modernization programs and prospects for continuing congressional support in the coming years

INTRODUCTION

The United States is currently recapitalizing each delivery system in its "nuclear triad" (land-based ballistic missiles, sea-based ballistic missiles, and heavy bombers) and refurbishing many of the warheads carried by those systems. The plans for these modernization programs have raised a number of questions, both within Congress and among analysts in the nuclear weapons and arms control communities, about the costs associated with the programs and the need to recapitalize each leg of the triad at the same time. The Obama Administration offered strong support for these programs, with Secretary of Defense Ashton Carter referring to the nuclear arsenal as "the bedrock" of U.S. security and the Pentagon's "highest priority mission." Secretary Carter argued that all the recapitalization programs are necessary because "most of our nuclear weapon delivery systems have already been extended decades beyond their original expected service lives."[1]

While the Trump Administration has not yet completed its Nuclear Posture Review or outlined its specific goals for the nuclear triad, current Pentagon officials have also offered strong support for the full scope of modernization programs. In testimony before the House Armed Services Committee in May 2017, Robert Soofer, the Deputy Assistant Secretary of Defense for Nuclear and Missile Defense Policy stated that "Nuclear weapons remain a foundational element of U.S. strategy for the deterring strategic attacks and large scale war and for assuring U.S. allies." He noted that "After decades of deferred modernization, replacement programs must proceed without delay if we are to retain existing deterrent capabilities."[2]

---

* *The views expressed are those of the author and do not reflect on either the Congressional Research Service or the Library of Congress.*

Nuclear Weapons and Related Security Issues
AIP Conf. Proc. 1898, 020004-1–020004-7; https://doi.org/10.1063/1.5009209
Published by AIP Publishing. 978-0-7354-1586-7/$30.00

Analysts outside government have questioned the Administration's plans, however, arguing that the United States can sustain deterrence with fewer numbers of warheads or fewer types of delivery vehicles. While cost estimates for the nuclear programs vary, most indicate that the Pentagon would likely spend around $35 billion per year to these programs during the 2020s, as many of the nuclear modernization programs advance to the procurement phase. Thus, the costs would rise from a current 3-4% of the Pentagon budget to 6%-7% of the budget which, critics claim, could force trade-offs with other programs reaching the procurement phase at the same time. As a result, they note that these modernization programs could detract from U.S. efforts to procure advanced conventional weapons.

This paper covers four distinct issues. It begins with a brief review of the planned modernization programs, then addresses questions about why the United States is pursuing all of these modernization programs at this time. It then reviews the debate about how much these modernization programs are likely to cost in the next decade and considers possible changes that might reduce the cost. It concludes with some comments about congressional views on the modernization programs and prospects for continuing congressional support in the coming years.

THE NUCLEAR TRIAD

Rationale for the Triad

Throughout the Cold War, the United States deployed a "triad" of nuclear weapons delivery systems that consisted of long-range land-based and sea-based ballistic missiles and heavy bombers equipped with gravity bombs and cruise missiles. The United States first developed these three types of nuclear delivery vehicles, in large part, because each of the military services wanted to play a role in the U.S. nuclear arsenal. However, during the 1960s and 1970s, analysts developed a more reasoned rationale for the nuclear "triad." They argued that these different basing modes would enhance deterrence and discourage a Soviet first strike by complicating Soviet attack planning and ensuring the survivability of a significant portion of the U.S. force in the event of a Soviet first strike.[3]

The systems also had complementary strengths and weaknesses that would strengthen the credibility of the U.S. deterrent strategy. For example, ICBMs eventually had the accuracy and prompt responsiveness needed to attack hardened targets such as Soviet command posts and ICBM silos, SLBMs had the survivability needed to complicate Soviet efforts to launch a disarming first strike and to retaliate if such an attack were attempted, and heavy bombers could be dispersed quickly and launched to enhance their survivability, and they could be recalled to their bases if a crisis did not escalate into conflict.

The United States has maintained the triad since the end of the Cold War, with successive Administrations outlining the rationale for this force structure after conducting Nuclear Posture Reviews. While this rationale continued to note that the mix of forces offered the United States a range of capabilities and flexibility in nuclear planning and complicated an adversary's attack planning, it also added the argument that the triad of delivery systems and warheads allowed the United States to hedge against unexpected problems in any single delivery system. This latter issue became more of a concern as the United States retired many of the different types of warheads and missiles that it had deployed over the years, reducing the redundancy in its force.

The Obama Administration highlighted this combined rationale in its report on the 2010 Nuclear Posture Review, noting that

> "Each leg of the Triad has advantages that warrant retaining all three legs at this stage of reductions. Strategic nuclear submarines (SSBNs) and the SLBMs they carry represent the most survivable leg of the U.S. nuclear Triad.... Single-warhead ICBMs contribute to stability, and like SLBMs are not vulnerable to air defenses. Unlike ICBMs and SLBMs, bombers can be visibly deployed forward, as a signal in crisis to strengthen deterrence of potential adversaries and assurance of allies and partners."

The report went on to state that "retaining sufficient force structure in each leg to allow the ability to hedge effectively by shifting weight from one Triad leg to another if necessary due to unexpected technological problems or operational vulnerabilities."[4]

Current Force Structure

The United States currently deploys 400 Minuteman III land-based intercontinental ballistic missiles (ICBMs) at three Air Force bases in North Dakota, Wyoming, and Montana. It has recently completed the process of reducing the force from 450 to 400 deployed missiles to meet the limits under the 2010 New START Treaty, but plans to retain all 450 silo launchers and to keep the missiles spread among all three bases. It is developing a new Ground-based Strategic Deterrence (GBSD) to replace the Minuteman III missiles, with deployment beginning in the late 2020s. The Air Force plans to acquire 642 new missiles to support testing and the deployment of a force of 400 missiles.

The United States currently deploys 14 Ohio-class ballistic missile submarines at two bases in Bangor, Washington and Kings Bay, Georgia. Each submarine was initially equipped to carry 24 Trident II (D-5) missiles, but the Navy has converted all the submarines so that they will carry only 20 missiles under the New START Treaty. The Navy plans to replace the 14 Ohio-class submarines with 12 new Columbia-class submarines, with deployment beginning in 2031. The Navy is also extending the life of the Trident II (D-5) missiles, and the National Nuclear Security Administration (NNSA), a semi-autonomous agency in the Department of Energy, is extending the life of the W76 warheads and beginning an alteration program on the W88 warheads, both of which are carried by D-5 missiles.

The Air Force currently has 20 B-2 bombers deployed at Whiteman Air Force Base in Missouri; these can carry both B61 and B83 nuclear gravity bombs. All of these bombers will remain nuclear-capable under the New START Treaty. It also maintains 76 B-52H bombers at Minot Air Force Base in North Dakota and Barksdale Air Force Base in Louisiana. Most are equipped to carry nuclear or conventional air-launched cruise missiles but the Air Force is converting some to carry only conventional weapons. The Air Force plans to retain 40 nuclear-capable B-52 bombers under the New START Treaty. The Air Force is also beginning the design and acquisition of new heavy bomber – the B-21 – that is expected to enter the fleet beginning in the mid-2020s. While the timing and size of this program (planning is for between 80 and 100 aircraft) is essentially determined by the requirements for a penetrating bomber equipped with conventional weapons, the bomber will also contribute to the nuclear mission as the B-52 and B-2 bombers retire in the 2040s.

The B61 bombs carried by B-2 bombers (and deployed in Europe in support of NATO's nuclear mission) are undergoing a life extension program (LEP) that will enhance the bomb's safety, security, and use control features. The Air Force is also designing a new tail kit for the B61 bomb, to improve its accuracy. The new model is scheduled to begin to enter the force in 2020. The B83 bomb is likely to be retired around 2025, after the completion of the B61 LEP. The Air Force is planning to replace the air-launched cruise missiles carried by B-52 bombers with a new advanced long range standoff (LRSO) cruise missile, with deployment beginning in the mid- or late-2020s. NNSA is also conducting a life-extension program on the W80 warhead to provide a warhead for the new LRSO.

PURSUING SIMULTANEOUS MODERNIZATION PROGRAMS

Analysts have offered a number of explanations for why the United States is pursuing multiple nuclear weapons modernization programs at the same time. Some have suggested that the United States has taken a "procurement holiday" since the end of the Cold War and must now address each of its aging systems at the same time because it neglected to do so over the past 25 years. Others have suggested that the United States must modernize its entire nuclear arsenal in response to Russian and Chinese nuclear modernization programs, or risk falling behind in an emerging nuclear arms race. Critics of the modernization programs offer a similar rationale, but in reverse; they argue that the United States, by pursuing all these programs at the same time, risks igniting an arms race with Russia and China at a time when all nations should seek to reduce, and eventually eliminate, their nuclear weapons.

Neither of these explanations captures the reality of the U.S. nuclear modernization cycle. While it is true that the United States has not designed or acquired new nuclear delivery systems during the past 25 years, it has not taken a holiday from maintaining and upgrading its existing systems. Most were expected to last well into the 2000s when they were first deployed and each has been subject to extensive life extension programs and modifications to ensure that they have remained safe and effective during that time. At the same time, though, many elements of the force are nearing the end of their service life, which will eventually introduce risk into their operations and effectiveness.

Moreover, to argue that the nuclear modernization programs either allow the United States to join, or cause the United States to initiate, a nuclear arms race presumes that the absence of nuclear modernization programs in other nations would allow the United States to forgo its own programs, or the absence of modernization programs in the United States would encourage other countries to abandon their efforts. Neither of these statements is true. Russia and China are both modernizing and, in some cases, expanding their nuclear forces. Russia's programs seek to replace aging systems, some dating from the Soviet era, and to incorporate new technologies. China, for its part, has pursued the slow and steady modernization of its nuclear capabilities for decades. While each may point to the United States as a source of its own insecurity, both nations' modernization efforts pre-date the U.S. commitment to recapitalization and neither nation would likely abandon its own programs if the United States slowed or stopped U.S. programs.

The United States also is not pursuing its modernization programs in response to Russian and Chinese programs, and those who suggest such a link ignore the fact that the United States likely *would not* abandon its efforts if these two nations slowed or stopped their programs. The United States is pursuing a wide range of modernization programs at this time because its systems are aging and the time has come to replace them. The United States first deployed the full triad of nuclear delivery vehicles in the early 1960s. It then replaced many of these systems in the late 1980s and 1990s. The current recapitalization represents a third cycle, which will bring new systems into the force around 2030 and retain them for another 30-50 years. According to officials in both the Obama and Trump Administrations, the United States delayed the programs a few years ago, after the Budget Control Act reduced Pentagon spending, and they are now "late to need," leaving little room for further delays or reductions in funding.

Analysts who question the current plans to recapitalize all three legs of the nuclear triad recognize that U.S. nuclear delivery vehicles are aging, but they dispute the conclusion that all need to be replaced at the same time. Some have suggested that the United States could delay some of the programs, and defer some of the costs, by further extending the service life of the current Minuteman III land-based ICBMs or air-launched cruise missiles. Pentagon officials in both the Obama and Trump Administrations have disputed this assertion, noting that, in some cases, the cost to maintain the aging systems could exceed the cost to replace them and that a decision to delay modernization could lead to the elimination of some U.S. nuclear capabilities.

Some have also questioned whether the United States needs to replace each of its current capabilities and retain, for at least another 50 years, a force structure identical to the force developed during the Cold War. They note that a force structure developed to deter and, if necessary, respond to the threat of global nuclear war with the Soviet Union may be excessive to U.S. needs in the coming decades, even if the United States faces numerous nuclear-armed regional adversaries. Others, however, have suggested that the United States not only needs each of its existing types of nuclear weapons systems, but may need to restore capabilities retired since the end of the Cold War, like sea-based cruise missiles or carrier-based nuclear capable aircraft, to address emerging regional challenges. This debate is likely to continue in the coming years if the Trump Administration reaffirms the U.S. commitment to recapitalize most elements of the U.S. nuclear force and a constrained budget environment continues to raise questions about whether the United States can, or should, finance all these programs at the same time.

THE COST OF NUCLEAR MODERNIZATION

Estimating the Cost

A number of different organizations, both inside and outside the government, have offered estimates of how much the U.S. nuclear weapons modernization programs will cost over the next 10, 20, or 30 years. Analysts outside government began raising the question in late 2011, when the programs began to take shape and Congress limited future defense budgets through the Budget Control Act. Congress has since joined the debate and has mandated that both the Congressional Budget Office and the Pentagon provide estimates of the future costs, and that the Government Accountability Office review the Pentagon estimates. These studies have offered a range of estimates – from around \$300 billion over 10 years to more than \$1 trillion over 30 years, due, in part to the use of different methodologies and different timelines, but also because they have calculated costs for different collections of programs.

In testimony before the House Armed Services Committee in July 2016, Robert Scher, then the Assistant Secretary of Defense for Strategy, Plans, and Capabilities, noted that the Department of Defense expected to spend \$350-\$450 billion over 20 years to recapitalize U.S. nuclear delivery systems and nuclear command and control systems. This total did not include the cost of nuclear weapons activities at the Department of Energy's National Nuclear Security Administration, nor, apparently, the cost to maintain and operate existing nuclear forces. In May 2017, Robert Soofer, now the Deputy Assistant Secretary of Defense for Nuclear and Missile Policy told the House Armed Services Committee that "DOD expects nuclear recapitalization costs to total approximately \$230 billion to \$290 billion over more than two decades." He offered no explanation for why his total differed so much from the total Assistant Secretary Scher provided a year earlier, but did indicate that the new estimate includes "only a portion of the B-21 bomber, which will have both conventional and nuclear roles." Moreover, the total did not include funding that the Pentagon had committed to providing to NNSA, to support its weapons activities accounts, as the Trump Administration plans to provide that funding directly to NNSA. Further, the total again seems to exclude the cost of maintaining and operating existing forces.

The Congressional Budget Office has published a number of reports that estimate the expected costs of U.S. nuclear forces – including both the operations of existing forces and the modernization programs – over a 10 year period. In the first edition, published at the end of 2013, CBO estimated that these programs would cost $355 billion between 2014 and 2023. This total included both DOD and NNSA programs, and allowed for a measure of cost growth based on historical data. At the time, CBO noted that, while it included estimates for the early years of the modernization programs, most of the procurement, and, therefore, higher costs, were likely to occur outside the dates of the study, in the latter part of the 2020s and 2030s. In its most recent edition, published in February 2017, CBO estimated that the 10-year cost of U.S. nuclear weapons programs would reach $400 billion between 2017 and 2026. The increase, from an average of $35 billion to $40 billion per year, reflects the growing procurement costs as the modernization programs mature during the 2020s.

Because the highest costs of the modernization programs are outside the 10-year window in recent studies, some studies have sought to estimate the costs through the 30-year time frame needed to complete most of the programs. For example, in 2014, the James Martin Center for Nonproliferation Studies (CNS) published a study titled "Trillion Dollar Nuclear Triad: U.S. Strategic Nuclear Modernization Over the Next Thirty Years." While this study's conclusion, and the catchy title, provided a frequently-quoted cost estimate for the modernization programs, some observers questioned its methodology, noting that anything could cost a trillion dollars if the estimate were stretched over long enough period of time. In addition, some argued that the study was deceptive in that it did not note that the magnitude of the defense budget over the same time period and the fact that nuclear weapons spending was likely to remain a small proportion of that total budget. Nevertheless, the 30-year estimate provided in that study was consistent with other studies, suggesting spending on the average of $30 billion per year. Moreover, while the methodologies likely differ, press reports indicate that a soon-to-be released study from the Congressional Budget Office will also provide an estimate of $1.2 trillion over 30 years to maintain and operate U.S. nuclear forces and pursue the full scope of planned modernization programs.

Managing the Costs and Affording the Programs

While these studies and several others completed during the past 5 years use different assumptions and methodologies to calculate the costs of U.S. nuclear weapons, all come to a similar conclusion. The United States will spend an average of $35-$40 billion per year, for the next 25-30 years, to operate its nuclear forces, recapitalize its nuclear weapons delivery vehicles, refurbish its nuclear warheads, and maintain the infrastructure needed to ensure a "safe, secure, and effective" nuclear deterrent. Some consider these costs to be excessive, and argue that the United States should reduce the size or scope of its nuclear programs so that it can divert the funds to higher priority conventional weapons programs. Others argue that the funding is reasonable, and affordable, in an environment where U.S. systems are aging and other nations are modernizing and expanding their own nuclear arsenals.

Each of these studies sought to answer the question of "how much do nuclear weapons cost?" The debate about the future of U.S. nuclear modernization programs is, however, more accurately characterized an attempt to answer the question of "how much are these weapons worth?" The debate is less about whether the United States can afford to commit resources to nuclear modernization programs than it is about whether it should choose to do so, in an environment where defense budgets are likely to be constrained and other priorities may lack necessary funding. The Pentagon's annual budget exceeds $600 billion per year. As many Pentagon officials have noted, spending on nuclear weapons currently accounts for about 3% of the Pentagon's budget; during the 2020s, as the procurement phase of the nuclear modernization programs peaks, it may rise to around 6%-7% of the Pentagon's budget.

But funds are limited, and demands for spending on new programs are likely to exceed available funds, even if the defense budget grows in the next decade. There will be trade-offs, and opportunity costs, if the United States pursues all of these programs at the same time.

While those who support the current plans to recapitalize the nuclear force argue that each of these programs is necessary and must be seen as the highest priority of Department of Defense, others argue that the Pentagon will not be able to sustain funding for all the programs when costs peak in the late 2020s. As a result, they argue that the Pentagon should find ways, now, to reduce and restrain costs, later, if it wants to ensure that budget restraints do not interrupt or interfere with the programs.

Some have suggested that the Pentagon reduce the eventual size of the U.S. nuclear force, and plan now to deploy fewer than 12 new ballistic missile submarines or fewer than 400 new land-based ballistic missiles. While this approach may eventually reduce the total cost of the procurement programs and long-term costs of operating the force, it is unlikely to relieve the fiscal pressure during the peak procurement period. This is because the highest costs associated with procurement are in the earlier years of the programs, in the 2020s, when greater numbers of systems are entering the force. Eliminating weapons systems that would be purchased in the latter years of the programs would do little to ease pressures at that time.

Others have suggested that the Pentagon slow the ongoing programs, and stretch procurement out over a longer period of time. This could relieve pressure on the budget in the mid-2020s, and lower the "bow wave" associated with increased procurement in all the programs. Critics of this approach have argued that the programs have already been slowed, and are now "late to need," and that further delays could undermine the capability and credibility of the U.S. nuclear force. At the same time, although such an approach might reduce annual costs for some years of the programs, it would increase the total cost of each program because a slower program would be less cost efficient.

Finally, many analysts have suggested that the Pentagon reduce the cost of the nuclear modernization programs by cancelling one or more of the planned new systems. This would reduce costs in both the near term and the long term. Many have focused, in particular, on the new long-range stand-off weapon (LRSO), arguing that it is not needed if the United States deploys a new bomber that can penetrate air defenses. Others have suggested the cancellation of the new land-based missile, arguing either that the United States no longer needs a land-based component for its nuclear triad or that the older Minuteman III missiles can be extended and maintained for a longer period. While these proposals are crafted and presented as a way to save money, they often reflect the view that these weapons add little value to U.S. national security, regardless of their cost. Those who support the programs dispute these conclusions, arguing that the LRSO and GBSD will provide added flexibility to the U.S. force and continue to complicate an adversary's ability to attack the United States

This again reveals that the debate over the nuclear modernization programs is as much, or more, about how much nuclear weapons are worth than about how much they cost.

CONGRESSIONAL PERSPECTIVES

In 2010, when the Obama Administration submitted the New START Treaty to the Senate for its advice and consent to ratification, it pledged support for the recapitalization of much of the U.S. nuclear enterprise. In response to a congressional mandate, it submitted a report, known as the 1251 report, that detailed expected levels of funding for both DOD and NNSA programs over the next decade. Most experts agree that the Administration's promise of support for modernization funding served as an essential ingredient in winning enough votes for the Senate to reach the two-thirds threshold needed to consent to ratification of the Treaty. At the same time, the promised ratification and implementation of the Treaty served as an important ingredient in winning support for continued funding for the nuclear modernization programs.

In the years since the ratification of the New START Treaty, Congress has continued to support the Obama Administration's (and now the Trump Administration's) requests for funding for nuclear modernization programs. The support has not been universal, and many Members have questioned both the funding levels and strategic rationale for some of the programs. Nevertheless, as remains true with both the House and Senate Mark-ups of the FY2018 National Defense Authorization Act (NDAA) a majority of Members in both the House and Senate continue to support the recapitalization of U.S. nuclear forces.

On the other hand, the broad consensus in support of these programs may be somewhat fragile. While most Members support most of the "program of record" outlined by the Obama Administration, changes that either add new programs or remove existing programs could undermine the broad consensus. For example, several Members have sought to reduce or eliminate funding for the LRSO. While it seems unlikely that the Trump Administration would request reductions in funding for the LRSO, it might seek to accelerate the program, which could not only embolden those seeking to restrict or eliminate the funding, but could also undermine the broader consensus. Of more concern is the possibility that the Administration will seek studies or funding in support of new types of nuclear warheads. While the NNSA continues to pursue Life Extension Programs to refurbish existing warheads, the United States has not added a new warhead to the force since the late 1980s and Congress rejected the Bush Administration's efforts to do so during the 2000s. Such an effort by the Trump Administration could spur significant resistance in Congress, particularly if it was seen as a step towards the resumption of explosive nuclear tests. Some believe that this resistance could undermine support, more broadly, for other elements of the recapitalization program.

The Trump Administration might also test the link between Senate consent to the ratification of the New START Treaty and congressional support for the nuclear modernization programs. While the reductions mandated by the New START Treaty must be complete by February 2018, the Treaty remains in force through February 2021, with the possibility of extension through February 2026. Press reports indicate that President Trump has questioned the value of the treaty and has argued that it was poorly negotiated and favors Russia. Some in Congress agree with this assessment and question the value, more broadly, of negotiated limits on U.S. nuclear weapons. Others, including many officials in the Pentagon, continue to support New START, both because it limits Russian nuclear forces and because it mandates monitoring and verification activities that provide the United States with insights into Russian force levels and deployments.

In a speech before to the Carnegie Nuclear Policy Conference in March 2017, Chris Ford, who is responsible for nuclear weapons and arms control issues at the National Security Council, indicated that the United States plans to fully implement the New START, but that its future, including its possible extension, is under review. It is not clear whether this statement offered support for the treaty through 2021, or only through 2018, when the reductions must be complete. What is clear, however, is that a U.S. decision to withdraw from or suspend participation in New START before 2021 could cause significant numbers of Members of Congress to reconsider their support for the Pentagon's nuclear modernization programs.

REFERENCES

1. U.S. Department of Defense, Remarks by Secretary Carter to troops at Minot Air Force Base, North Dakota, September 26, 2016. http://www.defense.gov/News/Transcripts/Transcript-View/Article/956079/remarks-by-secretary-carter-to-troops-at-minot-air-force-base-north-dakota

[2] U.S. Congress, House Armed Services Subcommittee on Strategic Forces. Hearing. on Nuclear Forces and Atomic Energy Defense Activities for Fiscal 2018 Priorities. May 25, 2017.

[3] U.S. Department of Defense. Annual Report to Congress, Fiscal Year 1989, by Frank Carlucci, Secretary of Defense. February 18, 1988. Washington, 1988. p. 54.

[4] U.S. Department of Defense, Nuclear Posture Review, Washington, DC, April 6, 2010, p. 20-22. http://www.defense.gov/npr/docs/2010%20Nuclear%20Posture%20Review%20Report.pdf

Global Strike Hypersonic Weapons

Mark J. Lewis

Director, Science and Technology Policy Institute
Institute for Defense Analyses
1899 Pennsylvania Ave NW
Washington DC 20006

Abstract. Beginning in the 1940's, the United States has pursued the development of hypersonic technologies, enabling atmospheric flight in excess of five times the speed of sound. Hypersonic flight has application to a range of military and civilian applications, including commercial transport, space access, and various weapons and sensing platforms. A number of flight tests of hypersonic vehicles have been conducted by countries around the world, including the United States, Russia, and China, that could lead the way to future hypersonic global strike weapon systems. These weapons would be especially effective at penetrating conventional defenses, and could pose a significant risk to national security.

BACKGROUND

Hypersonics refers to flight in the atmosphere in excess of approximately five times the speed of sound, beyond the speed of even the fastest operational manned aircraft. A hypersonic craft would cover more than a mile per second, flying at more than seven times the speed of a conventional airliner. A hypersonic airplane could fly from New York to London in less than an hour; a hypersonic missile could cover 1000 miles in about 15 minutes. Because they operate in the atmosphere, a hypersonic missile could also use aerodynamic forces to maneuver and change its trajectory.

After decades of research and a list of mostly failed programs, a number of recent successful hypersonic activities around the world have inspired a renewed interest in high speed weapons. Such weapons would introduce a new threat to defensive systems, complicating both detection and response. Particularly challenging will be responding to the combination of the combination of high speed and maneuverability afforded by hypersonic technology. Several recent studies have addressed the growing threat of these future high-speed maneuvering weapons (HSMWs), employing speeds in the hypersonic realm to offer rapid effects against which there may be limited defenses.

A recent report of the Air Force Studies Board of the National Academies of Science, Engineering, and Medicine explored the threat, and recommended responses to the development of these HSMWs.[1] This report was written in response to developments around the world, including efforts in The People's Republic of China and in the Russian Federation, to flight test hypersonic systems. In fact, those two countries are already publicly acknowledging their interest in high-speed systems that, if fielded, could endanger both forward-deployed U.S. forces and even the continental United States itself. At the same time, several efforts in the United States have also explored the combination of speed and maneuverability for a next generation of missiles and information-gathering platforms.

The combination of speed, maneuverability, and trajectories associated with HSMWs means that they may operate in the seams of current defensive systems. Put another way, while operational doctrine and command structures adequately address traditional atmospheric air attack or exoatmospheric ballistic missile attack, existing doctrine and organizational structure may not be adequate to address the cross-domain threat posed by HSMWs. These weapons may be used as either strategic or tactical systems, with conventional or nuclear

Nuclear Weapons and Related Security Issues
AIP Conf. Proc. 1898, 020005-1–020005-6; https://doi.org/10.1063/1.5009210
Published by AIP Publishing. 978-0-7354-1586-7/$30.00

warheads, introducing not only flexibility, but ambiguity of intent. At present, a ballistic missile's trajectory would be readily discernible; that would not be the case for an HSMW.

The applications of hypersonic technology can be sorted into three general categories: weapons, aircraft, and spacecraft. High speed weapons include hypersonic cruise missiles (HCMs), able to fly several hundred kilometers in a matter of minutes on jet power, or unpowered boost-glide hypersonic cruise vehicles (HVCs) that reach hypersonic speeds on a rocket booster, then glide at high-speed to a precise impact. Hypersonic cruise missiles would operate in a manner similar to existing subsonic cruise missiles, except they would travel in ten minutes about the same distance that a conventional missile would cover in more than an hour. Boost gliders resemble ballistic missiles, except they fly at significantly lower altitudes and are far more maneuverable, and thus far less predictable.

FIGURE 1. The USAF X-51 flight test vehicle. Forward (grey) part is the scramjet-powered test vehicle; rear (red) segment is a solid rocket motor for initial acceleration to scramjet speeds.

A type of engine known as a supersonic combustion ramjet, or scramjet would likely power HCM's operating above Mach 5. This is a class of engine first conceived in the late 1950's but only fully demonstrated in 2013. In contrast, HCV's do not require any new engine technology developments, but are instead boosted on conventional rockets to high speed, which is then bled off in an unpowered glide. A convenient way to think of these two systems is that boost gliders invest all of their propulsion energy up front, and then deplete that energy gradually, whereas cruise missiles release their energy at a steady rate until a short final terminal dive. Both airbreathing cruise missiles and boost glide weapons would be maneuverable, and operate at altitudes above those of conventional aircraft, but significantly below that of ballistic missiles. Because of their maneuverability, but HCMs and HCVs fall into the general category of HSMWs. Both types of systems operate below the classical ballistic missile trajectory and above typical low-speed cruise missile operating altitudes. The altitudes at which they fly also contribute to their lethality, as they would difficult to detect either from the ground, because of limited viewing angle, or from space because of background clutter. The USAF X-51, shown in Figure 1, is representative of an airbreathing HCM, though X-51 was a flight experiment that lacked essential components of a real weapon, including operational control and warhead systems. A representative boost-glide HCV is shown in Figure 2, which depicts the DARPA HTV-2 flight test vehicle as it would have appeared in its trans-pacific hypersonic glide.

In addition to flying at high speed through most of their trajectories, both categories of HSMWs will also likely decelerate, but still impact their targets at velocities in the high- supersonic (Mach 3.0-4.0) range. They could maintain significant maneuverability with precision even in the terminal phase, and thus be capable of engaging fixed or slow-moving targets, such as a runway, command and control facility, or seagoing vessel. Both categories of high-speed weapons may be capable of carrying conventional or nuclear warheads, thereby complicating strategic intent and posture as well as operational identification, response, and engagement.

Military applications of high-speed flight include more than just weapons. Fast penetrating intelligence, surveillance, and reconnaissance (ISR) platforms might also benefit from the technologies associated with this flight regime. This could include either manned or unmanned systems, supplementing or replacing space surveillance, or even operating as cueing systems for weapons.

FIGURE 2. DARPA HTV-2 boost glide flight test vehicle.

TECHNICAL CHALLENGES TO HIGH SPEED FLIGHT

Hypersonic flight is not new. Man-made objects have been flying at hypersonic speeds since February 24, 1949, when a two-stage American "Bumper" rocket, assembled from a captured German V-2 with a WAC Corporal sounding rocket as an upper stage, reached a speed of 2.3 kilometers per second. Bumper flew less than a decade after German researchers Eugen Sänger and Irene Bredt did studies of intercontinental hypersonic missiles concepts capable of ocean-spanning flights. With the start of the space age, hypersonic flight became practical, not only as a means of returning spacecraft from orbit, but also for the reentry of missile warheads. The class of weapons that are categorized as HSMWs represent a significant departure from previous hypersonic systems because they will be optimized for sustained atmospheric flight. This in turn requires that they have low drag and high lift, and they must handle the associated heating challenges.

Spacecraft and missile nosecones have typically been designed with blunt shapes that have high drag but can also handle the heating associated with high-speed flight through the atmosphere.[2,4] Because blunt shapes have significant drag they are fine for slowing down on return from space, but are not suitable for a vehicle designed to fly an extended distance through the atmosphere. To fly for sustained distances through the atmosphere requires an aerodynamic form which has low drag, which means sharp leading edges and a slender form.[4] Since the mid 1980's there has been renewed interest in such hypersonic shapes that offer both low drag and high lift; that combination not only allows for sustained flight through the atmosphere, but maneuvering using aerodynamic forces.

Heating and Drag

The term "hypersonic' is actually somewhat arbitrary, but it is generally applied to speeds at which a craft generates very strong shockwaves and in which skin friction yields temperatures that are high enough to produce chemical changes to the surrounding air. There are no sudden delineating changes in flow properties that occur as a craft enters the hypersonic realm; this is in strong contrast to the transition that occurs as a vehicle passes through

Mach 1, the speed of sound, where the fundamental behavior of air changes.

Some of the challenges of hypersonic flight are connected to the high temperatures and aerodynamic forces. Extreme surface temperatures, especially at leading edges, can destroy all but the most robust materials. As such, an understanding of high temperature materials is essential to the practical design of an HSMW. The shockwaves produced on a vehicle flying at any speed above that of sound, realized as sudden changes in pressure and temperature, can contribute significantly to drag and heating. There are still significant gaps in our basic knowledge of high speed flow; for example, it is uncertain when the air moving past a hypersonic vehicle will transition from a smooth laminar state to turbulence, with an associated increase in drag and heating. There are also significant challenges in accurately simulating the flight environment of a hypersonic vehicle on the ground, though various types of wind tunnels can be employed to capture parts of the performance envelope.

There are clearly also challenges in designing sensors and actuators that can operate effectively in a hypersonic flight environment, as well as warhead systems. Seekers that are not affected by the hot gases surrounding a vehicle in hypersonic flight must be developed, as well as control systems that are effective across a wide operating envelope.

Propulsion Options

Either rockets or so-called airbreathing engines can power high-speed weapons. Rockets are a good choice for either short range weapons, such as for air-to-air engagement, or extremely long range weapons such as ballistic missiles. Airbreathing engines can offer significant advantages over rockets for flight within the atmosphere, since oxidizer can be collected rather than carried on board in a fuel tank. This means that the vehicle can travel farther on a tank of fuel, or carry more useful payload for a given overall weight, clearly an advantage for a long range weapon. Airbeathing propulsion is however a special challenge at high Mach numbers. Traditional jet engines will not work in this regime, and thus a new type of engine is required.

The likely hypersonic engine of choice, the scramjet, functions by allowing the air that passes through the engine to remain essentially at flight speed. This keeps temperatures inside the engine at levels where fuel can still burn. Timescales in a scramjet become important: the engines that powered the X-51 craft had to swallow air through an inlet, inject and mix fuel into the air, burn that fuel with the air, and exhaust out the nozzle in about a thousandth of a second. This has been likened to lighting a match in a hurricane. As a result of shockwave physics, the engine on a hypersonic vehicle must actually be designed as part of the whole vehicle in one integrated package, creating one of the most difficult design and optimization problems in all of aerospace engineering.

Scramjets were first envisioned in the late 1950's, with groundbreaking work by Weber and McKay at the NACA Lewis Research Center.[5] Their analysis examined the speed limitations of current jet engines and considered theoretical systems that could operate up the supersonic Mach scale. Efforts to develop scramjet engines were a later part of the 1960's X-15 program, and continued as a primary element of the National Aerospace Plane Program of the mid-1980's and early-1990's. A research team at the University of Queensland flew what is generally believed to be the first airbreathing scramjet in July 2002, followed by two successful flights of NASA's X-43 in 2004 that proved that scramjet thrust could be greater than vehicle drag.

A milestone in the development of airbreathing HSMWs occurred on May 1, 2013, when a B-52H bomber launched the last of four X-51 vehicles at Edwards Air Force Base. The X-51 was an unmanned experimental research platform designed to probe the limits of highs-speed flight under scramjet power. X-51's engine was built by the Rocketdyne Corporation, and burned a relatively conventional high-temperature jet fuel. Before X-51, scramjets had long been seen as a key to sustained hypersonic flight in the atmosphere, and other important steps in their development had already occurred in ground test and in the air, including NASA's X-43 craft, and the Australian HyShot flights. But X-51's flight was definitive proof of the viability of scramjets as a practical operational hypersonic propulsion system. A key feature of the X-51 design is that it incorporated a scramjet engine in a missile-like configuration that could operate for long durations with a practical operational fuel, and that the scramjet could provide thrust for as long as it had available fuel.

IMPLICATIONS FOR THE UNITED STATES

In "Americas Air Force: A Call to the Future" the Air Force leadership detailed a 30-year plan for modernization, including the pursuit of five so-called game-changing technologies.[6] Each of the included technologies was chosen for its ability to contribute to the basic attributes of airpower: speed, range, flexibility, and precision. The first technology on that list was hypersonics.

As the vision statement explains, hypersonics holds the promise of significant advantages for future weapon systems, with a "profound impact" that would "revolutionize the way we approach…core missions in the future." This is a clear signal that leaders within the United States military view hypersonic weapons as worthy of investment. Flight test programs such as the DARPA Hypersonic Test Vehicle (HTV-2) boost glide, the U.S. Army Advanced Hypersonic Weapon (AHW), and scramjet programs such as X-51 and the U.S Navy HyFly all pointed the way to the development of HSMWs, even though each of those programs has since ended, and their record of success was mixed. In fact, HyFly attempted three tests, none of which even reached hypersonic speed, and the two flights of HTV-2 ended a few seconds into a planned ocean-spanning glide. Even X-51 had only one fully successful test out of four. These programs and associated position papers have clearly communicated a national intent to potential adversaries.

The nature of the U.S. air and space power advantage is captured in the Air Force's recognition of 'Global Vigilance, Reach, and Power' as its core contributions to national security. As the National Academies' report observed, through the awareness, access, and combat force made possible by U.S. Air Force systems, the United States has maintained a robust global presence sustained by air and sea lines of communication, and overseen by a rapidly responsive command and control infrastructure. The United States relies on this presence to exercise policy options that support its friends and allies, defend its own national security interests, conduct humanitarian relief, and promote other initiatives. HSMWs threaten that doctrine.

As the National Academies report also noted, at the present time, the United States military is facing an emerging class of HSMW threats in several countries that would severely challenge its ability to maintain global vigilance, would constrain its global presence, and impede its global power.[1] A potential adversary's use of HSMWs would render U.S. defensive capabilities inadequate, and would be an impediment to the application of global power in key regions of the planet. Both the People's Republic of China and the Russian Federation have flown examples of HSMWs, and other countries have shown interest in many of the underlying technologies for hypersonic flight.

As just one example of the challenges of trying to respond to an HSMW attack, the very nature of an HSMW flight profile will greatly compress decision and response timelines. Even if available, useful countermeasures would have to be deployed almost immediately. Another major challenge to organizational leadership is that the development of countermeasures to HSMW systems may fall outside of current DOD organizational mission responsibilities.

The threat of high speed global strike weapons could potentially be mitigated through arms controls and non-proliferation approaches. Pursuing such solutions in an integrated manner would provide policy developers a sense of direction, as well as offer the diplomatic community an informed base for integration into international regimes for arms control and deterrence. Further, a visible response to impending threats will demonstrate resolve to both our potential competitors and our own military personnel.

In summary, HSMWs may impede operations, global vigilance, maintenance, and supporting logistics. At a national strategic level, HSMWs could hold at risk the fundamental U.S. construct of global reach and presence.

ACKNOWLEDGEMENTS

The contributions of the members of the National Academies Air Force Studies Board Committee on Future Air Force Needs for Defense Against High-Speed Weapon Systems are gratefully acknowledged. Committee members included Dr. Thomas R. Bussing, Dr. Richard P. Hallion, Mr. Terry J. Jaggers, Mr. Eric D. Knutson, Mr. Gary O'Connell, The Honorable Malcolm R. O'Neill, Dr. David A. Whelan, Adm. (ret). Richard W. Mies, Dr. Brian R. Shaw, and Maj Gen (ret.) Suzanne Vautrinot. The contributions of Dr. Joan Fuller, Executive Director of the Air Force Studies Board, are also gratefully acknowledged.

REFERENCES

1. M. J. Lewis, and the Committee on Future Air Force Needs for Defense Against High-Speed Weapon Systems, *A Threat to America's Global Vigilance, Reach, and Power–High-Speed, Maneuvering Weapons* (Unclassified Summary) The National Academies Press, Washington D.C. 2016.
2. R. P. Hallion *The Path to the Space Shuttle: The Evolution of Lifting Body Reentry Technology*, U.S Air Force History Office, Washington D.C .1983.
3. Johnson, J. E., Starkey, R. P., and Lewis, M. J., "Aerothermodynamic Optimization of Reentry Heat Shield Shapes for a Crew Exploration Vehicle", *Journal of Spacecraft and Rockets*, Vol. 44, No. 4, July-Aug. 2007, pp. 849-859.
4. R. Starkey and M. Lewis, "Critical Design Issues for Airbreathing Hypersonic Waverider Missiles," *Journal of Spacecraft and Rockets*, Vol. 38, No 4, July-August 2001 pp. 1214-1221.
5. R. J. Weber and J. S. McKay, "An Analysis of Ramjet Engines Using Supersonic Combustion", TN-4386, National Advisory Committee for Aeronautics, Sept. 1958.
6. M. A. Welsh, *America's Air Force: A Call to the Future,* Office of the Secretary of the Air Force, 2014.

Unmaking the Bomb: Verifying Limits on the Stockpiles of Nuclear Weapons

Alexander Glaser

Nuclear Futures Laboratory
Engineering Quad, J207
Princeton University
Princeton, NJ

Abstract. Verifying limits on the stockpiles of nuclear weapons may require the ability for international inspectors to account for individual warheads, even when non-deployed, and to confirm the authenticity of nuclear warheads prior to dismantlement. These are fundamentally new challenges for nuclear verification, and they have been known for some time; unfortunately, due to a lack of sense of urgency, research in this area has not made substantial progress over the past 20 years. This chapter explores the central outstanding issues and offers a number of possible paths forward. In the case of confirming numerical limits, these include innovative tagging techniques and approaches solely based on declarations using modern cryptographic escrow schemes; with regard to warhead confirmation, there has recently been increasing interest in developing fundamentally new measurement approaches where, in one form or another, sensitive information is not acquired in the first place. Overall, new international R&D efforts could more usefully focus on non-intrusive technologies and approaches, which may show more promise for early demonstration and adoption. In the meantime, while warhead dismantlements remain unverified, nuclear weapon states ought to begin to document warhead assembly, refurbishment, and dismantlement activities and movements of warheads and warhead components through the weapons complex in ways that international inspectors will find credible at a later time. Again, such a process could be enabled by modern cryptographic techniques such as blockchaining. Finally, and perhaps most importantly, it is important to recognize that the main reason for the complexity of technologies and approaches needed for nuclear disarmament verification is the requirement to protect information that nuclear weapon states consider sensitive. Ultimately, if information security concerns cannot be resolved to the satisfaction of all stakeholders, an alternative would be to "reveal the secret" and to make available select warhead design information.

INTRODUCTION: WHAT IS TO BE VERIFIED?

Nuclear arms control will face qualitatively new verification challenges once there are bilateral or multilateral agreements that place limits on the total number of warheads in the arsenals of weapon states. Such a step is widely expected as the total number of warheads now far exceeds the number of warheads that are actually deployed, i.e., there are thousands of warheads in reserve, in storage, or in the dismantlement queue. This chapter explores and illustrates with examples two verification tasks that are likely to be central in this context: confirming numerical limits on declared nuclear warheads and confirming the authenticity of nuclear warheads before they are dismantled (Figure 1). Meeting these objectives would require new technologies and approaches. In particular, onsite inspections at new types of facilities, including warhead storage sites, may be required, which could put at risk highly sensitive information both related to military operations and warhead design. Weapon states may be reluctant to consider—or at least cautious to embrace—some of the anticipated procedures. To address these concerns, the discussion examines the potential of verification approaches that emphasize non-intrusiveness from the outset. Relevant examples include innovative tagging approaches and hashed declarations to confirm the correctness of warhead declarations and novel types of inspection systems to confirm the authenticity of nuclear warheads, while satisfying the different and sometimes conflicting requirements by the host and the inspector.

Nuclear Weapons and Related Security Issues
AIP Conf. Proc. 1898, 020006-1–020006-9; https://doi.org/10.1063/1.5009211
Published by AIP Publishing. 978-0-7354-1586-7/$30.00

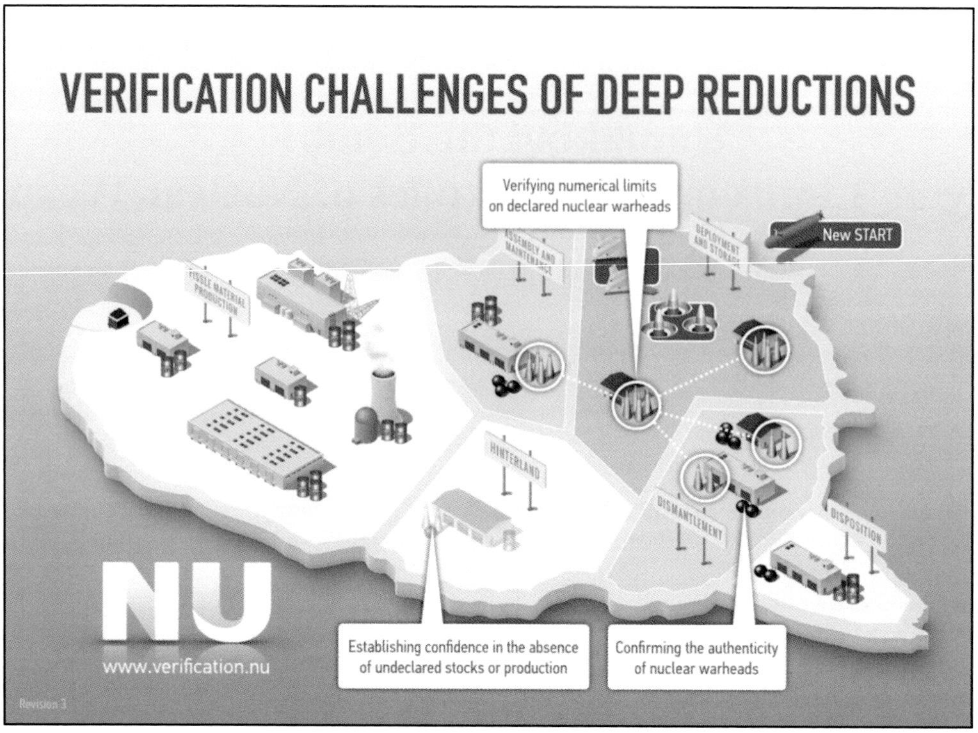

FIGURE 1. Verification Challenges of Deeper Warhead Reductions. Confirming numerical limits on declared nuclear warheads, including those in storage, is a challenging verification task because their locations and movements are generally considered sensitive and tagging may be difficult to implement. Prior to dismantlement, the authenticity of declared warheads needs to be confirmed, which requires trusted inspection systems that can protect sensitive design information. Development and demonstration of such systems has proven difficult. Finally, as with any other arms control agreement, there has to be confidence in the absence of undeclared stocks and production, which could be accomplished using a number of technical and non-technical approaches. The map is based on a fictional state model available at www.verification.nu, which can be used to illustrate various verification strategies.[1]

CONFIRMING NUMERICAL LIMITS

Procedures and techniques to confirm upper limits on the number of nuclear warheads would become a key verification objective should future arms-control agreements place limits on the *total* number of nuclear weapons in the arsenals. Verifying such agreements may then require the ability for inspectors to account for individual warheads rather than launchers. In principle, this can be accomplished by tagging treaty accountable items with unique identifiers (UIDs), which effectively transforms a numerical limit into a ban on untagged items.[2, 3] Direct tagging of warheads may be difficult to implement in practice, however, because the host may have safety, performance, or other concerns, and inspections would necessarily be highly intrusive. At the same time, the locations and movements of warheads and weapons in these categories may be considered sensitive information, and robust verification measures to ensure the authenticity and integrity of a declared treaty accountable item could put such information at risk. Two distinct approaches to confirm numerical limits while preserving high levels of non-intrusiveness and privacy are offered here for discussion.

Example 1: Buddy Tags

One approach to address this dilemma but still use tagging techniques is to separate the (sensitive) treaty accountable item from its tag, an idea first proposed by Sandia National Laboratories in the early 1990s as an option to confirm numerical limits on missile systems (Figure 2, left). This so-called "buddy tag" would act as a token, proving ownership of a treaty accountable item without having to present the item itself for close inspection.[4] Since

2016, Sandia together with Princeton has explored enhanced and miniaturized prototypes using modern software and hardware concepts to examine the viability and value of the concept for verifying future arms control agreements.[5]

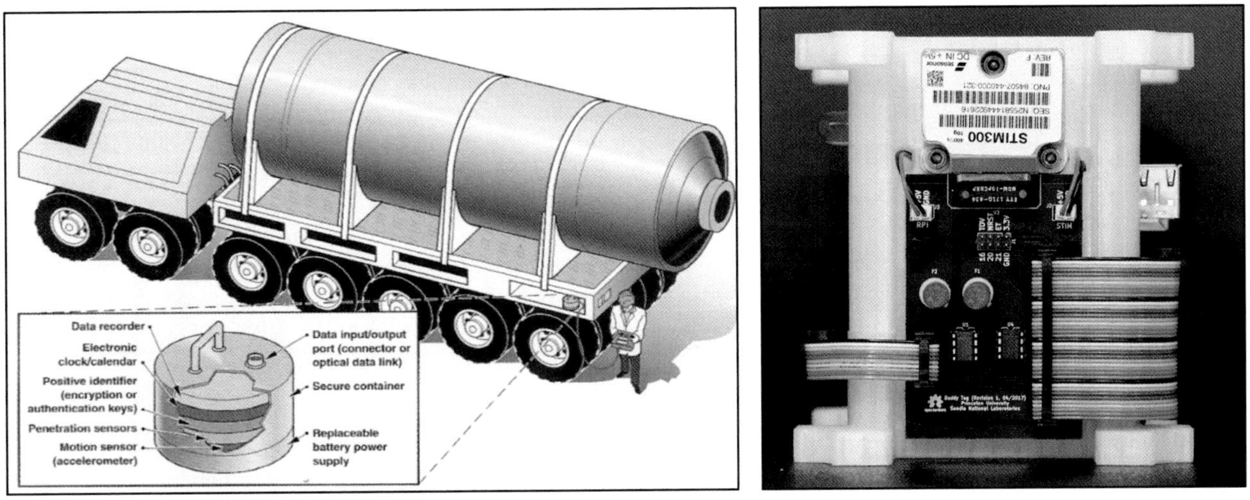

Figure 2. Artist's conception of the original Buddy Tag for confirming limits on numbers of treaty-accountable missiles. Shown on the right is the 2017 buddy-tag package developed as part of a joint project between Princeton University and Sandia National Laboratories. *Source James Fuller*[6] *and U.S. Department of Energy (left) and author (right).*

In a tagging regime using buddy tags, a party would declare a certain number of treaty accountable items and receive exactly one tag for each of these items. The basic idea is that, during a short-notice onsite inspection later on, the inspected party must be able to present one buddy tag for every declared treaty accountable item encountered at the site. Between inspections, buddy tags would be stored "near" the treaty accountable items, at the same site, perhaps in a dedicated and lower-security display area to which inspectors would have relatively easy and quick access. If a treaty accountable item is moved between sites, the buddy tag must go with it. Buddy tags do not have to be associated with a particular treaty accountable item, which could help protect operational patterns that the host party may consider sensitive.

Once a short notice inspection is called, the designated site goes into a stand-down, and the buddy tags present at the site are activated by the inspected party, ideally within minutes of the call. From then on, the inspected party may not move any treaty accountable items or buddy tags on or off site until the inspectors arrive at the site and the inspection can be completed. The stand-down on movements of treaty accountable items could be monitored through a combination of national technical means such as satellite reconnaissance and possibly also through portal-perimeter monitoring. Inter-site movements of buddy tags, which are much smaller and whose movements would be difficult to monitor otherwise, are to be detected by a motion-sensing subsystem in the tag (Figure 2, right). During a typical inspection, inspectors would count the number of buddy tags present, verify that the tags have not been moved since they have been activated, check the integrity of the tags, and confirm their authenticity, e.g. through a unique identifier that is attached to the tag. Inspectors would then be allowed to visually confirm that an equal number of treaty accountable items are present at the site and accept these declared items without further authentication. Inspectors might also be able to verify that additional items present at the site are in fact non-treaty-accountable and access other areas large enough to contain treaty accountable items to ensure the absence of such items in those areas. Over time, the numerical limit on accountable items could be gradually lowered, and a corresponding number of buddy tags would be jointly disabled or destroyed following specified procedures.

When implemented effectively, the buddy-tag approach makes it difficult for an inspected party to store undeclared treaty accountable items at a declared site. The party would be forced to hold such items at undeclared sites, which may be difficult to conceal from national technical means over time, especially when it is assumed that adequate security and a parallel support infrastructure would have to be maintained.

Example 2: Cryptographic Escrow and Hashed Declarations

A fundamentally different approach to confirm numerical limits could be based on declarations alone and avoid entirely the need for tagging. In particular, the concept of hashed declarations has been previously proposed as a strategy to protect information as part of a regime based on declarations.[7] Again, it is assumed that parties would be reluctant to provide the exact breakdown of storage locations for the treaty accountable items in their inventories. This could be due to security concerns or due to the fact that an adversary might be able to infer operational information that the host considers sensitive. Hashed declarations would avoid this concern and could also contain additional information that may or may not be revealed at a later time. The concept has recently been expanded to take advantage of modern cryptographic escrow schemes, which provide a way for multiple parties to commit, exchange, and potentially make computations on sensitive data without revealing the content of the data until the parties are ready to do so.[8] In the most elementary implementation of such an escrow scheme, a hash (or "message digest"), which is much shorter than the message itself, is shared and made public. The underlying cryptographic hash function is designed such that it is extremely difficult to find a valid message for a given hash or to construct two different messages that produce the same hash (collision resistance). Here, for illustrative purposes, a simplified step-by-step example is offered to explain how such a method could be used to confirm limits on stockpiles of nuclear weapons.

Under such a framework, treaty parties would regularly or even frequently (for example, electronically, every 48 hours) exchange hashed declarations with exactly one entry (row) for each treaty accountable item. These declarations could be made public as they do not contain any information besides the total number of declared items, i.e., the number of rows in these declarations (Figure 3, top insert). This aggregate information is generally not considered sensitive. For example, the United States has already declared the exact number of nuclear warheads in its arsenal. In September 2014, this number stood at 4,717. Similarly, the United Kingdom and France have given upper ceilings that are considered very close to the actual values. Making declarations as considered here would therefore not be without precedent and could reaffirm existing transparency measures.

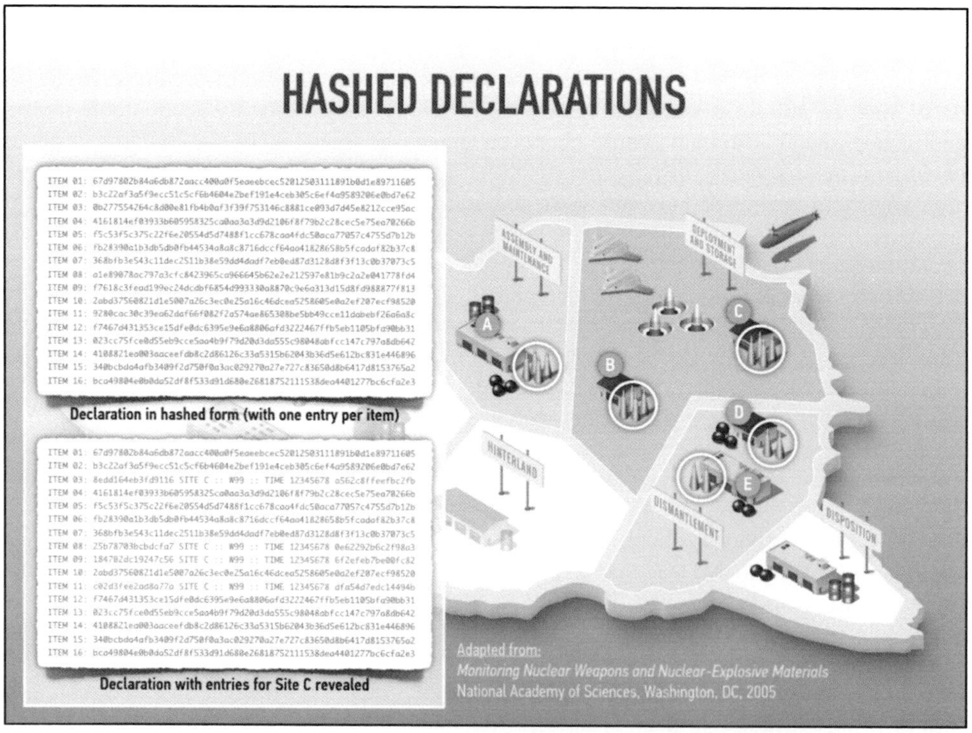

FIGURE 3. Hashed Declarations. The declaration contains one message digest for every treaty accountable item. The host reveals the cleartext for all items that are present at the site selected by the inspector. It is imperative that the host cannot reliably generate hash collisions, such that two different messages (e.g. showing two different storage locations for the same item) produce identical message digests; this could be used to conceal items exceeding the declared number. Shown message digests (generated with SHA-256) are for illustration purposes only; actual messages and digests may have to look differently to meet security requirements. *Map: verification.nu.*

In preparation for an onsite inspection and based on the most recent available hashed declaration, the inspecting party would first announce the storage or deployment location that it would like to inspect. At that point, a stand-down for the relevant site would take effect, which could be verified by national technical or other means. Alternatively, the inspecting party could simply pick one entry from the hashed declaration at random, i.e., without knowing the name or type of the site that will ultimately be inspected. Once the site is revealed, the host would then also reveal all other entries for the same site. Again, a stand-down for this site would take effect. The host party would then provide the cleartext for all treaty accountable items that are present at the selected site (Figure 3, bottom insert). The inspecting party can now confirm that the cleartext entries produce the correct message digests for all items. The cleartext would contain previously agreed basic information about each treaty accountable item. At a minimum, the cleartext would contain a trusted timestamp, a designator for the item, and the name of the site or simply the GPS coordinates that fall within the boundaries of the site. The cleartext would also contain some random-character segments to make brute-force, dictionary, or replay attacks impractical; it could also include information to prove the freshness of the entries, i.e., timestamps that demonstrate that the declaration has not been prior to a certain date.

The inspector team can now proceed to the site itself. Once the team arrives at the site, the host would present the declared warheads. Procedures would have to be available to confirm that additional objects that could be mistaken for nuclear warheads are in fact not treaty accountable. Inspectors may also be allowed to access other areas to gain confidence in the absence of undeclared treaty accountable items at the site. Note that, in this most basic scheme using hashed declarations, no tags whatsoever are required; and yet, hashed declarations could over time provide high confidence in the correctness of the warhead declarations made by the parties of a treaty. Hashed declarations could also include extra columns for additional information that is not necessarily revealed in the initial stages of an agreement (such as serial numbers or perhaps even the amounts of special nuclear material in particular warhead types). Disclosure of these entries in the future could further increase the confidence in the correctness and completeness of the declarations, potentially reaching back for years or decades.

CONFIRMING THE AUTHENTICITY OF WARHEADS

At some point prior to dismantlement, and even if verification arrangements seeking to confirm numerical limits on nuclear warheads have been in place for extended periods of time, the inspecting party will require reassurance that declared warheads are authentic and reductions in the arsenals credible. This is the second qualitatively new challenge of nuclear disarmament verification, and it is generally considered the more difficult one.

There are several plausible cheating scenarios that an inspecting party might worry about when the other party removes warheads from its balance sheet. First, the inspected party might present objects that are similar to genuine warheads, but prior to inspection some fissile material is removed or replaced by other (less "valuable") nuclear materials, for example, by natural uranium. The main objective of this strategy would be to withhold fissile material, presumably for future use in weapons. A second scenario would involve presenting objects that might or might not resemble real warheads, but they could be designed primarily to deceive the inspection system. These objects would presumably contain some fissile material in order to pass an inspection based on radiation measurements. The main objective of such a strategy would be to withhold real warheads or extra amounts of fissile material. A viable inspection system should be robust against many of these possible scenarios; more importantly perhaps, the limits of such a system ought to be well understood, i.e., there may be certain cheating scenarios that an inspection system cannot detect, but the parties may decide that they don't matter enough to address them.

Nuclear warheads contain kilogram quantities of plutonium or highly enriched uranium and often both. Even when placed inside a missile or a transport container, these materials emit detectable neutron and gamma radiation signatures that are unique for special nuclear materials, the respective amounts of these materials, and their configuration. Taking advantage of these signatures, two fundamental concepts have been proposed to confirm the authenticity of nuclear warheads: the attribute approach and the template (or template-matching) approach. In a nutshell, the attribute approach examines a set of properties that are considered characteristic for nuclear weapons; this can include qualitative criteria, such as the mere presence of a special nuclear material (e.g. the presence of plutonium), and quantitative criteria, such as meeting agreed threshold values for mass or isotopics (e.g. a maximum ^{240}Pu content). In contrast, the template approach does not seek to determine absolute or relative attributes of the inspected item; instead, it compares a unique radiation signature or "fingerprint" against a previously recorded template generated with a reference item that is known or believed to be authentic.

The radiation signatures acquired for both the attribute and the template method are considered highly sensitive and cannot be revealed to an inspecting party seeking to confirm the authenticity of a nuclear warhead. Primarily for this reason, warhead inspections would necessarily involve complex measurement techniques and procedures. To enable such measurements, the concept of an information barrier has been developed since the 1990s. An information barrier processes the acquired radiation signatures, but displays the outcome of the analysis in a simple pass/fail manner. The critical functional requirements for the barrier are threefold: First, the inspected party must be assured that classified information is protected so that under any circumstances, i.e., even when the equipment is malfunctioning or operated incorrectly, only non-sensitive information is presented to the inspecting party ("certification"); second, the inspecting party must be confident that the inspection system measures, processes, and presents the conclusion drawn from the data in an accurate and reproducible manner ("authentication"); and third, the system must reliably distinguish valid from invalid items ("confirmation"). Simultaneously certifying and authenticating information barriers has been the most serious obstacle to demonstrating the concept as a viable verification technology.[9, 10]

Joint development of the information barrier, including of both its hardware and software, has generally been considered a key strategy to facilitate mutual confidence and trust in the inspection equipment.[11] Even then, however, the question remains who supplies or has last custody of the equipment that will be used in an actual inspection. Project participants have usually concluded that the information barrier would *de facto* be host-supplied. This makes certification relatively straightforward, but authentication for the inspector extremely difficult. From the perspective of information security, host-supplied equipment appears preferable but, from the outset, the inspector is at a disadvantage given the difficulty of authentication equipment in a host-controlled environment. This task is particularly difficult, and perhaps impossible, given recent advances in stealthy insertion of Trojans at the hardware level, which could trigger hidden switches to produce non-genuine results during an inspection. There has recently been some interest in challenging the premise of host-supplied equipment in favor of equipment that would be provided by the inspector instead.[12]

Finally, both attribute and template systems face some additional challenges that are characteristic for each approach. In the case of attribute measurements, the question arises what types of attributes should be selected and what the exact threshold values for those attributes should be. A party may even refrain from proposing certain attributes worrying that the mere suggestion might reveal weapon features that the other party is unaware of. In general, the more representative the attributes are, the more robust the verification approach would be; but more information about the inspected warhead would necessarily be revealed also. Some experts have therefore suggested going back to very simple and attributes, essentially confirming the presence of fissile material (and perhaps high explosive) but nothing else.[13, 14] In the case of template measurements, qualitatively different challenges exist; these include: how to establish the authenticity of the template in the first place, how to protect sensitive design information that the template contains, how to account for differences between valid items (e.g. manufacturing tolerances, age of material) and how to store the template between measurements so that the inspector remains confident in its authenticity.

In all traditional approaches to nuclear warhead verification, the root cause for most (if not all) key problems is the acquisition and storage of sensitive information, which has then to be protected, removed, or destroyed. This requires an information barrier that both the host and the inspector can simultaneously trust, which may be an impossible task in the case of electronic equipment. Recognizing these difficulties, more recently, there has been increased interest in developing fundamentally different measurement approaches where, in one form or another, sensitive information is not acquired in the first place and where electronic equipment is avoided at critical stages of the data acquisition process. One such approach is based on neutron radiography using non-electronic bubble detectors combined with an interactive zero-knowledge protocol.[15, 16] In this particular case, if a genuine item is presented, detectors have a predetermined bubble count after the inspection; in other words, they do not contain any information, and nothing is learned in the process except that a genuine warhead was presented. Other approaches following a similar philosophy have been proposed since 2014, including a fast-neutron imaging detector that produces a "null image" from a positive and a negative image of two identical items, which cancel each other at all times;[17] a "single-pixel x-ray imager," where all image information from an x-ray image is compressed into a single pixel;[18] and a technique using a high-energy x-ray beam to generate a response from the inspected item based on the effect of nuclear resonance fluorescence, which is then combined with an encrypting foil to remove sensitive features from a transmission spectrum.[19]

None of these technologies and approaches are ready for deployment, and viable inspection systems for warhead confirmation are years away from field-testing with genuine warheads in the presence of international inspectors. This raises the question of what types of inspection tasks and systems would be most valuable and useful in the short and medium term. As one answer to this question, we have recently proposed to focus at least some research efforts on a set of different, more tractable questions, in particular, on measurements to confirm that an inspected item is non-treaty accountable, i.e., not a nuclear object.[20] Such measurements are already envisioned under New START and could help support confirming numerical limits on the total number of warheads in the arsenals discussed in the first part of this chapter. If inspection systems developed for this task were based on simple neutron or gamma radiation detection equipment and measurement types, they may not require information barriers.

WHAT'S NEXT?

This chapter has reviewed the main technical challenges of verifying deeper reductions in the nuclear arsenals: verifying numerical limits on the arsenals and confirming the authenticity of nuclear warhead prior to dismantlement. Due to a lack of sense of urgency, research in this area has not made substantial progress over the past 20 years. When setting new priorities for R&D efforts in this area, it is important to recognize that verifying numerical limits is more straightforward than verifying dismantlements of nuclear warheads, and this is where more research and development efforts could be usefully focused.

In general, new international R&D efforts could more usefully focus on non-intrusive technologies and approaches, which may show more promise for early demonstration and adoption. Ideally, these concepts could be demonstrated at selected sites or "test beds" and be designed in ways that allow future upgrades if desired. Efforts could and should involve non-nuclear-weapon states, non-governmental, and international organizations such as the International Atomic Energy Agency. If demonstrated, such non-intrusive verification approaches could be particularly important for moving forward discussions about expanding the scope of current agreements and facilitating discussions with weapon states that have so far not been part of formal nuclear arms control agreements and may currently have concerns about inspections that the United States and Russia are already used to. In parallel, immersive virtual reality (VR) has been proposed as a fundamentally new pathway to enable collaborations, especially at the government-to-government expert level, going beyond the traditional exchange of ideas at conferences and workshops [21]. VR exercises can be planned and carried out remotely, and VR offers a means to interact in flexible and realistic environments, allowing parties to explore new concepts and approaches and build confidence, laying a basis both for live exercises and for new policy initiatives (Figure 4).

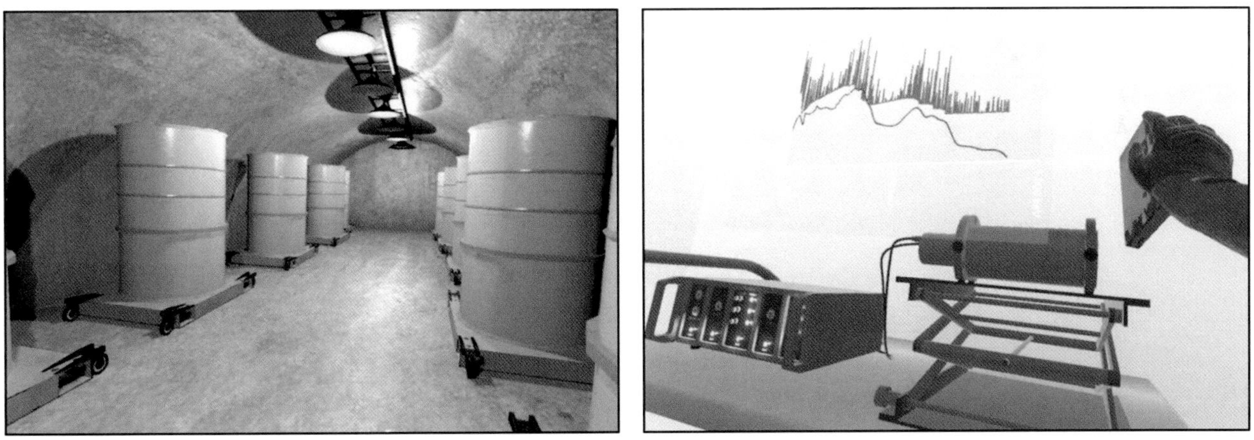

Figure 4. Inspections in the Matrix. While efforts on joint development of verification technologies and approaches are stalling, virtual reality may offer a new approach to support innovations in nuclear arms control. Immersive virtual environments in particular can offer levels of accessibility typically much more difficult to achieve in actual facilities. Networked environments involving host and inspector parties can allow for collaboration among research groups and governments working to find solutions to existing verification challenges. Real-time simulation of radiation fields, as shown on the right, enable the use of realistic inspection systems during virtual exercises.[21]

Ultimately, however, technologies and approaches have to be demonstrated in the field. Here, verified warhead dismantlement is generally considered the hardest task. Relevant work on information barriers ought to be expanded and place particular emphasis be placed on certification and authentication of the equipment. These efforts could rely on joint platforms for hardware and software development. Even then, however, it is far from certain that security concerns, which both host and inspector parties may have, will ever be overcome given that there are numerous pathways to attack and defeat electronic systems. This recognition has led a number of research groups to explore radically different verification approaches, for example, using zero-knowledge protocols.

In the meantime, and with viable inspections years away from deployment, nuclear warheads are continuously being dismantled or, depending on the practices established in different weapon states, regularly refurbished. Almost 90% of all nuclear weapons do not exist anymore today. Timely warhead dismantlements are highly desirable for many reasons, but from an arms-control perspective, they also create new challenges because these dismantlements have so far been unverified and will remains so, at least, for the medium-term future. Recognizing this dilemma, nuclear weapon states ought to begin now to document warhead assembly, refurbishment, and dismantlement activities and movements of warheads and warhead components through the weapons complex in ways that international inspectors will find credible at a later time. This includes generating and preserving appropriate records for all relevant transactions. Modern cryptographic techniques, such as escrowing and perhaps blockchaining, could help demonstrate the authenticity of these records in the future. Although these records may not be made public now, they would help establish the provenance of treaty-accountable items and drastically simplify the verification challenges of nuclear disarmament. Again, a first step would be to raise the awareness of this dilemma and to engage weapon states and non-weapon states in a debate on what to do about this issue.

Finally, it is important to recognize that the main reason for the incredible complexity of technologies and approaches needed for nuclear disarmament verification is the requirement to protect information that nuclear weapon states consider sensitive. This includes both weapon design information such as, for example, the mass of fissile material in a nuclear weapon and operational information such as locations and movements of warheads through the weapons complex. Ultimately, if information security concerns cannot be resolved to the satisfaction of all stakeholders, an alternative would be to "reveal the secret" and to make available select warhead design information, such as fissile material mass. If such a path were to be pursued, then many of the current challenges with defining nuclear weapons and verifying their reductions would go away.

ACKNOWLEDGEMENTS

This chapter is based on research that has been generously supported by the Carnegie Corporation of New York, the MacArthur Foundation, the U.S. Department of Energy's Consortium for Verification Technology under Award DE-NA 0002534, and the U.S. Department of State's Bureau of Arms Control, Verification and Compliance (AVC).

REFERENCES

1. Tamara Patton and Alexander Glaser, "Mapping Nuclear Verification," *58th INMM Annual Meeting,* Indian Wells, California, July 2017.
2. Thomas Garwin, Tagging Systems for Arms Control Verification, AAC-TR-10401/80, Analytical Assessment Corporation, Marina del Rey, CA, sponsored by the Office of Technology Assessment, Washington, DC, February 1980.
3. Steve Fetter and Thomas Garwin, "Using Tags to Monitor Numerical Limits in Arms Control Agreements," Chapter 3 in Barry M. Blechman (ed.), Technology and the Limitation of International Conflict, Washington, DC, 1989, pp. 33–54.
4. Sabina E. Jordan, Buddy Tag's Motion Sensing and Analysis Subsystem, Sandia National Laboratory, Albuquerque, New Mexico, 1991.
5. Ben Reimold, Jay Brotz, Sharon DeLand, Alexander Glaser, Michael Hepler, Andrew Kim, Thomas Schaffner, Heidi Smartt, and Dan Steingart, "Development and Demonstration of a Buddy-Tag Prototype," *58th INMM Annual Meeting*, Indian Wells, California, July 2017.
6. Jim Fuller, "US START TID Development Program: The Quest for Extreme Security Unique Identifiers (1986–1992)," April 2006.
7. *Monitoring Nuclear Weapons and Nuclear-Explosive Materials: An Assessment of Methods and Capabilities,* National Academy of Sciences, Washington, DC, 2005.
8. Sébastien Philippe, Alexander Glaser, and Edward Felten, "Cryptographic Escrow of Nuclear Warhead Inventories for Early Commitment and Non-intrusive Verification," *58th INMM Annual Meeting*, Indian Wells, CA, July 2017.
9. James Fuller, "Information Barriers," *41st Annual INMM Meeting,* New Orleans, Louisiana, July 2000.
10. Yan Jie and Alexander Glaser, "Nuclear Warhead Verification: A Review of Attribute and Template Systems," *Science & Global Security,* 23 (3), 2015.
11. *Trust in Verification Technology: A Case Study Using the UK-Norway Information Barrier,* United Kingdom Norway Initiative, Non-Paper, United Nations, New York, 2015.

12. Keith Tolk, "Authentication in Arms Control," *56th Annual INMM Meeting,* Indian Wells, California, July 2015.
13. James Fuller, "Going to Zero: Verifying Nuclear Warhead Dismantlement," Chapter 4 in Corey Hinderstein (ed.), *Cultivating Confidence,* Nuclear Threat Initiative, Washington, DC, 2010.
14. Danielle Hauck and Duncan MacArthur, "Benefits of a 'Presence of Fissile Material' Attribute for Warhead Confirmation in Treaty Verification," *54th Annual INMM Meeting*, Palm Desert, California, July 2013.
15. Alexander Glaser, Boaz Barak, and Robert J. Goldston, "A Zero-knowledge Protocol for Nuclear Warhead Verification," *Nature,* 510, 26 June 2014.
16. Sébastien Philippe, Robert J. Goldston, Alexander Glaser, and Francesco d'Errico, "A Physical Zero-knowledge Object-comparison System for Nuclear Warhead Verification," *Nature Communications*, 7, September 2016.
17. Peter Marleau and Rebecca Krentz-Wee, *Investigation into Practical Implementations of a Zero Knowledge Protocol,* Sandia National Laboratories, SAND2017-1649, February 2017.
18. Andrew J. Gilbert, Brian W. Miller, Sean M. Robinson, Timothy A. White, William Karl Pitts, Kenneth D. Jarman, and Allen Seifert, "A Single-Pixel X-Ray Imager Concept and Its Application to Secure Radiographic Inspections," *Nuclear Instruments and Methods in Physics Research A,* 861, 2017.
19. R. Scott Kemp, Areg Danagoulian, Ruaridh R. Macdonald, and Jayson R. Vavrek, "Physical Cryptographic Verification of Nuclear Warheads," *Proceedings of the National Academy of Sciences,* 113 (31), August 2016.
20. Alexander Glaser, "Ceci N'est Past une Bombe: Toward a Verifiable Definition of a Nuclear Weapon," *58th INMM Annual Meeting*, Indian Wells, California, July 2017.
21. Luke Petruzzi, Bernadette Cogswell, Alexander Glaser, Malte Goettsche, Tamara Patton, and Drew Wallace, "Nuclear Inspections in the Matrix: Working with Radiation Detectors in Virtual Reality," *58th INMM Annual Meeting*, Indian Wells, CA, July 2017.

Dick Garwin, the Government and Science and the Danger Ahead[1]

Joel N. Shurkin

Science Writer
2503 Manhattan Ave.
Baltimore, MD 21215, USA

The first major government-sponsored science in America was very early in the new country's history—the Lewis & Clark expedition, the Corps of Discovery, up the Missouri River and out to the West Coast. It was the brain child of Thomas Jefferson and was, for its time —1804-1806—the scientific equivalent of a moon shot. There is no greater example of government-sponsored science before nor after it until the Manhattan Project in WWII.

Presidents did have advisors, but there were few formal organizations until 1951 with the creation of the Science Advisory Committee by Harry Truman.

Few things encourage science-government cooperation like war. WWI was a chemist war. Men like Germany's Fritz Haber— a Nobel laureate who not only invented nitrogen fertilizer, but poison gas— altered the way wars were fought—how people got killed. WWII—on the other hand—was the physicists' war.

The story is often told. A letter signed by Albert Einstein in July 1939 was delivered to President Roosevelt with the help of Alexander Sachs, warning of the dangers of atomic weapons and the fear that Germany would get one first. There was no formal way to advise the president. While the cream of European physics had fled to the US and Britain—many of them were Jewish—some physicists remained, including Werner Heisenberg, creator of the Uncertainty Principle of quantum physics. Heisenberg had a first-class mind, perfectly capable off figuring out eventually how to build a fission device. If he had one, Adolf Hitler was perfectly capable of using it.

During the war, there still was no formal way of getting scientists face-to-face with top government officials like the President. But there was one man who was the exception, an MIT engineer named Vannevar Bush. Bush became the main conduit scientists used to get the government to do anything important. **Three years** after Sachs delivered the Einstein letter, the recommendation finally found its way through the bureaucracy and got to Bush. Bush immediately knew what buttons to push. He turned the project over to the U.S. Army Corps of Engineers' Manhattan district—yes, the New York Manhattan—who put a colonel, later general Leslie Groves in charge.

When I visited Los Alamos last summer, Groves was described as a "fat buffoon" by a scientist there. He wasn't. He was one of the most competent military officers in American history. He decided to give the project to Robert Oppenheimer, one of the stars of American physics, made sure Oppie had everything and everyone he needed—and was smart enough and humble enough to turn him loose. He stayed in Manhattan, out of Oppenheimer's way.

Nuclear Weapons and Related Security Issues
AIP Conf. Proc. 1898, 020007-1–020007-5; https://doi.org/10.1063/1.5009212
Published by AIP Publishing. 978-0-7354-1586-7/$30.00

Eventually tens of thousands of people were involved, including most of the world's best physical scientists. Any evening in Los Alamos you could put together a poker game with only Nobel laureates or future laureates playing. (You probably wouldn't want to play with some of those guys. The mathematician John von Neumann was so good at poker he was barred from Monte Carlo casinos.) Those not based there visited. Virtually no one turned down Oppenheimer's recruiting drive. It was a war of survival. Uncle Sam wanted us—one physicist said. It was easily the largest concentration of scientific genius in world history and probably still is. The space program was a bigger operation but truth be told, that was more engineering than science.

But when the first bomb exploded over Trinity site in New Mexico, July 16, 1945, things changed. The minute the bomb went off it stopped being mainly a scientific project and became a weapon. To use the common vernacular, it had been weaponized. Scientists around the country signed a petition urging Harry Truman to invite Japanese authorities to witness a bomb blast in hopes they would surrender before a bomb had to be used. Nothing happened to the petition—it either never made it to Truman or he rejected it.

After Trinity, Oppenheimer famously quoted the Bhagavad-Gita, "I am death. Destroyer of worlds." There is a less famous quote just as pertinent.

"In some crude sense—he wrote—which no vulgarity, no humor, no overstatement can quite extinguish, the physicists have known sin, and this is a knowledge they can't lose."

Oppenheimer sat in on target meetings where it was decided that the first target would be Hiroshima, the second one included several possibilities, including Nagasaki—places where hundreds of thousands of Japanese men, women and children—maybe a million—would be blown apart, fried, melted, vaporized or zapped and even genetically altered by radioactivity. Without the bomb, there would have to be an invasion of the Japanese home islands. The army ordered a million Purple Hearts.

Dick Garwin was not at the Manhattan Project. He was too young. But in 1951 Garwin—whom Enrico Fermi called the one true genius he had met—solved the problem that Edward Teller and Stanislaw Ulam could not solve, and designed the first hydrogen bomb. He did it in three weeks. He was 23-years-old. He did not—incidentally—stay around to see the H-bomb explode—October 31, 1951. The bomb—code named Ivy-Mike—destroyed an archipelago in the South Pacific. Garwin had returned home. He has never seen a nuclear explosion. "I have a good imagination," he told me. Teller was known as the father of the H-bomb but in the 1970s, after a heart attack, he admitted the main design was Garwin's. Even Garwin's family didn't know until Teller's admission was printed in the *New York Times.*

Garwin went to work for IBM, and at the same time for the government, as a consultant and advisor. He became one of the most political scientists in America. He was advisor to half a dozen presidents and served on so many committee's and agencies, only his brain can keep track of them all. He is still at it at the age of 89. He has testified before Congress dozens of times through the years.

The deal he made with IBM was that he would work 2/3 of the time for the company and 1/3 for the government. Most of what he did for the government is classified and not even James Watson knew where he was or what he was doing.

A few notes on his IBM work. Garwin has 60 patents and for a while ran several of IBM's research labs. His inventions include a device to wash mussels, the prototype of the first ATM, touch screens and the laser printer. A polymath, he has involved himself in everything from how to stop a flu pandemic, to a floating airport and air traffic control, and a solar-powered spaceship for traversing the solar system. He is responsible for the wide use of the Fast Fourier Transform. His testimony in Congress was responsible for killing the American supersonic airliner (SST) and Star Wars, Ronald Reagan's dingbat defensive system.

Most of his time, however, has been spent on disarmament. Every single treaty between the US and Russia has his fingerprints on it.

He has also been a front-row spectator at the demise of scientific influence in America.

After WWII, scientists emerged as heroes—gods. They had—through patriotism, hard work, and genius—ended the goriest war in history. Their influence on science policy was profound. The apotheosis of scientific influence was in the 50s and 60s, partly a reflection of two presidents. Remember the three years it took Einstein's letter to get atomic bomb research moving? That didn't happen again.

When the Soviet Union launched Sputnik in 1957, panic ensued and for several years, scientists like Dick Garwin had the government's ears. But that did not last.

Dwight Eisenhower as a five-star general but distrusted the military. It was he who coined the phrase the "military industrial-complex" and warned Americans about its influence. We clearly didn't listen.

John Kennedy was Harvard-educated and had great respect for intellect. When he was assassinated, Lyndon Johnson, decidedly not from Harvard, took over. He had much less respect. It's been downhill since then. After Garwin testified against the SST in 1971, Richard Nixon killed the Presidential Science Advisory Committee that Garwin was on.

By the time of the Vietnam War, many of the Los Alamos scientists had died or retired and the Sputnik boost was petering out. Many of those few left formed an organization called JASONs to better organize the influence science had with government. Remember the Oppenheimer quote about knowing sin? They also knew arrogance.

They had intervened in WWII and, to some extent saved the Allied nations. As the Vietnam War progressed and emitted the stench of failure, scientists in the JASONs—which included some who worked on the bomb—decided they better intervene again. The war was going badly. The Pentagon was made up of officers who mostly never read Sun Tsu's *The Art of War* and were losing to Ho Chi Minh, who had.

Robert McNamara, Secretary of Defense under Kennedy and Johnson, wanted to find an excuse to end Rolling Thunder, a huge bombing operation designed to cut off the Ho Chi Minh Trail, the main route through the jungles and mountains for getting men and supplies to the communist forces in the south. The JASONs had a plan:

They would build a wall across Vietnam. It would be partly an actual wall or fence but part of it would be electronic. They would drop tiny microphones into the grass and trees to pick up sounds indicating troop or truck movement. As soon as they heard the sounds, the Air Force would bomb the area. Sounds logical.

Garwin's specialty when confronted by a neat plan—especially from the military—was to find out what's wrong with it and to find a simpler and cheaper way to do the same thing. The military, however, does not generally want simple and cheap. Garwin's predictions that the Viet Cong would find ways to get through the wall were accurate. The Air Force turned out to be masterful at passive aggression. They had their own research establishment and weren't interested in anyone else's ideas.

When the listeners to the sound devices—who were based in Thailand—heard movement in the network the air force waited a day before sending the bombers. The enemy was nowhere near the site by the time the bombs came down.

Another sensor spotted urine. If a Viet Cong soldier decided to take a leak, the sensor would detect the chemicals in the urine and send the message to Thailand. Somehow, the Viet Cong figured it out and sent old women into the jungle with buckets of buffalo urine. The United States Air Force then sent multi-million dollar airplanes to bomb buckets of piss.

Essentially the wall was a failure in Vietnam although it became one of the first uses of electronic warfare and versions of it are in use today by US forces in Asia.

As the public turned against the military and the war, they also turned against the scientists. Scientists were participating in the war, the war was evil, the scientists were evil.

America loves to destroy its gods and that's what happened to science. Add to that the fact that science also keeps coming up with inconvenient truths, and you have the picture. In the years since, science has had less influence culminating in the present administration. Garwin has seen it firsthand.

To a degree, this was all part of the old tradition of anti-intellectualism in America. Unlike Europeans, we do not revere smart, educated people. This isn't new: in 1830, the French writer Alexis de Tocqueville commented on it. The world of the intellect exists here despite the American people, not because of it.

There were other factors. Scientists became more specialized and the federal government became larger and more decentralized. There was no Vannevar Bush to go see. There is in Washington—Frank Von Hippel told me—science for policy and policy for science and each federal department has its own science advisors, many their own scientists. Special interests have much greater influence. The National Rifle Association can—through its Congressional allies—block the Centers for Disease Control and Prevention from doing research on gun violence.

The scientific community shares some of the blame, Tom Postol of MIT told me. Turn on cable television and you will hear alleged experts spout all kinds of nonsense—fake scientific news—and this winds up diminishing respect from the public. And, in the real world, there are questions that even experts cannot answer. During coverage of the Three Mile Island disaster, people called my newspaper and demanded to know how serious it was and asked should they flee. No one—including the experts—knew and the callers got mad at us for not knowing either. Experts don't know everything although some claim they do on television.

There are scientists who—Postol says—will prostitute themselves, taking money to procure skewed research. How many climate change deniers do research funded by fossil fuel producers? That undermines the clear majority who don't.

And the biggest change may be that many of the brightest people in America no longer go into science. They can make such more money and have much less aggravation going to Wall Street or the financial industry. They used to go to law school—now they go to business school.

There is no sign it is going to get better soon—no sign that science will ever attain the influence it once had. We live in a world of alternative facts and facts—at least as we best know them—have lost their primacy.

Isaac Asimov—the science fiction writer who also was a biochemist—wrote:

"There is a cult of ignorance in the United States, and there has always been. The strain of anti-intellectualism has been a constant thread winding its way through our political and cultural life, nurtured by the false notion that democracy means that 'my ignorance is just as good as your knowledge."

[1] This summarizes some aspects of material in Joel Shurkin, "True Genius, the Life and Work of Richard Garwin, the Most Influential Scientist You've Never Heard Of" (Amherst, NY: Prometheus Books, 2017).

The Secret of the Soviet Hydrogen Bomb*

Alex Wellerstein[a] and Edward Geist[b]

[a]*Department of Science and Technology*
Stephens Institute of Technology
Hoboken, NJ
USA
[b]*Rand Corporation*
17776 Main St.
Santa Monica, CA 90407
USA

Abstract. Was the first Soviet thermonuclear device really a step in the wrong direction? No bomb design has been as much maligned or otherwise disparaged as the first Soviet thermonuclear weapon. Detonated in August 1953, the bomb, officially tested under the name RDS-6s but usually known as *Sloika* or "layer cake" (the name Andrei Sakharov coined for it), was nothing to sneeze at. Shown in Figure 1 and able to be dropped from aircraft, it released the explosive equivalent, or yield, of almost half a megaton of TNT. The result was a blazing fireball with 20 times the power of the bomb that leveled Nagasaki, Japan.

But when discussed today, the Sloika is almost immediately qualified by US experts as not a "true" hydrogen bomb. The downgrading is a curious reflex, one with interesting cultural and nationalistic origins. At one level, it is a technical determination: The bomb's design did not allow it to be scaled up to near unlimited explosive yields that true hydrogen bombs would allow, and it differed from what has become the foundation for all modern thermonuclear weapons, the famous US-developed Teller–Ulam design.

Historically, there was also a political reason to downplay the weapon. In the 1950s a lot was at stake for a US government official to say exactly when the Soviets first had true thermonuclear capability or that their first H-bomb was a militarily useless weapon and a dead end from a design standpoint. Decades later, the downplaying of the Sloika remains emblematic of how that phase in nuclear history is spoken of by US historians.

But a perusal of Soviet-era sources released in the past decade and of several new and obscure publications that have come from the Russian weapons establishment has convinced us that the disparaging view is incomplete. The new sources seem to point to two contrary views: First, the Sloika was, in fact, seen as a useful weapon in its own right, even if it was inefficient; it was not "cobbled together" just to make a statement, as some American scientists thought at the time. Second, it was also not as much of a nuclear dead end as it was perceived and is still claimed to be: Rather, it appears to have been absolutely crucial in helping to solve the riddle of how the Soviets developed their later, multistage, multimegaton thermonuclear weapon designs. Indeed, the Sloika is possibly the answer to the most curious question about the Soviet thermonuclear program: How did it develop a form of the Teller–Ulam design only a year after developing the Sloika?[1]

WHY DO WE CARE?

The origins of the reflexive diminishing of the Sloika are found, in part, in an earlier debate in the US over whether to develop an H-bomb at all. In the fall of 1949, after the first Soviet atomic bomb test (labeled Joe-1 by the US, RDS-1 by the Soviets), a polarizing debate took place among US scientists and policymakers about whether a crash program to develop the H-bomb was the appropriate response to the loss of the nation's nuclear monopoly. The debate became public by the end of the year, and the resulting publicity prompted President Harry S. Truman in late January 1950 to issue a mandate to the US Atomic Energy Commission to develop thermonuclear weapons. The problem was that nobody knew how to do such a thing. The H-bomb, which would use the power of nuclear fission

---

* Reproduced from *Physics Today*, Vol. 70, April 2017, p. 40–47 with permission of AIP publishing.

Nuclear Weapons and Related Security Issues
AIP Conf. Proc. 1898, 020008-1–020008-7; https://doi.org/10.1063/1.5009213
Published by AIP Publishing. 978-0-7354-1586-7/$30.00

to initiate substantial reactions of nuclear fusion, was then still an uninvented technology, an idea without an implementation.[2]

FIGURE 1. The Sloika, or "layer cake," is the informal name for the Soviet Union's first thermonuclear bomb. Although its casing was roughly similar in shape and size to Fat Man, the US atomic bomb dropped on Nagasaki, Japan, in World War II, the Soviet bomb was 20 times as powerful: It detonated with the explosive equivalent of 400 kilotons of TNT. (Nuclear Weapon Museum, Russian Federal Nuclear Center, VNIIEF.)

A few days after Truman's mandate, the UK announced that physicist Klaus Fuchs had confessed to being a Soviet agent. Fuchs had been a key member of the UK nuclear program and had worked at Los Alamos during and slightly after World War II, before the US had stopped all classified cooperation with the British. Newspapers reported that Fuchs may have passed on information about the H-bomb to the Soviets as well. Now the American narrative was a new one: The Soviets had stolen the design for an atomic bomb, not independently developed or reinvented it, and they were possibly galloping ahead. For Edward Teller and others who had argued in favor of developing the H-bomb, the story was a vindication.[3]

Thermonuclear research in the US revolved primarily around one design, later called the Classical Super. The Classical Super was Teller's idée fixe and posed difficult technical problems. The basic idea involved using a high-yield nuclear fission bomb to ignite fusion reactions in a mixture of deuterium and tritium, which would generate enough heat to propagate further fusion reactions. The appeal was that arbitrarily large explosions could be generated that way; just add more deuterium. The problem—aside from the fact that it would require large amounts of scarce tritium and a very large fission stage to set the bomb off—was that it didn't seem to work.

In the spring of 1951, US scientists Teller and Stanislaw Ulam, shown in Figure 2, made their famous breakthrough. Instead of a reaction that propagated relatively slowly down a tube of material, one fission bomb (the "primary") would detonate inside a heavy chamber that reflects its radiation to compress another, separate fusion capsule (the "secondary") inside a heavy chamber. That scheme, based on the concepts of staging (keeping the primary and secondary physically separated) and radiation implosion (the use of radiation to compress the secondary), became known later as the Teller–Ulam design. From the perspective of most weapons designers at the time, it was a radical departure from the approach taken with the Classical Super.[4]

The first prototype of the Teller–Ulam design was tested in November 1952 and was known as shot "Mike" of Operation Ivy. Yielding an equivalent of more than 10 million tons of TNT, the prototype vindicated the concept, though Mike was not designed to be used as a weapon. Its extensive cryogenic equipment, designed to keep deuterium in liquid form, meant that it weighed some 80 tons.

Just before the test, a fierce, secret debate about the importance of Fuchs's information began in the weapons laboratories. Fuchs's last contact with the US thermonuclear program was in 1946. Was information he could have gleaned from the program at that time valuable to the Soviets? On one side was Hans Bethe, who argued that the successful Teller–Ulam design differed so much from the original Classical Super design that anything Fuchs could have given them would be at best irrelevant and at worst completely misleading. Poised against him was Teller, who argued that the theoretical distance between the Classical Super and the Teller–Ulam design was not as large as

Bethe thought. Furthermore, he pointed out, Fuchs had himself been involved with working on certain lines of research that eventually proved crucial: Fuchs, along with John von Neumann, had worked on a hydrogen bomb design that involved a version of the concept of radiation implosion.[5] (See the articles by German Arsen'evich Goncharov, *Physics Today*, November 1996, pages 44, 45, 50, and 56.)

FIGURE 2. Edward Teller (left) and Stanislaw Ulam. In 1951 Ulam proposed a two-stage weapon design, in which a "primary" fission bomb compresses and ignites a "secondary" fusion capsule. Teller then proposed that the bomb's radiation, not explosive shock, be used for the compression. The Teller–Ulam design now serves as the foundation for all modern hydrogen bombs. (Teller photo courtesy of Lawrence Berkeley National Laboratory and the AIP Emilio Segrè Visual Archives. Ulam photo courtesy of the AIP ESVA, Ulam Collection.)

The content of the debate was technical, but the reasons for it were clearly political. If Teller was correct, then the US was potentially behind the Soviets in weapons development. In Teller's eyes, it was because people such as J. Robert Oppenheimer stood in the way of his work on the H-bomb in the years after World War II and squandered a potential lead. If Bethe was correct, though, then not only would Fuchs be unable to pass on useful information about later US H-bomb design (since he did not know anything about it), but Oppenheimer would be vindicated for not supporting Teller's early, wrong-footed schemes. That debate started in the spring of 1952, even before the Mike test, and versions of it arguably continue today in any history of the thermonuclear program.

The Soviet Sloika entered the story in the middle of the debate and before the US had tested deliverable versions of the Teller–Ulam design. On 8 August 1953, Soviet premier Giorgi Malenkov gave a speech to the Supreme Soviet in which he declared that "the United States has no monopoly in the production of the hydrogen bomb." On 12 August, the fourth Soviet nuclear test, dubbed Joe-4 in the US, was detonated over the Kazakh steppe. On 20 August, *Pravda* published a statement proclaiming that "within one of the last few days an explosion of one of a variety of hydrogen bombs was carried out for experimental purposes" and attributed its "great strength" to a "mighty thermonuclear reaction."[6]

It looked, then, as if the Soviets might be keeping pace with the US, if not beating it: If the Joe-4 test was of a deliverable thermonuclear bomb, then the Soviets could be seen as ahead in the H-bomb race in one sense, because the US Mike device was an experimental apparatus, not a weapon. More than US pride was in the balance: A leading Soviet program might be a vindication of those who said that the US program had been needlessly stalled.

By September 1953, however, there were reasons to doubt that the Soviets were, in fact, ahead in the race for a deliverable H-bomb. A panel consisting of physicists Bethe, Enrico Fermi, Richard Garwin, and Lothar Nordheim conducted an analysis of the fallout residues from the August test. Their full conclusions are still redacted more than six decades later, but from what has been released, we can see they found that the Joe-4 test used highly enriched uranium, not plutonium, and that it involved "a substantial thermonuclear reaction." They were able to estimate the amount of uranium in the device and the amount of energy release attributable to fusion reactions, and they could speculate on the geometry of the device. They concluded that it was not a Teller–Ulam design but a weapon that had achieved "a high-yield, high-efficiency [fission] reaction with the help of the boosting principle." Bethe would later call the device “a big boosted fission weapon” and “a glorified booster,” and he would say that it was clear from the analysis that it was a “single stage” weapon that involved “alternating layers of uranium and lithium deuteride.”[7]

The Soviet bomb was therefore not really an H-bomb, if by H-bomb one means something along the lines of the Teller–Ulam design. Instead, it shared characteristics with two other thermonuclear designs the US had pursued. One design, Booster, involved a fission weapon that had a small amount of deuterium–tritium gas injected into its core at the moment of its detonation, which generated enough fusion neutrons to cause extra fission reactions.

The other design was one Teller had proposed in 1946 as the Alarm Clock—a weapon that would use spherical layers of fissionable and fusionable fuel in a matryoshka-doll arrangement, one sphere inside the other. The design had serious negatives: Its fusion yield would necessarily be limited, primarily serving to enhance fission reactions, like Booster; the various layers would interact in complex ways that were extremely difficult to calculate with the computing technology of the time; and to increase the Alarm Clock’s yield to the megaton scale meant increasing a bomb’s radius so much that it would not fit inside a bomber. For Teller, the fact that Alarm Clock’s yield could not be increased indefinitely made it less interesting. Like the Booster, it was considered an auxiliary approach to what was then still the main show, the Classical Super.

If Joe-4, the Sloika, was just an Alarm Clock, then it wasn’t the main show. If it was a glorified Booster, it definitely wasn’t an H-bomb. But the Soviets saw it somewhat differently.

THE SOVIET VIEW OF SLOIKA

The Soviets called their H-bomb design the Sloika in reference to a layered Russian pastry similar to a napoleon. The code name hints at the bomb’s internal geometry: layers of highly enriched uranium, fusion fuel made of solid lithium deuteride, lithium deuteride tritide, and uranium tamper materials, all of which would be compressed by high-explosive lenses. Many details, such as the number of layers, their order, and their relative masses, remain classified.

Declassified documents and imagery indicate that the test device was roughly a sphere, 1.5 meters in diameter, and that it weighed about 4.5 tons. The test device apparently fit into the same casing as the original Soviet atomic bomb and differed from the production-line (military) version of RDS-6s mainly in that the latter used two to three times as much tritium and uranium-235 as the test version and thus would have likely had a substantially larger yield.[8] In terms of weapons design, that size is not extreme (see Figure 1)—it would be roughly the same shape and weight as the US Fat Man bomb dropped on Nagasaki, Japan, in World War II, though with a far more powerful explosion.

The basic problem with the Sloika, from a weapons designer’s viewpoint, is that chemical high explosives simply lack the power to compress the entire mass sufficiently for substantial fusion reactions to occur. Such a weapon would also be extremely expensive in terms of enriched uranium usage.

As tested in 1953, the Sloika detonated with an explosive yield of 400 kilotons, of which around 80% of the energy came from fission reactions and 20% came from nuclear fusion. That ratio of fission-to-fusion reactions is less useful in determining a true H-bomb than it might seem: The Mike design, like practically all US H-bombs, relied heavily on a final uranium fission stage to increase its yield, and it had the same fission/fusion ratio as the Sloika. As J. Carson Mark, one of the few US weapons designers not to quibble about the status of the Sloika, argued in an interview: “They managed to get 400 kilotons without going to an unreasonable or even a heavier size. And, they did it by using thermonuclear reactions. Want to call that a hydrogen bomb? Well, why not?”[9]

Where Sloika really loses is in terms of the yield-to-weight ratio, the amount of energy release divided by the total bomb weight, which is the preferred method by which weapons designers gauge weapon sophistication. At 0.08 kilotons of energy per kilogram of bomb weight, the Sloika was an order of magnitude better than Fat Man, but still an order of magnitude less efficient than the first deliverable US H-bomb designs.

Documents declassified in the past decade give us some insight into how the RDS-6s device was viewed by those who made it. Although privately the Soviet designers also would consider it a glorified booster, they had grand plans for the Sloika. Contrary to US analyses that insisted those weapons could never achieve yields much greater than the 1953 test, the Soviets originally envisioned it as a megaton-range weapon.

But the multimegaton Sloika proved more difficult to develop than Soviet nuclear scientists, including Andrei Sakharov, originally envisioned. Serious problems emerged because a Sloika of a particular diameter could only make efficient use of expensive materials such as uranium–235 and tritium up to a certain yield. By mid 1954 it became clear that within the 1.5-m radius dictated by the size of delivery vehicles, constructing a Sloika with a yield greater than 0.5–1.0 megaton without using costly tritium would be difficult.

Over the next year, some of the Soviet Union’s most brilliant technical experts devoted their attention to constructing a cost-effective, multimegaton Sloika. They also explored developing a design with a focus on economy rather than yield. That effort resulted in a new budget-friendly design that was the same size as the multimegaton Sloika but with a yield of only 350 kilotons. Soviet nuclear scientists, however, insisted that the seemingly inferior weapon exploited the full potential of the Sloika concept for maximizing yield while minimizing the need for scarce nuclear materials such as lithium-6, making it vastly more cost-effective.

In September 1953 the Bethe panel produced a lengthy analysis, in which it asserted that weapons on the Sloika principle would scale poorly even if "the yield they have achieved is certainly enough to cause concern." How are we to account for the contrast between Bethe's dismissal of the RDS-6s design and the Soviets' ambitious plans for the Sloika? One possibility is that Bethe's conjectural reconstruction of the weapon's internal geometry was in error, but we cannot be sure, as both Bethe's analysis and the design details of the Soviet device remain classified.

However, given the extreme difficulty the Soviets experienced developing the megaton-scale version of the RDS-6s (dubbed the RDS-6sd), it is also possible that their ultimate design differed very substantially from the weapon tested in 1953. For a predicted yield of 1.8 megatons, it may have incorporated a dissimilar internal geometry, different materials, and counterintuitive features that never occurred to Bethe or other US scientists who lacked hands-on experience with weapons of the Sloika type. Bethe's dismissal of the Sloika may also have been based on the analysis that had been done in the US on the Alarm Clock; although it shared a similarly layered design with the Sloika, it may also have differed in several ways.

By August 1955 both the RDS-6sd and its budget version were ready for testing, but the imminent arrival of a more advanced rival delayed their debut. After struggling to improve the Sloika, the Soviets had finally hit on their version of the Teller–Ulam design. They called it Sakharov's Third Idea, and gave the prototype the code name RDS-37. A two-stage weapon employing radiation implosion to produce a multimegaton yield, the RDS-37 used about a quarter of the nuclear explosive materials the RDS-6sd used and had the capacity for a much greater yield in a package that the Soviet Union's bombers and missiles could carry. Soviet leaders decided to wait to test the costly multimegaton Sloika until after the performance of Sakharov's new invention could be verified in a live test.

The successful airburst of the RDS-37 on 22 November 1955 sounded the death knell of the RDS-6sd and, in time, all other Sloikas. Deliberately detonated at half its total predicted power, the weapon fit into the same case as the Sloika but released 1.6 megatons of energy. Assuming its weight was similar to the Sloika's, it was a full order of magnitude more efficient and, more importantly, much more flexible for scaling weapon output both up and down. The handful of RDS-6sd devices were promptly dismantled so their precious lithium-6 and enriched uranium could be incorporated into more modern weapons. The Sloika had passed into history.

THE SLOIKA'S LEGACY

Was the Sloika merely a dead end? The Soviet records suggest not. For Soviet weapons designers, Sloika served as a means of exploring thermonuclear concepts while still producing deliverable weapons that though not as powerful as later developments were large enough to be considered serious city busters. Moreover, reading between the lines of the secret Soviet histories, there are reasons to suspect that Sloika was more important to their program than one might expect.

No doubt the Soviet fission bomb program owed much to espionage. The work of spies such as Ted Hall and David Greenglass allowed the Soviet Union to have a reasonably good understanding of what went into the construction of plutonium-implosion nuclear weapons, and its first fission bomb, RDS-1, was a "Sovietized" copy of the weapon dropped on Nagasaki.[10]

The Soviets also received some information on US thermonuclear work from Fuchs. Declassified documents from the Soviet archives show that Fuchs gave them extremely detailed accounts of the state of US work as of 1946 and of the work he did with von Neumann. The Soviets did have a research program for the Classical Super design, which they dubbed the *Truba*, or "Tube," that ran parallel with the Sloika work.[11, 9]

As noted earlier, the Soviets eventually hit upon the two-stage, radiation implosion design known in the US as the Teller–Ulam idea. Although the Soviets called it Sakharov's Third Idea (see figure 3), internally they noted that exact authorship was difficult to determine. As Lev Feoktistov, a scientist on the project, recalled, "New ideas dawned upon us suddenly like light in a dark kingdom, and it was clear that the instant of truth had come. Rumors ascribed these fundamental thoughts in Teller's spirit now to [Yakov] Zel'dovich, now to Sakharov, now to both, or to someone else, but always in some indecisive form: likely, possibly, and so on."[12] The first two ideas were, in order, the Sloika's layering scheme and the use of lithium deuteride as a fusion fuel; both had been well-documented by 1949.

Both the Third Idea and the Teller–Ulam design differentiate themselves from earlier H-bomb designs in their use of radiation energy as a means of achieving extremely high densities in a thermonuclear assembly (see the box on page 45). As Soviet designers drew it, the weapon was a heavy box with an atomic bomb at one end and the thermonuclear capsule at the other. In the earliest US Teller–Ulam designs, the capsule was a cylinder with multiple layers: on the outside a heavy tamper, then liquid deuterium or lithium deuteride, and in the center a "spark plug" of plutonium and tritium.[13]

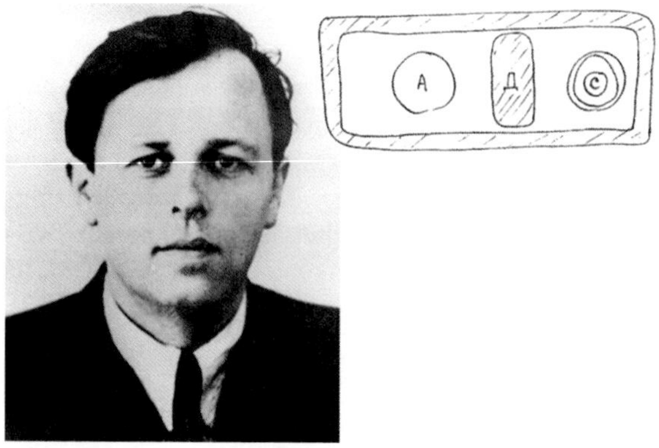

FIGURE 3. Andrei Sakharov is pictured next to the first formulation of the "Third Idea," a heavy box containing an atomic bomb (A), a neutron shield or diaphragm (Д, the Cyrillic letter for "D"), and a Sloika (C, the Cyrillic letter for "S"). (Russian Federal Nuclear Center, VNIIEF Museum and Archive)

Two approaches to thermonuclear weapons. In a basic implosion bomb, like the one dropped on Nagasaki, Japan, a solid-metal plutonium core is compressed using high-explosive lenses and an aluminum pusher to around 2.5 times its original density. In the deuterium-boosted design, a hollow core of plutonium is injected with deuterium–tritium gas at the moment of detonation, which causes a small number of fusion reactions. Those reactions produce high-energy neutrons that enhance the efficiency of the fission reactions in the core. In the *Sloika* design, alternating layers of lithium hydride and uranium-238 surround a uranium-235 core. The high-explosive lenses compress the entire core and set off of a fusion reaction that in turn compresses the fusion fuel. The fusion reactions produce high-energy neutrons that induce further fissioning.

These bomb designs are the basic ingredients that the US and the Soviet Union adapted into thermonuclear weapons—also known as H-bombs. In the original US Teller–Ulam design, a boosted fission bomb sits at one end of a heavy radiation case. At the other end sits the thermonuclear charge, a cylinder with a neutron shield on one end, liquid deuterium inside it, and a thin "spark plug" of plutonium mixed with tritium. At detonation, the radiation from the fission bomb reflects off the inside of the radiation casing and compresses the thermonuclear charge to many times its original density. The compression, in turn, begins a fission reaction in the "spark plug," which compresses the fusion fuel from the other side simultaneously. Thus compressed, the fusion fuel is primed for fusion reactions, which contribute significantly to the explosive yield. The reactions produce high-energy neutrons that induce further fissioning in a uranium-238 tamper. Sakharov's "Third Idea" adopts a similar scheme, except the neutron shield is integrated into the overall design, and the Sloika is stripped of its high-explosive components.

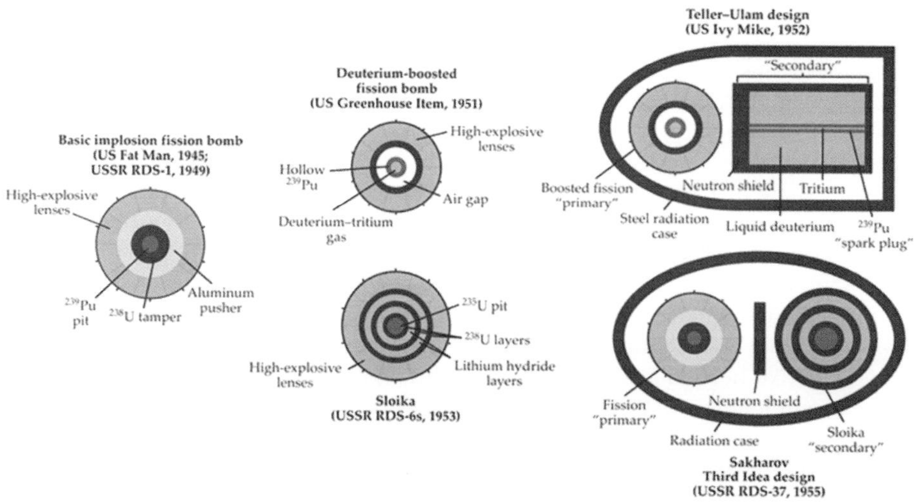

The first record of that idea from the Soviet archives dates from January 1954, a brief memo from Zel'dovich and Sakharov titled "On the use of a gadget for implosion of supergadget RDS-6s."[14] The memo describes a heavy box inside which an atomic bomb (labeled "A") sits at one end, a neutron shield (labeled the Cyrillic character for "D") sits in the center, and what looks like a Sloika (labeled with a Cyrillic "S") sits at the other end (see Figure 3). Along with the title of the paper, the sketch suggests a plausible genealogy of the Third Idea: The Sloika became a second stage of a two-stage thermonuclear weapon, the "supergadget" imploded by the fission "gadget."

Much remains missing in our knowledge of Soviet thermonuclear developments, but the path to the Third Idea may have been paved in part by the intensive work on the Sloika. The chief practical problem of that device is achieving the high compressions one needs for fusion. If high explosives can't cause them, what can? A new approach appeared to answer the question: using a fission bomb to compress the entire Sloika, first imagined as a compressive shock, later as radiation implosion. The Sloika, minus its high explosives and simplified a bit, is essentially a high-performance thermonuclear secondary: layers of fusion fuel, tamper, and fission material.

Over the decades many authors have asserted that the Soviet Union somehow learned of radiation implosion from the US rather than developing it independently. In more recent decades, a few accounts have asserted that the Soviets could not have discovered radiation implosion on their own and that a still-unidentified mole must have given away the secret of the H-bomb.[15]

Declassified Soviet documents contradict those views. They reveal that the thermonuclear information the Soviets got from spies was of limited value and not responsible for the work on either the Sloika or the later RDS-37 device. There is simply nothing to suggest that the Soviet scientists had insight into US weapons designs; even after they developed their own two-stage design, Soviet nuclear scientists remained uncertain whether the American bombs operated on the same principle.[16] And if the Russian security services could have taken credit for the Soviet H-bomb, which would serve to delegitimize the dissident Sakharov, it seems likely they would have done so by now.

If just one lesson were to be taken from the history of the Sloika, it may be that in the journey toward invention there is no single path to a right idea. Too often the American case is taken to be the default path of technological development, often because the US did it first and perhaps because it is much easier to document than other countries' programs. But the secrecy involved meant that each national program, to various degrees, reinvented the bomb, and finding some national variances should not be so surprising.

The Sloika, rather than just being a relic, sheds much light on alternative approaches toward a similar technological end.

REFERENCES

1. For more information on US scientists' perceptions of the Sloika, see R. Rhodes, Dark Sun: The Making of the Hydrogen Bomb, Simon & Schuster (1995), p. 524.
2. See, for example, G. Herken, Brotherhood of the Bomb: The Tangled Lives and Loyalties of Robert Oppenheimer, Ernest Lawrence, and Edward Teller, Henry Holt & Co (2002).
3. Ref. 2, chap. 12.
4. Ref 1, chaps. 23, 24; K. Ford, Building the H Bomb: A Personal History, World Scientific (2015), chaps. 9–14.
5. H. Bethe, "Memorandum on the History of the Thermonuclear Program" (28 May 1952), documents NV0318420 and NV0311423, Nuclear Testing Archive, Las Vegas, NV; E. Teller, "Comments on Bethe's History of the Thermonuclear Program" (14 August 1952), document NV0318421, Nuclear Testing Archive. For information on the Fuchs–von Neumann design, see J. Bernstein, Phys. Perspect. 12, 36 (2010).
6. New York Times (10 August 1953), p. 6; H. E. Salisbury, New York Times (20 August 1953), p. 1.
7. H. Bethe et al., "Analysis of Joe-4," (11 September 1953); H. Bethe, Bull. At. Sci. 46(8), 8 (1990), p. 8; H. Bethe, quoted in C. Hansen, The Swords of Armageddon: U.S. Nuclear Weapons Development Since 1945, 2nd ed., vol. 3, Chuckelea Publications (2007), p. 136; H. Bethe, quoted in ref. 1, p. 524.
8. See, for example, I. E. Tamm, A. D. Sakharov, Y. B. Zel'dovich, in L. D. Riabev, ed., Atomnyi proekt SSSR: Dokumenty I materialy, vol. 3, bk 2, VNIIEF (2009), doc. 5, p. 21.
9. Y. B. Khariton, V. B. Adamskii, Y. N. Smirnov, Phys. Usp. 39, 185 (1996); J. Carson Mark, interview by R. Rhodes (undated), transcript available at Voices of the Manhattan Project, www.manhattanprojectvoices.org/oral-histories/j-carson-marks-interview.
10. M. Gordin, Red Cloud at Dawn: Truman, Stalin, and the End of the Atomic Monopoly, Farrar, Straus, and Giroux (2009), chaps. 3, 4.
11. Ref. 4, Ford, p. 99.
12. L. Feoktistov, quoted in G. Goncharov, Phys. Usp. 40, 859 (1997), p. 864.
13. Ref. 4, Ford, p. 156.
14. R. B. Zel'dovich, A. D. Sakharov, note to I. B. Khariton, in ref. 8, doc. 56, p. 128.
15. D. Hirsch, W. G. Mathews, Bull. At. Sci. 46(1), 22 (1990); T. C. Reed, D. B. Stillman, The Nuclear Express: A Political History of the Bomb and Its Proliferation, Zenith Press (2009), p. 34.
16. A. P. Zaveniagin et al., to the Presidium of the Central Committee, in ref. 8, doc. 170, p. 394.

Sakharov, Gorbachev, and Nuclear Reductions*

Frank von Hippel

Program on Science and Global Security
Princeton University
Princeton, NJ 08544
USA

Abstract. Two years before his death in 1989, Andrei Sakharov's comments at a scientists' forum helped set the stage for the elimination of thousands of nuclear ballistic missiles from the US and Soviet arsenals.

The great Soviet dissident physicist Andrei Sakharov and Soviet Union leader Mikhail Gorbachev met for the first time in January 1988. That was a little more than a year after Gorbachev had given Sakharov permission to return to Moscow from the closed city of Gorky, to which Sakharov had been exiled for seven years.

When Sakharov returned, US–Soviet nuclear arms control was at an impasse; Gorbachev was insisting that the US commit to keeping its ballistic missile defense (BMD) program within the constraints of the 1972 Anti-Ballistic Missile (ABM) Treaty, and President Ronald Reagan was refusing to do so. Sakharov publicly argued that Reagan's program, ridiculed as "Star Wars" by its US critics, would never produce militarily significant capabilities and that Gorbachev therefore should seize the opportunity for nuclear arms reductions. Two weeks later that view was endorsed by the Soviet leadership and opened the path to deep cuts in Soviet and US nuclear forces.

In 1987 I coorganized a scientists' forum in Moscow on nuclear disarmament, where Sakharov went public with his views, and I was present at his first meeting with Gorbachev a year later. I also took part in a private discussion with Sakharov in his apartment before the forum. Five years later I received a copy of a partial transcript of that discussion, which had been delivered to Gorbachev by the head of the Soviet secret service, the KGB.

For the younger generation, many of whom may know little about Sakharov, I start this article with a brief summary of the man's remarkable career.[1] I then describe how he helped delink nuclear reductions from the issue of BMD at that key moment in the history of nuclear arms control; the meeting in his apartment; his meeting with Gorbachev; and finally how relinking arms reductions and BMD has now returned to hobble progress on nuclear disarmament.

ANDREI SAKHAROV

Sakharov (1921–89) was recruited into the Soviet Union's nuclear weapons program in 1948, a year after he completed his doctorate. In 1949 the US detected the first Soviet test of a fission bomb, and the two countries embarked on a desperate race to design a thermonuclear hydrogen bomb that was a thousand times more powerful. The race ended in a rough tie five years later. (See the article by Alex Wellerstein and Edward Geist on page 40 of this issue.)

Sakharov made key contributions to the Soviet effort and was awarded the honorary title Hero of Socialist Labor three times, in 1953, 1956, and 1962. During the same period, he coinvented the tokamak (see *Physics Today*, December 2005, page 15), a toroidal magnetic plasma confinement device that is still the main focus of international efforts to develop a fusion reactor.

Like his US counterparts, Sakharov justified his H-bomb work by pointing to the danger of the other country's achieving a monopoly. But also like some of the US scientists who had worked on the Manhattan Project, he felt a responsibility to inform his nation's leadership and then the world about the dangers from nuclear weapons.

Sakharov's first effort to influence policy was stimulated by his concern about possible genetic damage from

---

* Reproduced from *Physics Today*, Vol. 70, April 2017, p. 48–54 with permission of AIP Publishing.

Nuclear Weapons and Related Security Issues
AIP Conf. Proc. 1898, 020009-1–020009-8; https://doi.org/10.1063/1.5009214
Published by AIP Publishing. 978-0-7354-1586-7/$30.00

long-lived radioactive carbon-14 created in the atmosphere from nitrogen-14 by the enormous fluxes of neutrons released in H-bomb tests.[2] In 1961 he urged Soviet leader Nikita Khrushchev to maintain the bilateral Soviet–US testing moratorium that had begun in 1959. Khrushchev told him that tests were required to show the US that the Soviet Union could not be intimidated. In the subsequent final spasm of Soviet and US atmospheric testing, the Soviet Union exploded a 50-megaton bomb, by far the highest-yield nuclear explosion ever set off. (Yield refers to the amount of TNT needed for an equivalent explosion.) The next year the Cuban Missile Crisis sobered both sides, at least temporarily, and in 1963 they agreed on an atmospheric test ban.

In 1968 a friend suggested that Sakharov write an essay about the role of the intelligentsia in world affairs. Samizdat (self-publishing) was the method at the time for spreading unapproved manuscripts in the Soviet Union. Many readers would create multiple copies by typing with multiple sheets of paper interleaved with carbon paper. One copy of Sakharov's essay, "Reflections on Progress, Peaceful Coexistence, and Intellectual Freedom," was smuggled out of the Soviet Union and published by the New York Times. More than 18 million reprints were produced during 1968–69. I still remember my excitement on reading that call for cooperation between East and West coming from the heart of the closed Soviet nuclear weapons complex. It was also a call, as Sakharov put it, for "freedom to obtain and distribute information, freedom for open-minded and unfearing debate and freedom from pressure by officialdom and prejudices."

After the essay was published, Sakharov was barred from returning to work in the nuclear weapons program and took a research position in Moscow. With his political views known, however, dissidents began to ask him to lend his name to their appeals for more freedom—requests he could not refuse. His status as coinventor of the Soviet H-bomb protected him, but as other dissidents were sent to prison, he became more and more outspoken and joined vigils outside the courtrooms where they were being tried.

In 1980, after an interview with the New York Times in which he denounced the Soviet invasion of Afghanistan, the government's patience finally ran out. To put him beyond the reach of Western journalists, the Soviet Union exiled Sakharov and his wife, Elena Bonner, to Gorky (now known as Nizhniy Novgorod). There he undertook prolonged hunger strikes. The longest, which lasted for about a year, was to get permission for Bonner to go abroad for heart-bypass surgery. During that period he was repeatedly force-fed before the government finally relented. (See the article by Sidney Drell and Lev Okun, Physics Today, August 1990, page 26.) Under the stress, Sakharov suffered three heart attacks.

Foreign supporters—including Jeremy Stone, president of the Washington-based arms-control group the Federation of American Scientists—campaigned tirelessly to keep Sakharov's case in the spotlight of public and government attention. As chairman of the federation, I was a supporting character in that effort.

In March 1985 Gorbachev became general secretary of the Soviet Communist Party. More than a year and a half later, he persuaded the Politburo, the party's executive committee, to allow Sakharov and Bonner to return to Moscow. (For more details about Sakharov's life and work, see the special issue of *Physics Today*, August 1990.)

THE US–SOVIET NUCLEAR IMPASSE

Reagan's election in 1980 led first to the intensification of the nuclear arms race and then to the largest ever public uprising against it. A powerful advocacy group, the Committee on the Present Danger (CPD), had convinced Reagan that the US was falling behind in the nuclear arms race and was in mortal danger of a Soviet first nuclear strike. Many of its members obtained high-level positions in the administration, including in the Department of Defense, where they proposed to add almost 10 000 ballistic and cruise missile nuclear warheads to the US arsenal. The new weapons would threaten Soviet intercontinental ballistic missiles (ICBMs) in their hardened underground silos in the same way that the CPD claimed Soviet ballistic missiles already threatened US silos. (See the article by Harold Feiveson and me, *Physics Today*, January 1983, page 36.)

The proposed buildup and public statements by some middle-level DOD appointees that it might be possible to win a nuclear war led to huge public demonstrations in the US, Europe, and elsewhere against the nuclear arms race.[3]

In March 1983 President Reagan announced the Strategic Defense Initiative (SDI), which would focus on developing technology to make Soviet ballistic missiles "impotent and obsolete." The initiative communicated a less threatening image than the CPD buildup to the US public and allies. In Moscow, however, it suggested a scenario in which the US could threaten to destroy most of the Soviet Union's missiles in a first strike and then use its BMD to block the ragged counterattack by the surviving Soviet missiles.

In early November 1983 the Reagan administration created a severe nuclear crisis with a NATO exercise, Able Archer, that Soviet intelligence mistook for preparations for an actual nuclear attack. Two years later the situation finally began to diffuse. At the recommendation of UK prime minister Margaret Thatcher, Reagan in November

1985 had a get-acquainted meeting in Geneva with Gorbachev, who had just become general secretary after two elderly predecessors, in quick succession, had died in office.

In their legendary October 1986 Reykjavik summit, Reagan and Gorbachev agreed on the goal of nuclear disarmament. But they could not agree on a first tranche of cuts because Gorbachev insisted that Reagan commit to remaining within the constraints of the ABM Treaty for 10 years. The treaty limited both sides to 100 ground-based interceptors at a single site. At the time, the Reagan administration was focused on a concept in which hundreds of orbiting, high-power lasers would burn holes in Soviet booster rockets as they rose out of the atmosphere.

Arms controllers on both sides, meanwhile, had found each other and were brainstorming about how to end the nuclear arms race. Stone and I had become involved in such discussions in November 1983. We met with a group headed by Evgeny Velikhov, a vice president of the Soviet Academy of Sciences. After Gorbachev came to power in March 1985, we learned that Velikhov and his colleagues had been advising Gorbachev (see my article, *Physics Today*, September 2013, page 41.).

Gorbachev's first move, in August 1985, was to declare a unilateral moratorium on nuclear testing, now underground because of the 1963 atmospheric test ban. The Reagan administration disparaged the initiative, but the Democrats, who controlled the House of Representatives until 1995 and the Senate during 1987–95, were impressed. In 1992 they were able to force the Bush administration to end US testing on the condition that other countries not test either. In 1996 the Clinton administration negotiated the Comprehensive Nuclear-Test-Ban Treaty (CTBT). Although that treaty has not yet been ratified by the US, China, India, Pakistan, or North Korea, only North Korea has tested since 1998 (see the article by Pierce Corden and David Hafemeister, *Physics Today*, April 2014, page 41.).

In February 1987 Velikhov and I organized the scientists' forum on nuclear disarmament in Moscow. Sakharov's release from Gorky may have been timed so that his exile would not become an issue at the forum. Stone and I arrived with our wives in Moscow a few days early, and Stone arranged a meeting with Sakharov and Bonner in their apartment (see Figure 1). Now that he was able to speak out again, we knew the world was eager to hear what Sakharov had to say. We saw the forum as an opportunity for him to lay out his views on nuclear disarmament at that critical time.

FIGURE 1. Andrei Sakharov (right) and I speak in his Moscow apartment on 11 February 1987, three days before a scientists' forum at which he made public his proposal to delink Soviet–US negotiations on nuclear reductions from the Reagan administration's Star Wars program. (Photo by Jeremy Stone.)

Stone started the discussion by suggesting that Sakharov urge Gorbachev to ignore the SDI and seize the opportunity for arms reductions. He argued that Reagan's successors would abandon the program as unaffordable and that the US would not break the ABM Treaty if progress was being made on nuclear reductions. Sakharov responded that he had been thinking along the same lines.

My part of the discussion was not so easy. I told Sakharov that my colleagues and I were publishing estimates of the civilian consequences of US and Soviet nuclear "counterforce" attacks on each other's nuclear forces. We had found that the direct consequence would include tens of millions of civilian deaths. I argued that both sides should settle for "minimum deterrence" or what McGeorge Bundy, President John F. Kennedy's national security adviser, called "existential deterrence," a situation in which just the fact that a country has nuclear weapons instills caution—like a policeman's holstered gun—in other countries.

Sakharov argued that persuading both countries' military leaders to abandon counterforce strategies would be virtually impossible. In the two decades since our conversation, the numbers of warheads deployed on Russian and US strategic missiles have come down by a factor of about five, to less than 2000 each, but the two missile forces' highest-priority targets remain each other.

During our discussion, the apartment doorbell rang every 10 minutes or so. Bonner told us to ignore it, saying, "It's just the KGB." It turned out that the KGB was also recording our conversation.

THE KGB TRANSCRIPT

In 1992, after the collapse of the Soviet Union, the new Russian government made public a collection of Communist Party documents, including a KGB transcript of parts of our conversation in Sakharov's apartment. Matthew Evangelista, a historian at Cornell University, obtained a copy and shared it with me.

Certain passages of the transcript were underlined. If that was done by Gorbachev—or by the KGB for his benefit—they indicate that he was interested in learning more about the nuclear balance. Among the underlined parts were my statements that Soviet warheads had higher explosive yields than US warheads and that the Soviet Union was about five years behind the US in reducing the weight of its warheads per unit of explosive power. That information may have been relevant to the debate within the Soviet leadership over Gorbachev's unilateral nuclear testing moratorium, which was about to end.

Sakharov's statement that silo-based ballistic missiles accounted for a much larger fraction of Soviet than US strategic warheads also was underlined, as was his response when Stone conveyed an invitation from Senator Edward Kennedy for Sakharov to visit the US. Sakharov said he would not be allowed to travel abroad in the absence of "very strong pressure" from "foreign political leaders and organizations," and he added that an effort of such magnitude would be "disproportionate to the goal."

DELINKING MISSILE DEFENSE FROM NUCLEAR REDUCTIONS

Sakharov argued at the scientists' forum that the Soviet Union should delink its objections to Reagan's SDI from the issue of bilateral nuclear cuts. He was heavily criticized for that by some Soviet nuclear strategists at the meeting. Later, at an event at the Kremlin, in a speech summarizing the conclusions of the scientists' forum for Gorbachev and a large audience, I presented Sakharov's recommendations.[4] In his book Perestroika, published later that year, Gorbachev recounted,

> At the Moscow International Forum "For a Nuclear-Weapon-Free World and the Survival of Humanity"—a meeting unprecedented in the number of participants and their authority—I had the opportunity to feel the moods and hear the thoughts and ideas of an international intellectual elite. My discussions with them made a great impression on me. I discussed the results of the congress with my colleagues in the Politburo and we decided to make a major new compromise—untie the Reykjavik package and separate the problem of medium-range missiles in Europe from the other issues.[5]

Sakharov was not the only one arguing for delinking. The day before the crucial 26 February 1987 Politburo meeting, Alexander Yakovlev, a close adviser, sent Gorbachev a memo arguing passionately for the delinking; his argument was based primarily on an analysis of European and US public opinion. Also, Velikhov's group had convinced Gorbachev that any SDI system could be handled with countermeasures. As Gorbachev said it, "A tenth of the US investments would be enough to create a counter-system to frustrate SDI."[6]

Later that year Gorbachev and Reagan signed the Intermediate-Range Nuclear Forces Treaty, which eliminated about 2700 medium- and intermediate-range nuclear missiles. The two men also agreed in principle on 50% cuts of strategic warheads, which laid the basis for the Strategic Arms Reduction Treaty (START) that entered into force in 1994. In the meantime, with both houses of Congress under solid Democratic control, Senator Sam Nunn, the chairman of the Senate Armed Services Committee, informed Reagan that if his administration reinterpreted the ABM Treaty to allow testing of a space-based BMD, funding for SDI would be cut deeply.

MEETING WITH GORBACHEV

Sakharov's meeting with Gorbachev in January 1988 came about because of another Velikhov initiative, the establishment of an independent, international foundation in Moscow to work on global problems. Velikhov invited Sakharov to be on the foundation's board, along with several foreign luminaries and me. The foundation's creation was announced at the scientists' forum, and the board was invited to meet with Gorbachev in the Kremlin.

I sat with Sakharov on the bus to the Kremlin. He told me about Gorbachev's call to Gorky to inform him that he was free to return to Moscow. Sakharov said that his immediate response to Gorbachev had been that freeing him was not enough. It was necessary to free all political prisoners.

At the Kremlin, there was first a reception outside Gorbachev's office. Gorbachev greeted Sakharov, who thanked him for "restoring my freedom and responsibility" (someone translated the exchange for me). Gorbachev responded that he was happy to hear Sakharov connect those two words.

Then the foundation board members sat down with Gorbachev around a conference table, and each of us had an opportunity to address him. According to my recollection, when it was Sakharov's turn, he started, "Mikhail Sergeyevich, when we spoke during your call, I raised the issue of other political prisoners. Today I have brought with me a list." Gorbachev responded, "Andrei Dmitrievich, we can't go too quickly. Remember what happened with the Red Guards in China," referring to the chaos that resulted when Mao Zedong unleashed young activists on China's establishment during the late 1960s. Gorbachev did, however, have an aide take Sakharov's list. A year later Sakharov could say, "The majority of prisoners of conscience have been freed."[7]

The foundation operated for a few years, which gave me an opportunity to get to know Sakharov better. He was absolutely uncompromising, starting with the foundation's name. He insisted that it be called the International Foundation for the Survival and Development of Humanity. I commented that was a rather long name but Sakharov responded, "What do you want to leave out? Humanity? Development? Survival?" I surrendered. On another occasion, he suggested that members of the board personally pay half of their travel expenses to make sure that we weren't being motivated by the opportunity to travel. That idea did not attract support from any other board member.

I had the opportunity to see another side of Sakharov when I accompanied him to a lunch with Sweden's ambassador to the Soviet Union. The subject was Raoul Wallenberg, a heroic Swedish diplomat who in 1944 saved thousands of Jews in Budapest from being shipped to Nazi extermination camps. After the Soviet Union occupied Hungary, Wallenberg was sucked into the KGB's prison system. The KGB said that he died in 1947, but from time to time released prisoners reported having seen him. Sakharov had come to discuss the latest rumor. I was moved to see how, in the midst of his battle for democracy in the Soviet Union, this great man was still pursuing the cases of individual political prisoners.

Sakharov was elected as an opposition member to the Soviet Congress of People's Deputies in 1989. Later that year he had a heart attack and died in his apartment. He left behind a draft of a new Soviet constitution that emphasized democracy and human rights. In a poll taken shortly thereafter, Sakharov was found to be the most revered person in Soviet history.[8]

Sakharov stood up for the principles that he had enunciated in "Reflections on Progress, Peaceful Coexistence, and Intellectual Freedom," and for that he was recognized with the 1975 Nobel Peace Prize. The European Parliament honored him by establishing the Sakharov Prize for Freedom of Thought in 1988. The American Physical Society created a Sakharov Prize in 2006 for physicists who uphold human rights.

RELINKING

US presidents George H. W. Bush and Bill Clinton did not have the same enthusiasm for BMD as Reagan did. But weapons programs are difficult to kill, and its funding continued at about $5 billion per year, as shown in Figure 2.

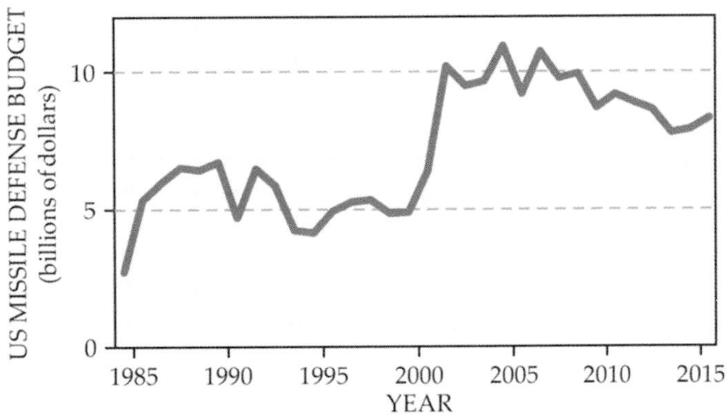

FIGURE 2. Budget of the US Missile Defense Agency and its predecessors, the Strategic Defense Initiative and Ballistic Missile Defense Organizations. The initial rise followed President Ronald Reagan's Star Wars speech in 1983. The second major increase followed President George W. Bush's announcement in 2002 that he was taking the US out of the 1972 Anti-Ballistic Missile Treaty, which had limited ballistic-missile defenses. (Data from ref. 15, converted to constant 2016 dollars.)

In 1996 a new Republican majority in both houses of Congress established the Commission to Assess the Ballistic Missile Threat to the United States, chaired by Donald Rumsfeld. The commission reported back in 1998 that within five years Iran and North Korea could have intercontinental ballistic missiles armed with weapons of mass destruction. It also said that Iraq could do the same within 10 years or, if it used ship-based ballistic missiles, "within a very short time" and that those capabilities might emerge with little warning.[9]

After his election in 2000, President George W. Bush appointed Rumsfeld to be his secretary of defense. In 2002 they took the US out of the ABM Treaty and committed to fielding missile defenses by the end of Bush's first term in 2004. The annual budget for BMD was quickly doubled to $10 billion (in 2016 dollars), a level from which it dropped only slightly during the Obama administration. Currently the US has 30 land-based interceptors deployed in Alaska and California and 33 Aegis cruisers and destroyers equipped with missile-detection radars and launchers for interceptor missiles (see figure 3). The most advanced Aegis interceptor, the Standard Missile 3 Block IIA, which was first tested in 2015, has sufficient speed, if launched from ships near the continental US, to intercept intercontinental ballistic missiles sent from Russia or China. The same system is being deployed on land in Romania and Poland. Designed to intercept missiles above Earth's atmosphere, the systems could be defeated by lightweight decoys and other countermeasures.

FIGURE 3. The launch of an interceptor missile from the rear vertical launch magazine of a US Aegis cruiser. The cruiser's phased-array radars can detect incoming warheads at a range of about 300 kilometers. To defend the US against long-range missiles, the Aegis missiles would have to be given their interception points by more powerful early-warning radars. (Courtesy of the US Navy.)

The ABM Treaty came about in part because US physicists in the late 1960s explained the many obvious countermeasures to both Congress and their Soviet counterparts. Sakharov's 1968 essay cites a key article that informed the debate:

> … the practical impossibility of preventing a massive rocket attack. This situation is well known to specialists. In the popular scientific literature, for example, one can read this in an article by Richard L. Garwin and Hans A. Bethe in the Scientific American of March 1968.

When Garwin and Bethe wrote their article, the proposed US interceptor missiles were nuclear tipped. That was one reason for public interest in the issue. Suburbanites did not want nuclear-armed interceptors in their backyards. Today the interceptors are terminally guided with IR sensors. But the problem of decoys and other countermeasures remains.[10]

FIGURE 4. On 23 March 1983 President Ronald Reagan announced his Strategic Defense Initiative on television. The first three-quarters of the speech was spent justifying dramatic increases in the US defense budget, "to make America strong again after too many years of neglect and mistakes." The photograph behind the president shows a Soviet MiG-23 fighter-bomber base in Cuba. (Courtesy Ronald Reagan Presidential Library/Carlos E. Santa Maria)

Both Russia and China have expressed concern about the US deployments. Russia cites US BMD as a principal reason why it is not interested in negotiating further reductions in strategic nuclear weapons.[11] And although it would make it easier for Russia and the US to further reduce their arsenals if China committed to not build up, China is increasing its small force of intercontinental ballistic missiles, in part because of its own concerns that US BMDs could neutralize its deterrent.[12]

The original justification by the Bush administration in 2002 for deploying a BMD was the imminent threat of weapons of mass destruction carried by missiles launched by Iran, Iraq, and North Korea. Today there are no such threats from Iran or Iraq. But the US BMD program goes on, including the provocative deployments in Poland and Romania. A North Korean threat of a nuclear-armed intercontinental ballistic missile has materialized, but there are alternative, potentially more effective defenses that would not threaten the deterrents of large countries such as China and Russia. Specifically, North Korean missiles still in their boost phase could be within reach of interceptors based off the country's shores or to the north in China or Russia.[13]

Gorbachev had the wisdom to ignore the Reagan administration's fantasies about space-based BMD. But this time, the US should take responsibility for weighing the questionable advantages of exo-atmospheric missile defense against the obstacles that it poses to further nuclear reductions. Physicists could again make an important contribution by explaining the technical issues in the debate.

REFERENCES

1. This part of the story is primarily based on A. Sakharov, Memoirs, R. Lourie, trans., Knopf (1990), and Moscow and Beyond, 1986–1989, A. Bouis, trans., Knopf (1991); for Elena Bonner's account of their time in Gorky, see E. Bonner, Alone Together, A. Cook, trans., Knopf (1986).
2. A. Sakharov, At. Energy 4, 6 (1958), reprinted in Sci. Global Secur. 1, 175 (1990).
3. L. Wittner, Confronting the Bomb: A Short History of the World Nuclear Disarmament Movement, Stanford U. Press (2009).
4. F. von Hippel, Bull. At. Sci. 43(4), 12 (1987).
5. M. Gorbachev, Perestroika: New Thinking for Our Country and the World, Harper & Row (1987), p. 153.
6. Ref 5, p. 234. see also P. Podvig, Sci. Global Secur. 25, 3 (2017).
7. Ref 1, Moscow and Beyond, 1986–1989, p. 44.
8. Ref 1, Moscow and Beyond, 1986–1989, p. x.
9. D. H. Rumsfeld et al., Executive Summary of the Report of the Commission to Assess the Ballistic Missile Threat to the United States (15 July 1998).
10. For an update, see A. M. Sessler et al., Countermeasures: A Technical Evaluation of the Operational Effectiveness of the Planned US National Missile Defense System, Union of Concerned Scientists (2000).
11. V. Putin, meeting with heads of international news agencies, St. Petersburg Economic Forum (17 June 2016), http://en.special.kremlin.ru/events/president/transcripts/52183.
12. L. Bin, Arms Control Today 45, 8 (December 2015).
13. R. Garwin, Arms Control Today 30, 8 (September 2000).
14. B. Whitmore, "Revolutions of '89: The Moscow Spring," Radio Free Europe/Radio Liberty (12 November 2009).
15. US Missile Defense Agency, "Historical Funding for MDA FY85-16" (February 2016).

Lab-to-Lab Cooperative Threat Reduction

Siegfried S. Hecker

Center for International Security and Cooperation
Stanford University
Stanford, CA 94305-6165

Abstract. It is difficult to imagine today how dramatically global nuclear risks changed 25 years ago as the Soviet Union disintegrated. Instead of the threat of mutual nuclear annihilation, the world became concerned that Russia and the other 14 former Soviet states would lose control of their huge nuclear assets – tens of thousands of nuclear weapons, more than a million kilograms of fissile materials, hundreds of thousands of nuclear workers, and a huge nuclear complex. I will describe how scientists and engineers at the DOE laboratories, with a focus on Los Alamos, Lawrence Livermore and Sandia national laboratories, joined forces with those at the Russian nuclear weapon institutes for more than 20 years to avoid what looked like the perfect nuclear storm – a story told in the two-volume book *Doomed to Cooperate*[1] published in 2016.

INTRODUCTION

On December 25, 1991, the Soviet Union dissolved into 15 independent states. The Soviet breakup was the first time in history that a country with the capacity to destroy the world was thrust into chaos—its political system collapsing, its economy in free fall, the institutions that regulated the daily lives of Soviet people crumbling, and their future uncertain. In Washington and other capitals around the world, the nuclear dangers took an unexpected turn, with nations no longer feeling threatened by the enormous nuclear arsenal in the hands of the Soviet government, but rather by the prospect that the huge former Soviet nuclear assets—weapons, materials, and experts—most of which were inherited by the Russian Federation, would no longer remain in the government's control.

The specter of another Chernobyl, of nuclear weapons or nuclear materials winding up in Iraq, Iran, or North Korea, or of a nuclear environmental disaster haunted Washington and much of the international community. The West feared for the fate of Soviet nuclear weapons, the security of its nuclear materials and facilities, the potential leakage of nuclear weapons expertise abroad, and the possibility of dangerous nuclear exports to raise revenues. These fears collectively can be called the four "loose nuclear dangers" – namely, loose nuclear weapons, loose nuclear materials, loose nuclear experts, and loose nuclear exports. These concerns were exacerbated by early reports of economic hardship, corruption, and the rise of criminal elements in Russia, as well as disarray in Russia's vaunted security services and institutions that protected the nuclear complex. It appeared to be a perfect storm for disaster in the Russian nuclear complex, which had been largely opaque to the western world.

The attempted coup against Soviet President Mikhail Gorbachev in August 1991, during which he reputedly lost control of the nuclear launch codes, triggered alarm of a Soviet breakup and the potential looming nuclear dangers among some professionals in the US government, nongovernmental organizations, and academia. A group of university and non-governmental intellectuals in the security field recognized the dangers as well as the opportunity to fundamentally change the nuclear relations between our two countries. They brainstormed cooperative security initiatives that in a remarkably short order became codified in the visionary Nunn-Lugar Cooperative Threat Reduction legislation.

The Soviet Union was estimated to have roughly 39,000 nuclear weapons near the time of its dissolution, spread across the 11 time zones of its republics. Most of the weapons were in the Russian Federation, but significant stockpiles were left in the now independent countries of Ukraine, Kazakhstan, and Belarus. The huge Soviet nuclear complex of some 60 sites with 10 dedicated closed cities and hundreds of facilities had produced nearly 1.4 million kilograms of fissile materials, namely, plutonium and enriched uranium, that could fuel nuclear weapons. The combined military and civilian nuclear complex employed one million people.

David Hoffman tells the story of how the Cold War arms race came to an end to leave a legacy of peril in his Pulitzer Prize winning book, *The Dead Hand.*[2] He points out that the worst fears during the last days of the Soviet

Nuclear Weapons and Related Security Issues
AIP Conf. Proc. 1898, 020010-1–020010-15; https://doi.org/10.1063/1.5009215
Published by AIP Publishing. 978-0-7354-1586-7/$30.00

Union were that loose nukes, fast money, and a weak state would come together to present Russia and the former Soviet states with "an inheritance from hell." To Washington, everything in Russia's nuclear complex posed only dangers and offered no benefits. This was the predominant American view and, quite frankly, it was believed to be an inheritance from hell for the rest of the globe as well.

Yet, none of the West's fears came to pass. In the nearly 25 years since the Soviet dissolution, there has been no perfect storm—no disaster in Russia's nuclear weapons complex, no serious nuclear accidents, very limited loss of nuclear material, limited leakage of nuclear weapons know-how or brain drain, and no nuclear weapons stolen, sold, or diverted.

How did this happen? In *Doomed to Cooperate,* I argue that the extraordinary professionalism, dedication, and patriotism of the Russian nuclear weapons workers and leaders, combined with an extraordinary and timely assist from the United States through innovative government programs and scientific cooperation, saved the world from potential disaster. Numerous firsthand accounts from Russian scientists and their leaders, published for the first time in the book, capture the extraordinary actions of the Russians at the frontlines of the nuclear dangers, actions enabled and enhanced by direct scientific cooperation between the Russian and American nuclear laboratories.

In retrospect, in spite of the current political acrimony between Moscow and Washington, cooperative threat reduction, including scientific cooperation, served the United States well because it helped avoid a nuclear disaster of unknown proportions. Cooperation also enabled the rapid implementation of a series of nuclear weapons treaties, which over 30 years resulted in a nearly 80 percent reduction in each country's nuclear arsenal. It brought long-needed relief from Cold-War tensions and fears of annihilation.

Cooperation benefitted Moscow as well. Most importantly, it allowed Russia to safely reduce its nuclear arsenal, which involved the difficult tasks of transporting record numbers of nuclear weapons, disassembling them, and storing the fissile materials safely and securely. Bilateral cooperation greatly enhanced security in Russia's nuclear complex and its military sites, allowing them to be upgraded to meet the new challenging security environment resulting from the dissolution of the Soviet Union. Cooperation between Russian nuclear scientists and engineers and their American counterparts facilitated successful resolution of many of the most sensitive nuclear security challenges and helped to prevent a brain drain, that is, the leakage of technical expertise outside the Russian nuclear complex.

Doomed to Cooperate tells the stories of lab-to-lab collaboration between Russian and American nuclear weapon lab scientists. This article also focuses on that collaboration and how it helped to avert the nuclear dangers that arose with the breakup of the Soviet Union.

SETTING THE STAGE FOR LAB-TO-LAB COOPERATION

The political changes that eventually led to the lab-to-lab initiative began in 1985 when Mikhail Gorbachev became the General Secretary of the Soviet Communist Party. The nuclear arms race was at its peak, President Reagan had labeled the Soviet Union "the evil empire," and the Soviet economy was failing. Gorbachev decided to take his country in a new direction. He introduced measures to reduce military expenditures, demilitarize foreign policy, and use the resources to shore up the economy. He also introduced policies of *glasnost* and *perestroika,* aimed at creating a more open society and loosening government controls over the economy.

President Reagan quickly saw the value of Gorbachev's initiatives and pursued a policy of *détente*. At the Reykjavik Summit in October 1986, Presidents Reagan and Gorbachev came close to agreeing to eliminate US and Soviet nuclear arsenals. Although they did not succeed, they set in motion a number of arms control agreements. They eliminated an entire class of nuclear-tipped missiles with the Intermediate-Range Nuclear Forces (INF) Treaty in December 1987. They charged their nuclear weapons experts to develop verification measures to enable ratification of the Threshold Nuclear Test Ban Treaty (TTBT). And, on July 31, 1991, Gorbachev and President Bush signed the Strategic Arms Reduction Treaty (START-I), the first treaty calling for reductions in strategic nuclear arsenals.

During that period Gorbachev loosened the Soviet hold over the Eastern Bloc nations. He did nothing to resist the unification of East and West Germany in November 1989, and after initially blocking the independence of the Baltic States, he allowed them to exit the Soviet Union. However, the Soviet hardline opposition rebelled, and less than three weeks after the 1991 summit, they attempted the coup against Gorbachev. Although Russian Republic President Yeltsin helped Gorbachev regain control, the events of August proved to be the beginning of the end of the Soviet Union.

The coup triggered widespread concern in the West about potential nuclear chaos during the last months of 1991. The ensuing disintegration of the Soviet empire posed unprecedented nuclear security dangers and immediate concern about the safety of nuclear weapons – that is, how to safely transport and dismantle a record number of nuclear weapons from the former Soviet Union and how to safely operate or decommission different parts of the huge nuclear infrastructure spread across the vast territory of the defunct empire.

President Bush helped open the door to collaboration with the Presidential Nuclear Initiatives (PNIs),[3] following shortly after Gorbachev's return from house arrest. In a phone call to Gorbachev on September 27, 1991, Bush informed him that he would take several unilateral actions, such as taking all nuclear weapons off US surface ships and taking some of the weapon systems off alert. In a nationwide address on October 5, Gorbachev announced that he would take reciprocal actions. The Russian side followed up immediately through nongovernmental US channels. By October 18 and 19, 1991, Viktor Mikhailov, who at that time was deputy head of the Ministry of Atomic Power Industry, was in Washington, DC, at a meeting organized by the Federation of American Scientists and the National Resource Defense Council to discuss several joint security initiatives. At that meeting, Thomas Neff of the Massachusetts Institute of Technology explored his idea of a highly enriched uranium purchase deal with Mikhailov for the first time.[4] This idea led to the successful 20-year bilateral program known as the Highly Enriched Uranium (HEU) purchase agreement.

The American nongovernmental scientific community had been instrumental in promoting nuclear cooperation, particularly since Gorbachev's initial openings to the West in 1985.[i] The US National Academy of Sciences Committee on International Security and Arms Control, the Carnegie Corporation, the National Resources Defense Council, and the Federation of American Scientists, as well as several university programs, such as those at Princeton, Massachusetts Institute of Technology, Harvard Kennedy School's Belfer Center, and Stanford University's Center for International Security and Arms Control (CISAC), promoted cooperation from the early days of the Gorbachev administration. The legislative branch of the US government also joined in through its Senate Committee on Arms Control.[5]

Most of these organizations worked closely with the Soviet Academy of Sciences, in which Academician Evgeny P. Velikhov played the most influential role. He was particularly effective because he also served in the Gorbachev administration and was president of the Kurchatov Institute.[ii] Other Soviet academicians, such as Roald Sagdeev, Vladimir Fortov, and Gennady Mesyats[iii] also reached out to the American scientific community. Sergey Rogov and Andrey Kokoshin of Moscow's USA-Canada Institute were also influential in reaching out in late 1991 to brief US senators on nuclear control issues.[6]

One of the most influential contributions of the NGO community was the Belfer Center's study and prescient report on how to manage the nuclear impacts of a Soviet breakup, "Soviet Nuclear Fission: Control of the Nuclear Arsenal in a Disintegrating Soviet Union,"[7] published in November 1991. It found fertile ground in the office of Senator Nunn.

Senator Nunn had met with Gorbachev in Moscow following the failed August coup and Gorbachev's release from house arrest and had become deeply concerned about the fate of Soviet nuclear assets. He then collaborated with Congressman Les Aspin to add a last-minute amendment to a November 1991 authorization bill targeting aid to the Soviet nuclear complex. There was sufficient opposition to the amendment that Nunn and Aspin withdrew it.

Shortly thereafter Senator Lugar joined forces with Nunn to champion nuclear security assistance to the Soviet Union. It was in the Senate offices on November 19 that Ashton B. Carter, the director of the Belfer Center study, presented the results to Nunn, Lugar, their staff, and a small group that included William J. Perry, former Department of Defense Assistant Secretary, Director of Defense Research and Engineering, and then co-director of Stanford University's CISAC; David Hamburg, president of the Carnegie Corporation, which had funded the Belfer study; and John Steinbruner of the Brookings Institution. In Carter's words, "The study predicted that the breakup of the Soviet Union posed the biggest proliferation threat of the Atomic Age and outlined a new form of 'arms control' to stop it: joint action by the two former Cold War opponents against the common danger."[8]

---

[i] The role of US nongovernmental individuals and organizations, working in a Track II—that is, nonofficial, nongovernmental—role is captured nicely in David A. Hamburg's *A Model of Prevention: Life Lessons,* Paradigm Publishers, Boulder, London (2015). Scientific cooperation through the US – Soviet Union was promoted through the National Science Foundation in the United States and the Russian Academy of Sciences in the Soviet Union. An appraisal of the 1972-82 intergovernmental agreement on scientific and technical exchanges is described by Catherine P. Ailes and Arthur E. Pardee, Jr. in *Cooperation in Science and Technology: An Evaluation of the U.S.-Soviet Agreement.* Boulder, Colorado: Westview Press, 1986.

[ii] E. P. Velikhov's efforts during the Gorbachev era are described by Hoffman in the *Dead Hand* and by David A. Hamburg in *A Model of Prevention: Life Lessons,* Op. cit.

[iii] V. Fortov, G. Mesyats and R. Sagdeev were three prominent member of the Soviet Academy of Sciences. Fortov recently served as the president of the Russian Academy, Mesyats was vice president for many years, and Sagdeev was a pioneer in the Soviet space program and later moved to the University of Maryland. Fortov was one of the first Soviet scientists to regularly collaborate with several LANL scientists.

Carter's briefing backed up by the report, *Soviet Nuclear Fission,* lent empirical and analytic weight to Nunn's arguments and validated the urgency to adopt policy to confront the security dangers posed by the political turmoil in the Soviet Union.[9] The impact was immediate. The Nunn-Lugar legislation, titled the "Soviet Nuclear Threat Reduction Act of 1991," passed as an amendment to an unrelated bill and was signed into law by President Bush on December 12, 1991. The legislation did not appropriate new funds to the Department of Defense (DoD), but instead the DoD was given authority to transfer or reprogram $400 million from existing accounts.

The Bush administration's efforts to turn the legislation into a real program were half-hearted at best. It took an enormous effort by Nunn and Lugar in 1992, including several congressional delegation visits to the former Soviet Union to begin to develop such a program. These efforts finally got off the ground under the Clinton administration, when William Perry became Deputy Secretary of Defense and eventually Secretary, and Ashton B. Carter became Assistant Secretary. They engineered cooperation between US and Russian defense ministries and militaries under Cooperative Threat Reduction (CTR), which came to be known as the Nunn-Lugar program. Their efforts solved one of the most immediate threats, namely that of safely bringing the nuclear weapons left in the former Soviet republics of Ukraine, Kazakhstan, and Belarus back to Russia for dismantlement and disposition.

It would have been a deadly blow to the nonproliferation regime and an enormous safety and security risk had these countries kept the Soviet nuclear arsenals stationed on their soil.[iv] The predecessor agency to the Defense Threat Reduction Agency in the DoD also initiated several urgent projects that provided secure transportation of Soviet nuclear warheads, initiated planning and construction for a centralized fissile material storage facility, and helped incapacitate the nuclear testing infrastructure at the former Soviet Semipalatinsk Test Site in Kazakhstan. In addition, the DoD also orchestrated the incredible feat of airlifting nearly 600 kg of HEU from the former Soviet nuclear industrial city of Ust-Kamenogorsk (by then in the Republic of Kazakhstan) to the United States.[10]

These efforts were just the beginning of the Nunn-Lugar CTR program. Senators Nunn and Lugar, later joined by Senator Pete Domenici, spent much of the next two decades in various capacities personally promoting and supporting this entirely new relationship with our former adversary.[11]

While these formal government-to-government programs addressed many of the most immediate, crucial threats, they were not sufficient to meet the whole spectrum of nuclear safety and security concerns. Many of these issues were so sensitive that they required a much greater level of trust and cooperation than was possible to achieve in the political sphere at that time. They also required a level of technical knowledge far beyond what was available to diplomatic representatives of each side. Moreover, the enormous differences between how the governments of the United States and Russia viewed the threats posed by Russia's nuclear weapons complex during these chaotic times hindered taking some of the most important actions necessary to avert these threats.

THE BEGINNING OF LAB-TO-LAB COOPERATION

Understanding the Russian perspective was crucial to tailoring nuclear cooperation to achieve Russia's aspirations and to reduce what Americans viewed as the four nuclear dangers. But that required close collaboration and American boots on the ground in Russia's nuclear complex, which was so wrapped in deep secrecy during the Cold War that nuclear sites existed in cities that were not even on Soviet maps. It was Gorbachev-Reagan détente that opened the door at the October 1986 Reykjavik Summit, which led to the Joint Verification Experiment (JVE) of 1988.[v]

The JVE cooperation helped overcome a stumbling block related to verification procedures needed to ratify the 1974 Threshold Nuclear Test Ban Treaty (TTBT). Soviet and American nuclear weapons scientists were asked to conduct parallel nuclear-explosion yield measurements at testing grounds in Nevada and Semipalatinsk, located in what is now Kazakhstan. The 1988 nuclear tests enabled the two sides to sign a new ratification protocol in Geneva in June 1990, leading to the TTBT entering into force in December 1990.

Several of the Soviet and American scientists who participated in the JVE strengthened these relationships when, over the next two years, they served in Geneva as technical experts in developing the verification protocols for the TTBT. In 1990, while they were in Russia for technical verification discussions and a conference, Mikhailov, who led the Soviet technical delegation in Geneva, invited Los Alamos and Lawrence Livermore lab scientists to visit the secret nuclear cities. This was the first time an American was permitted to step foot into the Soviet nuclear complex.

[iv] See Nunn-Lugar scorecards, Defense Threat Reduction Agency. http://www.dtra.mil/Missions/Partnering/NunnLugarCTRScorecards.aspx

[v] The US NGO community, particularly the efforts of Tom Cochran and Chris Paine, played a critical role in promoting the idea of the JVE and in orchestrating the off-site seismic measurements.

The US laboratory scientists returned with proposals from the Russian nuclear labs to collaborate on fundamental science projects with Los Alamos. In follow-up correspondence, the leadership of the Russian nuclear weapons labs proposed visits to the US labs and invited me, then director of the Los Alamos National Laboratory, to visit the Russian labs. These invitations came as a surprise, and it took until late 1991, just prior to the dissolution of the Soviet Union, before I was able to convince the US government to accept the proposals for research collaboration and exchange visits. Admiral James D. Watkins, then secretary of energy, approved my proposal for an exchange visit of US nuclear weapons lab directors with the Soviet (soon to be Russian) counterparts.

On February 23, 1992, little more than three years after the JVE and less than two months after the dissolution of the Soviet Union, John Nuckolls, director of the Lawrence Livermore National Laboratory (LLNL), and I touched down in an Aeroflot Turboprop on the tarmac in Sarov, Russia's secret city and birthplace of the Soviet nuclear bomb. Sarov was the Soviet analogue to Los Alamos, America's formerly secret city where Manhattan Project pioneers built the world's first atomic bomb. We were greeted by Academician Yuli Borisovich Khariton, who—like America's J. Robert Oppenheimer—had led the building of his country's first atomic bomb. Almost five decades later, Khariton, at age 88, had just stepped down from his decades-long role as scientific leader of VNIIEF,[vi] Russia's first nuclear weapons laboratory, known as Arzamas-16 in Soviet times. This visit marked the beginning of a new and unprecedented scientific collaboration between members of the Russian and American nuclear weapons laboratories.

FIGURE 1. On February 23, 1992, Yuli B. Khariton greets Siegfried S. Hecker and a US delegation in Sarov.

The visit to Sarov and the continuation to the city of Snezhinsk in the Urals, the home of the second nuclear weapons laboratory, VNIITF, shattered our Cold War preconceptions of the Soviet nuclear program. Contrary to expectations, their scientific and technical staff was outstanding. Although they lacked modern computers and electronics, their computational achievements were remarkable, and their experimental facilities were innovative and functional. We found their dedication to the mission deeply patriotic and their attention to nuclear weapons safety reassuring. During the briefings and tours, their scientists described leading-edge research in the fundamental science that underpins the nuclear weapons program. It was these visits that convinced me that we not only should collaborate with the Russian nuclear institutes to help solve immediate problems, but that we would benefit scientifically.[vii]

Most surprising, however, was that their readiness to cooperate directly with us, lab to lab and scientist to scientist, far exceeded what we were prepared for. They, more than we, realized that the end of the Cold War would require collaboration rather than confrontation. They had lived through declining political support for their nuclear complex during the Gorbachev era, but that changed when the Russian government, only two months old, chose Mikhailov to head the new Ministry of Atomic Energy (Minatom). He was one of their most respected nuclear scientists and our colleague during the JVE and in Geneva. After the Soviet breakup, Mikhailov continued to

---

[vi] The full name of the institute (we also refer to it as laboratory in this article) was All-Russian Research Institute of Experimental Physics, abbreviated as VNIIEF. The analogue to LLNL was called the All-Russian Research Institute of Technical Physics, abbreviated as VNIITF. During the Cold War, the nuclear centers VNIIEF and VNIITF were referred to as Arzamas-16 and Chelyabinsk-70, respectively. The Institute of Automatics (VNIIA), the analog to the US Sandia National Laboratories, not known in the West during Soviet times, was located in Moscow.

[vii] I am even more amazed today as I look back at the notes I took in the journal that they gave us at the beginning of the visit. The Russian scientific achievements were truly remarkable. These areas became areas of intense scientific collaboration over the next two decades.

orchestrate Russia's nuclear dialogue with the United States, and he approved the lab directors' visits on the Russian side, marking the beginning of lab-to-lab collaboration

On the American side, members of our delegation had already experienced positive collaboration with the Russians in the JVE and in Geneva. Also, the Los Alamos scientists who visited Russia's secret nuclear labs strongly encouraged me to accept the early Russian proposals for scientific collaboration. Fortunately, Nuckolls and I, as well as our entire delegation, were employees of and reported to the University of California. We enjoyed greater freedom to explore scientific collaboration with our Russian colleagues than would have been possible for US government employees. So, we quickly seized the opportunity, and in the office of VNIITF Chief Engineer Boris V. Litvinov, under the portraits of Vladimir Lenin and Igor Kurchatov (the first head of the Soviet atomic project), Russian and US counterparts hammered out a joint protocol describing areas of potential scientific and technical cooperation. The document had astonishing depth and breadth and included sensitive areas that directly addressed the nuclear dangers arising from the recent breakup of the Soviet Union.[viii]

It turned out that we scientists were far ahead of what the US government was prepared to authorize at the time. We heard that when members of the National Security Council staff, which coordinated interagency government issues with Russia, received a copy of the protocol, they declared it did not exist and threw it in the waste paper basket. When we presented the protocol to Secretary of Energy, Admiral James D. Watkins, we received approval to proceed, though only in fundamental science. But that was sufficient to begin collaboration.

Scientific cooperation got off to a fast start in a large part because it tapped into the most basic interests of scientists and engineers on both sides, namely, to create new knowledge and new technologies. Science is fundamentally an interactive, cooperative pursuit, which requires exposing results to review, critique, and ultimately improvement by others. The common language of science allowed us more easily to cross cultures and borders. We found science, unlike politics, to be a unifying force – one that allowed us to build trust through collaboration.

Through hundreds of scientific visits following our 1992 lab directors exchange, American participants in the scientific cooperation witnessed firsthand that the nuclear weapons complex in the Russian Federation, although severely stressed financially, remained remarkably well organized, primarily because nuclear experts such as Mikhailov, who had years of experience in the nuclear complex, ran their government's ministry. In the new government of the Russian Federation, Mikhailov had direct access to President Boris Yeltsin and used his position to keep the nuclear ministry intact and give it the most favorable treatment even under the country's dire economic conditions.

We quickly discovered that the Russian view of its nuclear enterprise could not have been more radically different than the prevailing American view. Our Russian colleagues and the nuclear ministry viewed their nuclear weapons and nuclear complex as the crowning achievement of Soviet times and essential to Russia's political, scientific, and economic future. With Russia's conventional forces having atrophied, nuclear weapons were viewed as their guarantor of sovereignty, assuring Moscow a seat at the international table. Nuclear materials were seen as an economic resource and a national treasure. Nuclear experts and their scientific institutions were seen as an engine for the recovery of the national economy.

On the American side, the road to scientific nuclear cooperation was more complicated. In fact, it would never have happened without the opening of three independent policy windows. First, the US presidents set the political tone that enabled a cooperative effort with the Russian nuclear complex. Second, the US nongovernmental (NGO) community, including universities and other private organizations, not only reached out to the Soviets well before the fall of the Soviet Union and then to the Russians but also anticipated the nuclear dangers that lay ahead as the Soviet Union was disintegrating. And third, Senators Sam Nunn and Richard Lugar, heeding those NGO voices, launched the transformative congressional Nunn-Lugar Cooperative Threat Reduction (CTR) program.

To launch and carry on scientific cooperation, particularly what we came to call lab-to-lab cooperation—that is, predominantly lab initiated cooperation—scientists and engineers also needed champions in the US government who were willing to take a risk, persevere through the endless interagency stalemates, and help promote our efforts. We particularly credit the courageous and farsighted support of Charles Curtis, Rose Gottemoeller, Victor Reis, Victor Alessi, Robert Gallucci, Anne Harrington, Ernest Moniz, Linton Brooks, and Daniel Poneman. In addition, we acknowledge the strong support and encouragement we received from many others in the government who worked behind the scenes.

[viii] The areas of cooperation in the protocol included enhancing security and safety of nuclear weapons during reduction and dismantlement; preventing proliferation of nuclear weapons knowledge and promote conversion/diversification of institutes and plants, preventing proliferation of nuclear weapons to non-nuclear states and terrorists; developing joint mechanisms for nuclear emergency response; enhancing safety of nuclear arsenals of both states; preventing unauthorized use of remaining weapons in both stockpiles; and promoting the protection and cleanup of the environment at nuclear weapons facilities.

LAB-TO-LAB COOPERATION IN ACTION

The scientific camaraderie and common ground forged between the nuclear weapons laboratories came none too soon. The multiple exchange visits by lab scientists and joint research projects that followed provided us with unparalleled access to each other's nuclear facilities and people. We developed trust and one-on-one relationships that led to unprecedented cooperation over the next 20 years.

The initial focus of Mikhailov's visits to Washington was to ask for assistance in what was called safe, secure dismantlement (SSD), specifically to get help to protect the security of nuclear weapons being returned to Russia for disassembly and for a fissile materials storage facility for safe and secure storage of the fissile materials from the weapons. The US government enlisted the technical support of nuclear scientists and engineers from the three nuclear weapons laboratories. The initial effort of advising the government on these technical issues in late 1991 turned into one the most enduring and successful cooperative efforts, the Warhead Safety and Security Exchange (WSSX), which helped keep Russian nuclear weapons safe and secure during chaotic political times.

The State Department launched the International Science and Technology Center (ISTC) program in early 1992, designed to prevent a potential brain drain of nuclear weapon scientists from an economically desperate nuclear complex. The Russian nuclear weapon scientists were supported directly by ISTC funds and often assisted by their American laboratory counterparts.

And concurrently, we at the nuclear weapons laboratories developed joint scientific collaboration with the Russian nuclear weapons labs, which helped to build personal relationships and trust that served all US government programs well over the years. These relationships opened the path to the WSSX program and the immensely successful nuclear materials protection, control, and accounting (MPC&A) program that helped Russia secure and safeguard the huge inventory of fissile materials. These relationships also facilitated the participation of US and Russian nuclear scientists over the entire spectrum of government nuclear cooperative programs. I describe here the most salient attributes of scientific lab-to-lab cooperation that guided all of these efforts.

A View from the Inside

With boots on the ground at each other's nuclear facilities, we gained a better understanding of one another's nuclear weapons complexes. For us at the US labs, that was especially important since so little was known about the Russian complex. Lab-to-lab scientific collaboration provided the first close look at Russia's people and facilities dedicated to nuclear weapons, particularly those at the three weapons institutes. We found their scientific expertise comparable to ours. Surprisingly, we found their attention to nuclear weapons safety also to be comparable, a big surprise since we were inclined to think we would find the same alarming disregard that led to the Chernobyl disaster in the civilian nuclear sector.

Through the multiple visits and early collaborations we came to know the inner workings of the Russian nuclear weapons complex. For example, we found that the dedication and patriotism of their nuclear workers was unshakable and, hence, the threat of brain drain was much less than what Washington anticipated. However, we also saw the severe economic and psychological stress endured by the nuclear workers in the 1990s. Their salaries were miniscule, and we often found them to be living on grocery store credits between paychecks that at times arrived six months late. We found the US brain drain programs, those aimed to relieve the plight of the Russian nuclear workers, to be much more effective than was reported by US government audits: we could see how these programs provided crucial financial support to nuclear workers, along with giving them hope for the future.

The stress on the nuclear workers was of particular concern since we found that safety and security in Russia's nuclear complex relied primarily on the nuclear workers themselves and less on technology than in the US complex. Hence, despite the unshakable patriotism and dedication of the nuclear workers, the quiet confidence that Russia's nuclear leadership had in nuclear safety and security was no longer realistic or justifiable. Our visits to their nuclear facilities and subsequent collaboration convinced us that they were not fully prepared for the post-Soviet world in which both safety and security of their complex and its assets would be greatly challenged by the chaotic new economic, political, and security environment. Our Russian colleagues maintained resolute confidence,[ix] but the fact

[ix] Although the Russian leadership maintained confidence, some Russian analysts expressed their concerns. For example, V. Orlov, R. Timerbaev, A. Khlopkov, in "Nuclear Nonproliferation in U.S.-Russian Relations." PIR Center Library Series, 2002, assess the situation in the 1990s as follows: "Russian politicians, military and experts did not immediately understand the scale of the problems of maintaining the security of nuclear munitions and materials which emerged as a result of the collapse of the Soviet Union. ...the threat of the external enemy disappeared, and the threat of the internal enemy emerged, be it a terrorist, or an unreliable employee at a nuclear facility."

that we had seen their situation from the inside and that they had witnessed how we use technology to back up human security measures made them receptive to cooperation, and we progressed in facing the new challenges together.

Cooperate to Do Good, Not Just Prevent the Bad

Lab-to-lab cooperation differed in another significant manner from the governmental programs that were directed almost exclusively at preventing bad things from happening. Most scientists want to make good things happen, not just prevent the bad. Lab-to-lab cooperation was focused on both from the beginning. Scientists and engineers want to create new knowledge, develop new technologies, and build things. Doing so together in early lab-to-lab projects built professional respect, which, in turn, made it easier to establish trust. We formed true partnerships that benefitted both sides. We treated each other as equals, not as donors or recipients of US aid, as was often the case in governmental circles.

Creating things together through scientific cooperation became the hallmark of our relationship. We joined forces to create new science—for example, the highest magnetic fields ever generated in experiments, advances in the quest to create pure fusion, a better understanding of the vexing instabilities of plutonium (the most complex of all elements), and the application of new parallel computing algorithms for modeling instabilities in explosive experiments or for modeling bio-molecular processes. Russian and US scientists and engineers jointly pursued peaceful applications of the technologies residing in their nuclear weapons laboratories. These ranged from kidney dialysis applications to artificial prostheses, computer software for industrial applications, and much more.

Unfortunately, little came of cooperation in civilian nuclear energy. It was the area in which the Russian laboratories and the ministry showed the greatest interest in collaborating over the years, but Washington showed little interest. This difference demonstrates the fundamentally different views of the two governments. Washington's focus was on preventing bad things from happening; Moscow wanted good things to result from cooperation. To the Russian nuclear complex, civilian nuclear energy was the long-term solution to global energy and climate change problems. To Washington, Russia's intense interest in the next generation of fast breeder reactors was considered a step in the wrong direction (primarily because of the perceived nonproliferation challenges of breeder reactors) or, at best, premature by half a century.

FIGURE 2. VNIIEF's Sasha Bykov lifts Steve Younger of LANL off the ground in a Russian bear hug in jubilation of the success of doing something positive together in a 1993 joint pulsed-power experiment.

Stockpile stewardship was critically important to both sides. President Clinton's Presidential Decision Directive-47 declared sustaining "the scientific competence of individuals responsible for ensuring confidence in the Russian and US nuclear stockpiles" a national priority.[12] This policy was quite remarkable in that it made maintaining the expertise of Russian nuclear weapons scientists, our Cold War adversaries, to be in the US interest. The ensuing collaboration focused primarily on nuclear weapon safety and security.

Quick, Flexible, and Concrete Results

In the early days of lab-to-lab collaboration, we at the nuclear laboratories communicated by phone, fax, and e-mail rather than through diplomatic cables. We explored, analyzed, and planned our cooperation together before seeking approval from our respective governments for the collaboration. Thanks to dedicated and creative financial and administrative experts at the laboratories on both sides, we managed to cut through government red tape to get funds to the Russian side quickly and to accelerate cooperation in the governments' effort toward safe, secure dismantlement of Russian warheads. The labs managed to send some of their discretionary funds to rapidly promote collaboration and achieve what the Russians liked to call concrete results.[x] These kinds of collaborations helped build the relationships and trust, which, in turn, allowed us to greatly accelerate and expand governmental efforts in sensitive nuclear security areas.

A True Partnership – Side by Side As Equals and Step by Step

Our Russian counterparts embraced the lab-to-lab approach. They appreciated the partnership, the focus on doing positive things together, and the creativity, flexibility, and speed of these efforts. American scientists and engineers came as collaborators, not as inspectors or spies, although there was considerable lingering suspicion in government circles on both sides. Steve Younger of LANL and Academician Yuri A. Trutnev of VNIIEF developed two principles that would guide future lab-to-lab cooperation during dinner in Sarov in 1993: *side by side as equals*[xi] and *step by step*. The first principle, "side by side as equals," implied that Los Alamos was not at VNIIEF to pick up technologies at bargain prices, which was a big concern throughout the Russian science and technology sector at the time. Younger convinced Trutnev that Los Alamos was committed to combining the best of both laboratories for joint scientific advances. The second principle, "step by step," explicitly recognized the sensitive nature of interactions between nuclear weapons laboratories. It meant that even if we could begin to work quickly, we needed to take deliberate and transparent steps one at a time to keep each other and our governments informed.

Secrecy and classification issues were challenging, of course, especially since the two countries had different classification guidance. Some of the projects that were explored often ended with one side or the other stating, "We can't talk about that." However, both sides managed the secrecy and classification issues well through the years. In fact, in the 1990s, Russian security services in the secret cities often supported the lab-to-lab efforts directly because both sides viewed the work as being urgent and essential.

Lab-to-lab was not a one-way street. Russian scientists and engineers came to the US labs as well as vice versa, and collaborations benefitted US scientists as well as science more generally. One such project conducting pulsed-power experiments required scientists from both sides to spend weeks at a time at VNIIEF and at Los Alamos. This remarkable scientific collaboration produced some 400 publications and presentations over 20 years. Scientists from VNIIEF and VNIITF made extended visits to Sandia National Laboratories and its Cooperative Monitoring Center and visited LLNL to collaborate on numerical algorithms, as well as other projects.

We invited the leadership of the Russian nuclear laboratories and Minatom to visit the three weapons laboratories so they could see US security and safety practices firsthand, as well as to better appreciate our scientific and technical capabilities. I will never forget the bewildered look on the face of Gennady Khandorin, the director general of the Siberian Chemical Combine,[xii] when I had to show my DOE security badge to the guard at the Los Alamos Plutonium Facility that we were entering in the summer of 1998. He exclaimed, "Don't they know you—after all aren't you the director of this laboratory?" He had difficulty understanding that this was required by our security culture. During the 1990s, most of the top leadership of Russia's nuclear complex visited Los Alamos and several of the other US nuclear sites.

This type of cooperation naturally required governmental agreements and approvals. We were fortunate to have the support of Victor Reis and his DOE Defense Programs Office. Both in this case and in other programs, lab-to-lab

---

[x] As one might imagine, the legal, political, and public relations aspects of sending money to the Russian nuclear weapons institutes early in our collaboration took a monumental effort. Thanks to creative financial staff at the US laboratories and solid support in DOE it was accomplished with remarkable speed – and, over the years, had an excellent track record of accountability.

[xi] This mantra was captured beautifully in Necia Grant Cooper, Los Alamos Science, "Russian-American Collaborations to Reduce the Nuclear Danger," No. 24. 1996

[xii] After visiting the Siberian Chemical Combine or Tomsk-7 as it was known in Soviet times in November 1997 (this huge site housed facilities similar to the US Rocky Flats, Savannah River and Oak Ridge Y-12 plants combined), I invited director general Gennady Khandorin to visit Los Alamos and several other sites. We hosted a similar visit for Valery Lebedev, the director general of the Mining and Chemical Combine, also known as Krasnoyarsk-26.

efforts worked best when they were aligned with the vision of farsighted government officials who made it possible to work the two hand in hand.

An Additional Diplomatic Tool – from "Trust but Verify" to "Trust and Benefit"

The attributes of the lab-to-lab approach described above created an effective parallel pathway to governmental diplomacy. It opened new doors and allowed us to accelerate and expand the governments' programs. In effect, the lab-to-lab approach added another diplomatic tool to the governments' diplomacy kit—namely, scientific diplomacy. Scientists in NGOs and universities had for decades played a seminal role in informing and sometimes guiding government actions in the nuclear arena. However, scientists at the US nuclear laboratories had been typically restricted to collaboration in the fundamental sciences, mostly through international conferences, or to provide support to governmental arms control initiatives. I should add that scientists were typically consulted only after such agreements were reached and mostly to help monitor or verify the agreements.

We were asked to play that role for the JVE to fulfill President Reagan's famous dictum "trust but verify," which he is said to have borrowed from a popular Russian saying "doveryai no proveryai." This role garnered mutual respect, admiration, and trust between Russian and US nuclear weapons scientists. The JVE not only provided the tools to verify the TTBT, but it helped launch a relationship that went far beyond verification. The stories in this book demonstrate how the ensuing cooperation between Russian and US nuclear scientists created new science, applied new technologies, benefitted both sides by improving the safety and security of nuclear weapons and materials, and helped Russia avert the four "loose nuke" problems.

Continuity of Effort – a Catalyst and a Holistic Approach

One of the most important features of cooperation between Russian and US nuclear scientists is that the relationships built early in the lab-to-lab programs endured for more than 20 years in spite of changing political relations between Moscow and Washington. Even now that cooperative programs have been largely eliminated because of the political impasse over Ukraine, we remain in contact. During these 20-plus years, Russian and US nuclear scientists and laboratory leaders have made many visits to each other's nuclear facilities. As an example, as of summer 2017, I have made 53 trips to Russia since the inaugural 1992 lab directors exchange.

These continuing and close relationships have allowed us to contribute across the entire spectrum of nuclear issues. We have developed and retained the best direct pipelines into each other's nuclear complexes. We have contributed broadly across both government- and lab-directed cooperative programs. Our governments have called on our relationships numerous times to help solve particularly thorny diplomatic problems. To provide just one example, the close relations developed over many years allowed Russian and American scientists to convince their respective governments to tackle one of the most dangerous Soviet nuclear legacies, that of fissile materials left behind in what now is the Republic of Kazakhstan when the Soviet Union disintegrated.

Russian and US laboratory scientists also acted as catalysts for a variety of government programs, such as the Nuclear Cities Initiative and the Initiative for Proliferation Prevention. Also, Russian scientists called on US lab scientists to be collaborators on their government-sponsored ISTC grants. And, quite often, we were asked by our Russian colleagues to send a message to US government agencies, typically regarding what they considered an ill-advised US effort. For example, my Russian colleagues told me right at the beginning of the government's plutonium disposition program that, in spite of strong US government insistence, Russia would never burn its plutonium in light-water reactors because they considered that a waste of a precious resource. Indeed, after more than 10 years of trying to convince the Russians, the US government agreed that Russia could burn its plutonium in fast reactors.[xiii]

The scientists' wide-ranging involvement in the cooperative programs also allowed us to take a holistic rather than a program-specific approach to nuclear cooperation. Success in one program opened up opportunities in another. For example, scientific cooperation built the trust necessary for cooperation in nuclear materials protection, control, and accounting (MPC&A). It helped overcome a two-year diplomatic stalemate that prevented progress in the US government's effort to help Russia secure and safeguard its vast stockpile of nuclear materials.

[xiii] It is interesting to note that in February 2016, more than 20 years after the plutonium disposition programs were initiated, the US government has decided to halt its own program to burn plutonium in light-water reactors. Future options are still being debated. See Derrek Asberry, "Obama's budget request nixes MOX; S.C. sues DOE over missed deadline," Aiken Standard, Feb. 9, 2016.

Finally, the relationships between the scientists engendered close relations between the nuclear cities, Sarov and Los Alamos. It was the close collaboration between scientists and the visits to each other's cities and homes that led to close relations between medical doctors, teachers, students, city officials, first responders, and law enforcement agencies.

A Window of Opportunity

The lab-to-lab approach also benefitted from good timing. During the Cold War, the US nuclear weapons laboratories were viewed as essential national security institutions by the administration and had great bipartisan support in Congress. The contractual arrangement for managing the laboratories was based on the remarkably successful GOCO or "government owned, contractor operated" principles established during the Manhattan Project. The University of California managed LANL, LLNL, and AT&T Bell Laboratories managed Sandia National Laboratories on a not-for-profit basis. The laboratories were considered as essential partners to the DOE.

I was personally encouraged by my boss, David Gardner, president of the University of California, to do what was in the best national interest during the path-breaking days of building relations with the Russian nuclear weapons laboratories. That encouragement combined with the sense of partnership with the US government allowed the laboratories to step up to the new challenge of how to work with the Russian nuclear complex. The three US nuclear weapons labs pursued cooperation with the Russian labs because it was in our national interest, not because of the prospect of increased revenues for our labs. In fact, with approval from DOE's Defense Program office, the labs tapped their own programmatic funds to support some of the initial activities.

On the Russian side, it was fortunate that in 1992 President Yeltsin decided to keep the military and civilian sectors together in one ministry and select Mikhailov, who was greatly respected in Russian scientific circles, to lead it. Moreover, Yeltsin supported and protected the ministry as best he could during these difficult times. And, Mikhailov had strong support from Yeltsin to promote collaboration with Americans.

The early lab-to-lab effort also benefitted from uncertainties and chaos in the US political system as it scrambled to address the tectonic political and military shifts resulting from the breakup of the Soviet Union. In *Agendas, Alternatives, and Public Policies*, W. Kingdon[xiv] posited that the epistemic community—a network of knowledge-based experts, scientists, and engineers in this case—is most effective during chaotic political periods. It is during these times that policy windows open and governments facing uncertainty are most receptive to new ideas. Unfortunately, for the time being that window has closed.

THE HUMAN DIMENSION

The human dimension, that is, the personal relationships, were an essential element of lab-to-lab cooperation. Although nuclear weaponry is mostly hidden in secrecy, nowhere is the human dimension less documented, yet more crucially important. The end of the Cold War lifted the veil of secrecy and allowed collaboration. *Doomed to Cooperate* tells the remarkable story of how Russian and American nuclear scientists and engineers, who formerly designed and built the world's deadliest weapons targeting each other's countries, embraced and joined forces to overcome the nuclear dangers resulting from the end of the Cold War and the breakup of the Soviet Union. We found time and again that the personal relations between the scientists based on respect and trust, along with a sense of shared global responsibility to manage the risks and benefits of nuclear energy, helped to overcome the new nuclear dangers and the bureaucratic hurdles to cooperation.

The book tells a story of courage, brotherhood, and perseverance, especially on the Russian side—courage to overcome geopolitical rivalry and reach out to a former adversary in a way that could have resulted in grave personal consequences several years or even just months before; brotherhood rooted in sharing the legacy of great schools of European physics and the common language of science; and perseverance to overcome obstinate barriers to collaboration presented by the chaotic political environment and the legacy of Cold War confrontation. It is a story of how scientists and engineers brought us from confrontation to cooperation.

[xiv] John W. Kingdon, *Agendas, Alternatives, and Public Policies.* (Boston: Little, Brown, 1984). p 166. Emanuel Adler and Michelle Dover demonstrated these principles in their analysis of different times in the Soviet and Russian system. See: Emanuel Adler, "The Emergence of Cooperation: National Epistemic Communities and the International Evolution of the Idea of Nuclear Arms Control," International Organization 46, no. 1 (1992). Michelle E. Dover, "Hard Problems, Soft Answers: Impact of Epistemic Communities on US-Russian Nuclear Cooperation," American University, 2013. http://aladinrc.wrlc.org/handle/1961/14915

During these 25 years we discovered that we not only shared common scientific bonds but also the enormous sense of responsibility we had for nuclear weapons. The scientists and engineers of weapons laboratories on both sides considered themselves the stewards of the nuclear weapons. We conceived them, we designed them, we helped build them, we gave custody to the military, and finally we took them back for disassembly. We had cradle-to-grave responsibility for the weapons and could not rest until they were dismantled.

This job was so important and so difficult that each side created world-class scientific institutions that housed some of each nation's best technical talent. Each side did its job during the Cold War providing the technical basis for nuclear deterrence, which we believe helped avoid a major military confrontation between the two superpowers. We also realized, however, that the fruits of our labor produced the most destructive weapons ever created by mankind and that the nuclear deterrent brought at best an uneasy peace. This common bond provided a strong incentive to work together to mitigate the new nuclear dangers.

CHANGING TIMES AND SLOWED COOPERATION

Times changed by the end of the 1990s. The end of the Cold War brought into question the role of the US nuclear weapons laboratories. The congressional consensus of support for these labs forged during the Cold War, when they were considered essential to hold off the Soviet Union, dissolved. The US government lost the sense of partnership with the laboratories, which was essential to the GOCO (government-owned, contractor-operated) governance model, resulting in an arms-length contractor relationship. Instead of the laboratory leadership blazing the path to cooperative programs with Russia, the laboratories were relegated mostly to serve at the discretion of DOE project managers who ran the Russian cooperative programs.

That problem was exacerbated in 2000 when Congress decided to create the National Nuclear Security Administration (NNSA), designed to be a semi-autonomous agency with the DOE but in fact added another layer of federal bureaucracy. In addition, by that time Charles Curtis and Victor Reis, two of DOE's staunchest supporters of the lab-to-lab approach, had left the department. Although the labs continued to enjoy the support of several high-level officials, such as DOE's Rose Gottemoeller, Ernest Moniz, and Linton Brooks, the federal bureaucracy made it increasingly difficult to cooperate with the Russians on a lab-to-lab basis. In 2005, this situation deteriorated further when, in response to Congressional pressure, the DOE/NNSA replaced the six-decade GOCO-style management of the laboratories with a for-profit structure, which further reduced the role of the laboratory scientists and leaders in developing and executing cooperative programs.

Ironically, the strong financial support from the Clinton administration and Congress for Russian cooperative threat reduction programs actually constrained the effectiveness of the lab-to-lab approach in these programs because it led to more control from DOE's program managers, more congressional oversight, and countless audits from the Government Accountability Office. These changes are to be expected for a program that began with a few million dollars and grew to hundreds of millions annually, but they naturally bring more bureaucracy, more conservatism, and less innovation.

The rapid budgetary growth also attracted many additional organizations to participate, including most of the other DOE laboratories. Although these labs often brought special talent and great enthusiasm to the cooperative efforts, over time this diversification reduced the lab-to-lab nature of the programs. There were so many DOE labs competing for funding that they were no longer able to shape and guide the cooperative programs. Instead, that role fell mostly to DOE program managers, who often had little technical competence and little understanding of the Russian nuclear complex. Moreover, the competition among the US labs confused and often overwhelmed our Russian counterparts. They were inundated with requests for visits to their nuclear sites. A common complaint that we began to hear from the Russian labs and from Moscow headquarters was that their complex suffered from American nuclear tourism.

By the late 1990s, times also changed in Russia. Mikhailov's tenure as minister ended in 1998 and, much like on the American side, more and more of the leadership responsibilities for cooperative programs was delegated to Moscow's bureaucracy, rather than Russia's scientific leadership. Our Russian laboratory colleagues complained as much about Moscow's bureaucracy as we complained about Washington's. Beginning in the early 2000s, the Russian security services, which in the nuclear cities were initially essential partners in some of the most sensitive lab-to-lab work, placed much greater restrictions on US access to Minatom's nuclear facilities. Although the greater access restrictions were imposed partly in response to the enormous expansion of requests for visits from the American side, it was mostly an indication that they believed the American presence in the Russian nuclear complex was too intrusive and the time had come to scale it back. This attitude reflected Moscow's view that Russia and the nuclear complex had survived the difficult 1990s and that now that the financial situation had improved markedly, it no longer needed help from the American side.

In light of all of these changes, after a decade of remarkably successful collaborations, which kept nuclear weapons and nuclear materials safe and secure and led to radical improvements on all nuclear security fronts in the Russian nuclear complex, cooperation began to fade.[13] Nunn-Lugar programs continued to fund some important cooperative work between the Russian and US nuclear complexes for more than another decade, but that work was quite limited and not nearly as effective as it could have been. It was fortuitous then that in the 1990s when Russia's nuclear complex faced its most serious challenges, the conditions were right to allow the nuclear scientists and the lab-to-lab approach to make great progress.

LOOKIN BACK AND LOOKING FORWARD

As we look back at the nearly 25 years of nuclear cooperation with the Russian nuclear complex, we take comfort in the fact that the nuclear crisis of the early 1990s has passed. There was no perfect storm, no nuclear disaster in the ensuing 20-plus years. The Russian complex rebounded following the financial and political crises. For the time being, steps necessary to provide for the safety and security of its nuclear assets have been taken. It did not necessarily have to turn out this way. It took the confluence of far-sighted political action on both sides to pave the way for the cooperation that prevented the potential perfect storm. And in the United States, relentless pursuit of cooperation by some NGOs and university groups galvanized the government's actions. All of this made it possible for nuclear weapons scientists on both sides to take the concrete actions described in *Doomed to Cooperate* that averted some of the most dangerous risks.

US-Russian nuclear cooperation has declined dramatically during the past decade. Beginning in 2005, Moscow opted not to renew most of the cooperative programs. The remaining efforts were largely government directed, with scientists and engineers providing a supporting role. In June 2013 the Russian government allowed the umbrella CTR agreement to expire. At the last minute, a weak substitute was signed that allowed limited nuclear cooperation. In September 2013, the DOE and Rosatom attempted to resurrect cooperation by signing a new cooperative nuclear agreement. However, the Ukraine crisis in February 2014 – compounded by Russia's actions in Syria, its military maneuvers in the Baltics, cyber hacking, and alleged interference in the 2016 US presidential election – has led Washington and Moscow to terminate or suspend almost all nuclear cooperation.

That list is long. In the fall of 2016, Moscow suspended the US-Russian agreement on the disposition of excess weapons plutonium, terminated an agreement on converting research reactors from highly enriched to low-enriched uranium, and ended a 2013 agreement to cooperate on nuclear- and energy-related scientific research and development. These actions followed Moscow's termination of most initiatives under the Nunn-Lugar Cooperative Threat Reduction Program, including cooperation on nuclear materials security, which ended in December 2014. And in 2016, Russian President Vladimir Putin chose not to have his country participate in the last of the Nuclear Security Summits organized by President Barack Obama. The termination of any one of these programs on its own would have been lamentable, but Moscow's wholesale withdrawal from nuclear cooperation is alarming. Even more alarming is Putin's insistence that Washington undo what he considers to be the egregious actions inflicted upon Russia over the past two decades before cooperation can be resumed. For example, he wants to curb NATO's military presence in countries that joined the alliance after 2000, wants Washington to repeal the Magnitsky Act, which imposes visa bans and financial sanctions on certain Russian officials, and insists that Washington makes reparations for economic sanctions imposed on Russia since 2014.

The sharply different views Washington and Moscow have of a new world order today and how to respond to current international crises imperil future nuclear cooperation. Many in the United States now ask whether or not the nuclear cooperation of the past 25 years was in our country's best interest. Would it not have been better in the 1990s to leave Russia alone to deal with its economic troubles and deteriorating nuclear complex? My answer is a resounding, "No!" The situation in the Russian nuclear complex in the 1990s was the most dangerous in nuclear history. Although the likelihood of a nuclear confrontation between our countries declined dramatically, the potential use of a nuclear weapon somewhere in the world increased because of the possibility of the theft or diversion of nuclear weapons or nuclear materials and the potential proliferation of nuclear weapons expertise from an economically stressed Russian nuclear complex.

Moscow looks at past nuclear cooperation through the lens of prolonged post Cold-War grievances against Washington – such as NATO expansion, withdrawal from the anti-ballistic missile treaty, indiscriminate use of military force around the world, development of precision global strike munitions, and attempts to dominate the space and cyber domains.[xv] It views the desperate 1990s as a lost decade, one in which Washington took advantage of Russia's weakness to advance its own agenda.

In the nuclear arena, though, these claims are unfounded. First, they assume an incorrect premise. Unlike the Russian economy and military, the nuclear complex did not collapse. It was not embarrassed. No secrets were lost. In spite of severe financial, political, and personal hardships, it avoided nuclear catastrophe and positioned itself to rebound in the ensuing decade. Second, the actual record of US-Russian cooperation in the nuclear sphere during the past 25 years does not warrant the narrative that Washington took advantage of a challenged Russian nuclear complex. In *Doomed to Cooperate* dozens of articles by both Russian and American nuclear experts and leaders point to the mutual benefits of cooperation. It tells the story of how they cooperated to successfully avoid the four great post-Cold War nuclear dangers—loose nukes, loose nuclear materials, loose nuclear experts, and loose exports.

STILL DOOMED TO COOPERATE

The current political acrimony prevents Russia and the United States from confronting the greater dangers that threaten the security of the world today, those that arise from failed states and the disintegration of state power, which require cooperation by the major powers. Moreover, it takes attention away from collaborating on other serious issues that threaten mankind's existence and which require scientific cooperation, such as global climate change; a sustainable supply of clean, affordable energy that includes nuclear fission and fusion; and the protection of the earth from catastrophic asteroid impacts.

Whereas these issues will take time to resolve and get our two countries back on track to a sensible future, the nuclear issues cannot wait. Nuclear is different, both because of its enormous destructive power and its potential to electrify the world. The devastation wrought by the two atomic bombs in Japan in 1945 was so horrific that neither atomic bombs nor their more powerful kin, hydrogen bombs, have been used in anger since. Likewise, nuclear power plants are such efficient producers of electricity that a mere 400 nuclear power plants around the world have supplied nearly 15 percent of the world's electricity with minimal carbon production in the past couple of decades. However, the nuclear accidents at Three Mile Island, Chernobyl, and Fukushima are grim reminders that nuclear safety is a never-ending quest for the international nuclear community. Preventing the use of nuclear weapons and supplying safe and secure nuclear electricity require cooperation, not isolation, on both state and technical levels.

Our governments must again find common ground to promote nuclear cooperation as they did at the end of the Cold War. Their cooperation for the Iran deal offers hope. The development and the implementation of the Joint Comprehensive Plan of Action[14] would not have been possible without close cooperation between Russia and the United States. The two nations put aside political differences in the interest of nuclear cooperation that benefits both sides.

Washington and Moscow should do the same to sustain the enormous improvements in nuclear safety and security of Russia's nuclear complex and the corollary benefits accrued in the US complex as a result of lab-to-lab cooperation. The authors in *Doomed to Cooperate* overwhelmingly call for continued scientific and technical cooperation because ensuring nuclear safety and security, be it for weapons, nuclear facilities, or nuclear power, requires continued vigilance, cooperation, and exchange of best practices and lessons learned. These efforts necessitate close collaboration between the nuclear stewards, which was the hallmark of lab-to-lab cooperation.

In *Doomed to Cooperate*, the nuclear weapons scientists on both sides also make a convincing case that their knowledge and continued technical cooperation could play a vital role to prevent the proliferation of nuclear weapons to states or terrorist organizations. Likewise, pooling their expertise will be essential in responding to nuclear threats or nuclear accidents as well as achieving the maximum benefit from civilian nuclear power worldwide.

Fortunately the requisite talent still exists at the Russian and US nuclear labs today. The reservoir of good will accumulated over the past 25 years still exists, but it must be nurtured. Formal intergovernmental agreements signed in the past five years, but not exercised, provide the necessary legal foundation. What is needed now is for our governments to encourage and task the laboratories to jointly address these problems to the benefit of each nation and the world.

[xv] These have been a common theme of President Vladimir Putin's complaints about the US, starting with 2007 Munich speech. "Putin's Prepared Remarks at 43rd Munich Conference on Security Policy in 2007," Courtesy Munich Conference on Security Policy, *The Washington Post,* February 12, 2007. http://www.washingtonpost.com/wp-dyn/content/article/2007/02/12/AR2007021200555.html

That sentiment is captured in the comments of Lev D. Ryabev, the former First Deputy Minister of the Atomic Energy of the Russian Federation, in an interview for the book, "Our job is to convince our respective leaderships that we must keep working together. The only alternative is suspicion, return to an arms race, and increased tension. We must not go in that direction." I concur with his assessment. In the nuclear activities that I have been engaged in around the world, I have found that cooperation is essential, whereas isolation may lead to nuclear disaster.

ACKNOWLEDGMENTS

Doomed to Cooperate and this article are dedicated to the Russian and American scientists and engineers who contributed to Russia-US nuclear cooperation during the past 25 years. Thanks go to my Stanford University research team for assistance with the book, the support of the National Nuclear Security Administration and the Los Alamos National Laboratory, and the support of the Carnegie Corporation of New York and the John D. and Catherine T. MacArthur Foundation. The professional support of Bathtub Row and the Los Alamos Historical Society in the publication of *Doomed to Cooperate* is likewise greatly appreciated.

REFERENCES

1 S. S. Hecker, ed., *Doomed to Cooperate: How American and Russian Scientists Joined Forces to Avert Some of the Greatest Post-Cold War Nuclear Dangers.* Bathtub Row Press, Los Alamos Historical Society, June 2016.

2 D. E. Hoffman, *The Dead Hand, The Untold Story of the Cold War Arms Race and its Dangerous Legacy,* Doubleday, 2009, p. 478.

3 S. J. Koch, *The Presidential Nuclear Initiatives of 1991–1992. Case Study Series 5*, National Defense University Press, Washington, D.C., September 2012.

4 T. L. Neff, "A Grand Uranium Bargain." *New York Times*, October 24, 1991.

5 P. S. Corden, D. W. Hafemeister and P. D. Zimmerman, ed., *Nuclear Weapons in the 21st Century,* American Institute of Physics, Conference Proceedings, Conference dates: 2–3 November 2013, Volume 1596. ISBN: 978-0-7354-1230-9, May 9, 2014.

6 S.A. Nunn and R. G. Lugar, "The Nunn-Lugar Initiative: Cooperative Demilitarization in the Former Soviet Union," in Allen E. Goodman (ed.), *The Diplomatic Record 1992–1993* (Boulder, CO: Westview Press, 1995), p. 144.

7 A. B. Carter, K. M. Campbell, S. E. Miller, and C. A. Zraket. *Soviet Nuclear Fission: Control of the Nuclear Arsenal in a Disintegrating Soviet Union.* Cambridge, Mass.: Center for Science and International Affairs, Harvard University, November 1991.

8 A. B. Carter and W. J. Perry, *Preventive Defense: A New Security Strategy for America* (Washington, DC: The Brookings Institution, 1999), 70.

9 P. I. Bernstein and J. D. Wood, "The Origins of Nunn-Lugar and Cooperative Threat Reduction," National Defense University. Center for the Study of Weapons of Mass Destruction, 2010, https://www.hsdl.org/?view&did=22899

10 D. E. Hoffman, S. Savranskaya, and T. S. Blanton. *Project Sapphire 20th Anniversary: More than a Half-ton of Weapons-grade Uranium Removed from Kazakhstan in 1994.* Issue 491, National Security Archive Electronic Briefing Book. 2009.

11 M. B. Nikitin and A. Woolf, The Evolution of Cooperative Threat Reduction: Issues for Congress, Congressional Research Service, 7-5700, R43143, June 13, 2014. https://www.fas.org/sgp/crs/nuke/R43143.pdf
Also see: http://www.nti.org/newsroom/news/celebrating-20-years-nunn-lugar-program/

12 W. J. Clinton, Presidential Decision Directive PDD/NSC-47, issued 21 March 1996; http://www.fas.org/irp/offdocs/pdd/pdd-47.pdf.

13 S. S. Hecker, "Thoughts About an Integrated Strategy for Nuclear Cooperation with Russia," *Nonproliferation Review* (Summer 2001), pp. 1-24.

14 UN Security Council, *Security Council resolution 2231 (2015) [on Joint Comprehensive Plan of Action (JCPOA) on the Islamic Republic of Iran's nuclear programme]*, July 20, 2015.

Reviews of the Comprehensive Nuclear-Test-Ban Treaty and U.S. Security

Raymond Jeanloz

Departments of Earth and Planetary Science and of Astronomy, and Miller Institute for Basic Research in Science, University of California, Berkeley, CA 94720, USA

Abstract. Reviews of the Comprehensive Nuclear-Test-Ban Treaty (CTBT) by the National Academy of Sciences concluded that the United States has the technical expertise and physical means to i) maintain a safe, secure and reliable nuclear-weapons stockpile without nuclear-explosion testing, and ii) effectively monitor global compliance once the Treaty enters into force. Moreover, the CTBT is judged to help constrain proliferation of nuclear-weapons technology, so it is considered favorable to U.S. security. Review of developments since the studies were published, in 2002 and 2012, show that the study conclusions remain valid and that technical capabilities are better than anticipated.

BACKGROUND

The U.S. National Academy of Sciences was requested both in 2000 and in 2009 to examine technical aspects of the Comprehensive Nuclear-Test-Ban Treaty (CTBT), a treaty banning all nuclear explosions around the world. The first occasion was motivated by the Senate voting on October 13, 1999 not to support CTBT ratification, despite the U.S. having been among the nations leading international negotiations of the Treaty. The focus of the 2000 study was to determine from a technical perspective whether the U.S. could maintain its nuclear weapons capability under the CTBT, and adequately monitor global compliance with the Treaty; it also addressed the impact on U.S. security of the CTBT coming into force [1]. The conclusion at the time was that the Stockpile Stewardship Program could assure U.S. expertise for sustaining its nuclear weapons as well as for other essential activities, from arms control and non-proliferation to counterterrorism in the nuclear domain. In addition, the combination of the International Monitoring System (IMS) being deployed by the Comprehensive Nuclear-Test-Ban Treaty Organization (CTBTO)[1] and U.S. National Technical Means (NTM)[2] promised to be sufficient for monitoring the Treaty.

Because the Stockpile Stewardship Program was relatively new (it was initiated in 1996) and the IMS was only starting to be deployed at the time, the 2000 study largely described technical capabilities based on plans and limited initial experience [1]. This fact motivated a second study in 2009, intended to determine whether operational experience with maintaining U.S. nuclear-weapons expertise and with global nuclear-explosion monitoring in fact validated the conclusions of the earlier report [2].[3] Broadly speaking, the 2009 study confirmed the results of the 2000 study.

The purpose of the present article is to briefly examine the degree to which the primary conclusions of the 2009 study, hence also the 2000 study, have stood the test of time. It is worth emphasizing that both 2000 and 2009 studies only address technical capabilities; they do not assure political support for maintaining those capabilities, which require sustained leadership and funding, and depend on national priorities that can shift over time.

---

[1] Prior to the CTBT's entry into force, the CTBTO is known as the Preparatory Commission for the Comprehensive Nuclear-Test-Ban Treaty Organization. Entry into force requires ratification by 44 nations deemed to have significant capabilities in nuclear technologies ("Annex 2" states), of which 36 have ratified as of July 15, 2017. The CTBTO is affiliated with the United Nations, with headquarters near the International Atomic Energy Agency in Vienna, Austria. Further details are given in [1, 2], as well as https://www.ctbto.org/ (accessed July 15, 2017).

[2] National Technical Means refers to all of a nation's capabilities for observation and monitoring, including from its intelligence community. The CTBT allows for evidence from any and all sources to be considered in evaluating whether a nuclear explosion has taken place: examples include open academic research, and monitoring for other purposes (e.g., recording of earthquakes, volcanic eruptions and other natural phenomena).

[3] The two studies were primarily carried out in 2000 and in 2009-2010, so are labeled with those dates in order to clarify the vintage of the information that was considered. They were published in 2002 and 2012, respectively, after technical and other reviews had been completed. Both were carried out under the auspices of the National Academy of Sciences (NAS), including the National Research Council (NRC).

Nuclear Weapons and Related Security Issues
AIP Conf. Proc. 1898, 030001-1–030001-7; https://doi.org/10.1063/1.5009216
Published by AIP Publishing. 978-0-7354-1586-7/$30.00

SUMMARY OF 2009 STUDY

The 2009 study concludes that the U.S. "has the technical capabilities to maintain a safe, secure and reliable stockpile of nuclear weapons into the foreseeable future without nuclear explosion testing" and "will retain effective protection against technical surprises" under the same conditions [2, pp. 1-2]. Therefore, ratification of the CTBT does not amount to U.S. nuclear disarmament, and essential expertise can be maintained for important applications in arms control, counter-proliferation and the like.

The study further documents that the International Monitoring System[4] can detect nuclear explosions in the subsurface, underwater and in the atmosphere to below yields of 1 kiloton (kt) worldwide, and that "the IMS provides valuable data to the United States both as an augmentation to the U.S. NTM and as a common baseline for international assessment and discussion of potential violations when the United States does not wish to share NTM data" [2, pp. 1-2]. In addition, the global monitoring capabilities "will reduce the likelihood of successful clandestine nuclear-explosion testing, and inhibit the development of new types of strategic nuclear weapons." The study therefore recommends that the U.S. "support both the completion of the IMS and its operations, training and maintenance" [2, p. 2].

Appendices D and E of [2] address additional details about seismic monitoring, including the possibility of surreptitious or evasive testing: that is, underground nuclear-explosion tests conducted in such a manner ("decoupled") as to muffle the seismic waves that are propagated into the ground. The study finds that whereas "the 90 percent confidence levels for IMS seismic detection are well below 1 kiloton worldwide for fully coupled explosions," inclusion of regional monitoring and better characterization of background signals mean that "an evasive tester in Asia, Europe, North Africa, or North America would need to restrict device yield to levels below 1 kt (even if the explosion were fully decoupled) to ensure no more than a 10 percent probability of detection by the IMS" (i.e., a 90 percent probability of successfully evading the IMS) [2, p. 1].

Monitoring of the oceans and atmosphere is sensitive to nuclear explosions down to about 0.001 kt and 1 kt, respectively, based on the IMS prompt-detection capabilities offered by hydro-acoustic and infrasound sensors. However, radionuclide monitoring is in any case sensitive to debris from explosions far below 1 kt levels in both oceans and atmosphere.

The 2009 study was intended to provide technical assessments that can be used in considering the merits and limitations of the CTBT. It was not intended to recommend for or against ratification, which is a political decision, and several of the study conclusions are identified as valid whether or not the CTBT enters into force.

However, for the following pragmatic reasons the technical conclusions of the study form a strong basis for arguing that U.S. national-security interests would be well served by Treaty ratification and entry into force. First, the study concluded that the CTBT can help curb horizontal and vertical proliferation of nuclear-weapon capability.[5] Yet there can be no treaty-based response to any apparent detections of a nuclear explosion before CTBT entry into force, whether inspections of the suspected explosion site or international sanctions in response to a proven Treaty violation. In this sense, the IMS is a capability without purpose before the CTBT enters into force.

Second, sustaining the IMS and CTBTO over the long term, with maintenance and upgrade of hardware, software and communications infrastructures, along with recruitment and retention of the requisite teams of experts, seems implausible if the underlying treaty regime does not exist. Some nations, the U.S. included, have ongoing interest in global nuclear-explosion monitoring, but the international consensus that provides access to locations for stations (e.g., for regional monitoring) and expertise is linked to the treaty-based CTBTO and its monitoring capability.

Third, the study concluded that the threats it contemplated can be addressed whether or not the CTBT is in force; that is, without the U.S. returning to nuclear-explosion testing. Specifically, sustaining a nuclear arsenal – for as long as needed – and responding to development of "weapons with lower capabilities, such as those that might pose a local or regional threat, or that might be used in local battlefield scenarios… would not require the United States to return to [nuclear-explosion] testing" [2, p. 2]. If, in contrast, threats were to develop beyond those contemplated in the 2009 study, "the United States could invoke the supreme national interest clause and withdraw from the CTBT."

---

[4] The IMS consists of networks of seismic, hydro-acoustic and infrasound sensors for detection, location and characterization (as explosive) of events in Earth's crust, oceans and atmosphere, and of radionuclide stations for confirming that the event is a nuclear explosion. It also consists of a global communication system that transmits data from the sensors distributed worldwide to the International Data Center (IDC) at the CTBTO. Along with the hydro-acoustic and infrasound networks, the seismic stations also record many events in the atmosphere and oceans.

[5] E.g., "inhibit the development of new types of strategic nuclear weapons" [2, p. 2]. Horizontal proliferation refers to nuclear weapons being obtained by states that did not previously have them, and vertical proliferation describes new, more advanced nuclear weapons being developed by states that already have nuclear-weapons capability.

These technical considerations indicate how CTBT entry into force is in the interest of U.S. national security and underpins the Treaty's constraints on nuclear proliferation.

ASSESSMENT OF STUDY FINDINGS

Stockpile Stewardship

The technical capability sustained by the U.S. Department of Energy and National Nuclear Security Administration's (DOE/NNSA) Stockpile Stewardship Program can be assessed in a number of different ways, from documenting the major facilities that have been established to reviewing progress in large programs such as the computing initiatives or maintenance of the nuclear arsenal.

The Dual-Axis Radiographic Hydrodynamic Test facility at Los Alamos National Laboratory (DARHT, LANL), Microsystems Engineering and Systems Applications complex at Sandia National Laboratory (MESA, SNL) and National Ignition Facility at Lawrence Livermore National Laboratory (NIF, LLNL) represent significant investments in research and production infrastructure for the U.S. nuclear-weapons complex since 1995, for instance. As another example, implementation of numerical simulation at Terascale in the 1990s and Petascale in the 2000s, with continued progress since then, illustrates successful exercising of relevant capabilities [2].[6]

It is worth noting that the benefits of these developments are not limited to the nuclear-weapons complex, but spin off important capabilities for society more broadly. Thus, as the DOE laboratories have invested in it, high-performance computing has similarly advanced rapidly in the business and academic communities. As another example, requirements for NIF prompted the development at industrial scale of glass manufacturing having unprecedented quality (i.e., density of flaws orders of magnitude lower than previously achieved). Advanced manufacturing is being developed vigorously in the NNSA laboratories and production facilities at the present time, with engagement of businesses around the country. These are but a few illustrations of cutting-edge technology development having practical applications that enhance societal capability and competitiveness.

As a different indication of practical capability, stockpile surveillance, Life-Extension Programs (LEPs) and related activities support the maintenance and refurbishment of the U.S. nuclear arsenal.[7] One weapon-type LEP had been completed by the time of the 2000 study, but there was experience with several more LEPs by 2009 and the bulk of the arsenal is currently being refurbished (completion due by 2020).

Maintaining a nuclear-weapons stockpile might not be seen as progress in arms control, but can be considered as such in the following sense. First, to the degree that the CTBT is an important instrument for countering nuclear proliferation,[8] it deserves support in its own right, independent of arms reductions among nuclear-weapon states; in particular, linking CTBT to disarmament is politically problematic for the U.S. in the present era. Second, the Stockpile Stewardship Program was successful enough to give the United States the confidence to greatly reduce its nuclear arsenal, which is now smaller than it has been in more than 60 years.[9] That is, preparations for CTBT have *indirectly* supported arms reductions. Third, it is retaining capability in nuclear-weapons technology that is especially important: without that knowledge and understanding, described here as technical capability, the deterrent is of limited credibility, inevitably vanishing in utility over time. Also, it is ongoing technical knowledge that is needed for essential efforts in arms control, non-proliferation, counter-terrorism and related domains (e.g., training inspectors and developing inspection technologies for countering proliferation).

Sustaining the creativity and quality of technical capability for the long term is especially important, and is proven by the quality of research performed in science and engineering.

Among accomplishments in the more applied research domains is a much deeper understanding of energy-balance in nuclear weapons, based on both simulation and experiment; also, the properties of plutonium have been significantly clarified, with aging of the 239 isotope due to self-irradiation and related processes being characterized

---

[6] Terascale and Petascale refer to computing at speeds of a trillion (10^{12}) and thousand-fold faster (10^{15}) floating operations per second [e.g., https://en.wikipedia.org/wiki/Computer_performance_by_orders_of_magnitude (accessed July 15, 2017)]. Software development and implementation are as important as advances in hardware in making these increased computing speeds useful for practical applications.

[7] For simplicity, I do not distinguish here between Life Extension, Alteration and Modification programs (LEPs, ALTs and MODs), nor between repair, refurbishment and replacement of components. These are important distinctions within the Stockpile Stewardship Program, but not essential for the present summary of technical capabilities. See https://nnsa.energy.gov/ourmission/managingthestockpile/lifeextensionprograms and https://nnsa.energy.gov/content/2016-year-review, for example (both accessed July 15, 2017).

[8] See Footnote 5 and associated text.

[9] E.g., https://2009-2017.state.gov/t/isn/npt/statements/241165.htm (accessed July 15, 2017). Another way to characterize evolution of the U.S. stockpile of nuclear warheads is that it was reduced by half after dissolution of the Warsaw Pact and Soviet Union, and then was cut by half again after establishment of Stockpile Stewardship.

at a level that informed major decisions about Pu production over the coming years [e.g., 3]. It is remarkable that just a few years' "accelerated aging" experiments using material spiked with ^{238}Pu (87.7 year half-life) could document changes that take place in ^{239}Pu (24,110 year half-life) over periods exceeding 200 years (in detail, samples have to be studied over a range of temperatures in order to characterize the crystal-structure annealing that helps mitigate the radiation-induced damage) [e.g., 4]. Neither of these breakthroughs was anticipated when Stockpile Stewardship was initiated. In addition, work at LLNL's NIF has documented important steps toward controlled thermonuclear fusion in the laboratory [5].

At the same time, more basic research has been pursued aggressively using the major facilities at the NNSA laboratories. For example, there have been significant breakthroughs at the SNL Z facility and elsewhere in understanding the properties of hydrogen and other elements at high pressures and temperatures, with implications ranging from improving quantum mechanical simulations to understanding the constitution and evolution of planets [6, 7]. Exquisitely sensitive methods for measuring the elasticity of materials have been refined [8], and improved measurements of opacity have had significant astrophysical applications [9]. These few examples highlight applications closely related to Stockpile Stewardship, so are not intended to reflect the broader array of basic and applied research within the NNSA complex.

Ironically, rather than cessation of nuclear-explosion testing crippling U.S. nuclear-technology capability, it allowed – through the Stockpile Stewardship Program – a much deeper level of understanding than previously possible, while also enabling new kinds of research in basic science and engineering. It is this level of demonstrated technical knowledge that lends significance to the major accomplishment of Stockpile Stewardship, providing for an assessment that for more than twenty years has allowed the Secretaries of Defense and Energy to attest annually to the safety and reliability of the U.S. nuclear stockpile without the need for nuclear-explosion testing.[10]

Nuclear-Explosion Monitoring

The International Monitoring System does not have a pre-determined sensitivity, but is instead expected to improve over time as new technologies and analytical approaches are developed. By 2013, Kvaerna and Ringdahl empirically showed that 3 or more primary stations of the IMS detect, with 90 percent probability, explosions across all continents and oceans around the world down to yields of 0.1 kt or less[11] ([10] describes the seismic detection limit, and the high sensitivity of the hydro-acoustic network is shown in Fig. 2-10 of [2]). Evasive testing in the subsurface raises this yield limit, and if the logic is to provide 90 percent assurance of successful evasion (analogous to the 90 percent successful detection expected for the IMS) then the sensitivity limit may be degraded to several kilotons for muffled explosions worldwide and to under 1 kt for regions of interest (e.g., Table 4-2, Fig. 4-1 and associated discussion of evasive testing in [2]).[12] A comparable detection sensitivity of about 0.1-1 kt for an explosion in the atmosphere anywhere around the world was established nearly a decade ago for the IMS infrasound network (Fig. 2-13 of [2], based on work presented in 2009).

Both for the atmosphere and the subsurface (in oceans or crust), the sensitivity being discussed here is for prompt detection and location of an event, with signals being recorded within minutes to tens of minutes, the time required for acoustic waves to travel from the event to the closest sensors. The CTBTO International Data Center transmits the data worldwide[13] from the primary seismic stations in near-real time, and the information is processed within tens of minutes to determine i) that an event occurred; ii) the magnitude and location of the event, typically to within 10-30 km spatial resolution;[14] and iii) the nature of the displacement at the source ("source mechanism"), whether indicative of an explosion or not.

Rapid detection of an event and its location are important, both in the context of accumulating additional evidence (e.g., calling attention to the event and marshaling added sensors for more complete characterization) and in making a case for on-site inspection. It takes time to implement an on-site inspection, and field evidence of a

[10] See https://2009-2017.state.gov/t/avc/rls/202013.htm, for example (accessed July 30, 2017).

[11] The tip of S. America is monitored only down to about 0.15 kt under these same criteria.

[12] There has been considerable controversy about how effectively seismic signals can be muffled in different geological media and locations, but evidence that decoupling is less effective for higher- than lower-frequency seismic signals shows that recording at "regional" distances, under 1000-1500 km from the source (at short enough distances that geometric spreading, scattering and attenuation of seismic waves are minimized), is especially beneficial for monitoring decoupled as well as coupled explosions (see Chapter 4 and Appendix E of [2]).

[13] All state parties to the CTBT have access to the IDC data, as do others recognized by the CTBTO.

[14] The location of an event must generally be known to determine its magnitude, in that a small event closer to sensors can produce a signal of similar amplitude to that of a larger event farther away, although other characteristics may be determinative of location (e.g., amplitude, polarization and frequency content of the 3-dimensional displacements recorded by seismic sensors). The CTBT specifies that on-site inspections must be limited to an area of 1000 km^2, corresponding to a region roughly 30 x 30 km.

nuclear explosion decays over periods of days to weeks, so it is helpful to identify a suspect event as soon as possible.

The additional step of characterizing an explosion as nuclear in origin depends on gases and particulate debris being transported through the atmosphere or oceans to the IMS radionuclide stations.[15] Depending on distance, this can take several days. Also, for an underground event, days or even weeks may be needed for the telltale gases to escape to the surface, if they escape at all: there can be no measureable radionuclide evidence for an underground explosion, if it does not vent to the surface [2].

Unlike the case for underground nuclear explosions (potentially), evidence of an event in the atmosphere or oceans being a nuclear explosion is all but certain to be documented by the IMS radionuclide stations, and likely also by additional radionuclide-monitoring systems fielded by several nations (e.g., capability to collect at additional ground stations or by aircraft). Prompt detection for both atmosphere and oceans is assured through a combination of the IMS (infrasound and hydro-acoustic networks), national and commercial satellites, and related technologies.

The Chelyabinsk bolide of February 15, 2013 provides a good example of the effectiveness of current capabilities in detecting and characterizing an unexpected atmospheric event as an explosion. The 500-kt event was recorded within tens of minutes of its occurrence (the time required for acoustic propagation of the signal), and rapidly identified as an explosion due to a meteorite hitting Earth's atmosphere [11]. It is sobering to realize that bolides cause 10-kiloton-scale explosions in the atmosphere every year or two, as now established by IMS and related observations [12].[16] This was not known throughout most of the Cold War.

All nuclear-explosion testing has been conducted underground in recent decades, rather than in the oceans or atmosphere. Therefore, underground events receive special attention for purposes of monitoring nuclear-explosion testing around the world, especially because of the added challenges presented by the possibility that no radioactive particles or gases are released (containment) or that the explosion was decoupled (evasive testing). Clutter also plays a role, in that the rate of natural events increases exponentially with decreasing magnitude.[17]

The problem of identifying a small explosion within a large number of natural events is not resolved by simply developing more sensitive seismometers. However, monitoring for underground nuclear explosions benefits from the fact that seismometers are widely deployed for reasons of recording earthquakes and other natural phenomena.

In addition, data analysis has dramatically evolved so as to take advantage of the large and growing number of seismic stations around the world [e.g., 13, 14]. Groups of stations are treated as virtual antennas, with coherent processing of data being used to identify direction and velocity of wave-fronts crossing these sensor arrays. Repeat events (earthquakes, as well as mining explosions and other events) are used to characterize crustal velocity structure (i.e. refraction and focusing properties) around the world, amounting to developing matched filters for explosions in each region. Even the background noise of seismic records can be used to determine crustal velocity structure.

Zhang and Wen have used such approaches, for example, to identify an event that took place in the vicinity of the DPRK nuclear-testing site on May 12, 2010 [15]. They did a remarkable job of extracting a small signal from the background noise, although their inference that it was due to a nuclear-explosion test is not widely accepted [13, 16]. Interpretations of this event are ongoing, but the important point is that its yield was less than 0.001 kt (1 ton), far smaller than a typical truck bomb.[18] Although the capability is not global, this does illustrate the sensitivity of contemporary explosion monitoring down to truck-bomb levels in at least certain locations of interest. As more monitoring stations and improved analysis tools are deployed, it seems inevitable that a similar level of detection can be achieved for many if not all regions studied around the world.

---

[15] The nation where the event occurred might in the meantime have announced that it conducted a nuclear-explosion test, in which case the IMS plays a role of validating that claim.

[16] 10, 100 and 1,000 kiloton events are estimated to occur roughly every 2, 20 and 150 years, respectively [12].

[17] As captured by the Gutenburg-Richter relationship, the frequency of earthquakes increases by roughly one order of magnitude for each unit decrease in magnitude (e.g., https://www.iris.edu/hq/inclass/fact-sheet/how_often_do_earthquakes_occur accessed July 25, 2017). Global rates increase from about 10,000 events per year above magnitude 4 (0.4 kt) to 100,000 events per year above magnitude 3 (0.04 kt), with parenthetic values giving the equivalent yield for well-coupled explosions [2].

[18] A limit of 0.25 ton is suggested in [16] for the May 12, 2010 event, assuming it was located at the DPRK nuclear-testing site, and suggests it could have been due to an earthquake or industrial blast. For comparison, truck bombs typically have yields of 2-5 tons or more [17]: e.g., p. 4-7 of FEMA 428 (https://www.fema.gov/pdf/plan/prevent/rms/428/fema428_ch4.pdf) and the NIST description of the 1995 Oklahoma City bombing (https://www.nist.gov/el/oklahoma-city-bombing-1995) both accessed on July 27, 2017.

Required Sensitivity: Implications of Underground Nuclear-Explosion Testing

An overarching question for nuclear-explosion monitoring is: How sensitive do the networks need to be, and what is sufficient for effective monitoring? Table 4-3 of the 2009 report attempts to address this question, and it is arguably the most uncertain assessment of that study [2].

The atomic bomb that destroyed Hiroshima on August 6, 1945 demonstrated that a 15-kt weapon can be successfully produced without nuclear-explosion testing, and one has to assume that such a feat can be replicated far more readily now and into the future. Constraining nuclear-explosion testing therefore cannot be expected to block all development of nuclear weapons, including by a state that has no prior experience with the underlying technologies. Instead, the objective of the CTBT is to increase the time and effort required for development of more advanced nuclear weapons, whether the more deliverable and powerful weapons of the Cold War or weapons that are new in other ways.

Expert judgment informed the creation of Table 4-3 [2] and its predecessor (Table 3-1 of [1]), both of which highlight a threshold of 0.1-1 kt as being a key yield range for original (i.e., unaided) development of more advanced weapons of this type. One concern is that this estimate assumes no help from a state having nuclear-weapons capability, yet there is historical evidence of nuclear-weapons technology being proliferated:[19] How much lower does the threshold need to be, in case there has been some proliferation of knowledge and materials?

In principle, all of the requisite technology might be transferred to a country to establish a new nuclear-weapon capability there, in which case deliverable and powerful weapons could be produced without nuclear testing.[20] That is, although the CTBT can increase the effort and time required for establishing and proliferating knowledge, it does not limit the consequences of such proliferation. Similarly, it is conceivable that with advances in technology, new alternatives will emerge for producing the materials or developing the designs required for nuclear weapons. The estimates in Table 4-3 would accordingly need to be revised downward.[21]

The point is that explosions having yields much lower than this 0.1-1 kt threshold can be detected across much of the globe, notably including in regions of special interest,[22] when one takes into account the combined capabilities of IMS and non-IMS monitoring systems (e.g., non-IMS seismometers, commercial satellite imagery and other sources of information available to the research community and public at large). Therefore, it is likely that a set of nuclear-explosion tests at yields sufficient to have military significance would be detected, perhaps even if those tests are conducted evasively. While proliferation of knowledge or materials may lower the required sensitivity threshold in the future, the monitoring capability is itself improving dramatically and can – from a technical perspective – become better over time relative to the level needed for effective monitoring.

Conclusions

The nearly 10 years since the work informing the 2009 NRC study show that its conclusions are robust in documenting both the success of Stockpile Stewardship and the capability of nuclear-explosion monitoring [2]. In fact, there have been unexpected advances in both domains, and in this sense the report was appropriately conservative in its conclusions. Continued need for expertise motivates ongoing support for activities in nuclear technologies, and this view is unlikely to change in the foreseeable future. Monitoring is also likely to continue improving, if for no other reason than that it provides societally important information about natural hazards and the environment, including changes in the environment [18]. From a technical perspective, the U.S. abides by the constraints of the CTBT and would therefore benefit from entry into force as soon as possible.

---

[19] This horizontal proliferation of knowledge or materials violates the Nuclear Nonproliferation Treaty (NPT), but not all states are party to (or always abide by) the NPT and some forms of transfer can be inadvertent (see the U.S. State Department's description of one proliferation network, for example: https://2009-2017.state.gov/t/isn/115913.htm, accessed July 30, 2017). In addition, Table 4-3 indicates that countries with extensive experience in nuclear-weapons technology can make some advances with nuclear-explosion tests of lower yields (vertical proliferation). Such tests would be correspondingly more difficult to detect, but the overriding conclusion is that they do not change the strategic balance that already exists today between experienced nuclear-weapons states: these states are already deterred by each being able to cause unacceptable damage to another. The CTBT would, nevertheless, constrain vertical proliferation by limiting development of new designs requiring higher-yield nuclear-explosion testing.

[20] The same can be said about nuclear-weapon proliferation to a subnational group. In either case, designs might be comparable to early-Cold War technology, and in this sense "primitive." Nevertheless, such weapons would be dauntingly powerful.

[21] The ability to create and update such information as contained in Table 4-3 of [2] illustrates the need for ongoing capability in nuclear-weapons technology, regardless of the sizes or even existence of nuclear-weapons arsenals.

[22] Regions of past nuclear-weapons development, or potential nuclear-weapons development in the near future, represent different levels of interest for different countries.

ACKNOWLEDGMENTS

Work supported by the DOE and University of California. I thank M. L. Adams, P. Corden, M. D. Lowenthal and J. Zucca for helpful discussions.

REFERENCES

1. National Academy of Sciences (NAS), *Technical Issues Related to the Comprehensive Nuclear Test Ban Treaty*, Washington, D C: National Academy Press, 2002, 84 pp.
2. National Research Council (NRC), *The Comprehensive Nuclear Test Ban Treaty, Technical Issues for the United States*, Washington, D C: National Academy Press, 2012, 204 pp.
3. A. Heller, *S&TR*, Livermore, CA: Lawrence Livermore National Laboratory, August, 2015, pp. 6-14.
4. A. J. Schwartz, *J. Alloys Comp.* **444-445**, 4-10 (2007).
5. O. A. Hurricane, et al., *Nature* **506**, 343-348 (2014).
6. M. D. Knudson, et al., *Science* **348**, 1455-1460 (2015).
7. R. F. Smith, et al., *Nature* **511**, 330-333 (2014).
8. A. Migliori and J. D. Maynard, *Rev. Sci. Instrum.* **76**, 121301 (2005).
9. J. E. Bailey, et al., *Nature* **517**, 56-59 (2015).
10. T. Kvaerna and F. Ringdal, *Bull. Seismol. Soc. Am.* **103**, 759-772 (2013).
11. A. Le Pichon, et al., *Geophys. Res. Lett.* **40**, 3732-3737 (2013).
12. P. G. Brown, et al., *Nature* **503**, 238-241 (2013).
13. P. G. Richards, *Nonprolif. Rev.* **23**, 287-300 (2017).
14. M. E. Pasyanos, et al., *Trends in Nuclear Explosion Monitoring Research & Development – A Physics Perspective*, Los Alamos National Laboratory Report LA-UR-17-21274, (2017). Accessed June 15, 2017 at http://permalink.lanl.gov/object/tr?what=info:lanl-repo/lareport/LA-UR-17-21274
15. M. Zhang and L. Wen, *Seismol. Res. Lett.* **86**, 1-8 (2015).
16. S. R. Ford and W. R. Walter, *Seismol. Res. Lett.* **86**, 1160-1170 (2015).
17. K. D. Koper, et al., *Bull. Seismol. Soc. Am.* **92**, 527-542 (2002).
18. R. Jeanloz, et al., *Science* **339**, 761-762 (2013).

On-Site Inspections Under the Comprehensive Nuclear-Test-Ban Treaty[1]

Edward Ifft

Security Studies Program
Walsh School of Foreign Service
Georgetown University
37 and O Streets, NW
Washington, DC, 20057 USA

Abstract. On-site inspections (OSI) are a key aspect of the verification regime for the Comprehensive Nuclear-Test-Ban Treaty (CTBT). Preparations to be able to conduct such inspections are going well. In particular, realistic major field exercises in Kazakhstan in 2008 and Jordan in 2014 enabled the Preparatory Commission of the Comprehensive Test-Ban-Treaty Organization (CTBTO) to learn valuable lessons regarding both the technical and human factors involved in being able to successfully implement the treaty's provisions.

NUCLEAR TESTING

The world has experienced more than 2,000 nuclear explosions. All countries except North Korea have voluntarily refrained from conducting nuclear explosions since 1998. The following table summarizes the history of world nuclear testing.

TABLE 1. World Nuclear Testing by Nation.

World Nuclear Testing

Country	First test	Last Test
U.S.	1945	1992
USSR/Russia	1949	1991
U.K.	1952	1992
France	1960	1996
China	1964	1996
India	1974	1998
Pakistan	1998	1998
North Korea	2006	2016

Nuclear Weapons and Related Security Issues
AIP Conf. Proc. 1898, 030002-1–030002-9; https://doi.org/10.1063/1.5009217
Published by AIP Publishing. 978-0-7354-1586-7/$30.00

Underground testing is generally carried out either in a vertical emplacement hole or in a horizontal tunnel. Below are museum exhibits showing the situation before and after vertical and horizontal nuclear tests.[2]

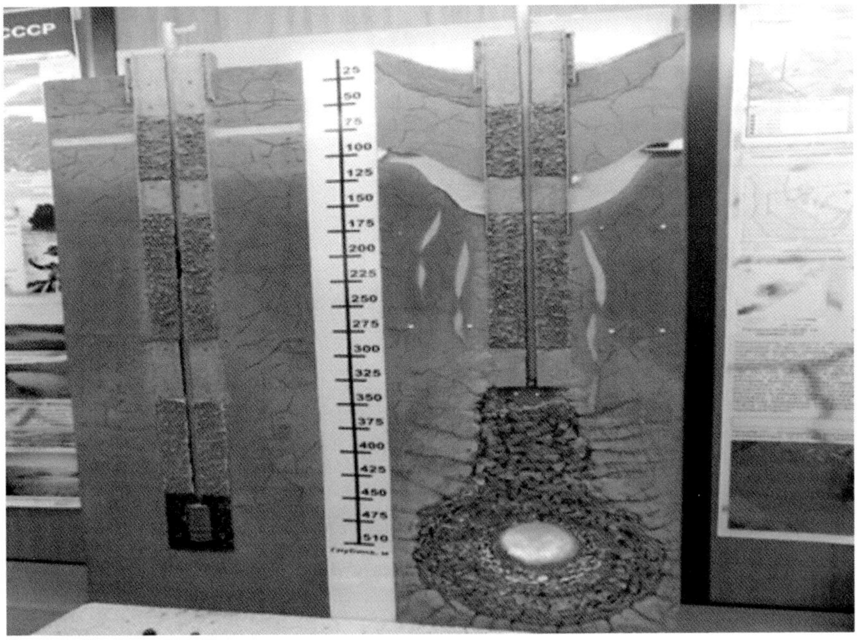

FIGURE 1. Museum exhibit showing the situation before and after a vertical nuclear test.

FIGURE 2. Museum exhibit showing the situation before and after a horizontal tunnel nuclear test.

An inspection team may be able to detect aftershocks caused by rocks in the rubble zone falling into the cavity. They may also be able to detect radioactive particles or radionuclide noble gases escaping from the area. Both of these possibilities are more likely the sooner the team arrives on-site. The treaty negotiators recognized this urgency and specified that the team should arrive at the inspection area within about 10 days from the time the request for an inspection is received. Below is a typical preparation scene for a U.S. vertical test.

FIGURE 3. Typical preparation scene for a U.S. vertical test.

TREATY PROVISIONS

From the earliest days of interest in a CTBT, it was generally accepted that OSI would be essential for effective verification. This was because the remote sensing systems would have false alarms and probably be unable to distinguish between nuclear and chemical explosions. In addition, an underground nuclear explosion would leave unambiguous and long-lasting evidence and be very difficult to conceal. Thus, in the CTBT negotiations from 1994 to 1996, it was readily agreed that provisions for OSI needed to be part of the treaty. The details, however, were controversial and agreement on derails was difficult to reach. This negotiating process has been recorded in several authoritative accounts.[3]

The basic provision is found in Article 4, paragraph 34, which states, "Each State Party has the right to request an on-site inspection in accordance with the provisions of this Article and Part II of the Protocol in the territory or in any other place under the jurisdiction or control of any State Party, or in any area beyond the jurisdiction or control of any State." The text further specifies that the sole purpose of an OSI shall be to clarify whether a nuclear explosion has been carried out in violation of the treaty and, to the extent possible, to gather any facts which might assist in identifying any possible violator. The details of how requests for an OSI are to be dealt with and the rights and obligations of both the inspecting and inspected parties have been described elsewhere.[4]

EXPERIENCE WITH OSI

The world has a great deal of experience with OSI in the context of arms control agreements, as shown in the following table (PPCM refers to Perimeter and Portal Continuous Monitoring)

TABLE 2. World experience with on-site inspections for nuclear tests.

World Experience with OSI

Agreement	Number
NPT (IAEA)	8,000+ per year
CWC (OPCW)	6,000+
CFE	6,000+
INF	850+ also PPCM
START	1,100+ also PPCM
New START	18 per year
Open Skies flights	1,000+ (100 per year)
Iraq (UNSCOM, UNMOVIC, IAEA)	hundreds
ABACC	hundreds

All of this successful experience is quite valuable, but, of course, CTBT OSI would have unique features. Many of the other arms control regimes have regular inspections subject to a quota. Many involve confirming the elimination or conversion of certain items or facilities. There is nothing to eliminate or inspect on a regular basis under the CTBT—all OSI would be of the "challenge" variety. Thus, CTBT OSI is very likely to be relatively rare.

In some respects, CTBT OSI would be more intrusive than such activities under other arms control regimes. The inspection teams would be large—up to 40 people. These teams could be on-site for a long period of time—up to 130 days. They would also be taking multiple scientific measurements and collecting samples. On the other hand, in other important respects, CTBT OSI would be less intrusive. As noted above, inspections would almost certainly be infrequent events. They would probably be conducted in remote areas, not at sensitive military installations or commercial facilities, as is the case for START and CWC inspections. Further, again in contrast to other regimes, CTBT inspection teams would generally have no need to enter buildings.

TOOLS OF OSI

In accordance with the treaty, the inspection team has at its disposal a formidable array of tools. These include:

- Overflight to confirm the boundaries of the inspection area and to identify areas of possible interest.
- Ground-based visual observation.
- Environmental sampling and analysis of solids, liquids and gases above, at and below the surface.
- Local seismic arrays for passive monitoring for aftershocks.
- Gamma surveys, ground, underground and air.
- Magnetic and gravitational field mapping, ground-penetrating radar and electrical resistivity measurements at the surface and from the air.
- Resonance seismometry and active seismic surveys to search for underground anomalies, including cavities and rubble zones.
- Drilling.

Some of these may be used in the "initial" period (up to 25 days), while others may be used only during the "continuation" period (up to 60 days) or the "extension" period (up to 130 days). Drilling would be the most invasive and expensive technique and would be considered as a last resort.[5]

The rights of the inspected party are also spelled out in detail in the treaty text. These include the right to declare "restricted-access sites." Such a site may be no larger than four square kilometers. If multiple such sites are declared, their total area may not exceed 50 square kilometers and individual sites must be separated by at least 20 meters. As is the case in other recent treaties involving OSI, the inspected party may also employ "managed access."

EFFECTIVENESS OF OSI

An authoritative 2012 study by the U.S. National Academy of Sciences gave a positive assessment of the ability of OSI teams to detect clandestine nuclear explosions.[6] Assuming that the inspection team is given adequate location information, the study judged that it would have a high likelihood of detecting explosions greater than 0.1 kt. The inspection area within which the team may operate is up to 1,000 square km (roughly the area inside the Washington beltway). Location information would most likely come from the International Monitoring System (IMS), which has provided very accurate information on the five North Korean tests. Such information could also come from the National Technical Means of the States-Parties, technical data from outside the IMS—e.g., national seismic stations outside the IMS-- whistleblowers or even open sources.[7]

FIELD EXPERIENCE

The CTBTO conducts a very active program to test the equipment which would be used for OSI, and its use by inspectors under field conditions. Two large-scale field exercises particularly stand out.

The 2008 exercise in Kazakhstan (IFE08) occurred at a former Soviet test site (Semipalatinsk). This was particularly useful since real nuclear explosions had taken place there and there were real testing artifacts and real residual radiation. A tent camp was constructed in a remote area. There were stressful weather conditions, some unplanned equipment malfunctions and other events which provided valuable lessons. The exercise especially shed light on the importance of logistics. The scenario was a suspicious seismic event and the release of Cs-137. It was "compliance" scenario, meaning that the evidence pointed to the conclusion that there had been no nuclear explosion.

The 2014 exercise in Jordan (IFE14) took place just east of the Dead Sea. There was no history of nuclear activity here, but there was a history of significant seismic activity, being located in the Great Rift Valley. This time, the inspectors lived comfortably in a Jordanian Dead Sea resort. Thus there were no issues with lodging, toilets, meals, etc. and the inspectors could focus on their scientific field activities. The scenario was that a suspicious seismic event of 4.0 magnitude (1 kt) had taken place, along with the release of a radioisotope of xenon. This was a non-compliant scenario, meaning that the evidence pointed to the occurrence of a nuclear explosion. IFE14 was even more elaborate and realistic than IFE08, involving about 200 experts and the transport of 150 tons of equipment.[8]

Both exercises employed the full 1,000 square km inspection area. Both compressed the inspection period of months into weeks. Each employed nearly the full range of inspection techniques and equipment, but not drilling. Both were realistic in that they involved unanticipated problems, as well as suspicious artifacts and data built into the scenarios. Both were successful, in that they demonstrated the capabilities of the entire OSI mechanism, along with identification of areas for improvement. Both the Kazakhstanis and Jordanians played very constructive roles as hosts.

The author participated in both exercises and some of his informal amateur photos below provide some insights into OSI activities:

IFE08 IN KAZAKHSTAN

FIGURE 4. (a) left and (b) right.

FIGURE 5. (a) and (b).

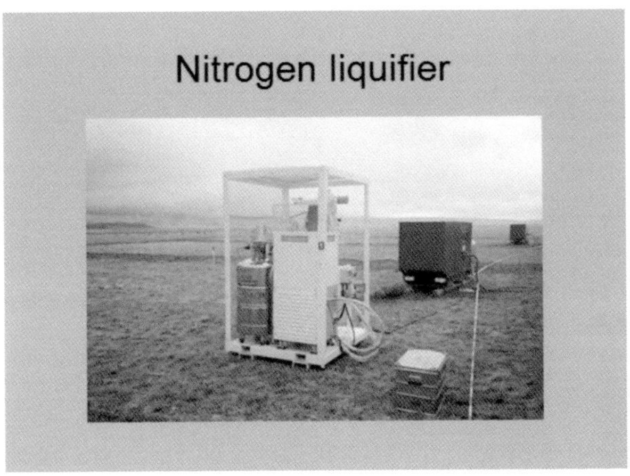

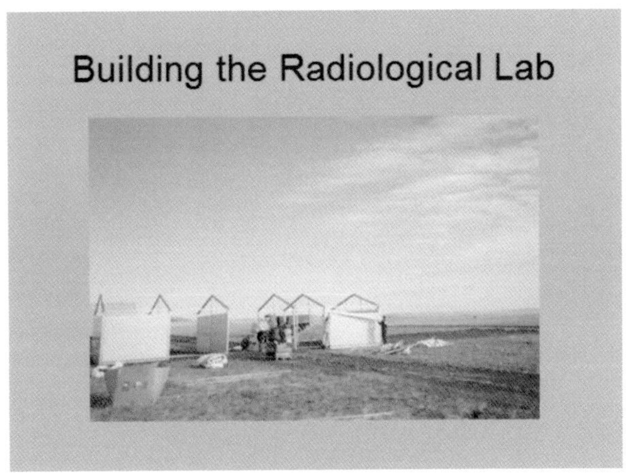

FIGURE 6. (a) and (b).

FIGURE 7. (a) and (b).

IFE14 IN JORDAN

FIGURE 8. (a) and (b).

FIGURE 9. (a) and (b).

FIGURE 10. (a) and (b).

FIGURE 11. (a) and (b).

FIGURE 12.

CONCLUSION

Unfortunately, entry into force of the CTBT appears to still be some distance in the future. The IMS functions now and is providing very useful information every day. However, OSI can only be employed after the treaty is in force, unless a state voluntarily invited an inspection to address a problematic situation. Nevertheless, work continues to prepare for the day when OSI will be a powerful tool at the service of an effective CTBT verification regime and significant progress has been made. The CTBT Science & Technology Conference held in Vienna June 26-30, 2017 demonstrated once again the considerable interest in the world scientific community in these developments and issues. Results of this conference will be published by the CTBTO.

REFERENCES

[1] The views expressed are those of the author and do not necessarily reflect the policies of the U.S. Department of State or Georgetown University.

[2] This and the following photograph are of exhibits in the Atomic Museum at Kurchatov City, Kazakhstan.

[3] See, for example, Jaap Ramaker, Jenifer Mackby, Peter D. Marshall and Robert Geil, *The Final Test: A History of the Comprehensive Nuclear-Test-Ban Negotiations,* published by the Provisional Technical Secretariat of the Preparatory Commission for the Comprehensive Nuclear-Test-Ban Treaty Organization, 2003; see also Keith Hanson, *The Comprehensive Nuclear Test Ban Treaty; An Insider's Perspective*, Stanford University Press, 2006.

[4] The full text of the Treaty may be found at www.state.gov or www.armscontrol.org.

[5] For further information from the CTBTO, go to www.ctbto.org; multiple articles on the CTBT are contained in *The Nonproliferation Review*, June-July, 2016, Vol. 23, Nos. 3-4, published by The James Martin Center for Nonproliferation Studies.

[6] *The Comprehensive Nuclear Test Ban Treaty: Technical issues for the United States*, U.S. National Academy of Sciences, National Academies Press, 2012.

[7] For further considerations regarding OSI see J. J. Zucca, "CTBT On-Site Inspections," *Nuclear Weapon Issues in the 21st Century*, American Institute of Physics Proceedings, AIP Publishing, 2014; see also Edward Ifft, "Occasional Paper 1, On-Site Inspections under the 1996 Comprehensive Nuclear Test Ban Treaty: Modalities," and Occasional Paper 2: "Technical Considerations," The Verification Research, Training and Information Centre (VERTIC), London, 2009; see also Ola Dahlman, Jenifer Mackby, Svein Mykkeltveit, and Hein Haak, *Detect and Deter: Can Countries Verify the Nuclear Test Ban*? Springer, 2011.

[8] For a detailed discussion of IFE14, see Jenifer Mackby, "Special Report: Did Maridia Conduct a Nuclear Test Explosion? On-Site Inspection and the CTBT," *Arms Control Today*, January-February, 2015; see also "Major Exercise Tests CTBT Verification Regime," International Institute for Strategic Studies, *Strategic Comments* 46, December, 2014.

The Big Science of Stockpile Stewardship*

Victor Reis[a], Robert Hanrahan[b] and Kirk Levedahl[b]

[a]*Office of the Secretary*
US Department of Energy
Washington, DC

[b]*National Nuclear Security Administration*
Department of Energy
Washington, DC

Abstract. In the quarter century since the US last exploded a nuclear weapon, an extensive research enterprise has maintained the resources and know-how needed to preserve confidence in the country's stockpile.

At the height of the Cold War, the US possessed more than 30 000 nuclear warheads and exploded, on average, about one of them per week at a desert complex known as the Nevada Test Site. In 1992, after the Cold War ended, President George H. W. Bush halted production of all nuclear weapons and signed an authorization bill containing the Hatfield-Exon-Mitchell Amendment, which instituted a nine-month moratorium on nuclear explosive tests. His successor, Bill Clinton, extended the moratorium, and the US hasn't exploded a nuclear weapon since.

Two decades later the US nuclear deterrent is none the weaker. How has the American defense apparatus retained confidence in its ability to explode nuclear weapons—and how has it demonstrated that ability to the world—without actually exploding nuclear weapons?

A QUESTION OF CONFIDENCE

When President Clinton took office in January 1993, he directed a National Security Council review of the nuclear weapons testing program to focus on two specific questions: How many nuclear explosive tests should the nation conduct, and should the administration support a comprehensive test-ban treaty? Underlying those questions was the broader issue of the role of nuclear weapons and their institutions in the post–Cold War world.

On 3 July 1993, after a vigorous interagency debate, he stated in a radio address,

> I have therefore decided to extend the current moratorium on United States nuclear testing at least through September of next year…. I therefore expect the Department [of Energy] to maintain a capability to resume testing. To assure that our nuclear deterrent remains unquestioned under a test ban, we will explore other means of maintaining our confidence in the safety, the reliability, and the performance of our own weapons.[1]

His statement was codified into a Presidential Decision Directive, a form of executive order, and became law with the signing of the National Defense Authorization Act of 1994 (section 3138). That act included a specific direction to "ensure the preservation of the core intellectual and technical competencies of the United States in nuclear weapons."[2]

The steps taken by Presidents Bush and Clinton drove extraordinary changes in nuclear weapons strategy and its

---

* Reproduced from *Physics Today* Vol. 69, August 2016, p. 46–53 with the permission AIP Publishing.

Nuclear Weapons and Related Security Issues
AIP Conf. Proc. 1898, 030003-1–030003-9; https://doi.org/10.1063/1.5009218
Published by AIP Publishing. 978-0-7354-1586-7/$30.00

underlying institutions. During the Cold War, the US produced approximately 70 000 nuclear warheads of more than 70 different types and conducted more than 1000 nuclear explosive tests. That nuclear buildup was accomplished within an industrial complex that stretched from Hanford, Washington, to Pinellas County, Florida. It employed more than 100 000 skilled workers in nine states. At the heart of the enterprise were two fiercely competitive nuclear weapons design laboratories—Los Alamos National Laboratory (LANL) and Lawrence Livermore National Laboratory (LLNL)—and a third lab, Sandia National Laboratories (SNL), focused on weapons engineering.

Around the time President Bush left office in 1993, the US retained some 11 000 nuclear weapons,[3] including two types of intercontinental ballistic missiles (ICBMs), five types of gravity bombs, two types of cruise missiles—one for the US Air Force and one for the US Navy—and two types of submarine-launched ballistic missiles. To comply with President Clinton's directive, the nuclear weapons enterprise would have to shift away from the Cold War model of aggressive competition between laboratories, continuous design and production of new nuclear weapons, and extensive underground nuclear explosive testing. By what other means could the US maintain "confidence in the safety, the reliability, and the performance" of its weapons while preserving "the core intellectual and technical competencies" in nuclear weapons?

THE END OF "BUILD AND TEST"

A new strategy was first envisioned by a small leadership team from the three weapons labs and the Department of Energy's Office of Defense Programs. The original team members were John Immele and John Browne from LANL, George Miller and Dick Fortner from LLNL, and Roger Hagengruber and Gerry Yonas from SNL. As the assistant secretary of energy for defense programs at the time, one of us (Reis) chaired the effort.

The team assumed that the nuclear stockpile would be reduced to between 1500 and 5000 weapons over the next 20 years, with similar weapon types. (That assumption proved prescient.[3]) The goal would be to attain sufficiently detailed scientific understanding of the nuclear explosive process to discover, comprehend, and correct any anomalies that might occur during the lifetimes of the stockpile weapons; advise the president on whether a return to nuclear explosive testing might be required; and carry out such tests, if needed.

FIGURE 1. The US Department of Energy's Stockpile Stewardship Program makes use of (clockwise from top left) subcritical plutonium implosive testing equipment at the Nevada National Security Site, supercomputers such as Lawrence Livermore National Laboratory's Sequoia, electron accelerators at Los Alamos National Laboratory's Dual-Axis Radiographic Hydrodynamic Test, and an inertial confinement fusion chamber at LLNL's National Ignition Facility. (Images courtesy of the US Department of Energy.)

To that end, the team recommended a new major project at each lab—the National Ignition Facility (NIF) at LLNL, the Dual-Axis Radiographic Hydrodynamic Test (DARHT) facility at LANL, and what became the Microsystems and Engineering Sciences Applications Complex at SNL. Each was a large-scale extension of an ongoing laboratory project and would require greatly enhanced diagnostic capabilities. Each lab would have a dedicated high-performance computer.

The design, fabrication, and operation of the proposed new experimental and computational tools would serve a dual purpose: to equip the laboratories' scientists and technicians to solve anticipated issues with the current stockpile and to demonstrate their continued nuclear competence, adding to the nation's long-term deterrent posture. Crucially, the three DOE labs would share a common vision, shifting from an explosive-test-based confidence model to one of science-based validated simulation. For the US to maintain testing expertise, plutonium experiments at the Nevada Test Site would continue. But they would be subcritical, to avoid igniting a nuclear explosion. Figure 1 shows an apparatus used to conduct the subcritical experiments, along with other images of stockpile stewardship facilities at LLNL and LANL.

The team estimated that the cost for the new weapons strategy would be roughly $4 billion to $5 billion per year for 10 years—about the same as had been required to fund the nuclear testing program. There would be no so-called peace dividend.

The team's concept was developed by the US nuclear weapons communities over the following months. It was then reviewed and endorsed by the administration's myriad national security agencies, Congress, and technical advisory groups, and it was adopted as official US policy.[4]

Thus a strategy of "build and test" was replaced with one that relied on a detailed understanding of nuclear warheads, enhanced technical surveillance of the stockpile, and new experiments and computer simulations. In essence, the viability of the weapons in the US nuclear stockpile was to be entrusted to the scientific competence and integrity of DOE's weapons labs. Thus the program's name: Science-Based Stockpile Stewardship (SBSS).

MUCH ADO ABOUT YIELD

After a series of intense interagency and international negotiations—and the resumption of testing by the French government—President Clinton announced on 11 August 1995 that the US would support a "zero yield" comprehensive test-ban treaty, which would forbid nuclear explosions outright. His spokesman added,

> I am assured by the secretary of energy and the directors of our nuclear labs that we can meet the challenge of maintaining our nuclear deterrent under a [comprehensive test ban] through a science-based stockpile stewardship program without nuclear testing.[5]

One year later the Comprehensive Nuclear-Test-Ban Treaty (CTBT) was adopted by the United Nations General Assembly and signed by President Clinton. In the ratification package that he sent to the US Senate, the president included an assurance that the US's acceptance of the CTBT was conditional on

> the conduct of a Science Based Stockpile Stewardship program to ensure a high level of confidence in the safety and reliability of nuclear weapons in the active stockpile, including the conduct of a broad range of effective and continuing experimental programs.[6]

He also assured the Senate that under the treaty the US would maintain modern nuclear laboratory facilities, maintain the basic capability to resume nuclear testing, continue to improve its treaty-monitoring capabilities, and continue developing its nuclear-intelligence-gathering apparatus. As a final safeguard, he stated that

> if the President of the United States is informed by the Secretary of Defense and the Secretary of Energy (DOE) … that a high level of confidence in the safety or reliability of a nuclear weapon type which the two Secretaries consider to be critical to our nuclear deterrent could no longer be certified, the President, in consultation with Congress, would be prepared to withdraw from the CTBT under the standard "supreme national interests" clause in order to conduct whatever testing might be required.[6]

Despite those reassurances, when the Senate finally took up ratification of the CTBT in October 1999, there was considerable skepticism as to the viability of stockpile stewardship. Brent Scowcroft, Henry Kissinger, and John Deutch evinced a typical concern in their open letter to the Senate Foreign Relations Committee:

> But the fact is that the scientific case simply has not been made that, over the long term, the United States can ensure the nuclear stockpile without nuclear testing… . But even with adequate funding—which is far from assured—the Stockpile Stewardship Program is not sufficiently mature to evaluate the extent to which it can be a suitable alternative to testing.[7]

The Senate did not consent to ratify the treaty and has not to this day.

At the time of the CTBT ratification debate, the skepticism of the Senate and some of its advisers was arguably justified. But in the decades that followed, the scientists at the national labs tasked with carrying out stockpile stewardship would allay the sceptics' reservations.

A BIG SCIENCE APPLICATION

During the Cold War, underground nuclear testing at the Nevada Test Site was the centerpiece of the development and deployment of the US nuclear stockpile. The tests were "final exams" not only for the nuclear warheads themselves but for the laboratory scientists and engineers, who were continually experimenting with new ideas, apparatus, and diagnostics. The program was the quintessential Big Science endeavor, applied to an existential national security mission.

Replacing that program, and doing so under a demanding time schedule, represented a continuing set of challenges for the weapons labs. To address the challenges, the architects of the new nuclear strategy developed three major research programs: nuclear primary hydrodynamics, high-energy-density physics, and plutonium metallurgy. Each subject area would require its own analysis and computer models, which would ultimately be integrated into a series of high-fidelity simulations to assess performance and safety.

NUCLEAR PRIMARY HYDRODYNAMICS

A thermonuclear weapon functions through a sequence of events, beginning with the primary-stage implosion, which ignites a fissile material—most often Pu—and thereby releases a massive amount of x-ray radiation. That radiation is trapped inside a radiation case and provides the energy to trigger the thermonuclear secondary.

Understanding how the fissile Pu pit is compressed during the primary-stage implosion is fundamental to nuclear weapons design. By simulating the pit's compression using nonfissile Pu surrogates, researchers at LANL's DARHT facility are able to generate a variety of two-dimensional, full-scale images that then serve to inform weapons primary design. At the Nevada Test Site, now known as the Nevada National Security Site (NNSS), primary implosions are performed with fissile Pu, but material quantities are reduced sufficiently to ensure that the assembly remains subcritical at all times, in compliance with the CTBT.

Advanced diagnostic tools both at the DARHT and at the NNSS allow implosions to be analyzed with far greater precision than was possible during the era of explosive testing. The precise diagnostics provide stringent tests of fabrication, designs, materials models, computer codes, the abilities of the designers to predict and understand results, and the ability of the researchers to execute the complex experiments. The Pu experiments at the NNSS also exercise many of the skills required to return to full-scale nuclear explosive testing, if such testing were ever required.

HIGH-ENERGY-DENSITY PHYSICS

From almost the beginning of nuclear weapons research, the Atomic Energy Commission and its successor agency, DOE, supported research on inertial confinement fusion (ICF), the ongoing effort to ignite a fusion reaction by compressing a target fuel with energy from lasers or other pulsed-power drivers. Work on ICF has attracted interest not only because of its intimate connection to nuclear design and performance but also because of its potential as a long-term source of energy.[8]

Well before 1992 DOE appreciated that ICF could provide an alternative to underground tests for exploring radiation hydrodynamics—the study of the flow of matter under circumstances where strong coupling with electromagnetic radiation (in this case, x-ray radiation) significantly affects the observed phenomena. Experiments with ICF could also help maintain essential scientific, technical, and design skills. As part of the stockpile stewardship strategy, NIF, the Z machine at Sandia, and the Omega facility at the University of Rochester in New York now conduct studies that integrate radiation hydrodynamics with focused physics experiments.

The collective challenge of building, maintaining, and operating complex integrated facilities; improving the power of pulsed optical and electrical systems; designing and fabricating novel experimental targets; and upgrading the state of the art in diagnostic instrumentation was similar in scale, complexity, and technical difficulty to the challenge of underground testing. The ICF experiments have elucidated fundamental properties of high-energy-density materials, atomic opacities, equations of state, kinetic effects in dense plasmas, and nonequilibrium processes—all with much greater precision than ever before. Many of the findings have become important components of radiation-hydrodynamics models, but they are also relevant to astrophysics, planetary physics, high-pressure science, and other fields of study.[9]

Of the SBSS facilities, NIF stands out in terms of its size, technological complexity, and breadth of scientific interest. At the time NIF was planned, the best radiation-hydrodynamics codes suggested that ignition of an ICF capsule was feasible at scales attainable in the laboratory.[10] Despite considerable effort—and much public hoopla related to fusion energy—ignition has not yet been achieved.

Recent work enabled by vastly improved instrument resolution and full 3D computational capabilities helps explain why. Studies such as the one in figure 2 demonstrate that, inter alia, laser drive asymmetries, engineering features, and imperfections in the target capsule distort implosions to a far greater degree than was initially appreciated.[11] As shown in Figure 3, even a plastic support membrane just 110 nm thick can substantially perturb an implosion.[12]

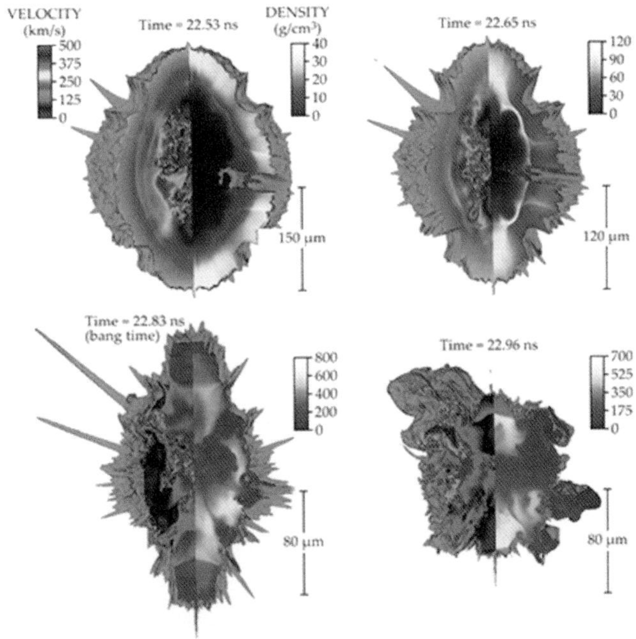

FIGURE 2. Inertial confinement fusion has proven elusive in part because tiny asymmetries and imperfections in the fuel-filled target capsule can produce vastly nonuniform implosion fronts. In this series of snapshots taken near a simulated implosion's "bang time"—the anticipated moment of ignition—spike-like perturbations arise from localized defects in the thin plastic capsule (green). In each panel, the cutaway surface shows fluid speed at left and fluid density at right. Over the course of the implosion, a hot spot develops at the upper left of the target. (Adapted from ref. 11.)

Run by DOE's National Nuclear Security Administration (NNSA), the ICF facilities provide the opportunity for scientists to gain experience applying new physical models and increasingly powerful computers to ignition and high-energy-density experiments—problems that closely parallel physical phenomena relevant to the performance and safety of the nuclear stockpile. Because much of the work performed for the ICF program is publishable, scientists responsible for stockpile stewardship have the opportunity to interact with the academic community, which in turn helps maintain public confidence in the stockpile scientists' abilities.

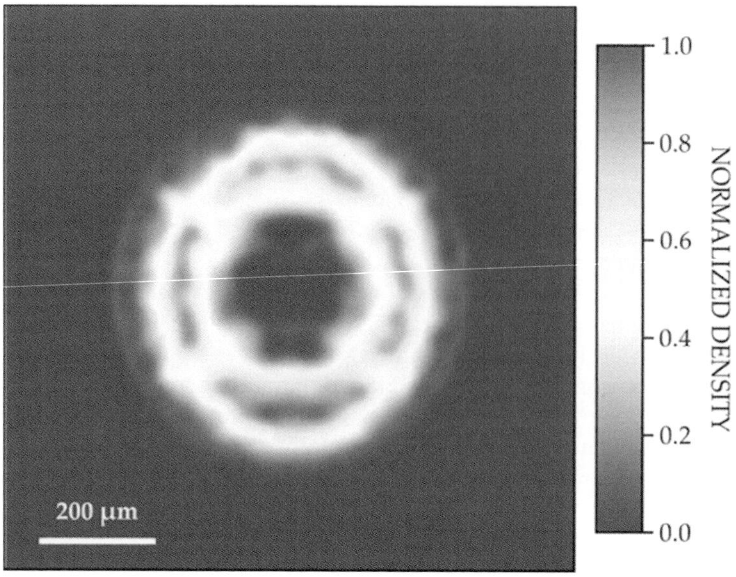

FIGURE 3. A capsule of nuclear fuel undergoing a laser-powered implosion shows density defects at four positions along its outer ring, as seen in this x-ray image. The defects occur where the capsule support membrane contacts the target surface. (Adapted from ref. 12.)

PLUTONIUM AGING

When stockpile stewardship began, American scientists understood little about how the material properties of Pu evolve with age. (Barely 50 years had passed since the first Pu metal was produced.) The scientists needed to either learn enough about aging to be able to certify the performance of existing Pu pits or build replacement pits from scratch. The latter task would have been formidable, given the closing of the US's pit-manufacturing facility at Rocky Flats, Colorado. So LANL and LLNL focused on three aspects of Pu aging: radioactive decay, helium ingrowth, and the microstructural changes those two processes induce under conditions ranging from near ambient to those of a nuclear detonation.

FIGURE 4. Aging weapons-grade plutonium develops helium bubbles, visible as white spots in this transmission electron micrograph. Theorists initially hypothesized that those bubbles would coalesce into large voids that could compromise the material's viability as a nuclear fuel. Such void swelling, however, has yet to be observed in Pu. (Graphic by Mark Wall/Lawrence Livermore National Laboratory.)

Materials scientists and nuclear scientists attacked the problem with a robust suite of tools: computational modeling to predict radiation effects at the atomic and lattice levels; continuum modeling to predict behavior at the grain scale; metallurgical studies on naturally aged materials from the stockpile; and studies on artificially aged alloys created by adding fast-decaying ^{238}Pu to the more slowly decaying, weapons-grade Pu alloy.[13]

Initially, many experts thought aging would be dominated by the onset of so-called void swelling, in which He bubbles agglomerate into voids in the Pu, causing a relatively rapid decrease in material density and strength under ambient conditions. The photomicrograph in Figure 4 shows the distribution of He bubbles in aged Pu. The bubbles increase in number as the Pu ages, but to date, void swelling has not been observed. Rather, growing bubbles appear to become pressure stabilized: At an equilibrium size determined by the strength of the Pu matrix, the pressure of He inside the bubble equals the yield strength of the Pu alloy, and there is no strain field around the bubbles.

Other properties of Pu were found to vary relatively smoothly with age. In other words, under experimentally accessible conditions, Pu appears to age gracefully. Based on the aging work, scientists at LANL and LLNL were able to report initial estimates of pit lifetimes in October 2006. For the majority of weapons types, most Pu pits had estimated lifetimes of 85 years or more.[14]

Subsequent work on Pu has ranged from developing techniques to study its behavior under subcritical implosion conditions at the NNSS to measuring its crystal structure under high pressures at NIF. But the key caveat, in 2006 and still to this day, is that we do not have complete data for aged Pu under the high-explosive-drive conditions of a pit. Obtaining those data will require diagnostics that can measure the hydrodynamic behavior of Pu through all the conditions obtained in a subcritical implosion. Aging Pu will continue to be a stockpile stewardship research area for many years to come.

HIGH-PERFORMANCE COMPUTING

Perhaps no aspect of the SBSS program demanded more resources and technical know-how than the high-fidelity simulation of the nuclear explosive process. Faced with weapons materials aging in a radioactive environment and the impending retirement of the few remaining nuclear weapons experts experienced in underground testing, the program's planners estimated in 1993 they had only 10 years to ramp up computation speeds by a factor of 10 000 over the highest performing computers at the time, equivalent to a factor of 1 million over computers routinely used for nuclear calculations. The feat would require an annual speedup about twice that of the already optimistic expectation from Moore's law, which predicts a doubling in computing speed every 18 months.

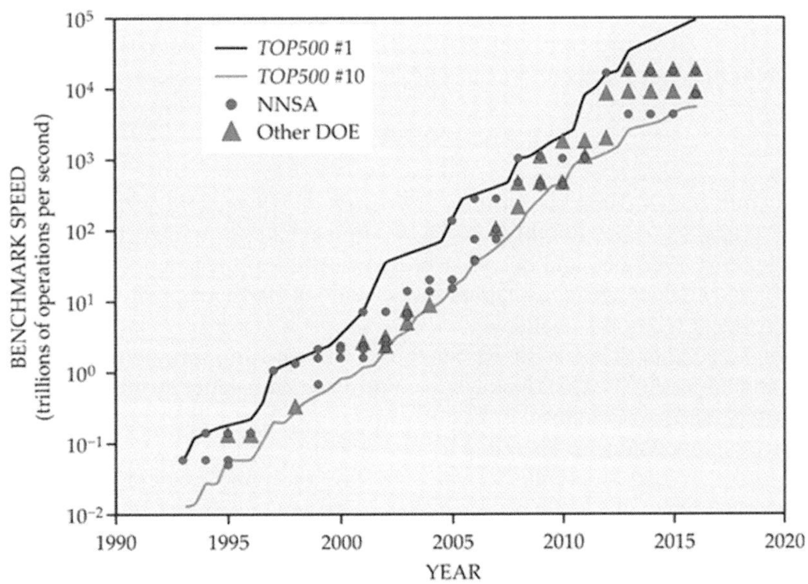

FIGURE 5. Supercomputers commissioned by the US Department of Energy's National Nuclear Security Administration (NNSA) for the purpose of stockpile stewardship have consistently ranked among the fastest in the world, as have other DOE computers. The black and gray lines, respectively, indicate the theoretical peak speeds of the world's fastest and 10th-fastest computers during a given year, according to *TOP500*. (NNSA data from www.top500org.)

To meet the goal, the DOE laboratories had to engage the computer industry in massively parallel processing, a technology that was just becoming available, to develop not just new hardware but new software and visualization techniques. It was a risky proposition,[4] but one built on a historic partnership. From that effort sprang a new dedicated program: the Accelerated Strategic Computing Initiative, now called the Advanced Simulation and Computing Program.[15] In addition, DOE adopted a procurement strategy wherein each laboratory took turns buying a new machine that was larger—and capable of solving harder problems—than the last.

Despite the inevitable obstacles, the program met its milestones and now exceeds its 10-year goal for computing speed (100 teraflops) by a factor of 200. Meanwhile, massively parallel processing has become the global technology standard for high-performance computing and an integral part of stockpile stewardship and other DOE laboratory programs. Figure 5 shows how the computation speeds of DOE computers have stacked up with the world's leading computers over time, as reported annually by *TOP500*. The LLNL supercomputer Sequoia, which contains over 1.5 million processor cores, currently operates at 17 petaflops.

EXTENDING THE STOCKPILE LIFETIME

The most direct application of the SBSS program to maintaining confidence in the US nuclear stockpile is the Life Extension Program (LEP), in which warheads are refurbished and modernized in order to prolong their useful lives.[16] Because of the deep understanding of the nuclear explosive process gained through the SBSS—particularly through studies of Pu aging—the lifetimes of the warheads in the current stockpile will be extended for another 20–30 years. (See *Physics Today*, November 2013, page 27.)

The US Air Force's W87 ICBM warhead has completed its LEP. The life extension of the US Navy's W76–1 warhead, a submarine-launched ballistic missile, is more than 50% complete. The LEP for the B61-12, a gravity bomb that will replace four existing versions of the B61, is completing development. And the LEPs for the navy's W88 ICBM warhead and the air force's W80 cruise missile warheads are in the early development stages.

In parallel with the LEP, scientists and engineers have had to upgrade the nuclear weapons' electronics. Electronic systems for nuclear weapons must be highly integrated, reliable, and secure and must be designed to operate in unique, high-radiation environments. Working with Intel Corp, SNL developed the Permafrost Application Specific Integrated Circuit, and its Microsystems and Engineering Sciences Applications facility has produced some 3000 units for the navy.

In 2005, Congress directed the NNSA to investigate whether the SBSS's tools could be used to develop a new, high-confidence nuclear warhead and a responsive nuclear weapons production infrastructure. The NNSA sponsored a vigorous conceptual design competition between LANL and LLNL that generated two concepts for a reliable replacement warhead.[17] (Both labs partnered with SNL.) Congress did not authorize any further development of either warhead, but much of the associated analysis has proven directly applicable to ongoing LEPs.

PASSING MARKS

Science-Based Stockpile Stewardship sits at the confluence of US nuclear weapons, science, and technology policy. The effort didn't always run smoothly; it has at times had to withstand turbulent political climates, organizational reshuffling, cost overruns, and delays. Yet by prioritizing the scientific understanding of the nuclear explosive process, DOE, its nuclear weapons laboratories, and its partners at the Department of Defense have successfully answered President Clinton's challenge.

In the two decades since the Science-Based Stockpile Stewardship program became official US nuclear weapons policy, every annual stockpile assessment has culminated in a letter from the secretaries of defense and energy certifying the safety and performance of the stockpile. All the major tools recommended by the SBSS planning team have been developed and are operating at near full capacity.

In reviewing the results of the Stockpile Stewardship Program, the Congressional Advisory Panel on the Governance of the Nuclear Security Enterprise stated in March 2014,

> To date, Science-Based Stockpile Stewardship has succeeded in sustaining confidence in the U.S. nuclear deterrent. Unmatched technical innovation on the part of NNSA's scientists and engineers has produced a dramatically increased understanding of the country's aging nuclear weapon stockpile.[18]

Looking forward, the nuclear landscape remains uncertain and dangerous. Although the Cold War has ended, the governments of Russia and China are modernizing their nuclear weapons. Nuclear proliferation and nuclear terrorism loom large. Our national response, as exemplified by the recent Joint Comprehensive Plan of Action to curtail Iran's nuclear program, is heavily dependent on DOE's laboratory system. (See *Physics Today*, December 2015, page 26.)

In his 30 March op-ed in the Washington Post, President Obama wrote, "Even as the United States maintains a safe, secure and effective nuclear arsenal to deter any adversary and ensure the security of our allies, I've reduced the number and role of nuclear weapons in our national security strategy." Underpinning presidential confidence in the nuclear arsenal is the confidence in the integrity and competence of our national labs. That is what is meant by "science based."

The views and opinions expressed by the authors in this article do not necessarily state or reflect those of the US government or any of its agencies.

REFERENCES

1. The American Presidency Project, "The President's Radio Address, July 3, 1993," www.presidency.ucsb.edu/ws/?pid=46803.
2. GovTrack.us, "Text of the National Defense Authorization Act for Fiscal Year 1994," www.govtrack.us/congress/bills/103/hr2401/text.
3. See "Stockpile Numbers, End of Fiscal Years 1962–2015," http://open.defense.gov/Portals/23/Documents/frddwg/2015_Tables_UNCLASS.pdf.
4. See, for example, S. Drell et al., Science Based Stockpile Stewardship (rep. prepared for the US Department of Defense), MITRE Corp (November 1994).
5. See "Comprehensive Nuclear Test Ban" (11 August 1995), video, at the 7:09 minute, www.c-span.org/video/?66686-1/comprehensive-nuclear-test-ban.
6. Federation of American Scientists, "Comprehensive Test Ban Treaty Safeguards," http://fas.org/nuke/control/ctbt/text/ctbsafeguards.htm.
7. 145 Cong. Rec. S24603 (8 October 1999) (letter by B. Scowcroft, H. A. Kissinger, J. Deutch).
8. For a detailed discussion of the history of inertial confinement fusion, see J. H. Nuckolls, Contributions to the Genesis and Progress of ICF, rep. no. UCRL-BOOK-219136, Lawrence Livermore National Laboratory (2006).
9. B. A. Remington, R. E. Rudd, J. S. Wark, Phys. Plasmas 22, 090501 (2015); https://doi.org/10.1063/1.4930134 , J. E. Bailey et al., Nature 517, 56 (2015).
10. J. Lindl, Phys. Plasmas 2, 3933 (1995); https://doi.org/10.1063/1.871025 , J. Lindl et al., Phys. Plasmas 21, 020501 (2014).
11. D. S. Clark et al., Phys. Plasmas 22, 022703 (2015).
12. R. Tommasini et al., Phys. Plasmas 22, 056315 (2015).
13. J. C. Martz, A. J. Schwartz, J. Miner. Met. Mater. Soc. 55(9), 19 (2003).
14. R. J. Hemley et al., Pit Lifetime (rep. prepared for the US Department of Energy, National Nuclear Security Administration), MITRE Corp (January 2007).
15. A detailed history of the Accelerated Strategic Computing Initiative is available at https://asc.llnl.gov/content/assets/docs/DeliveringInsightASCI.pdf.
16. A full discussion of the Life Extension Program is available at https://nnsa.energy.gov/ourmission/managingthestockpile/ssmp.
17. J. Medalia, The Reliable Replacement Warhead Program: Background and Current Developments (rep. prepared for the US Congress), Congressional Research Service (December 2007).
18. N. Augustine, R. Mies, Interim Report of the Congressional Advisory Panel on the Governance of the Nuclear Security Enterprise, testimony before the US Senate Committee on Armed Services, 26 March 2014, www.armed-services.senate.gov/imo/media/doc/Augustine-Mies_04-09-14.pdf.

A New Era of Nuclear Test Verification*

Matthias Auer and Mark Prior

International Monitoring System
Comprehensive Nuclear-Test-Ban Treaty Organization
Vienna, Austria

Abstract. The global network of sensors commissioned to monitor compliance with the Comprehensive Nuclear-Test-Ban Treaty has proven capable of that task and more.

In recent months considerable speculation has appeared in the international media about the prospects of a fourth nuclear test by North Korea. Although we hope, of course, that no such test comes to fruition, the hype is remarkable in itself: It shows how rare nuclear testing has become nowadays.

The credit for the slowdown in nuclear testing goes largely to the de facto moratorium that has been in place since the Comprehensive Nuclear-Test-Ban Treaty (CTBT) opened for signature in September 1996. (See Physics Today articles by Jeremiah Sullivan, March 1998, page 24; Kai-Henrik Barth, March 1998, page 34; and Pierce Corden and David Hafemeister, April 2014, page 41.) The CTBT bans nuclear explosions by everyone everywhere and in all environments—underground, underwater, and in the atmosphere. Its no-test norm, freezing the global status quo in nuclear weapons development, effectively hampers both the first-time development and the upgrading of nuclear weapons.

To date, the CTBT has been signed by 183 states and ratified by 162 of them, which makes it one of the most widely supported treaties in history. However, it will not enter into force until it has been ratified by all of the 44 states listed in its Annex 2—the states that possessed nuclear power or research reactors when the treaty was being negotiated. The eight Annex 2 states that have yet to ratify the CTBT are China, Egypt, India, Iran, Israel, North Korea, Pakistan, and the US.

Although the ban on nuclear tests is not yet legally binding, the effect of the CTBT has already been tangible. In the half century before the treaty opened for signature in September 1996, more than 2000 tests were conducted worldwide. Since then, all of the CTBT's signatory states have adhered to the no-test norm, and the seven nuclear tests that have been conducted—by India, Pakistan, and North Korea—were denounced worldwide. Part of the no-test norm's strength derives from the treaty's unique verification regime, which has proven its ability to detect even small events. When North Korea conducted nuclear tests in 2006, 2009, and 2013, the CTBT's monitoring system detected them immediately and with high precision; it narrowed down the 2013 blast's origins to within about 180 km2. The first data pertaining to the magnitude and location of that test were made available to CTBT signatory states almost within an hour—even before North Korea declared that it had carried out a test.

A GLOBAL NETWORK

To monitor compliance with the treaty, the Preparatory Commission for the Comprehensive NuclearTest-Ban Treaty Organization (CTBTO) in Vienna established the International Monitoring System (IMS), a global network of sensors designed to ensure that all nuclear tests are detected and located. When complete, the network will comprise 337 stations and laboratories in places as varied as sub-Saharan Africa, Alaska, the Australian outback, and the South Pole. As of August 2014, almost 90% of the network was operational.

* Reproduced from *Physics Today*, Vol. 67, September 2014, p. 39–44 with permission of AIP Publishing.

Nuclear Weapons and Related Security Issues
AIP Conf. Proc. 1898, 030004-1–030004-8; https://doi.org/10.1063/1.5009219
Published by AIP Publishing. 978-0-7354-1586-7/$30.00

The IMS uses three types of waveform technologies—seismic, hydroacoustic, and infrasound monitoring—complemented by radionuclide detection technology. Each station's type and location is prescribed by the treaty. Operations at most stations are automated. Each station has a state, usually the nation in which it is located, that is responsible for equipment maintenance and upkeep of the site.

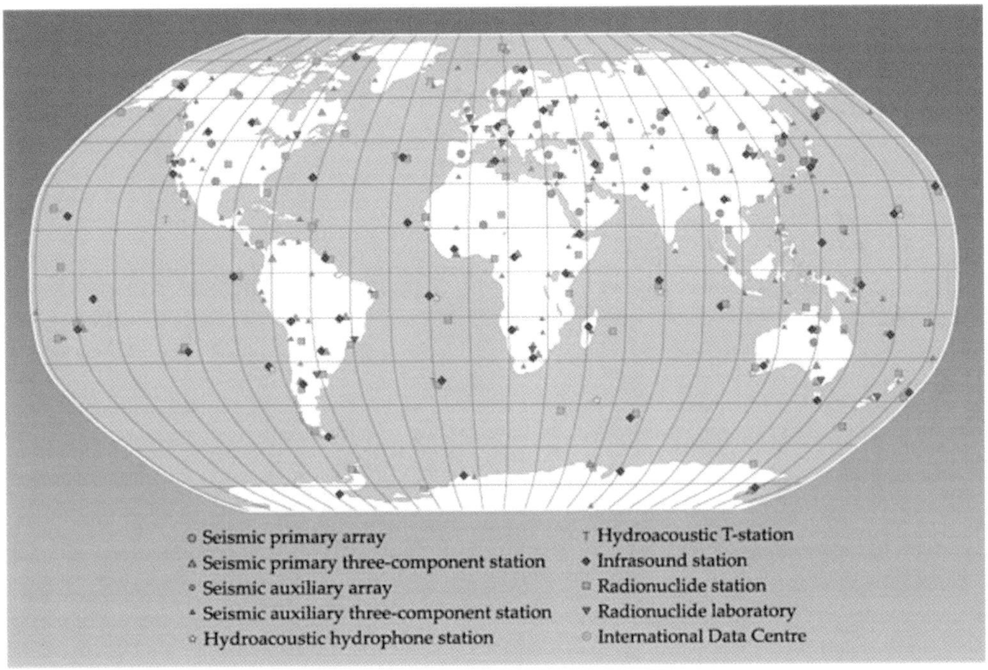

FIGURE 1. The International Monitoring System, once complete, will comprise 337 facilities: 170 seismic stations, 60 infrasound stations, 11 hydroacoustic stations, 80 radionuclide stations, and 16 radionuclide laboratories. As of May 2014, about 90% of the network was operational. (Image courtesy of the Comprehensive Nuclear-Test-Ban Treaty Organization.)

BY LAND

At the core of the IMS are its seismic monitoring stations. A seismic station may consist of a single three-component seismometer or of an array containing dozens of sensors spread over an area from a few to several hundred square kilometers. The three-component seismometers detect vertical, north–south, and east–west movements in Earth's crust; those data can be processed to give the polarization and propagation direction of seismic waves. When seismic sensors are arranged in an array, their data can be combined using beam-forming algorithms to accurately estimate a wave's direction, speed, and time of arrival.

Of the 170 seismic stations in the IMS, 50 are designated as primary seismic stations, which make the initial detection of an event, and 120 are specified as auxiliary stations, which refine estimates of the event's time and location. The two kinds of stations have the same capabilities and instrumentation; the main difference is that auxiliary seismic stations send data only on request, whereas primary seismic stations transmit data continuously.

North Korea's first nuclear test, on 9 October 2006, was detected by 22 seismic stations, including one in San Ignacio de Velasco, Bolivia, more than 17 000 km away. The event was similar in size to a magnitude 4.1 earthquake, and its location was calculated to a precision of better than 20 km. The country's second nuclear test, a magnitude 4.5 event, was declared on 25 May 2009. Due to the size of the explosion and an increase in the number of operating IMS stations, 59 stations detected the event, including one 18 000 km away in Paso Flores, Argentina. The trends of larger tests and detection by an increasing number of IMS stations continued with North Korea's third test, on 12 February 2013. Registering a magnitude of 4.9, the explosion was detected by 94 seismic stations—the farthest being in Paraguay.

FIGURE 2. Workers install a primary seismic station in Torodi, Niger. It is one of 50 such stations that continuously transmit measurements to the Comprehensive Nuclear-Test-Ban Treaty Organization's headquarters in Vienna. (Photo courtesy of the CTBTO.)

BY SEA

The IMS uses two types of hydroacoustic stations: hydrophone and T-stations. At the six hydrophone stations, seabed cables run from a shore facility to an offshore location where three sensors are deployed in a horizontal, triangular configuration suspended beneath subsurface buoys. Time delays between signals registered by the sensors are used to calculate the direction of passing acoustic waves. The five T-stations, by contrast, use coastal seismometers to detect the vibrations caused by waterborne waves that hit the coast and convert to seismic waves.

FIGURE 3. At a hydroacoustic station on an island in Chile's Juan Fernández Archipelago, workers lay underwater cable that will connect offshore sensors with the shore-based facility. (Photo courtesy of the Comprehensive Nuclear-Test-Ban Treaty Organization.)

Most hydrophones are positioned a kilometer or so below the surface, in the deep sound channel, a region of low sound speed that arises due to the cumulative effects of temperature and density. The speed of sound in seawater increases with temperature and pressure. Near the surface, temperature falls sharply with increasing depth and, therefore, so does the speed of sound. Eventually, the temperature levels off and, due to the gradual rise of pressure, the sound speed begins to increase. The depth at which the sound speed reaches its minimum marks the axis of the deep sound channel.

The deep sound channel behaves like a waveguide: It bends sound by refraction into paths that cycle around the channel axis and allow sound to travel long distances without hitting the sea surface or seabed. The phenomenon allows a single hydrophone station to monitor an entire ocean basin; in seismic surveys, waves emitted off the coast of Japan have been detected by IMS hydrophone stations as far away as the Juan Fernández Archipelago in Chile.

T-stations are less sensitive than hydrophone stations, largely because of the signal distortion and attenuation that occurs when waves cross the coastline. (See the article by Bill Kuperman and James Lynch, *Physics Today*, October 22004, page 55.)

BY AIR

The planned 60 infrasound stations detect airborne pressure waves at frequencies below the limit of human hearing. (See the *Physics Today* article by Alfred Bedard and Thomas Georges, March 2000, page 32, and the Quick Study by Curt Szuberla and Ken Arnoult, April 2011, page 74.) Arrays comprising between 4 and 15 ultralow-frequency microphones, or microbarometers, are used to sample the acoustic field over an area of around 10 km2. Each microbarometer sits at the center of a rosette of pipes that reduces the influence of wind noise. Choosing sites with vegetation cover can further mitigate wind noise, but that isn't always an option, given that some stations are located in high deserts and Antarctica.

FIGURE 4. Infrasound stations such as this one in Qaanaaq, Greenland, use ultralow-frequency microphones to sample the acoustic field over areas of several square kilometers. Each microphone is surrounded by a rosette of pipes that helps filter local wind noise. (Photo courtesy of the Comprehensive Nuclear-Test-Ban Treaty Organization.)

The infrasound stations are designed to detect signals in the frequency range of 0.02–4 Hz, corresponding to wavelengths between a few hundred meters and a few kilometers, at distances of up to thousands of kilometers from the source. At the CTBTO's International Data Centre (IDC), the wave attributes, including the arrival direction and wave speed, are calculated by correlating signals from the different microbarometers in an array. The processing method can efficiently extract low-amplitude coherent signals from noncoherent noise such as local wind.

North Korea's third nuclear test was the first to be detected by infrasound stations; it was registered by one station in Isumi, Japan, and another in Ussuriysk, Russia. Both stations are located not far from the test site—just 1200 km and 400 km, respectively. The infrasound signals were the result of pressure waves in the air generated by the shaking of mountains in the region of the test site.

Although it is tempting to think that the three waveform technologies are each responsible for detecting a different type of nuclear test—that seismic stations detect underground tests, hydroacoustic stations detect underwater tests, and infrasound stations detect airborne tests—the signal processing used by the CTBTO is sophisticated enough to detect events that cross from one medium to another. Underground events can be detected by infrasound stations when the vibrating ground generates airborne waves or shakes microphones. Underwater explosions couple energy into Earth's crust that can be detected with land-based seismometers. Large airborne sources can generate seismic waves when their shock waves hit Earth's surface. When the Chelyabinsk meteor exploded over the Ural Mountains on 15 February 2013, for instance, it was registered by 20 CTBTO infrasound stations—including one in Antarctica and one in Alaska—for three days as waves traveled twice around Earth. But the first signal detected from the explosion was a seismic wave that arrived at a nearby station earlier, before the slower airborne waves. (For more on the Chelyabinsk meteor and its detection, see the article by David Kring and Mark Boslough on page 32 of this issue; see also A. Le Pichon et al., *Geophysical Research Letters*, volume 40, page 3732, 2013, and P. G. Brown et al., *Nature*, volume 503, page 238, 2013.)

SMOKING GUNS

The network of seismic, hydroacoustic, and infrasound sensors is complemented by 80 stations that detect the radionuclides that are produced during a nuclear explosion. Most radionuclides condense quickly after formation and, if released into the atmosphere, attach themselves to atmospheric aerosol particles. Those aerosol-borne radionuclides are often referred to as particulates. All IMS radionuclide stations are equipped with particulate sampling and measurement systems.

Although particulate systems are well suited to monitoring explosions in the atmosphere, they are less likely to detect an underground or underwater nuclear explosion; unless a major vent of fission products occurs, only volatile isotopes—mainly noble gases and, to a lesser extent, iodine—are likely to escape into the atmosphere in significant quantities. Those isotopes typically don't condense or form particulates, but they can be analyzed using specialized analysis techniques. Whereas all of the IMS's radionuclide stations have particulate systems, only half are being equipped with noble-gas systems prior to the CTBT's entry into force. (Later, member states may decide to equip the remaining stations with noble-gas systems as well.)

Of the noble-gas isotopes, xenon-131m (here, "m" indicates a metastable nuclear configuration), xenon-133, xenon-133m, and xenon-135 are of the greatest interest for monitoring nuclear explosions. Produced directly by fission of uranium or plutonium and indirectly by the decay of indium, tin, antimony, tellurium, and iodine isotopes, ^{133}Xe, ^{133m}Xe, and ^{135}Xe are among the highest-yield products of uranium or plutonium fission, and their half-lives (5.25 days, 2.2 days, and 9.17 hours, respectively) are long enough that they can be detected at large distances from their source. Although ^{131m}Xe has a significantly lower yield, its longer half-life (11.8 days) means that it can be detected weeks after a nuclear explosion. Under very favorable conditions, differences between the yields of uranium-235 and plutonium-239 fission may allow discrimination between the two, based on the number of Xe isotopes detected and the time at which they were detected.

Some Xe isotopes decay into particulate nuclides that can be detected by the particulate systems. The most significant among them is ^{140}Xe, which decays into barium-140. With a half-life of 13.60 seconds, ^{140}Xe is sufficiently long-lived to vent from an underground or underwater test before decaying.

Other noble gases created in nuclear explosions are less relevant for the IMS, either because their yield is too low compared with their atmospheric background concentration, as in the case of krypton-85, or because we lack the technology to measure them at the required sensitivity, as in the case of argon-37. Measuring for ^{37}Ar, which has a half-life of 35 days, can be performed during an on-site inspection—the CTBT's final verification measure. Such inspections, however, can be requested only after the treaty has entered into force.

Particulate and noble-gas samples are acquired by pumping air through filters at a rate of at least 500 m^3/h and 0.4 m^3/h, respectively, for a period of up to 24 hours. For quality assurance and in order to verify the results, samples are taken regularly and reanalyzed at one of the CTBTO's 16 radionuclide laboratories. When combined with atmospheric transport modeling (ATM), a technique that uses meteorological data to simulate the dispersion of atmospheric radionuclides, measurements made at radionuclide stations and laboratories can provide information on the timing and location of an event. The ATM calculations can be done either in a backward mode, to indicate possible source regions of radionuclide emissions, or in a forward mode, to assess the area affected by the release. The detection of radionuclides also provides significant evidence as to whether an event detected by one of the waveform technologies was a nuclear explosion.

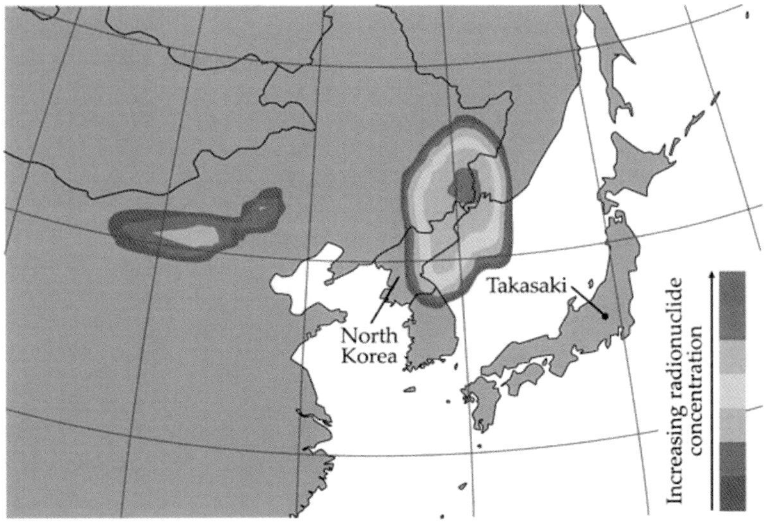

FIGURE 5. Atmospheric transport modeling (ATM) uses meteorological data to calculate how substances—for example, radionuclides—that are released into the atmosphere disperse over time and where and when they are likely to be detected. Backtracking ATM identifies where and when a release may have occurred based on the location of a detection. Shown here are the results of a backtracking calculation for an air sample containing radioactive xenon isotopes (131mXe/133Xe) collected at the International Monitoring System's radionuclide station in Takasaki, Japan, on 8 April 2013. The colored regions represent the most likely areas for the isotopes' release into the atmosphere, had it occurred one day earlier on 7 April 2013 between 03:00 and 06:00 UTC. The measured ratios of the Xe isotopes indicate that they were the product of a nuclear fission event that occurred 50 days earlier, consistent with the time of the 2013 North Korean nuclear test. (Image courtesy of the Comprehensive Nuclear-Test-Ban Treaty Organization.)

When North Korea announced its first nuclear test in 2006, only a few noble-gas systems had been installed, none of them in the East Asian region. Nonetheless, the IMS radionuclide station in Yellowknife, Canada, detected increased levels of ^{133}Xe, consistent with a leak from an underground nuclear explosion on the Korean peninsula. That scenario was also supported by detections of ^{133}Xe and ^{133m}Xe at a mobile station deployed in South Korea.

No radionuclide releases were detected in the aftermath of North Korea's second test. But seven to eight weeks after the third test, the two stations closest to the test site—one at Takasaki, Japan, and another at Ussuriysk, Russia—collected samples with $^{131m}Xe/^{133}Xe$ ratios consistent with release from a nuclear test on 12 February 2013. The hypothesis that the emissions were from a nuclear test was supported by the fact that the ratios and concentrations differed significantly from previous values obtained at the same sites. The CTBTO was the only group worldwide to detect radioactivity that could be attributed to the event.

DATA ON THE GO

The IMS network is unique not only in terms of its low detection thresholds and its global coverage but also because of its state-of-the-art equipment and the exceptional quality of the data it generates. Those data are time-stamped using GPS clocks and then transmitted to the IDC in Vienna by way of a network of satellites known as the Global Communications Infrastructure. Data are transmitted in near real time—about 90% of the stations transmit data to the IDC within 30 seconds of measurement. Hardware and software devices guarantee the data's integrity. The Global Communications Infrastructure is also used to distribute processed and raw data from the IDC to signatory states.

After they have been received by the IDC, seismic, hydroacoustic, and infrasound data are entered into an automatic processing pipeline. Signals that stand out from the background noise are identified, and then the type of signal and the path by which it is traveling are calculated from the waveform shape and wave speed. Once signals have been detected in the waveform data, the next stage of processing is to identify the events that generated them. Seismic stations detect vibrations not just from explosions but from natural tremors and quarry blasts. Hydroacoustic stations detect waves trapped in the ocean sound channel that have crossed entire ocean basins, but they also detect local noises made by whales and passing ships. Infrasound stations detect signals that, among other possible paths, travel up to the thermosphere and cross the stratosphere, where wind speeds can reach one-third of the speed of sound. Signals can also originate from volcanic eruptions, meteoroids, earthquakes, and space vehicle launches.

Events are identified using approaches taken from the field of global seismology; lists of detection times at IMS stations are converted to lists of hypothesized event times and locations. Because seismic, hydroacoustic, and infrasound waves travel at very different speeds, signals from even a small event may arrive at the IMS stations over the course of minutes to hours. To accommodate the staggered arrival of information, IDC processing runs in a repeated loop, with each iteration refined using new data as they become available. It is not unusual for the first estimate of an event's location to be produced while vibrations from the event are still traveling through the ground, ocean, or atmosphere.

After all signals have arrived, the automatic processing finishes and the results are sent to analysts who refine the estimates and produce the Reviewed Event Bulletin (REB), a list of the events that were detected by IMS stations. The bulletins are produced and distributed to CTBT signatory states daily and typically contain more than 100 events. All three nuclear tests by North Korea were included in an REB.

The scale and scope of the data collection and processing needed to produce the REB is remarkable. Data describing ground motion in Mongolia are combined with measurements of air pressure in Greenland to help determine the time and place of earthquakes in Asia; seismic signals from Scandinavia record quarrying in Siberia; infrasound sensors at high latitudes record very low frequency waves generated by the aurora borealis and aurora australis; underwater sensors in the tropical Atlantic record ice-breaking noises from Antarctica. Every day, gigabytes of data are recorded, transmitted across the world, and processed to produce a snapshot of global seismic activity, among other results.

THE SECOND WAVE

The processing of radionuclide data is substantially different from that of waveform data. Radioactive particles and noble gases may not arrive at a monitoring station until days or even weeks after an event—long after the waveform data for the same event have been analyzed.

Every day, more than 90 decay-energy spectra from radionuclide stations are transmitted to the IDC, where they are analyzed for relevant fission and activation products. For each sample, a reviewed radionuclide report is produced listing the nuclides that have been detected, with each analysis categorized according to the number of relevant isotopes detected. It takes approximately three days from the time sample collection is finished to complete the report.

One of the challenges in interpreting radionuclide data is distinguishing between radionuclides stemming from a nuclear explosion and those originating from other manmade sources. Noble-gas emissions from industrial facilities such as nuclear power plants and medical-isotope production plants are particularly common. When more than one relevant isotope is detected, nuclear explosions can be distinguished by their isotopic ratios, which are different from those of most civil nuclear activities. For example, nuclear explosions yield higher ^{133m}Xe/^{133}Xe and ^{135}Xe/^{133}Xe ratios than nuclear reactors. If the ratios are unobtainable or inconclusive, a nuclear test can often be ruled out if emissions can be traced by ATM back to known civil sources. To minimize interference with the CTBT verification regime, six producers of medical isotopes have pledged since June 2013 to reduce their emissions and to share information on emission levels.

The value of the CTBTO's global monitoring system extends beyond detecting and deterring nuclear tests. In March 2011 all radionuclide stations in the Northern Hemisphere and some in the Southern Hemisphere picked up traces of radioactive emissions from the stricken Fukushima Daiichi power plant in Japan. The data were provided to signatory states and scientific institutions, which were then able to assess the health and environmental risks caused by radiation. The CTBTO subsequently joined the Inter-Agency Committee on Radiological and Nuclear Emergencies, which coordinates the preparation and response of international intergovernmental organizations to nuclear and radiological emergencies.

After the devastating tsunami caused by the Indian Ocean earthquake on 26 December 2004, CTBTO data were made available to numerous tsunami warning centers, initially for a trial period. Since then, centers in 11 countries reached agreements with the CTBTO to receive data from around 110 IMS stations. The monitoring data enable the centers to issue more timely warnings.

A NEW NORMAL?

Over the past 17 years, the CTBT's verification system has grown both quantitatively and qualitatively. As its monitoring network approaches completion, considerable scientific and technological progress has been made in the monitoring technologies. With debates over the value of arms control agreements frequently focusing on verifiability and whether would-be violators will be detected, the CTBT's verification regime has already demonstrated capabilities that meet and even exceed the parameters defined by its negotiators.

Once the CTBT enters into force, a member state will be able to request an on-site inspection to establish whether a nuclear explosion has been carried out. On-site inspections complement the other elements of the verification regime and will serve as an additional deterrent to any potential violator of the CTBT. The CTBTO has been training extensively in preparation for the treaty's entry into force and will carry out field exercises in Jordan to simulate an on-site inspection in November 2014.

One can argue that the CTBT, with 183 signatures and 162 ratifications, is already strongly embedded in the international nonproliferation and security regimes. Entry into force will provide a strong and verifiable final barrier to developing and enhancing nuclear weapons, an issue of vital importance for addressing present-day security challenges.

FIGURE 6. (Photo courtesy of the Comprehensive Nuclear-Test-Ban Treaty Organization.)

REFERENCES

1. M. B. Kalinowski, A. Becker, eds., Recent Advances in Nuclear Explosion Monitoring, vol. 2, Birkhäuser/Springer, Basel, Switzerland (2014).
2. National Research Council, The Comprehensive Nuclear Test Ban Treaty: Technical Issues for the United States, National Academies Press, Washington, DC (2012).

Understanding the Conventional Arms Trade*

Rachel Stohl

Director – Conventional Defense Program
The Stimson Center
1211 Connecticut Avenue, NW (8th Floor)
Washington, DC 20036

Abstract. The global conventional arms trade is worth tens of billions of dollars every year and is engaged in by every country in the world. Yet, it is often difficult to control the legal trade in conventional arms and there is a thriving illicit market, willing to arm unscrupulous regimes and nefarious non-state actors. This chapter examines the international conventional arms trade, the range of tools that have been used to control it, and challenges to these international regimes.

INTRODUCTION

In March 2015, Saudi Arabia and coalition partners intervened militarily in Yemen in an effort to restore the Hadi government. Now in its third year, the ongoing campaign is largely characterized by air strikes and attacks against Houthi rebels. The campaign has been heavily criticized for repeated strikes against civilian targets – destroying schools, factories, markets, and hospitals, among other targets. Saudi Arabia's campaign is supported by arms sales from major global exporters, including the United States and the United Kingdom and has prompted a growing debate about the risks of providing more conventional weapons to an ongoing conflict amid civilian harm and an increasingly dire humanitarian crisis.

While arms sales to Saudi Arabia have repeatedly generated international headlines, conventional arms deals and their relevance to larger strategic and international security issues are often ignored. However, conventional weapons are responsible for significant challenges to the global order, causing hundreds of thousands of deaths a year, facilitating the continuation of deadly conflicts, and contributing to instability and insecurity worldwide. The global conventional arms trade is a multi-billion-dollar business that affects every country in the world in some way.

Every country is involved in some aspect of the international arms trade. Although the risks and consequences of the conventional arms trade impact countries in a variety of devastating ways, they are often essential components of national–security apparatuses, and thus their control is often challenging. Because conventional arms have legitimate military, police, and civilian uses, their outright ban is not an option. Governments must therefore be creative in trying to regulate and control the proliferation and use of conventional arms.

This chapter examines the international conventional arms trade and the range of tools that are used to control it. Conventional arms range from guns to sophisticated fighter aircraft and naval ships, in other words, the conventional weapons of war.

THE CONVENTIONAL ARMS TRADE

There are five aspects of the conventional arms trade that help explain the challenges of regulating global conventional arms transfers.

First, conventional arms are profitable. Conventional arms transfer agreements worldwide amounted nearly $80 billion in 2015, according to the most recent Congressional Research Service Report.[1] This total only accounts for

* Parts of this chapter appear in my book, *The International Arms Trade*, co-authored with Suzette Grillot (Polity Press, 2009) with the permission of the publisher.

Nuclear Weapons and Related Security Issues
AIP Conf. Proc. 1898, 030005-1–030005-8; https://doi.org/10.1063/1.5009220
Published by AIP Publishing. 978-0-7354-1586-7/$30.00

the legal trade in arms. There is a thriving black market trade in arms, as well as a robust grey market. The global trade in small arms and light weapons (SALW) is an important sub-category of conventional weapons – weapons which, from a control perspective in particular, are dealt with separately from their heavier counterparts. Small arms and light weapons are those that can be carried by one or two people, mounted on a vehicle, or carried by a pack animal. The majority of fighting in the post-Cold War world has been conducted with small arms. According to the Small Arms Survey (SAS), a Geneva-based NGO that monitors global flows of SALW, nearly 875 million small arms are in circulation around the world.[2] Small arms can be purchased in legal markets from an increasing number of suppliers. More than 1,200 companies in over 90 countries produce 8 million new SALW annually.[3] While the small arms trade is nowhere near the value of the heavy conventional weapons trade, its value is estimated to be $7.1 billion annually. There is also more than $1 billion in illicit small arms sales annually. In short, the global conventional arms trade, including small arms and light weapons, can impact national and global economies.

Second, the supply and demand for conventional weapons, both legal and illegal, ebbs and flows as international crises emerge and/or are resolved. Significant shifts in the conventional arms trade are marked by major world events, such as World Wars I and II, the Cold War, the 1991 Gulf War, and the post 9/11 War on Terror. Moreover, national and international political and economic issues more generally affect the international trade in arms and numerous factors influence weapons supply and demand.

Third, in many cases, short-term priorities outweigh long-term interests in arms transfer decisions, and national security considerations supersede human security concerns. The resulting unrestrained trade of conventional weapons leads to significant consequences. From deaths and injuries, to the undermining of human security, the uncontrolled conventional arms trade has put peacekeepers in danger, diminished national and multinational business opportunities, impeded the ability of humanitarian and relief organizations to conduct their efforts, hampered sustainable development, and overall negatively affected global peace and security.

Fourth, the control of and trade in conventional weapons are more complicated than that for other weapons systems. Unlike weapons of mass destruction, conventional arms do not primarily serve a deterrence function, but are tools that can be legitimately used by governments, militaries, police forces, and civilians. Indeed, Article 51 of the Charter of the United Nations recognizes the inherent right of all States to individual or collective self-defense and thus the right to manufacture, import, export, transfer, and retain conventional arms toward that end. Thus, regulating and controlling the trade in conventional arms poses additional challenges for States and the international community.

Fifth and finally, conventional arms controls are underdeveloped and face significant challenges. One of the most immediate challenges is that some of the primary exporters have not supported nor participated in proposed and existing controls.

CONVENTIONAL ARMS CONTROL OF THE PAST

While existing control mechanisms for regulating the conventional arms trade are relatively new (that is, within the last 25 years), informal controls on the arms trade date back centuries and formal controls were first considered in the 1890s in the context of regulating the slave trade. Yet the Cold War brought renewed interest in pursuing meaningful conventional arms controls.

One of the first international organizations to focus strictly on trade control issues emerged in 1949 when NATO members (with the exception of Iceland) and Japan created the Coordinating Committee for Multilateral Export Controls (COCOM) to control the sale of strategic goods to Communist bloc countries.[4] After the Cold War ended, COCOM was replaced by the Wassenaar Arrangement on Export Controls for Conventional Arms and Dual-Use Goods and Technologies – the only multilateral organization that today focuses on conventional weapons and critical dual-use items.[5]

Most significant were the Conventional Arms Transfer (CAT) talks of 1977 and 1978 between the United States and Soviet Union. Although the talks failed, they led to progress in controlling conventional arms in the 1980s.

In 1980, UN Member States adopted the UN Convention on Prohibitions or Restrictions on the Use of Certain Conventional Weapons Which May Be Deemed to Be Excessively Injurious or to Have Indiscriminate Effects (also known as the Inhumane Weapons Convention or CCW). The Convention currently has five annexed protocols. The goal of the CCW is to ban or restrict the use of those weapons that cause unnecessary or unjustifiable suffering to combatants or affect civilians indiscriminately – in and outside of situations of international armed conflict. There are protocols on non-detectable fragments, the use of mines, booby traps and other devices, and the use of incendiary weapons, the use of blinding laser weapons, and responsibilities surrounding the clearance, removal, or destruction of explosive remnants of war.

One of the most significant conventional arms control treaties of the 20th century emerged near the end of the Cold War. The Treaty on Conventional Forces in Europe (the CFE Treaty) was signed in November 1990 by 22 NATO and Warsaw Pact countries. The agreement was meant to bring parity, enhance transparency, and increase stability regarding conventional forces in Europe. Covering the area from the Atlantic Ocean to the Ural Mountains, the CFE Treaty outlined limits for five categories of conventional weapons systems, including armored combat vehicles, battle tanks, combat aircraft, attack helicopters, and large artillery. The limits were set at 30,000 armored vehicles, 20,000 tanks, 20,000 pieces of artillery, 6,800 aircraft, and 2,000 helicopters. Limits were set for each individual country within the Treaty area, and no one state could possess more than 1/3 of the total limit in each category. All weaponry in excess of these limits were to be destroyed within 40 months of the Treaty entering into force, and a rigorous verification and inspection regime was established to oversee the process. Not all of the weapons that exceed Treaty limits, however, have been removed and destroyed. Disagreements between state parties have emerged and former Soviet weapons in Moldova and Georgia, for example, remain today. Moreover, due to Russian concerns about the deployment of a U.S. missile defense system and the placement of U.S. military bases in Central and Eastern Europe, the Russian Duma voted unanimously in November 2007 to suspend the country's implementation of the CFE Treaty, leaving the future of this ground-breaking conventional arms treaty in question.[6]

21ST CENTURY CONVENTIONAL ARMS CONTROL

In the 21st Century, governments have attempted many different types of conventional arms trade controls, with varying levels of success. First are controls relating to transparency. With transparency regimes, the belief is that greater understanding at a global level of which country is selling what weapon system to whom (and even for what purpose) can help increase security of the international arms trade. When governments and experts have such knowledge, they can take steps to secure their arms transfers and verify the conditions of transfers. For example, such procedures can contribute to better weapons tracking and facilitate identification of potential circumvention of controls already in place. Transparency also provides important assurances that weapons are being used as intended and, in turn, enhances accountability for individual arms transfers.

Second are controls related to the export of conventional weapons. At a national level, governments establish procedures that identify the rules about which recipients can acquire which weapons systems and through what processes. Such controls have the added benefit of assisting industry in ascertaining the requirements for specific arms transfers and can create accountability for governments to follow specific procedures when making arms transfer decisions.

Third are outright limits and/or bans of conventional arms. In some cases, certain categories of conventional weapons – such as landmines and cluster munitions – have been banned. In other cases, governments develop appropriate limits for arms transfers that can prevent abusive end-users from obtaining weapons technologies or relevant parts and components. Putting clear limits on state provision of conventional arms makes clear what behaviors and uses will not be tolerated and contributes to the development of global norms related to the arms trade.

Fourth are confidence- and security- building regimes. Such arrangements often encourage end-users to provide limits for acquiring certain weapons or a certain number of a given weapons systems in order to facilitate greater confidence in a region or with the international community.

Fifth are punitive measures that are put in place to address violations of conventional arms agreements or national obligations. In order to deal with violations of any national, regional, or international standards, States employ punitive measures for irresponsible or illegal arms transfers. However, without clear enforcement and civil, criminal, and administrative punishments, violators will continue to act with impunity.

Sixth are regimes focused on humanitarian concerns, including the protection of civilians. As policy makers try to address the direct, indirect, and consequential impact of the weapons trade, the focus has turned to "humanitarian arms control" to address enduring weapons threats to civilians. The human security framework has focused on lessening the impact of particular weapons systems on the lives and livelihoods of civilians.

Since the end of the Cold War, governments and civil society organizations have realized that there is no one solution to regulating the global conventional arms trade. Instead, several approaches to controlling conventional arms proliferation and misuse need to be undertaken simultaneously. National governments, local communities, regional organizations, and international organizations all have a role to play. National, regional, and global actors can develop and implement their efforts in a coordinated fashion.

NATIONAL ARMS CONTROL

National arms control efforts are ultimately the first line of defense in preventing the unchecked spread of conventional arms. But, despite the need for common practices to harmonize national laws and regulations, national controls have been implemented haphazardly. Some States have robust arms control systems and fully comply with internationally accepted standards, while others lack even basic control measures. There are several elements that make up a national control system, ranging from specific arms trade legislation to customs authorities and border controls.

Elements of a National Arms Control System

- Arms Trade Legislation: Legal instruments, in the form of laws or decrees, serve to provide government authorities the legal basis to register and license weapons producers, brokers, traders, transporters, and owners – as well as to authorize individual weapons transaction.
- Licensing Procedures: Licensing procedures work to control the export, import, re-export, transit, and transshipment of weapons into, out of, and through national territory, including brokering activities.
- Export Criteria and Control Lists: Arms control lists detail those weapons and weapons systems that are subject to licensing procedures and must be authorized by government authorities. Lists of controlled goods also include lists of proscribed destinations, such as those subject to an arms embargo. Export criteria based on international norms and standards help to determine whether a weapons transfer should be authorized. End-user criteria such as respect for human rights, involvement in armed conflict, involvement in terrorist activities, and level of democratic development are often considered when making licensing decisions.
- Interagency Coordination: Multiple parties within government work together to manage weapons licensing activities and jointly consider and approve arms transfers. Open communication and coordination among relevant agencies allow for an effective licensing process.
- Customs Authority and Border Controls: Border control agents and customs officers work to provide security at points of entry and exit, as well as to inspect and verify weapons moving into, out of, and through national territory. These officials need to be granted the authority to investigate arms activities and enforce breaches of the law.
- Verification Documentation: Arms transfers are verified with the use of standardized documentation. Import certificates, end-use and end-user statements, and post-delivery checks can all be required and their authenticity verified in order to confirm and validate the legality of weapons transactions.
- Penalties and Enforcement: Appropriate civil and criminal penalties can prevent illicit weapons activities and punish illegal actions. Appropriate authorities must enforce the law, investigate violations, arrest and prosecute offenders, and ultimately punish those found guilty of illicit arms transfers.
- Transparency and Oversight: Information regarding arms production, possession, trade and control is shared and reported domestically and internationally. Transparency in the arms trade allows for greater oversight in order to prevent destabilizing and dangerous weapons transfers.
- Marking and Tracing: Weapons producers stamp their wares with identifying marks in order to allow for the tracing of weapons that have been diverted and misused.
- Stockpile Management: Weapons storage facilities are secured and effectively managed to prevent loss and theft.
- Collection and Destruction: Unnecessary surplus weapons are collected from among the public (particularly after conflict has ended and reconstruction begins) as well as from official stockpiles and destroyed in order to prevent additional build-ups, loss, and theft.
- Regional and International Cooperation: Because the international arms trade is a trans-national issue, countries are engaged in regional and international cooperative activities regarding weapons transfers and controls. Collective, consistent, comprehensive, and compatible arms trade controls require multilateral engagement, guidance, and governance.

REGIONAL CONTROL EFFORTS

Regional organizations also provide a forum for conventional arms control, particularly because it is often easier for States with similar experiences to reach consensus and address region-specific issues. Several regions of the world that have been affected by armed conflict in the post-Cold War era – from the Americas to Europe to West Africa – have developed regional agreements and local mechanisms for addressing the arms trade. Moreover, these regional agreements have often served as models for state arms control behavior worldwide, as well as for arms trade discussions in global forums, such as the United Nations. The Organization of American States set a global example in 1997 when it adopted the first legally binding regional agreement on illicit firearms trafficking. The European Code of Conduct on Arms Exports is one of the most often cited arms control documents outlining standards of appropriate arms trade behavior, and the EU's Common position on the Control of Arms Brokering helps ensure compliance with UN, EU, and OSCE arms embargoes. West African States, which have been significantly affected by the spread, accumulation, and misuse of small arms and light weapons, established the ECOWAS Convention on Small Arms and Light Weapons, Their Ammunition, and Other Related Materials, which is intended to control and prevent the accumulation of small arms, promote trust and the exchange of information among member states, and to build the capacities of ECOWAS and its member states to control the spread of small arms.

INTERNATIONAL CONTROL EFFORTS

Globally, several measures have been undertaken to control conventional arms. The Mine Ban Treaty and Convention on Cluster Munitions both ban the use of specific conventional systems and were negotiated by like-minded governments.

For non-ban conventional arms controls, the United Nations has been the primary international governmental organization that has actively implemented international arms trade controls. In 1991, in the wake of the Gulf War, for example, the permanent members of the United Nations Security Council (United States, Britain, France, China and Russia) established guidelines for conventional arms transfers, including advance notification of arms sales, considerations of human rights consequences, and concerns for destabilizing arms build-ups. These guidelines, however, were never fully implemented because China opposed U.S. arms sales to Taiwan and refused to abide by the guidelines until such sales ceased.[7] Future discussions about the role and purpose of arms transfer guidelines continued, however, with the UN Disarmament Commission reiterating the importance of restraint in its 1996 report.[8] Perhaps the most significant conventional arms control development of the 1990s was the creation of the UN Register of Conventional Arms in December of 1991, a global voluntary transparency mechanism where States annually report on their arms exports and imports.

On small arms, the UN adopted the Protocol against the Illicit Manufacturing of and Trafficking in Firearms, Their Parts and Components and Ammunition (known as the Firearms Protocol) in 2001. The Firearms Protocol supplements the United Nations Convention against Transnational Organized Crime and is the first international legally binding small arms control agreement. Ratifying States commit to adopting certain crime control measures and implementing domestic legislation to 1) make the illegal manufacturing or trafficking of firearms a criminal offence; 2) establish a governmental arms licensing system; and 3) create a system for marking and tracing firearms.[9]

The United Nations also adopted the UN Programme of Action to Prevent, Combat and Eradicate the Illicit Trade in Small Arms and Light Weapons in All Its Aspects in 2001. The PoA establishes a voluntary process for Member States to take action on small arms at the national, regional, and global levels. UN Member States have pledged to enact at the national level small arms control measures such as export licensing procedures, brokering controls, and stockpile security practices; promised to create regional networks to share information and promote arms trafficking controls; and at the international level pledged to work with the UN to enforce embargos, circulate data, and encourage international laws governing the arms trade.

Most notably, the first globally binding treaty on conventional arms is the Arms Trade Treaty, or the ATT. The ATT was adopted in 2013 and aims to regulate the cross-border trade in conventional arms, ranging from fighter aircraft and warships to small arms and light weapons. The treaty establishes common international standards for the trade in conventional arms that States must incorporate into their national control systems, and provides for oversight of the global arms trade by enhancing transparency and facilitating accountability, where States are responsible for ensuring their arms sales meet global standards and norms. For the first time in history, there is an outright ban on arms shipments that would be used to commit the horrors of genocide, war crimes, and attacks on civilians. The ATT prevents irresponsible arms trading by stigmatizing arms transfers to war criminals; requires exporters to seriously take human rights into account before selling arms to dealers and countries; mandates signatories to close down safe havens exploited by rogue arms dealers to transfer arms to war criminals with impunity; and requires States to be transparent in their arms trade decisions.

There are also two significant multilateral conventional arms control regimes. The first is the Wassenaar Arrangement on Export Controls for Conventional Arms and Dual-Use Goods and Technologies (WA), which was created in the early 1990s as a successor to COCOM. The Wassenaar Arrangement is the first and only multilateral organization that focuses on both conventional weapons and sensitive dual-use goods and technologies.[10] The organization seeks to prevent the accumulation of arms by promoting greater transparency and responsibility in the trade of conventional arms and dual-use goods and technologies. While participating States have final say over their arms transfers, they have pledged to create national legislation to not allow arms transfers that would create international instability or insecurity. Forty-one States make up the Wassenaar Arrangement, including most major global arms exporters with the notable exceptions of Belarus, China, Israel, and South Africa.[11] Wassenaar's objective is to prevent countries or regions from amassing large stockpiles of weaponry and sensitive materials, or what Wassenaar members describe as "destabilizing accumulations." There are no specific countries that Wassenaar members are prohibited from selling arms and dual-use equipment to. Yet members are expected to work together to stop weapons stockpiling by countries or regions judged to be of "serious concern" because of their behavior or activities.

The Wassenaar munitions list comprises conventional weapons, such as tanks and fighter aircraft, as well as military explosives, toxicological agents, biocatalysts, and other specialized military equipment. It covers exports that are almost exclusively designed for war fighting purposes. Dual-use goods and technologies appear on a separate list, which is divided into two tiers of "basic" and "sensitive" with a subset of "very sensitive." Software, computers, machine tools, and radar technologies are the types of items that appear on the dual-use goods and technologies list.

The second multilateral conventional arms control regime is the Missile Technology Control Regime (MTCR). It is presently the only multilateral effort dealing with missile nonproliferation, but has conventional application. It is not a treaty, but a suppliers' regime founded by the G-7 States in 1987. The MTCR is a politically binding, voluntary association of states sharing common interests in the nonproliferation of missiles, unmanned air vehicles, and related technologies. MTCR members undertake not to sell certain technologies to non-member states.

The MTCR focuses on ballistic and cruise missiles capable of delivering a 500-kilogram payload to a range of 300 kilometers, the parameters required for a missile to deliver a nuclear weapon. While nuclear-capable delivery systems were the initial target of the regime, the MTCR members extended the guidelines to cover delivery systems involving all weapons of mass destruction (WMD) in January 1993. This means that certain unmanned aerial vehicles, or drones, are covered by the MTCR.

The MTCR consists of a Guidelines List and an Equipment and Technology Annex. Missiles and their delivery systems fall into one of two categories, with differing levels of protection regarding sales. Category I items include ballistic and cruise missiles and space launch vehicles. Even amongst MTCR-member states themselves, the export of Category I items is subject to "strong presumption of denial." Category II items include missile-related components; these can be exported with the approval of the individual MTCR-member government who is acting as supplier. Sales are at the discretion of the exporting members, who have agreed to restrict transfers for "acceptable end uses," provided end-use guarantees can be secured from the recipient state, or, in some cases, under government-to-government assurances.

Currently, the MTCR has 34 member states, which meet for discussions at least once per year.[12] Partners have equal standing, and all MTCR decisions are taken by consensus. Some non-member states such as Israel, Romania and Slovakia have agreed to abide by MTCR Guidelines. China pledged in November 1991 to adhere to the MTCR's original Guidelines, but later refused to adhere to the updated Guidelines implemented in January 1993. Consequently, many countries, such as the United States, do not consider China an adherent to the regime. At the 13th plenary meeting in October 1998, the MTCR members repeated their invitation to China to join, which China declined.

ROLE OF STAKEHOLDERS

The regimes and instruments discussed in this chapter are for governments to implement. And, although States remain the primary actors involved in arms control activities, other stakeholders have become increasingly important in the control and regulation of conventional arms. Numerous nongovernmental organizations working individually and collectively have managed to inform the public and pressure governments to develop more responsible weapons practices, such as the International Campaign to Ban Landmines, the Cluster Munitions Coalition, and Control Arms. NGOs play a role in encouraging states to adopt more rigorous arms transfer controls, raising awareness about irresponsible arms trade activities, and facilitating the development of international standards and agreements that limit state arms trade behavior. Industry has also become a key player in the international effort to regulate conventional arms. They play varying roles depending on the weapon systems relevant to the international agreement.

CHALLENGES TO CONVENTIONAL ARMS CONTROLS

Despite the many international, national, and NGO activities that have been undertaken regarding conventional arms control in recent years, numerous challenges and obstacles remain in enhancing arms control measures. The first challenge is the legitimacy of the weapons. Conventional weapons, unlike weapons of mass destruction, serve many purposes. They provide for the national defense and support policing activities – and small arms are regularly and legitimately used for sport and hunting activities. It is undesirable and unlikely to ban the majority of conventional weapons. The challenge, then, is to limit and constrain the trade in conventional weaponry to prevent destabilizing build-ups and misuse.

Because conventional arms serve legitimate purposes, a second challenge arises due to the role of arms producers and exporters. Many countries produce weapons, but the role of the major weapons producers and exporters cannot be understated. The United States, Russia, China, Britain, and France hold the lion's share of the global arms market – with the United States leading the way. Hesitance on their part to enhance conventional arms trade controls has a significant impact on the will and capability of the entire international community to engage in stricter arms trade practices.

A third challenge to the development of stronger arms trade controls goes beyond great power and major supplier concerns. Existing ideas and norms of state sovereignty, national self-defense, self-determination, and territorial integrity have an impact on political will more generally in the international community. Many recent international agreements on weapons control have, in fact, reiterated the norms of sovereignty, national self-defense, self-determination, and territorial integrity, which are grounded in the UN Charter itself.[13] Moreover, Article 223 of the Treaty of Rome, which established the basis for the European Community and today's European Union, specifies that national governments have exclusive control over national arms industries, arms sales, and arms control decisions, providing additional hindrance to state interests in multilateral arms trade controls in Europe.[14]

Ultimately, the fourth and perhaps the primary challenge facing arms control today is the sheer complexity of the issue. The conventional arms trade is comprised of weapons both big and small and characterized by a multitude of issues such as supply and demand. Many factors affect demand and sources of supply to provide the arms that states, non-state actors, and individuals seek. It has become increasingly difficult to control the flow of weapons in a growing global economy, and even more difficult to reach all relevant arms trade actors and limit or constrain their behavior. Today, cross-border mergers and acquisitions spread weapons technology and governments become less able to control arms-producing corporations. Weapons are increasingly assembled in more than one country, making it hard to control the end product. Moreover, developing countries have improved their own weapons industries, lessening their reliance on traditional trading partners if avenues to arms are closed. In addition, a growing illicit trade significantly hampers the development and implementation of arms control measures. Furthermore, the economic benefits of arms sales are increasingly considered more important than the possible political consequences.

Despite the many challenges, there are also many prospects for the future of international arms trade controls. The past several years have indicated that increased attention and activity has focused on better regulating the international arms trade. Ideally, a more cooperative environment will emerge and states, NGOs, and relevant arms trade actors will work together to find common ground and agree to certain limitations on the arms trade that will meet everyone's needs in forums that are not necessarily based on consensus rule. The international community may also continue to use economic tools such as development assistance and trade, investment, and technology transfers as enticements when negotiating arms controls in the developing world, and individual governments can also use persuasion, through political dialogue, to help enact arms controls in a bilateral agreement.[15] Moreover, other international, regional, and national measures regarding the protection of human rights, reform of security sectors, enhancement of the rule of law, and good governance will also have an impact on the international arms trade as safer and more secure environments emerge.[16]

ACKNOWLEDGMENTS

This chapter draws heavily, and in some places takes entire sections from, my book, *The International Arms Trade* (Polity Press, 2009), co-authored with Suzette Grillot. I thank her for her continued partnership and collaboration. I would also like to thank my Stimson colleagues Shannon Dick and Eli Enos, for their help in updating my work for this endeavor.

REFERENCES

1. C. Theohary, "Conventional Arms Transfers to Developing Nations, 2008-2015," Washington, DC: Congressional Research Service, 2016, pp. 2.
2. Graduate Institute of International Studies (Geneva), *Small Arms Survey 2007: Guns in the City*, London: Oxford University Press, 2007, p. 39.
3. Graduate Institute of International Studies (Geneva), *Small Arms Survey 2003: Development Denied*, London: Oxford University Press, 2003, p. 13.
4. J. Stanley and M. Pearton, *The International Trade in Arms*, New York: Praeger, 1972, p. 14. For more on COCOM, see: M. Mastanduno, *Economic Containment: CoCom and the Politics of East-West Trade*, Ithaca: Cornell University Press, 1992; and R. T. Cupitt and S. R. Grillot, "COCOM Is Dead, Long Live COCOM: Persistence and Change in Multilateral Security Regimes," British Journal of Political Science 27, July 1997, pp. 361-389.
5. M. Lipson, "The Reincarnation of CoCom: Explaining Post-Cold War Export Controls," The Nonproliferation Review/Winter 1999, p. 1.
6. See P. Fedynsky, "Russia's Parliament Suspends Arms Control Treaty Compliance," Global Security.org, November 7, 2007, available at http://www.globalsecurity.org/wmd/library/news/russia/2007/russia-071107-voa01.htm.
7. J. A. Lewis, "Looking Back: Multilateral Arms Transfer Restraint: The Limits of Cooperation," Arms Control Today, November 2005, available at http://www.armscontrol.org/act/2005_11/NOV-LOOKINGBACK.asp.
8. United Nations Report of the Disarmament Commission, A/51/42 (1996).
9. For more on the Firearms Protocol see http://www.unodc.org/newsletter/en/200504/page005.html. For a list of states that are or are not a party to the Firearms Protocol, see IANSA "UN Firearms Protocol – Has Your Country Ratified?" available at http://www.iansa.org/un/firearms-protocol.htm.
10. R. Smith and B. Udis, "New Challenges to Arms Export Control," in Paul Levine and Ron Smith, *Arms Trade, Security and Conflict*, London and New York: Routledge, 2003, pp. 94-110.
11. Wassenaar's 41 members are: Argentina, Australia, Austria, Belgium, Bulgaria, Canada, Croatia, Czech Republic, Denmark, Estonia, Finland, France, Germany, Greece, Hungary, Ireland, Italy, Japan, Latvia, Lithuania, Luxembourg, Malta, Mexico, Netherlands, New Zealand, Norway, Poland, Portugal, Republic of Korea, Romania, Russian Federation, Slovakia, Slovenia, South Africa, Spain, Sweden, Switzerland, Turkey, Ukraine, United Kingdom, United States.
12. The 33 current members of the MTCR are: Argentina, Australia, Austria, Belgium, Brazil, Bulgaria, Canada, Czech Republic, Denmark, Finland, France, Germany, Greece, Hungary, Iceland, Ireland, Italy, Japan, Luxembourg, Netherlands, New Zealand, Norway, Poland, Portugal, Republic of Korea, Russian Federation, South Africa, Spain, Sweden, Switzerland, Turkey, Ukraine, United Kingdom and United States.
13. See the Charter of the United Nations at http://www.un.org/aboutun/charter/. Also see the UN Program of Action at http://disarmament.un.org/cab/poa.html for a reiteration of these norms.
14. A. Pierre, Cascade of Arms: Managing Conventional Weapons Proliferation, Washington, DC: Brookings Institution Press, 1997, pp. 166.
15. Ibid., pp. 344-347, 350
16. I. Davis, "Implementing and Deepening the OAS Agenda on Small Arms and Light Weapons," May 2002, http://www.basicint.org/WT/smallarms/OAS-IDpres-0502.htm.

Space Weapon Technology and Policy

Theresa Hitchens

School of Public Policy
University of Maryland
College Park, MD 20742 USA[1]

Abstract. The military use of space, including in support of nuclear weapons infrastructure, has greatly increased over the past 30 years. In the current era, rising geopolitical tensions between the United States and Russia and China have led to assumptions in all three major space powers that warfighting in space now is inevitable, and possible because of rapid technological advancements. New capabilities for disrupting and destroying satellites include radio-frequency jamming, the use of lasers, maneuverable space objects and more capable direct-ascent anti-satellite weapons. This situation, however, threatens international security and stability among nuclear powers. There is a continuing and necessary role for diplomacy, especially the establishment of normative rules of behavior, to reduce risks of misperceptions and crisis escalation, including up to the use of nuclear weapons. U.S. policy and strategy should seek a balance between traditional military approaches to protecting its space assets and diplomatic tools to create a more secure space environment.

I. INTRODUCTION

Outer space is recognized by all nations as "the province of mankind" not subject to national boundaries or appropriation via both treaty – especially the 1967 Outer Space Treaty[1] – and by the practice of nation states. Since the dawn of the space age, the use of satellites has become integral to the global economy, including providing communications, weather services, mapping, precision timing and navigation services for shipping, secure cross-border banking, and Internet connectivity. Every state has both an interest in making use of space, and reason to deal with its use by other states, because the activities in space by one actor have the potential to impact all others, for good or for bad. In addressing international and national security, and nuclear security in particular, the space environment has played a role of great importance from almost the beginning of the nuclear age. The first satellites launched by the Soviet Union and the United States were oriented toward seeking information on what was transpiring in areas controlled by the other, and to verify bilateral arms control agreements. While in short order space systems also were integrated to the offensive uses of long-range delivery systems by providing photographic information about potential targets, strategic space systems were during the Cold War widely viewed as stabilizing the Superpower nuclear competition.

The use of space for military purposes has continued into the present era, with increasing capabilities to take advantage of large segments of the electromagnetic spectrum for acquiring intelligence, communicating globally, and generally supporting ways of using nuclear weapons both for deterrence, and, should deterrence fail, use of those weapons against an adversary. Most of the nuclear weapon possessing states operate satellites for these purposes. Perhaps as importantly, space systems over the last two decades have become integral to the tactical warfighting ability of many modern states – a situation that has complicated the status of space systems as strategically stabilizing. Indeed, the growing use of space by many countries to achieve victory on the battlefield has increased both the vulnerability of militaries to attacks on their space systems and has, at the same time, increased their value as potential targets in a war.

Over the past 50 years, the Soviet Union, the United States, and China have carried out experiments in or aimed at the outer space environment – mostly the area close to the atmosphere in Low Earth Orbit (LEO) – that show the capability to destroy a satellite, or to disrupt its functions. The specter of space warfare for many years has, among other negative consequences, raised concerns that a state's nuclear retaliatory capability could be compromised. This concern also applies more generally, of course, to an ability to disrupt communications functions for other military, or civilian, purposes.

In the 1980s, there was a period when the United States, and perhaps others, explored whether systems based in space could be used to destroy an adversary's intercontinental ballistic missiles, or their payloads. The so-called Star Wars program under the Reagan Administration envisioned the deployment of a system of satellites that would seek

Nuclear Weapons and Related Security Issues
AIP Conf. Proc. 1898, 030006-1–030006-5; https://doi.org/10.1063/1.5009221
Published by AIP Publishing. 978-0-7354-1586-7/$30.00

to destroy the missiles/warheads launched at the United States. One technology explored envisioned detonating a nuclear explosive to generate a beam of x-rays that would put out of commission the adversary's warhead. Thus far, such technologies have not succeeded in playing a role in the nuclear-weapon situation globally. However, the U.S. descendant of the Star Wars program – currently limited to conventionally equipped, ground- and sea-based missile defense interceptors with limited capability against a full-blown nuclear attack – continues to stress nuclear deterrence and stability between the United States and Russia, as well as China, which maintains a much smaller nuclear arsenal than the Cold War adversaries. However, recent missile experiments by China have demonstrated the vulnerability of the geosynchronous equatorial orbit (GEO), where many hundreds of satellites are "parked" carrying out communications and other functions, including nuclear weapons support systems and spy satellites.

II. INCREASED THREATS INVOLVING OUTER SPACE

Since the first satellites were launched in the 1950s by the Soviet Union and then the United States, the Russian Federation, the United States, China, India, Japan, and other states have, without much coordination, launched so many satellites into space into various orbits and at various altitudes that there is currently a strong risk of both congestion and competition. There is no global regime for regulating outer space activities. The Outer Space Treaty of 1967, to which all the launching states, and most others, are party[2] mandates that outer space be used solely for peaceful purposes, and prohibits the stationing of nuclear or other weapons of mass destruction in that environment. (The Treaty does *not* prohibit the transit of nuclear weapons, e.g. as a payload on a submarine-launched ballistic missile, *through* outer space; furthermore under common law practice, defensive military activities are tolerated as compliant with "peaceful purposes.") The Outer Space Treaty, however, makes it clear that states are responsible for their own space activities, and compliance with international law. And while there are a number of other space-related treaties, UN principles and voluntary agreements managed by various UN and multilateral bodies, a nation's activities in space are largely regulated by that nation alone. There is no international legal requirement for any one state to coordinate its satellite launches or maneuvers with others.

Environmental Threats: Crowding and Debris

Some 1,500 operational satellites are now in orbit, owned by more than 80 states or other entities. These states and entities have varying levels both of proficiency and of knowledge of the established laws and rules affecting space. In the radio frequency band of the electromagnetic spectrum, interference is rising, especially in the GEO regime. Some of this interference is deliberate, undertaken for political purposes, despite the fact that deliberate interference is one of the few legally binding restraints in the international space arena[3]. The evolution in satellite technology has led to the wider use of smaller satellites, including so-called "Cubesats," that can be deployed in constellations, especially in LEO. The number of operational satellites is expected to rise to many thousands within the decade. LEO, in particular, is becoming incredibly crowded with satellites, making tracking of on-orbit objects extremely difficult. Furthermore, many small satellites have no ability to maneuver to avoid collisions with other satellites and space debris.

The half-century of using space has resulted, from the breakup of satellites and other activities, in a considerable amount of on-orbit debris – including satellites no longer in use, parts of satellites that have broken up, launcher stages, nuts and bolts, and debris from the deliberate destruction of satellites. The United States and others track some 23,000 orbiting pieces with a diameter of greater than 10 cm. This debris is especially dangerous if a satellite or transiting vehicle collides with a piece, since the closing velocity of such a collision on-orbit is very high – some 7.5 kilometers per second (faster than a bullet) in LEO. Worse yet, even very small debris, most of which cannot be detected much less tracked, can destroy an operational satellite; it is estimated that some 500,000 to one million pieces of debris smaller than 10 centimeters exist on orbit.

It is widely agreed that new international measures to better coordinate space activities are required to ensure that the space environment is sustained. In 2007, the United Nations Committee for the Peaceful Uses of Outer Space (COPUOS) in Vienna, Austria, agreed on a set of guidelines for the mitigation of space debris, which are slowly being implemented by many space-faring states. It may be that such measures will eventually require removal of debris from orbit, as the decay of debris from space into the atmosphere where it burns up (or falls on Earth) is a very long-term prospect, taking as much as 25 years in LEO. Sadly, the lifetime of debris in GEO, like diamonds, is practically forever. COPUOS currently is working on a set of recommended best practices to ensure the "long-term sustainability of space." COPUOS has a 2018 deadline to finish this work; however, there is already discussion of follow-on effort that may include international guidelines for debris removal.

Increasing Military Tensions in Space

In the geopolitical sphere, compared with the period following the breakup of the Soviet Union, the current decade is witnessing increased tensions between the United States and Russia, and between the United States and China. The geopolitical situation in space has been further eroded by the proliferation of experimentation with and/or deployment of dual-use technologies with "counterspace," i.e. satellite attack, capabilities. As noted above, China, Russia and the United States all have tested (or in some cases deployed) such technologies in both LEO and GEO. The United States continues to have an advantage in military space capabilities, but its edge is eroding as China and Russia dedicate more resources.

Most technologies involved in sustaining systems in orbit are dual-use, but certain specific activities are raising suspicions about potential intended weapons use. The capability to maneuver satellites is particularly relevant. Russia placed a satellite called Luch/Olymp in GEO that maneuvered or drifted over a considerable range, and at several points in 2015 came extremely close to commercial satellites owned by Intelsat.[4] Intelsat called the move "irresponsible," but their request for information from Russia went unanswered. The maneuvers further prompted concern at the U.S. Defense Department about the satellite's mission, which has not been revealed by Moscow. The United States also has carried out programs in GEO that could have potential weapons capabilities. For example, the PAN, an acronym for Palladium at Night, is a classified program apparently dealing with communications platforms, and perhaps providing other capabilities.[5] The Geosynchronous Space Situational Awareness Program (GSSAP) is a U.S. military satellite constellation that also maneuvers in orbit, designed, according to the Pentagon, with the objective of inspecting other satellites orbiting in GEO. Such activities are known as Rendezvous and Proximity Operations (RPO), and have a number of benign applications such as satellite refueling, inspection and repair. Russia is carrying out other such experiments in LEO, as are China, the United States, Japan and Sweden. The commercial applications of maneuvering satellites are also increasing.

Among the number of more directly identifiable counterspace technologies now available, the most widespread are ground-based radio-frequency jammers, which can be used to disrupt satellite communications and operations. In addition, there are efforts to develop lasers for disrupting or degrading systems based in space. Russia, China and the United States have also carried out projects involving terrestrially based missiles carrying anti-satellite payloads. The United States as early as the 1980s launched missiles from an F-15 fighter jet with this objective. A 2007 Chinese test, involving the destruction of a non-functional Chinese weather satellite in LEO, released a considerable quantity of debris. The United States subsequently launched a missile from an Aegis cruiser that was advertised to have the objective of destroying a satellite in a decaying orbit, but this did not prevent speculation that the mission also had the objective of demonstrating a similar capability to that of China. Over decades, the U.S. missile defense program has also heavily relied on the space environment, for early warning, for communications, and as a place for engaging and destroying hostile systems. Noted above is the Reagan Administration's "Star Wars" program, pursued with the idea of creating a "shield" against intercontinental ballistic missiles.

The harder-line rhetoric that has been employed in recent years also has had an inevitable impact of raising tensions. The United States has pivoted from an approach of "strategic restraint" to one emphasizing "warfighting."[6] In particular, the budgets for providing resiliency in space systems and counterspace capabilities have been increasing. At the same time, Russian accusations that U.S. activities have a hostile objective, and its responses to U.S. representations, have become shriller. Russia has called the anti-ballistic missile system SM-3 2A an anti-satellite weapon, while touting its own objectives for acquiring anti-satellite capabilities. In 2013, China tested a missile, the Dong Ning-2, which appears capable of reaching satellites in GEO. Chinese military space activities lack transparency, but it seems clear that such activities include the objective of being able to exercise counterspace actions. Most troubling, there has been a lack of serious dialogue among these Big Three states.

Multilateral Efforts to Reduce Risks

For many years, a direct approach to concerns about the potential for weaponizing space (space has been militarized since the dawn of the space age, but so far cannot be said to have been weaponized) has been debated within the United Nations, as well as at the Conference on Disarmament in Geneva. The Russian-Chinese co-sponsored initiative, on the Prevention of an Arms Race in Outer Space, has been on the agenda of the Conference on Disarmament since 1985, and under that agenda item Moscow and Beijing have proposed a treaty to ban weapons in space.[7] However, the Conference has been all but immobilized by wider disagreements since that time; and the United States remains firmly opposed to the proposed treaty.

There have been a number of efforts to set norms of behavior in space in order to guard against misunderstanding and conflict in space. Most recently, the 2013 UN Group of Governmental Experts (GGE) on Transparency and Confidence-Building Measures in Outer Space Activities released a set of recommended initiatives for states to implement, including improved communications about objects in orbit.[8] Unfortunately, little work has been done since to implement the recommendations, either at the multilateral level or by individual states. However, the United States, Russia and China have recommended that the UN Disarmament Commission, based in New York, and the deliberative body on arms control issues, take up the question of implementation of the GGE recommendations. While the initial proposal has been received favorably, a decision regarding whether to put the issue on the Commission's formal agenda will not be made until Fall.

III. POLICY QUESTIONS FOR THE UNITED STATES

In view of the increased uncertainties affecting the use of outer space, particularly in the area of international security, the United States needs to address several issues with some urgency.

First, what is the appropriate mix of resiliency measures to apply in the coming years? A subsidiary question in this regard is what is an appropriate role for commercial providers? And should the U.S. military switch to constellations of small satellites for some national security missions? The budgetary implications of achieving objectives, and establishing appropriate requirements, are important components of pursuing this mix. And there is the inevitable bureaucratic overlap between the Department of Defense and the Intelligence Community. Such "turf" issues require constant attention lest they adversely impact on the fulfillment of national, *vice* institutional, objectives. Lengthy acquisition programs put systems at risk of becoming obsolescent earlier than they would otherwise become outdated.

As part of this latter issue, the United States will need to consider what reforms are needed in the acquisition process, and related organizational arrangements. The integration of Department of Defense and Intelligence Community programs and activities is inevitably a delicate matter; it will require especial focus from the White House, in particular as resiliency is now being embedded into the requirements for acquisition of new systems.

A more far reaching issue is how best to strike a balance between the defensive aspects of counterspace and the offensive aspects. And integral to addressing this balance is the impact of U.S. options to respond to hostile space activities on the stability of the strategic/nuclear relationships: U.S.-Russia, U.S.-China, and a large number of other such relationships involving the nuclear-weapon-possessing states. If "arms racing" resumes, or, in the case of India and Pakistan, continues, how will the use of space, specifically for counterspace activities, impact on these races, and vice-versa? Will there be a deterioration in nuclear deterrence? Will an offensive strategy involving the targeting of an adversary's nuclear-related satellites emerge? These are questions that beg answers in the near-term, as budgetary and policy decisions are being made.

It is also important to consider the role of diplomacy in dealing with international security for outer space. Diplomacy, in the form of both self-restraint and in reassurance of potential adversaries regarding intentions, has been a part of the tool kit for managing competition in space from the beginning of the space age. Can effective "rules of the road" be further developed? The limited success, but slow pace, of multilateral efforts should not be seen as failure, however. Diplomacy is a difficult business, often characterized by a "one step forward, one step back" dynamic. There is some optimism to be found in the ongoing COPUOS effort, which while a slightly sideways approach, will have positive impacts on international security if successful. While the Disarmament Commission has little power, the advent of discussions there would provide a much needed multilateral forum for addressing the security issues for space given the decades-long impasse at the Conference on Disarmament. Finally, one should not overlook the value of bilateral diplomacy, particularly among the Big Three space powers.

Further work will also be needed to regulate the proliferation of technologies in the commercial sector. This will likely involve export control, and measures for the management of "traffic" in space (STM). However, care must be given to weigh national security concerns against the needs of commercial industry to thrive in the international marketplace. There is a tendency in the national security community to try to "close the barn door after the horses have escaped" that must not be indulged in the space domain, given the reliance of the national security sector on commercial capabilities and technological innovation.

IV. THE NEED FOR A "TIME OUT"

To date, no state is deploying dedicated anti-satellite weapons. Testing of capabilities does not a program make. That said, the trend lines are currently negative and require both time and analysis to mitigate. It would be irresponsible for the United States, or any other country, to leap to conclusions about the "inevitability" of all-out war in space. A balanced strategy, which combines resiliency, deterrence, and diplomacy will be required to protect national security and ensure international security. While development of some anti-satellite capabilities for potential future use may be wise, a run-away space arms race is not desirable for any party. It may be that a viable modus vivendi could be a situation of "implied deterrence:" i.e., the development of dual-use technologies with inherent weapons capabilities in a transparent manner so as to provide the knowledge to others that, if pushed, anti-satellite weapons could be deployed.

And despite the difficulties to date, the prospect of the multilateral establishment of norms shows some possibility of promise. This involves the implementation of recommendations by the Group of Governmental Experts discussed above; of the COPUOS LTS (long-term sustainability) best practices work making progress by 2018; the successful efforts to codify the legal regime that are underway (e.g., those at McGill University in Montreal), and perhaps the UN Disarmament Commission addressing TCBMs in 2018. These efforts must be given a chance to ripen, however much frustration is involved in the processes.

It can perhaps be helpful to think of the world as being surrounded on all sides by a large fishbowl, of indefinite dimensions in the outward direction, with the atmosphere at the intersection between "outer" space and the land and waters below. Looked at in this way, human activities in outer space have little room to be confined to a single state: the world as a whole is impacted by those activities. Accordingly, when dealing with outer space, traditional concepts of absolute roles for state sovereignty must inevitably be modified to serve the objectives of global peace, security and stability. Whether this reality will at some point lead to an appreciation that reliance on force, nuclear weapons in particular, cannot play the role in space that it does on the Earth, remains to be seen.

REFERENCES

[1] "Treaty on the Principles Governing the Activities of States in the Exploration and Use of Outer Space, including the Moon and other Celestial bodies" (Outer Space Treaty) entered into force in 1967. Text of the treaty is found here: http://www.unoosa.org/oosa/en/ourwork/spacelaw/treaties/introouterspacetreaty.html

[2] As of July 2017, there are 107 parties to the Outer Space Treaty, with 23 signatories awaiting ratification.

[3] Radio-frequency interference is governed by the International Telecommunication Union, which is a treaty-based body and to which almost every space-faring state is a party.

[4] Mike Gruss, "Russian Satellite Relocates – Next To Another Intelsat Craft," Space News, Oct. 16, 2015, http://spacenews.com/russian-luch-satellite-relocates-next-to-another-intelsat-craft/

[5] Justin Ray, "Clues Emerge Over Mystery Satellite Soon After Launch," Spaceflight Now, Sept. 9, 2009, https://www.space.com/7258-clues-emerge-mystery-satellite-launch.html

[6] See: Theresa Hitchens and Joan Johnson-Freese, "Toward a New National Security Space Strategy: Time for a Strategic Rebalancing," Atlantic Council Strategy Paper No. 5, 2016, http://www.atlanticcouncil.org/publications/reports/toward-a-new-national-security-space-strategy-time-for-a-strategic-rebalancing

[7] "Treaty on the Prevention of the Placement of Weapons in Outer Space, the Threat or Use of Force against Outer Space Objects," introduced to the Conference on Disarmament by the Russian Federation and the People's Republic of China, Feb. 29, 2008; updated June 12, 2014, https://documents-dds-ny.un.org/doc/UNDOC/GEN/G14/050/66/PDF/G1405066.pdf?OpenElement

[8] For text of the GGE Report, see: http://undocs.org/A/68/189

Ballistic Missile Defense Effectiveness

George N. Lewis

Judith Reppy Institute for Peace and Conflict Studies
Cornell University
Ithaca, New York

Abstract. The potential effectiveness of ballistic missile defenses today remains a subject of debate. After a brief discussion of terminal and boost phase defenses, this chapter will focus on long-range midcourse defenses. The problems posed by potential countermeasures to such midcourse defenses are discussed as are the sensor capabilities a defense might have available to attempt to discriminate the actual missile warhead in a countermeasures environment. The role of flight testing in assessing ballistic missile defense effectiveness is discussed. Arguments made about effectiveness by missile defense supporters and critics are summarized.

INTRODUCTION

While the debate on whether or not the United States should deploy ballistic missile defenses, in particular, a national missile defense (NMD) system, is now largely settled, the important question of how effective such systems are likely to be in actual use remains open. The answer to this question is not simple and will depend on many factors, such as the type of defense, the nature of the attacking missiles, the circumstances of the attacks, and the standards by which the success or failure of the defense is judged. Ballistic missile defense (BMD) supporters argue that defenses have already proven themselves effective in tests and in (limited) actual use. Critics argue that many of these tests are highly unrealistic and that the actual use of BMD systems has involved only the shortest-range systems and has produced mixed results.

This chapter will first briefly discuss the effectiveness of short-range within-the-atmosphere BMD systems. Its main focus, however, will be on the long-range above-the-atmosphere systems that are the main subject of debate today. Particular emphasis will be placed on the problems of midcourse countermeasures and discrimination.

TERMINAL AND BOOST-PHASE SYSTEMS

A ballistic missile's trajectory is typically divided into three phases. First, the boost phase, when the missile is under powered flight using its rocket booster. Second, the midcourse phase, in which the missile or its warhead coasts on a ballistic trajectory through the vacuum of outer space (assuming the missile's range is long enough that it leaves the atmosphere). Third, the terminal phase, in which the missile or warhead re-enters the atmosphere and falls towards its target. Each phase presents different opportunities and problems for ballistic missile defenses. Although the main focus of this chapter is on midcourse defenses, it is useful as background to first briefly discuss terminal and boost-phase defenses.

Nuclear Weapons and Related Security Issues
AIP Conf. Proc. 1898, 030007-1–030007-15; https://doi.org/10.1063/1.5009222
Published by AIP Publishing. 978-0-7354-1586-7/$30.00

Terminal-Phase Within-the-Atmosphere Systems

These defenses operate within the atmosphere (endo-atmospheric) in order to counter missiles as they descend towards their targets (terminal phase). A number of such BMD systems have already been deployed and several have been used in actual combat. These systems include the U.S. Army's Patriot system (with PAC-2 and PAC-3 interceptors), the U.S. Navy's Aegis system (with SM-2 Block IV and SM-6 anti-aircraft missiles), Israel's Iron Dome and David's Sling systems, and Israel's Arrow system (with Arrow-1 and Arrow-2 interceptors). All of these systems use radar for guidance and homing and most are equipped with high-explosive warheads (PAC-3 being the exception). Such defenses use aerodynamic forces for interceptor maneuvering and can also make use of atmospheric filtering to remove light-weight countermeasures. They typically cover small areas, with dimensions of tens of kilometers. In the U.S. context, defenses of this type are usually classified as theater or regional systems. They could be employed as stand-alone systems or could form a second layer of defenses operating behind midcourse defenses. They are sometimes also capable of intercepting aircraft.

The U.S. Army's Terminal High Altitude Area Defense (THAAD) is a special case. Although officially classified as a terminal phase system, its method of operation is more similar to that of a midcourse system. THAAD use an infrared-homing kill vehicle with rocket divert thrusters that maneuvers to destroy its target in a direct collision above the atmosphere. However, the THAAD kill vehicle is aerodynamically shaped and is thus also able to perform intercepts in the upper atmosphere, perhaps as low as about 40 km.

In the 1991 Gulf War, Patriot PAC-2 interceptors attempted to intercept about 44 600-km range Iraqi modified Scud missiles. Initially, the U.S. Army claimed a 96% success rate in destroying these Scuds, but after criticism from several sources, eventually reduced this claim to about 61%, where it remains today.[1] However, detailed outside technical analysis demonstrated that the actual successful intercept rate was very low, possibly zero.[2] In the 2003 Iraq War, the U.S. Army claimed that Patriot destroyed all nine short-range ballistic missiles it targeted (two using the new PAC-3 interceptor).[3]

Iron Dome was first deployed in Israel in 2011. It is a very short-range system, and is frequently described as a counter-rocket, artillery and mortar weapon. It has reportedly destroyed over 1,500 targets with an 85% success rate.[4]

The most straightforward method available to defeat endo-atmospheric interceptors is the use of vigorous atmospheric maneuvering, particularly if high missile accuracy is not needed. In 1991, the Iraqi Scuds broke apart before or on atmospheric re-entry, with the warhead section cork-screwing through the lower atmosphere – maneuvers that the Patriot PAC-2 interceptor could not cope with. The subsequent PAC-3 interceptor was designed to be much more maneuverable, using small rocket thrusters along the missile body to generate very rapid attitude changes and aerodynamic forces. It does not appear that the Iron Dome targets maneuvered significantly. How current terminal BMD systems would fare against targets similar to the 1991 Scuds is unclear, as there is no publicly available data (such as test results) that could be used to assess their likely performance.

Boost-Phase Defenses

Boost-phase defenses attempt to destroy missiles while they are still in powered flight. By destroying a missile's rocket booster before it has finished burning, its warhead will fall short of its intended target, although it is possible it could still explode. A boost-phase defense could use hit-to-kill interceptors, or because rocket boosters are much more fragile than missile warheads, beam weapons such as lasers.

Boost-phase defenses have some significant advantages relative to midcourse defenses. First, since destroying a missile during boost phase eliminates all of its warheads, boost-phase defenses are desirable against missiles equipped with multiple warheads. Boost-phase defenses are the only type of ballistic missile defense (leaving aside pre-emptive measures) that can counter conventional, chemical, or biological submunitions. Large numbers of submunitions released immediately after the end of boost phase will overwhelm any midcourse or terminal defense.

Second, the problem of countermeasures appears to be much less severe for boost-phase defenses than it is for midcourse defenses. While some types of countermeasures – such as various forms of trajectory modification (against homing interceptors) or reflective coatings (against lasers) – are feasible, their scope and effectiveness appear to be more limited than that of the wide range of potentially very effective midcourse countermeasures.

The greatest difficulty facing boost-phase defenses is that the boost phase of a ballistic missile is at most a few minutes long. This means that a boost-phase interceptor must have a very high speed and acceleration, and must also be based relatively close to an attacking missile's launch location.[5] Even so, they are likely to be practicable only against liquid-fueled long-range missiles with long burn times.[6] The need to be close to the launch site limits their potential utility to use against relatively small countries, such as North Korea or possibly Iran. While space-based interceptors or lasers have been proposed to get around this limitation, they would be prohibitively expensive.

These considerations have so far prevented the deployment of any boost-phase systems. The United States cancelled its two most recent boost-phase development programs, the Kinetic Energy Interceptor, which was too slow, and the Airborne Laser, which had inadequate range among other problems, in 2009 and 2011, respectively. The Missile Defense Agency is now developing several types of electrically-driven lasers with the objective of deploying one of them as a boost-phase weapon on an unmanned aerial vehicle in the mid-2020s.

LONG-RANGE EXO-ATMOSPHERIC MIDCOURSE DEFENSES

Long-range midcourse defenses are the primary focus of this chapter. They operate above the atmosphere (exo-atmospheric) and use direct collisions by infrared-homing hit-to-kill kill vehicles to destroy their target missiles' warheads as they travel through the vacuum of outer space. The term "midcourse" refers to the part of the target's trajectory after its booster rocket has burned out but before it begins to re-enter the atmosphere. Midcourse defenses are in principle capable of defending large geographic areas, with dimensions of hundreds to thousands of kilometers. However, their exo-atmospheric mode of operation makes them potentially vulnerable to defeat by a variety of light-weight above-the-atmosphere countermeasures. A primary objective of such defenses is countering nuclear-armed missiles, and in this role their effectiveness must meet very stringent criteria. During the Cold War, such defenses would have been considered strategic defenses, although they now have regional roles as well.

The United States currently has two midcourse missile defense systems, the Ground-Based Midcourse Defense (GMD) NMD system and the U.S. Navy's SM-3 Aegis BMD system. The U.S. Army's THAAD system also has some limited midcourse capabilities, but it will not be considered further here.

The Ground-Based Midcourse Defense System

The GMD system is a national missile defense system; that is, it is intended to defend U.S. territory from long-range ballistic missile attacks. In principle, it is capable of covering all fifty states from missiles launched from North Korea or Iran. The interceptor of the GMD system is the Ground Based Interceptor (GBI), an ICBM-sized missile deployed in silos at Fort Greely in central Alaska and at Vandenberg Air Force Base in California. The first GBI was deployed in July 2004 and the GMD system was declared operational later that year. Current plans a call for a total of 44 GBIs to be deployed by the end of 2017 (40 of them Alaska). The deployment of an additional interceptor site in the eastern United States is under consideration.

The 16.6 m long GBI's purpose is to deliver a roughly meter long, 55 kg Exo-atmospheric Kill Vehicle (EKV) to an intercept trajectory for an incoming missile warhead. After release from the GBI, the EKV detects the target using an infrared sensor, and maneuvers itself to intercept it using four small rocket divert thrusters. There are several generally similar versions of the EKV currently deployed; the most recent, the CE-II Block I EKV, had its first intercept test (a success) in May 2017. Due to its hasty development process, the EKV's reliability is poor, and beginning in 2020 the U.S. Missile Defense Agency (MDA) intends to begin replacing the EKVs with new Redesigned Kill Vehicles (RKVs).

The GBIs are supported by a network of large phased-array radars at Shemya in the Aleutians; in California; in central Alaska; on Cape Cod; at Thule, Greenland; and at Fylingdales in Britain. Although these are large and powerful radars capable of tracking many targets simultaneously at long ranges, they have little capability to discriminate actual warheads from decoys or other objects. Much smaller forward-based X-band TPY-2 radars in Japan and Turkey and Aegis radars on BMD-capable Navy destroyers and cruisers also support the GMD system (X-band extends from 8 to 12 GHZ, and is often taken to mean about 10 GHz.).

The GMD system's primary discrimination sensor is a large X-band radar on a modified ocean-going oil drilling platform. The radar, known as the Sea-Based X-band (SBX) radar, operates out of Honolulu and has sailed west on several occasions to observe anticipated North Korean missile flight tests. It has significant operational limitations, most notably a very limited electronic-scan field-of view (about ±12°, compared to about ±60° for a typical phased-array radar). By 2020 the MDA plans to deploy a new S-band Long Range Discrimination Radar (LRDR) in central Alaska (S-band extends from 2 to 4 GHz). The GMD system is also supported by a highly effective space-based infrared-detecting missile launch early warning system.

The SM-3 Aegis Ballistic Missile Defense System

The U.S. Navy's Aegis BMD system is based on the Aegis combat system on its cruisers and most of its destroyers. The Aegis system includes the Aegis computers and software, the four-faced SPY-1 phased-array radar, and a large number of vertical launching tubes (90-96 in the destroyers, 122 in the cruisers). In addition to BMD interceptors, the vertical launchers carry a mix of weapons including SM-2 anti-aircraft missiles, SM-6 long-range anti-aircraft missiles (which can also attack surface targets and short-range ballistic missiles), Tomahawk land-attack cruise missiles, and anti-submarine weapons.

As of mid-2017, the Navy had 22 cruisers and 63 destroyers equipped with the Aegis system. For a number of years the Navy has been upgrading Aegis ships to give them ballistic missile defense capabilities. As of mid-2017 about 35 ships had received some form of BMD upgrade; only a few, however, have what the Navy calls the "advanced" BMD capability, which allows them to perform ballistic missile defense and air defense simultaneously. At present, the Navy is upgrading ships to the advanced capability at a rate of about two per year. In addition, all new construction Aegis destroyers will be delivered (also at about two per year) with the advanced BMD capability built in.

The midcourse BMD interceptors currently deployed on Aegis BMD ships are the SM-3 Block IA and SM-3 Block IB missiles. These missiles are kinematically similar, with a burnout speed believed to be about 3 km/second. After their booster burns out and they leave the atmosphere, the interceptors release a small infrared-homing kill vehicle that attempts to destroy its target in a direct high-speed collision. Although these interceptors work in exactly the same manner as the GBIs of the national missile defense system, their much lower speed prevents them from being able to cover enough territory to be useful for a strategic defense of U.S. territory.

In 2018, however, the United States plans to begin deployment of the new SM-3 Block IIA interceptor. This interceptor will have a much higher burnout speed of about 4.5 km/s. This higher speed will enable it to cover much larger geographic areas, particularly when deployed near the target areas, so that the interceptor has more flight time available. In order to achieve this large area coverage, Aegis BMD will need to use its engage-on-remote capability, which allows its interceptors to be guided by external sensors (such as those of the GMD system).

CLAIMS ABOUT DEFENSE EFFECTIVENESS

According to many missile defense proponents, effective ballistic missile defenses are already an established fact. In March 2013, in discussing the GMD NMD system, Secretary of Defense Chuck Hagel stated that "…the American people should be assured that our interceptors are effective."[7] Three months earlier, his predecessor, Leon Panetta, in response to a question about a potential North Korean missile attack, stated that "I'm very confident that American defense capabilities are able, no problem, to block a rocket like this one."[8] Missile defense supporters also cite Israeli claims that their Iron Dome system is already achieving success rates of as high as 85-90% against short-range rockets as proof that missile defenses can work.

Critics argue that ballistic missile defenses, particularly those that operate above the atmosphere, are vulnerable to defeat by a wide range of simple countermeasures, and that this vulnerability has been known for decades yet remains an unsolved problem. They point out that missile defense test programs are highly artificial and scripted, and reveal little about how well the defenses would work in the real world. In their view, the Iron Dome experience against short-range rockets within the atmosphere has essentially no relevance to midcourse defenses against longer-range missiles.

WHAT DOES IT MEAN TO WORK?

What does it mean to say a missile defense system can or does "work"? How effective must a defense be in order to be meet its objectives? What fraction of attacking warheads should it be expected or required to destroy? There is no simple answer to these questions, since any answer is highly circumstance-dependent.

Israeli experiences with missile defenses illustrate the complexity of this problem. The performance of Iron Dome in 2012-2014 has been widely hailed as a success for ballistic missile defenses, and in particular for reducing the pressure on the Israeli military to respond to the rocket attacks in other, potentially much more escalatory, ways. However, a similar argument could be made that Patriot in 1991 was successful in preventing potentially coalition-fracturing Israeli strikes against the Iraqi Scud launchers. Yet while Patriot was widely perceived during the Gulf War as being highly effective, its actual effectiveness in destroying Scuds was essentially zero.[9] Moreover, few would argue that the claimed Iron Dome failure rate of about 10-15% would be adequate against an attack by more than a few nuclear-armed missiles.

Statements from U.S. officials suggest, however, that a system intended to counter nuclear-armed missiles could be considered effective if its predicted effectiveness is greater than about 90%. On June 16, 2009, just a week after Defense Secretary Robert M. Gates told Congress that the current GMD NMD system was "fully adequate to protect us against a North Korean threat," General James Cartwright, Vice Chairman of the Joint Chiefs of Staff, told a Senate committee that he assessed the effectiveness of the GMD system against a North Korean missile as "ninety percent, plus."[10] A year later, amid continuing statements by U.S. officials about their confidence in the effectiveness of the GMD system, MDA Director Lt. Gen. Patrick O'Reilly told the House Armed Services Committee that the probability that the system could counter a single ICBM launched by Iran "would be well over into the high nineties."[11]

RELIABILITY AND OPERATIONAL EFFECTIVENESS

The fundamental issue with midcourse missile defense effectiveness is not whether the defenses can work in principle or how well they might work on a test range. It is their operational effectiveness – that is, how well they can be expected to work on the battlefield, where unexpected circumstances and enemy measures to defeat the defense must be expected.

It is important to distinguish the operational effectiveness of a midcourse defense system from its reliability. Current midcourse defense intercept tests are essentially highly-scripted demonstrations designed to verify computer simulations of the defense's performance. As such, their success or failure primarily reflects the reliability of the systems being tested. Determining the operational effectiveness of a defense is a much more complex problem, which at a minimum requires establishing the ability of the defense to handle the full range of conditions, including plausible enemy countermeasures, it could face in actual use. The operational effectiveness of a midcourse missile defense system is difficult to assess for several reasons.

First, there is no real-world experience with the use of midcourse ballistic missile defenses. The only actual uses of ballistic missile defenses are of terminal, endo-atmospheric defenses (Patriot and Iron Dome) against short-range missiles and rockets. Not only are the outcomes of these experiences mixed, but they likely have little relevance to exo-atmospheric defenses against longer-range missiles, which are the primary focus of the current debate over ballistic missile defenses. It is possible that there will never be a significant body of real-world experience with midcourse defenses against long-range ballistic missiles, particularly in a nuclear environment

Second, there is no consensus on the nature of long-range ballistic missile threats, particularly regarding the potential use and effectiveness of steps to defeat defenses (countermeasures). For example, official claims of the effectiveness of the U.S. GMD national missile defense system assume that the missile threat it faces is the one that could exist in the near future – one that does not include any effective countermeasures. Missile defense critics argue that future missile threats must be expected to include simple yet highly effective countermeasures that could be difficult for existing or planned defenses to defeat. Missile defense supporters counter that effective countermeasures are not easy to employ and that future technical development of missile defenses will enable them to defeat countermeasures as they emerge.

Third, midcourse missile defense tests are complex and expensive. The most recent GMD test, FTG-15 in May 2017, cost $244 million.[12] The cost and complexity of the tests limits the number and frequency of tests, making it difficult to test over the full range of circumstances that the defense might expect to encounter. In this situation, a single test failure can cause severe setbacks for a missile defense development program.

Finally, the technical nature of ballistic missile defenses and the secrecy associated with some aspects of them inhibit informed public assessment and discussion of the actual and potential effectiveness of midcourse missile defenses. For example, in May 2002 the Missile Defense Agency announced that all details about future GMD test targets would be classified.[13]

THE PROBLEM OF MIDCOURSE COUNTERMEASURES

Enemy steps intended to defeat a missile defense, generally referred to as "countermeasures," have been the most fundamental problem facing midcourse ballistic missile defenses from their beginning, and remain so today. The prospects for defeating countermeasures, often narrowly described as "discrimination," is the most controversial technical issue involving midcourse defenses.

Several aspects of operating in the exo-atmosphere make midcourse defenses particularly vulnerable to countermeasures. In the vacuum of outer space, heavy and light objects released from a missile will travel on (nearly) identical trajectories. The lack of wind resistance allows objects with little structural strength to retain their shape. Taken together, these two factors enable a wide range of lightweight countermeasures. In addition, objects in space gain or lose energy only through radiation, allowing manipulation of their surface temperatures. For example, a sunlit spherical balloon in Earth orbit with an aluminum foil (shiny side) surface coating will have an equilibrium temperature of 454 K, while one with a white epoxy paint coating will have an equilibrium temperature of 237 K.[14] Any equilibrium temperature between these two can be achieved by covering varying proportion of the aluminum foil surface with dots of white paint. Non-sunlit spherical objects have an equilibrium temperature of about 180 K for all surface coatings.

Two recent major U.S. government-sponsored technical studies highlighted both the importance and the difficulty of discrimination. A 2011 report by the Department of Defense's Defense Science Board stated that "The importance of achieving reliable midcourse discrimination cannot be overemphasized."[15] The report went on to conclude: "Yet discrimination in the exo-atmosphere is still not a completely solved problem. Robust research and testing of discrimination techniques must remain a high priority."[16] A 2012 study by the National Academy of Sciences (NAS) concluded that "The hard fact is that no practical missile defense system can avoid the need for midcourse discrimination – that is, the requirement to identify the actual threat objects (warheads) amid the cloud of material accompanying them in the vacuum of space. This discrimination is not the only challenge for midcourse defense, but it is the most formidable one, and the midcourse discrimination problem must be addressed far more seriously if reasonable confidence is to be achieved."[17]

Countermeasures are not just a hypothetical possibility. All five of the first nuclear-weapon states have developed and, at least in some cases, deployed countermeasures on their nuclear-armed intercontinental ballistic missiles (ICBMs) or submarine-launched ballistic missiles (SLBMs).[18] Eighteen years ago, a U.S. national intelligence estimate concluded that "We assess that countries developing ballistic missiles would also develop various responses to US theater and national defenses. Russia and China each have developed numerous countermeasures and probably are willing to sell the requisite technologies."[19] More specifically, the estimate stated that "Many countries, such as North Korea, Iran, and Iraq probably would rely initially on readily available technology – including separating RVs, spin-stabilized RVs, RV reorientation, radar absorbing materials (RAM), booster fragmentation, low-power jammers, and simple (balloon) decoys – to develop penetration aids and countermeasures" and that "These countries could develop countermeasures based on these technologies by the time they flight test their missiles."

There are many possible exo-atmospheric countermeasures. Examples include replica decoys, electronic decoys, booster fragmentation, enveloping structures or screens, radar stealth, jammers, infrared stealth (low emissivity coatings or cooled shrouds), obscurants (radar chaff or aerosols), midcourse maneuvers, nuclear detonations to create electromagnetic pulse effects, and direct attacks on ground-based defense components (such as radars). Multiple countermeasures can be used in combination to enhance their effectiveness.

Two Countermeasure Approaches

As an illustration of the potential difficulties facing a system attempting discrimination, consider two potential approaches to using decoys: booster fragmentation and enveloping balloons. One might think that a decoy should be designed to have an appearance and dynamics as similar to the real warhead as possible. Such decoys are known as replica decoys, and it is known that the United States has developed and tested such decoys. For example, in 1971 it flight-tested the Inflatable Exoatmospheric Object, a replica decoy for the Mark-12 reentry vehicles on Minuteman ICBMs.[20] Replica decoys were also developed as targets for missile defense tests.[21]

With a replica decoy, however, there is a risk that the defender might be able to exploit a small discrepancy between the decoy and the real warhead to identify the warhead. In addition, to closely mimic all the characteristics of the real warhead might result in the decoy mass being higher than desired. A potentially more effective alternative approach would be to instead make the warhead look like a decoy – an approach known as anti-simulation.

One such anti-simulation approach is booster fragmentation. In this approach, the final stage of the warhead's rocket booster would be cut up, using explosive cutting cords or other methods, into numerous pieces comparable to or larger in size than the warhead. In 1975, the U.S. Army conducted a missile flight test, designated Signature of Fragmented Tanks, in which the second stage of a Titan II ICBM was fragmented, as intended, into 28 pieces.[22] This test was used to test the ability of the Safeguard Anti-Ballistic Missile radar in North Dakota to distinguish between the fragments and a warhead (it does not appear to be publicly known if the radar was able to do so). If the warhead itself was then disguised to look like debris, for example, by attaching several crumpled sheets of aluminum foil to it, the defense would then be confronted by a large number of possible targets, none of which look like a warhead.

Another anti-simulation approach is to enclose the warhead in a balloon with a thin metal coating and release it along with a large number of similar but empty balloons. The metal coating would prevent radars from observing what was inside the balloons and their temperatures could be manipulated using coatings over the metallic layer and by varying the shapes of the balloons (or by putting a small heater in some of the empty balloons). Making the shape and surface coating different for each balloon prevents the defense from singling out one balloon – the one with the warhead – as different from all the others. A 1999 report discusses this enveloping balloon countermeasure in detail.[23] The Chevaline countermeasure system developed by Britain in the 1970s for its Polaris submarine-launched ballistic missiles reportedly enclosed both of its two warheads in metallized balloons and released them along with balloons containing decoy warheads and empty balloons.[24]

Sensors for Midcourse Discrimination

Current midcourse defenses have two types of data available with which to attempt to perform discrimination: measurements from the kill vehicle's infrared sensors as it closes in on the threat cloud and measurements from ground-based radars.

For almost the entire homing process, the kill vehicle's infrared sensors will see a warhead-sized target (about 2 m long) as only a single pixel. For a 256 x 256 array with a one degree field of view, the field-of-view of a single pixel will reach 2 m only at a range of about 30 km, when at a typical closing speed of 10 km/s the kill vehicle is only 3 seconds from impact. The actual shape of the target that the kill vehicle is homing in on will not become apparent until the kill vehicle is roughly a second or less from impact. Such rough information on the shape of an object may or may not be helpful in discrimination, depending on the nature of the countermeasures used by the attacker.

At longer ranges, where each target appears only as a single pixel, the kill vehicle's two-color infrared sensor can in principle be used to measure the temperature and emissivity-area product of each object in the threat cloud. This assumes that each object has a well-defined temperature (for example, a sunlit balloon painted with broad stripes might not) and that there is no more than one object per pixel. Again, as discussed above, the attacker has some control over the temperature of the objects the missile deploys, and it is at best unclear how these measurements could be used to distinguish between different object comparable in size to the warhead.

The MDA is interested in deploying infrared sensors on high-altitude unpiloted aerial vehicles for midcourse tracking. Such UAVs could potentially be positioned so as to provide infrared data well before a sensor on a kill vehicle could. However, given the range from the UAV to the threat cloud will almost always be at least several hundred kilometers, each object in the threat cloud will always appear as no more than one or two pixels and closely-spaced objects could not be resolved. Thus while such UAVs may be able to provide useful tracking data, they cannot be expected to contribute significantly to discrimination.

Given the limitations of the GMD system's infrared sensors, against any but the least effective countermeasures the burden of discrimination will fall on its radars. A key measure of the discrimination capability of a radar is its range resolution, which is typically orders of magnitude better than its resolution in directions perpendicular to the range axis. The range resolution ΔR of a radar is approximately given by $\Delta R \approx c/2\beta$, where c is the speed of light and β is the radar's bandwidth.[25] For a phased array radar, its bandwidth is typically limited to about 10% of the radar's operating frequency.

The large phased-array early warning radars that were upgraded to provide the initial core radar capability of the GMD system operate at between 420 and 450 MHz. As originally built, they had maximum bandwidths that varied by radar between 1 and 10 MHz.[26] It is possible that as part of the upgrade process to incorporate them into the GMD system, their bandwidth might have been expanded to span their full operating band of 30 MHz. Even so, their range resolution would only be about 5 m, not small enough to provide much information beyond the target's radar cross section and how it varied with time, and these radars thus had little if any discrimination capabilities.

The limited discrimination capabilities of the upgraded early-warning radars is described in a 2004 Government Accountability Office assessment of the first two of them to be integrated into the GMD system: "Neither the Cobra Dane radar nor the upgraded early warning radar at Beale is capable of performing rigorous discrimination, a function acheivable only by the X-band radar. Rather, both radars will utilize common 'target classification' software that enables them to classify objects as threatening or non-threatening. For example, debris would be classified as non-threatening, but objects like deployment buses and decoy replicas would be classified as threatening."[27]

As discussed above, the primary radar discrimination sensor of the current GMD system is the SBX, which operates in the X-band (8–12 gigahertz) of radar frequencies. The SBX is often assumed to have a center operating frequency of about 9.5 GHz, or a center wavelength of about 3.16 centimeters (1.2 inches). It has a bandwidth of about 1.0 GHz, and a range resolution of about 25 cm or somewhat better.[28] The much smaller TPY-2 radars, which operate as forward-based radars for the GMD as well as the radars for THAAD batteries, have similar frequency and bandwidth characteristics. These X-band radars can thus measure features along a target in the range direction with a resolution of about 25 cm, which, depending on the circumstance, may or may not be useful for discrimination.

In addition, if a target has rotational motion relative to the radar, radars such as the SBX and TPY-2 can use the Doppler shift produced by this rotation to obtain resolution in a direction perpendicular to the range axis. This resolution ΔR_A is given by $\Delta R_A = \lambda/2\theta_R$, where λ is the radar wavelength and θ_R is the angle rotated through during the observation period.[29] Thus for a radar operating at 9.5 GHz with a range resolution of 25 cm, a target rotation of 0.063 rad = 3.6° perpendicular to the radar line of sight is needed to get a cross-range resolution equal to the range resolution.

The NAS report concludes that by using multiple high-resolution radars, long observation times, and techniques such as range-Doppler imaging, "...an adequate level of discrimination performance can be achieved in the near term and that this approach has a reasonable chance of keeping the United States generally ahead in the contest between countermeasures and counter-countermeasures."[30] However, the detailed analysis supporting this claim is classified, as are the type and sophistication of the countermeasures considered.

TESTING

The most direct and visible indicator of the effectiveness of a BMD system is its performance on the test range. If a BMD system performs poorly on the test range, then it is unlikely to work well in actual use. On the other hand, a system that performs well on the test range may or may not work well in combat. The key issue here is how realistic the testing is. An example is the Patriot missile in the 1991 Gulf War. Prior to the war, Patriot was a perfect 17 for 17 in tests, but in the war it performed poorly, destroying few if any of the 44 Scuds it attempted to intercept. The Scuds the Patriot system encountered in the war were faster than the ones it was tested against and, unlike the test targets that flew on stable trajectories, they maneuvered vigorously as they fell.

How can the effectiveness of a midcourse BMD system be established? One way to look at this problem is to break it down into a series of increasingly difficult criteria the BMD system must demonstrate it can meet in order to establish that it will work in actual combat conditions. In actual practice, these steps are likely to overlap.

For example, for a midcourse system, the first criterion would be to demonstrate the feasibility of the basic principles of midcourse exo-atmospheric infrared-homing hit-to-kill interception. Although many intercept tests have failed, the United States has already completed about 40 successful midcourse intercept tests (as of July 1, 2017), so this criterion can be regarded as having been met. The second step would be for a specific system to demonstrate that it can perform reliably under simple and controlled conditions. As Tables 1 and 2 below show, the GMD system clearly has not yet met this criterion, while the Aegis BMD system may have, at least for the Block IA and Block IB interceptors.

TABLE 1. Intercept test record for the U.S. Ground Based Midcourse system, organized by the type of exo-atmospheric kill vehicle used in the test, as of July 1, 2017.

Kill Vehicle Type	Number of Tests	Successful Intercepts	Success Rate
Prototypes	10	5	50%
CE-I	4	2-3	25-50%
CE-II	3	1	33%
CE-II Block I	1	1	100%
Total	18	9-10	50-56%

TABLE 2. Intercept test record for the U.S. Navy's Aegis BMD system's SM-3 interceptor, organized by the version of the interceptor used in the test, as of July 1, 2017.

Interceptor Type	Number of Tests	Successful Intercepts	Success Rate
Block 0 – Block 1	7	5	71%
Block IA	19	16	84%
Block IB	9	7	78%
Block IIA	2	1	50%

Table 1 shows that the GMD system has only succeeded in about half of its intercept tests and, further, that its success rate is not improving over time. (The reason for the two values for GMD CE-I tests is that, in one test, the kill vehicle delivered a "glancing blow" to the target rather than a direct hit. Although the MDA classifies this intercept as a success, the Pentagon's Director of Operational Test and Evaluation scored it as "not a kill."[31])

On the other hand, the Aegis BMD system has a combined overall success rate of about 82% for its Block I interceptors. This might be considered to be high enough to establish reliable performance, although a higher success rate is both desirable and possible (THAAD is 14 for 14 in tests (as of July 15, 2017) since it resumed testing in 2006, although most of the tests were not exo-atmospheric).

The third step would be to demonstrate that the system could maintain this reliable performance over a wide range of potential operating circumstances. Even if it had demonstrated reliable performance, the GMD system would fall far short on this criterion. Although the GMD system is primarily a defense against ICBMs, it has only been tested once against an ICBM-range target, in May 2017. It has never been tested at night – all its targets have been sunlit. Although the GMD is intended to operate in a salvo mode, in which multiple interceptors are fired at the target before the outcome of the first intercept attempt is known, it has never been tested in salvo mode. Nor has it ever been tested against more than one near-simultaneous targets.

Aegis BMD appears to do much better on this third criterion. It has been tested both during daytime and at night. It has been tested against short-, medium-, and intermediate range targets (the SM-3 Block I interceptors are not intended to intercept ICBMs). It has been tested against both intact missiles and against separated warheads. It has been tested against two near-simultaneous targets.

The fourth, final, and most difficult step would be to demonstrate reliable performance against a wide range of plausible countermeasures. Although the secrecy associated with the countermeasures used in testing makes this criterion difficult to assess, as discussed in the next section, it seems likely that current testing has not progressed beyond the simplest of countermeasures.

GMD Test Targets

Has the testing of U.S. midcourse defenses been realistic? If, not, when can we plausibly expect it to be? The GMD is the U.S. midcourse system for which the most detailed information about its testing is available, and the nature of the targets it has been tested against is described below.

The GMD program's first intercept test took place in October 1999. Prior to that test, in 1997 and 1998, MDA conducted two GMD flyby tests. In a flyby test, the interceptor's kill vehicle observes and collects data on the target warhead and threat cloud as it flies by them but does not attempt to intercept the target. In each flyby test, the target threat cloud consisted of the warhead target plus eight decoys which included both round and conical objects.[32] At least one of the spherical objects had stripes on its surface, so it would produce a fluctuating signal. Although the Ballistic Missile Defense Organization (BMDO, the former name of the current Missile Defense Agency) claimed the flyby tests had been successful, and that the kill vehicles would have been able to distinguish the warhead from the decoys (the discrimination analysis was done on the ground following the tests), a whistleblower claimed that in the first flyby test the data was improperly manipulated to get the desired result.[33]

In 2000, the *New York Times* published a BMDO chart showing the actual and planned target sets for the two flyby tests and first 19 GMD intercept tests as of May 5, 2000.[34] As of the date of this chart, the two flyby tests and the first two intercept tests had been completed. The chart shows that after the flyby tests, no conical or striped objects were to be used, leaving as decoys only spherical, uniform objects that did not resemble the conical warhead targets.

In the first five of these tests, IFT-3 (October 2, 1999) through IFT-7 (December 3, 2001), the threat cloud included two other objects in addition to the conically shaped warhead target: a large (1.7–2.2 meter diameter) balloon decoy, and the final rocket stage used to deploy them.[35] (Early flight tests with interceptors using prototype or surrogate components were given an IFT designation; later tests using operationally-configured interceptors used an FTG designation.) Due to its larger surface area, the balloon had a much larger infrared emission signal than the warhead target, allowing the two to be easily distinguished from each other. According to Brigadier General Willie B. Nance, the program manager for national missile defense at the BMDO, the target warhead "was the least visible in the IR spectrum of all the elements in the target array, and the smallest of all the objects in the target array."[36] In the 6th intercept test, IFT-8 (March 15, 2002), two small balloons with infrared signals much smaller than the warhead target were also added to the threat cloud.

In May 2002, the MDA announced that details about GMD test targets would be classified for all future GMD tests.[37] The intercept test after that announcement, IFT-9 (October 14, 2002) apparently used a threat cloud similar to that of IFT-8, with one large and two small balloons in addition to the target.[38] The BMDO planning figure published in the *New York Times* in June 2000 indicated that subsequent intercept tests would use similar target sets as their threat clouds.[39] In any event, in the three remaining tests in the IFT series (IFT-10, IFT-13C, and IFT-14), the GBI interceptor or its launcher failed before it reached the point at which the kill vehicle would have been able to observe the threat cloud.

It should be evident that the tests described above are not demonstrating any real discrimination capability. Rather, they are only demonstrating the ability of the kill vehicle to distinguish between objects with different infrared brightnesses and to home in on the one with the relative brightness that matches the information preprogrammed into the kill vehicle. The *New York Times* chart shows that, at least for the first 19 intercept tests, the BMDO did not plan to move beyond such simple targets. (Although the chart shows the 19th intercept test occurring in May 2005, in actuality the 19th intercept test did not take place until May 2017.)

The secrecy associated with both the nature of the countermeasures expected and the countermeasures actually used in the tests allows military and industry officials to claim that tests against even the simplest of targets demonstrate a robust real-world capability. Recall that the target threat cloud in the first intercept test consisted of the target warhead, a two meter diameter spherical balloon, and the bus used to deploy the target and the balloon. While such a simplistic threat cloud was appropriate for an early test, John Peller, vice-president and NMD program manager at Boeing, stated that the "…target suite was equal to, if not more challenging than, the current projected rogue threat.[40] An unnamed Pentagon official claimed that this target suite was more than threat-representative of a rogue state.[41]

ARGUMENTS ABOUT EFFECTIVENESS BY MISSILE DEFENSE ADVOCATES

Missile defense supporters argue that developing and deploying countermeasures in actual practice is more difficult than critics claim, and in particular they cite past U.S. difficulties encountered in developing and deploying countermeasures. According to the 2012 NAS report "One should avoid overstating the ease with which countermeasures that are theoretically possible can actually be made to work in practice…" and "It is perhaps noteworthy that U.S. (and U.K.) experience with the development of high-confidence penetration aids during the Cold War was of mixed success."[42] Missile defense advocates also point out that deployment of countermeasures will impose penalties in terms of payload, range, and/or reliability on missiles that may already have only marginal capabilities.

Missile defense supporters also argue that countermeasures would require lengthy and visible testing programs that would allow time to develop counter-countermeasures. Responding to the 2000 *Countermeasures* report, MDA director Lt. General Ronald Kadish stated that a country "would not be capable of testing the chosen countermeasures without revealing telltale characteristics to the NMD system."[43]

Missile defense supporters further argue that the current missile threat does not yet include anything but the simplest countermeasures, and the capability of missile defenses to cope with countermeasures will increase as they are further developed. In particular, they emphasize the countermeasure against counter-countermeasure contest is an ongoing competition in which U.S. technological advantages will allow it to stay ahead of any country such as North Korea or Iran. Thus the 2012 NAS report concluded that "...an adequate level of discrimination performance can – in the committee's judgment – be achieved in the near term and provide a reasonable chance of keeping the United States generally ahead in the contest between countermeasures and counter-countermeasures over time, at least against emerging missile states like North Korea and Iran."[44]

Missile defense advocates argue that strategies for structuring and operating defenses will enhance effectiveness, for example, multiple defense layers, firing multiple interceptors at each missile, and operating strategies such as shoot-look-shoot, in which the outcome of a first intercept attempt is observed before possibly firing a second interceptor.

Missile defense supporters also argue that advances in sensor technologies – in particular, wide bandwidth X-band radars and advanced infrared imaging sensors – will enable a defense to collect information that can be used to discriminate warheads. The 2012 NAS report's conclusion about radar discrimination is that "by observation over the longest possible time by X-band radars," an "adequate solution of the problem [radar discrimination] is possible."[45] More generally, the report concludes that by using all possible sensor information over the longest time intervals possible, "an adequate level of discrimination performance can be achieved in the near term."[46]

Missile defense advocates also frequently claim that critics do not have access to classified information that shows that defenses will be effective. Responding to the 2000 *Countermeasures* report, MDA Director Lt. General Kadish argued that "I would like to emphasize the fact that many of the discrimination technologies and techniques the proposed NMD system relies on cannot be discussed in an open forum.[47] While the NAS report stated that the committee believed the problem of radar discrimination was solvable, the discussion justifying this conclusion was in a classified appendix.[48]

Finally, while missile defense supporters acknowledge the importance of realistic testing, they argue that with modern BMD systems, the primary purpose of testing is to validate computer simulations of the defense's performance, so extremely frequent testing is not required, nor is testing against all possible target configurations necessary.

ARGUMENTS ABOUT EFFECTIVENESS BY MISSILE DEFENSE CRITICS

Missile defense critics argue that the employment of potentially effective countermeasures must be expected. It is unrealistic to think that a country that is capable of building both a nuclear warhead small and rugged enough to be deliverable by a missile and an intercontinental ballistic missile and reentry system capable of delivering such a warhead will not be able to take relatively simple steps such as deploying the warhead inside a balloon along with a number of empty balloons. Given that nuclear weapons are likely to be an extremely limited and valuable resource to countries such as North Korea, and the known existence of ballistic missile defenses, such countries will have powerful incentives to take steps to enhance the likelihood that their nuclear-armed missiles can penetrate a defense.

Missile defense critics dismiss claims that the United States had significant problems developing countermeasures during the Cold War as irrelevant, since such countermeasures would have had to defeat a nuclear-armed defense, a far more difficult problem than today's or tomorrow's missile attacker would face. Moreover, even though limited to 1960s or 1970s technologies, potentially very effective countermeasures were developed. An example is the countermeasures package planned in the late 1950s for the British Blue Streak medium-range ballistic missile, which was ultimately never deployed. Each missile was expected to carry 20-40 decoys, which, together with electronic jammers, would be dispersed in a spherical cloud with a diameter of roughly 30 km.

According to one British defense scientist at the time, “as regards invulnerability it is so advanced that neither the U.S. nor ourselves can conceive of a counter to it.”[49]

In addition, missile defense critics point out that there are many possible types of countermeasures which can be used in many combinations. An attacker need only find one combination of countermeasures that it believes it will be effective, whereas if a defense is to be relied upon, it must be able to defeat all plausible countermeasures. Moreover, attacks by long-range, nuclear-armed missiles are likely to be extremely rare events. Thus not only must a defense work extremely well, but it must do so the first time it is ever used, a difficult criterion to meet.

Critics also argue that it is unrealistic to expect countries such as North Korea to require anything like the same level of testing, reliability, or confidence in their systems that the United States would require in military systems. They note that for many years the United States has regarded North Korean missiles as an imminent threat to U.S. territory, even though North Korea had never successfully flight-tested a ballistic missile with intercontinental range. (In July 2017 North Korea flew a missile on a lofted trajectory that might have had intercontinental range if flown on a maximum-range trajectory.)

Firing multiple interceptors at the threat cloud from each missile is not only possible, but will likely be required to get an acceptable level of effectiveness even against a missile not equipped with countermeasures. However, if a first intercept attempt fails because of a countermeasure – not because of an interceptor reliability problem – then the same countermeasure could well also be successful against a second and subsequent intercept attempts.

While a shoot-look-shoot strategy is likely be desirable if timelines permit it, it does not increase the defense’s effectiveness against small attacks, it only allocates interceptors more efficiently. If two interceptors are fired at a missile, the probability of a successful intercept will be same regardless of whether the interceptors are salvo-fired or fired using a shoot-look-shoot doctrine. However, if the first intercept is successful, the shoot-look-shoot approach will only expend one instead of two interceptors.

While multiple layers operating using different principles can in principle improve overall effectiveness, there are significant limitations in practice. No country has deployed a boost-phase layer, and a terminal within-the-atmosphere layer can only cover limited geographic areas relative to a midcourse defense. Thus while a geographically small country might be able to employ a multi-layered defense, this would be much more difficult for a large country such as the United States. The current GMD NMD system consists of only a single layer. Adding another wide-area defense system, such one based on the Navy’s new SM-3 Block IIA interceptor, while feasible, would only add a system with exactly the same vulnerabilities to countermeasures as the GMD system.

Finally, critics point out that the effectiveness of many types of countermeasures are based on fundamental physical principles (such the inability of radar waves to penetrated even extremely thin metallic layers), and thus their effectiveness does not rely on potentially classified engineering details. Discrimination requires making a measurement that unambiguously and uniquely identifies the actual warhead, and the objective of many countermeasures approaches is specifically to deny the defense the ability to do this. Thus while it is highly desirable for the defense to gather all available measurement data, doing so does not guarantee that discrimination can be successfully accomplished.

SUMMARY

The potential effectiveness of wide-area exo-atmospheric ballistic missile defenses remains a controversial subject. In particular, because they operate in the vacuum of outer space, such defenses are potentially vulnerable to a wide range of light-weight countermeasures. The issue of how well a defense will be able to discriminate the actual warhead in a countermeasures environment is the central question in the debate over the potential effectiveness of exo-atmospheric defenses. Supporters of such defenses argue that U.S. technological advantages will allow it to defeat any countermeasures that a country like North Korea might be able to develop. Critics point out that, after more than fifty years of developing such defenses, no solution to the countermeasure problem has appeared and that no testing against credible countermeasures appears to be taking place. The secrecy associated with countermeasures and counter-countermeasures will likely make it very difficult to reach any consensus on the potential effectiveness of these defenses.

REFERENCES

1. G. N. Lewis and T. A. Postol, Patriot Performance in the Gulf War: Lewis and Postol Respond, *Science and Global Security*, **8** (2000), pp. 315-356. Available from: http://scienceandglobalsecurity.org/archive/sgs08lewis.pdf [Accessed 16 July 2017].
2. Ibid.
3. W. Boese, Army Report Details Patriot Record in Iraq War, 2003. Available from: https://www.armscontrol.org/act/2003_11/Patriotmissile [Accessed 16 July 2017].
4. Raytheon Company, Iron Dome Weapon System, Fact Sheet, 2017. Available from: http://www.raytheon.com/capabilities/products/irondome/ [Accessed 16 July 2017].
5. American Physical Society Study Group on Boost-Phase Intercept Systems for National Missile Defense, *Report of the American Physical Society Study Group on Boost-Phase Intercept Systems for National Missile Defense: Scientific and Technical Issues*, Washington, D.C.: American Physical Society, 2003. Available from: https://www.aps.org/policy/reports/studies/upload/boostphase-intercept.PDF [Accessed 16 July 2017].
6. Ibid.
7. U.S. Department of Defense, *DOD News Briefing on Missile Defense from the Pentagon,* March 15, 2013. Available from: http://archive.defense.gov/Transcripts/Transcript.aspx?TranscriptID=5205 [Accessed 16 July 2017].
8. B. Clapper, U.S. Hesitant in Condemning North Korean Launch, The Associated Press. December 13, 2012.
9. G. N. Lewis and T. A. Postol, Video Evidence on the Effectiveness of Patriot during the 1991 Gulf War, *Science and Global Security* **4** (1993), pp. 1-64. Available from: http://scienceandglobalsecurity.org/archive/sgs04lewis.pdf [Accessed 16 July 2017].
10. Department of Defense Appropriations for Fiscal Year 2010, Hearing Before Defense Subcommittee of the Senate Appropriations Committee, 111th Congress, p. 28, June 9, 2009; Department of Defense Authorization for Appropriations for Fiscal Year 2010, Part 1, Hearing Before the Senate Armed Services Committee, 111th Congress, p. 741, June 16, 2009.
11. Status of Implementing the Phased Adaptive Approach to Missile Defense in Europe, Hearing before the Subcommittee on Strategic Forces, House Armed Services Committee, 111th Congress. p. 32, December 1, 2010.
12. U.S. Department of Defense, Department of Defense Off-Camera Press Briefing by Vice Admiral James Syring on Missile Defense, News Transcript, May 31, 2017. Available from: https://www.defense.gov/News/Transcripts/Transcript-View/Article/1198464/department-of-defense-off-camera-press-briefing-by-vice-admiral-james-syring-on/ [Accessed 16 July 2017].
13. K. Gildea, MDA Classifies Missile Defense Flight Test Target Countermeasure Data, *Defense Daily*, May 15, 2002.
14. American Physical Society Study Group, *Report of the American Physical Society Study Group*, p. 122
15. U.S. Department of Defense, Defense Science Board, *Defense Science Board Task Force Report on Science and Technology Issues of Early Intercept Ballistic Missile Defense Feasibility*, Washington, D.C.: Defense Science Board, September 2011, p. 5. Available from: http://www.dtic.mil/docs/citations/ADA552472 [Accessed 16 July 2017].
16. Ibid, p. 27.
17. U.S. National Academy of Sciences, National Research Council, Committee on Assessment of Concepts and Systems for U.S. Boost-Phase Missile Defense in Comparison to Other Alternatives, *Making Sense of Ballistic Missile Defense: An Assessment of Concepts and Systems for U.S. Boost-Phase Missile Defense in Comparison to Other Alternatives,* Washington, D.C.: The National Academies Press, September 2012, p. 10. Available from: http://www.nap.edu/catalog.php?record_id=13189 [Accessed 16 July 2017].
18. A.M. Sessler, J.M. Cornwall, B. Dietz, S. Fetter, S. Frankel, R. L. Garwin, K. Gottfried, L. Gronlund, G. N. Lewis, T. A. Postol, and D. C. Wright, *Countermeasures: A Technical Evaluation of the Operational Effectiveness of the Planned US National Missile Defense System,* Cambridge, Massachusetts: Union of Concerned Scientists /MIT Security Studies Program, 2000, pp 35-37, 145-148. Available from: http://www.ucsusa.org/sites/default/files/legacy/assets/documents/nwgs/cm_all.pdf [Accessed 16 July 2017].
19. National Intelligence Council, *National Intelligence Estimate (NIE): Foreign Missile Development and the Ballistic Missile Threat to the United States Through 2015*, unclassified summary, September 1999, p. 16. Available from: https://fas.org/irp/threat/missile/nie99msl.htm [Accessed 16 July 2017].
20. R. H. Speier, K. S. McMahon and G. Nacouzi, *Penaid Nonproliferation: Hindering the Spread of Countermeasures Against Ballistic Missile Defenses,* Santa Monica, California: RAND Corporation, 2014; Federation of American Scientists, Decoys. Available from: http://www.globalsecurity.org/space/systems/decoys.htm [Accessed 16 July 2017].

21. C. Desaris, P. Millner, C. Grabowsky and M. O'Dea, *The Ballistic Missile Defense Organization's Consolidated Targets Program*, Defense Technical Information Center, ADA329067, August 1997. Available from: http://www.dtic.mil/docs/citations/ADA329067 [Accessed 16 July 2017].
22. D. K. Stumpf, *Titan II: A History of a Cold War Missile Program*, Fayetteville, Ark.: University of Arkansas Press, 2000, pp. 200-201.
23. Sessler, et. al., pp. 59-79.
24. Ibid, pp. 147-148.
25. J.C. Toomay, *Radar Principles for the Non-Specialist, 2nd ed.,* Mendham, New Jersey: Scitech Publishing, 1998, p. 93.
26. G. N. Lewis, Space Surveillance Sensors: The Pave Paws and BMEWS Radars (April 12 2012), blog post. Available from: https://mostlymissiledefense.com/2012/04/12/pave-paws-and-bmews-radars-april-12-2012/#more-90 [Accessed 16 July 2017].
27. General Accounting Office (GAO), *Missile Defense: Actions being taken to address testing recommendations, but updated assessment needed*, GAO-04-254. Washington, DC: February 2004, p. 17. Available from: http://www.gao.gov/assets/250/241487.pdf [Accessed 16 July 2017].
28. P. Ingwersen, W. Camp, and A. Fenn, Radar Technology for Ballistic Missile Defense. *Lincoln Laboratory Journal,* **13** (2002), pp. 109-148.
29. E. Brookner, Radar Imaging for Arms Control, in K. Tsipis, D. W. Hafemeister and P. Janeway, eds., Arms Control Verification: The Technologies That Make It Possible, Washington D.C.: Pergamon-Brassey's, 1986, pp. 135-165.
30. U.S. National Academy of Sciences, *Making Sense of Ballistic Missile Defense,* p. 102
31. Fiscal Year 2013 National Defense Authorization Budget Request for Missile Defense, Hearing before the Subcommittee on Strategic Forces, House Armed Services Committee, March 6, 2012, p 152.
32. W. Broad, Antimissile Testing Is Rigged to Hide a Flaw, Critics Say, *The New York Times*, June 9, 2000, p. A1. Available from: http://www.nytimes.com/2000/06/09/us/antimissile-testing-is-rigged-to-hide-a-flaw-critics-say.html [Accessed 16 July 2017]. The figure is reproduced in D. Wright, *The Target Set for Missile Defense Intercept Test IFT-9*, Technical working paper, Cambridge, MA: Union of Concerned Scientists, October 11, 2002. Available from: http://www.ucsusa.org/sites/default/files/legacy/assets/documents/nwgs/ift9.pdf [Accessed 16 July 2017].
33. Broad, Antimissile Testing is Rigged.
34. Ibid.
35. D. Wright and L. Gronlund, *Decoys and Discrimination in Intercept Test IFT-8*, Working paper. Cambridge, MA: Union of Concerned Scientists. March 14, 2002. Available from: http://www.ucsusa.org/sites/default/files/legacy/assets/documents/nwgs/acfxoq64k.pdf [Accessed 16 July 2017].
36. "NMD Kill Vehicle Performed 'Very Well' in Flight Test, Officials Say," *Inside Missile Defense*, October 20, 1999, p. 1.
37. Gildea, MDA Classifies.
38. Wright, *The Target Set.*
39. Broad, Antimissile Testing is Rigged.
40. "NMD Kill Vehicle Performed 'Very Well'."
41. "Raytheon Kill Vehicle Destroys Target in Inaugural Test of NMD System," *Inside Missile Defense*, October 6, 1999, p. 19.
42. U.S. National Academy of Sciences, *Making Sense of Ballistic Missile Defense*, p. 134.
43. Sessler, et. al., *Countermeasures*; National Missile Defense: Test Failures and Technology Development, Hearing before the House Subcommittee on National Security, Veterans Affairs, and International Relations, Committee on Government Reform, 106th Congress (2000), p. 119.
44. U.S. National Academy of Sciences, *Making Sense of Ballistic Missile Defense*, p. 10.
45. Ibid, p. 134.
46. Ibid.
47. National Missile Defense: Test Failures and Technology, p. 116.
48. U.S. National Academy of Sciences, *Making Sense of Ballistic Missile Defense*, p. 136.
49. Richard Moore, *Nuclear Illusion, Nuclear Reality: Britain, the United States, and Nuclear Weapons, 1958-64.* Houndmills, England: Palgrave Macmillan, 2010, p. 111-112.

Bugs and Gas:
Agreements Banning Chemical and Biological Weapons*

Amb. (ret.) Robert P. Mikulak

Former United States Representative to the Organization for the Prohibition of Chemical Weapons
215 N. Highland St.
Arlington, VA 22201 USA

Abstract. The use of chemical or biological weapons, whether by a State or terrorists, continues to be a serious security concern. Both types of weapons are prohibited by multilateral treaties that have very broad membership, but both the Biological Weapons Convention and the Chemical Weapons Convention are facing major challenges. In particular, the continued use of chemical weapons in the Syrian civil war by government forces risks eroding the norm against the use of such weapons. This paper briefly explore the recent history of efforts to constrain chemical and biological weapons and outlines challenges for the future.

INTRODUCTION

Among the many different types of weapons, chemical and biological weapons have a unique status. Both are subject to a strong taboo against use, reflected in the long-standing injunctions against the use of poison or disease as weapons. Both are considered "weapons of mass destruction," along with nuclear and radiological weapons and both are subject to global bans. Chemical weapons may roughly be thought of as "poison gas," although most agents are liquids or solids that are disseminated as aerosols. Biological weapons may be roughly characterized as "germ weapons." Despite the substantial technical differences, the two types of weapons are often lumped together as "CBW."

A SECURITY ISSUE FOR THE LAST 100 YEARS

While extensive use of chemical weapons, as well as limited use of biological weapons, took place one hundred years ago, the threat of use of chemical and biological weapons, particularly by terrorists, is still a serious security concern today. Examples of the use of chemical and biological weapons in the last two decades include:

---

*Note: The views expressed in this paper are those of the author and do not necessarily reflect those of the Department of State or the United States Government.

Nuclear Weapons and Related Security Issues
AIP Conf. Proc. 1898, 030008-1–030008-8; https://doi.org/10.1063/1.5009223
Published by AIP Publishing. 978-0-7354-1586-7/$30.00

TABLE 1. Chemical weapons used in the past, with dates, locations, perpetrators and consequences.

Agent Used	Date	Place	Perpetrators	Consequences
Sarin nerve agent	20 March 1995	Tokyo subway	Aum Shinrikyo	12 killed; 50 severely injured; roughly 100 with vision problems
Anthrax	Sept-Oct 2001	Washington, DC; facilities in Connecticut, New York City, Florida	Individual	5 deaths; 17 non-fatal cases
Sarin nerve agent	21 August 2013	Ghouta, Syria	Syrian government forces	approximately 1500 deaths
Mustard agent	21 August 2015	Marea, Syria	ISIL forces	at least 1 killed; several injured
Chlorine	2014, 2015	Kafr Zita, Syria (18 April 2014) Qmenas, Syria (16 March 2015) Binnesh, Syria (24 March 2015)	Syrian government forces	number of deaths unclear; many affected
VX nerve agent	13 February 2017	Kuala Lumpur airport	Allegedly North Korean agents	Assassination of half-brother of North Korean leader
Sarin nerve agent	4 April 2017	Khan Sheikhoun, Syria	Syrian government forces	approximately 100 killed and more than 500 injured

Banning Poison Gas: World War I and Its Aftermath

Poison gas was first used on a large scale during World War I, notwithstanding efforts around the turn of the twentieth century to constrain the use of poison gas as a weapon. On the Western Front, German troops released large quantities of chlorine from stationary gas cylinders during the Battle of Ypres (Belgium) on April 22, 1915 in effort to break the stalemate that had developed. Defenses were quickly improvised, however, and both sides raced to develop more effective agents and means of delivery, culminating in the widespread use of the blister agent sulfur mustard.

The actual effectiveness of poison gas as a weapon appears uncertain and recent research suggests that its military utility may have been overstated. Nonetheless, chemical weapons were singled out during the war and afterwards as being especially cruel and odious. The peace treaties specified that the losers were not to possess chemical weapons and an unsuccessful effort was made to incorporate a ban on chemical weapons in post-war armaments agreements. In particular, an agreement reached at the 1921-22 Washington Naval Conference on banning use of chemical weapons failed to enter into force because France rejected the limitations it also contained regarding submarines.

The Geneva Protocol of 1925

Despite the failure of a number of post-World War I efforts to restrict the use of poison gas, the United States continued to raise the issue and eventually succeeded in 1925 at a conference in Geneva on control of the arms trade. Although the initial proposal dealt with restricting exports, participants agreed that this was impractical and settled instead on a ban on use of "asphyxiating, poisonous or other gases and of all analogous liquids, materials or devices," which has become known as the 1925 Geneva Protocol. The agreement extends the prohibition also to "bacteriological methods of warfare." As formulated, the 1925 Geneva Protocol contains two important qualifications; it applies only "in war" and only between parties to the agreement. While drafted as a complete ban on use, a number of states attached reservations allowing retaliation in kind if attacked with chemical (or biological) weapons, making the agreement in effect a ban only on first use. Subsequently, however, the threat of retaliation in kind helped to deter use.

From 1925 to 1969

In the years leading up to World War II, the Geneva Protocol was ratified by all of the great powers, except the United States and Japan. (The United States did not ratify until 1975, although for most of the intervening period it held to a "no-first-use" policy.) Since a number of countries retained the right to retaliate in kind, chemical weapons capabilities continued to exist and grow, with Germany, the Soviet Union and the Western allies all building up substantial stockpiles. Despite the existence of large quantities on both sides, chemical weapons were fortunately not used during World War II, apart from extensive use by Japanese forces in China. The reasons that such weapons were not employed are complex, but include both the restraining effect of the 1925 Geneva Protocol, the lack of military utility in the type of warfare taking place, and the threat of massive retaliation against civilian targets.

After World War II and the disclosure that Germany had developed a much more potent form of chemical weapons - nerve gases - military interest in chemical weapons once again grew. In particular, through the 1950's and 1960's, both the Soviet Union and the United States produced large quantities of nerve agents, in addition to the well-known sulfur mustard agent developed during World War I. A variety of tactical munitions, including artillery projectiles, land mines, short-range rockets and missiles, bombs and spray tanks were developed to deliver the agents.

In addition to chemical weapons, a few countries developed and produced biological (and toxin) weapons, notably the Soviet Union and the United States. Both lethal and so-called "non-lethal" chemical and biological agents were stockpiled.

Proposals to prohibit chemical weapons entirely sprouted after World War I but concerns about compliance blocked progress on a ban until the re-armament activities of the 1930's sidelines discussions altogether. The issue was only revived in the 1960's, when U.S. use of chemicals (the riot control agent CS, herbicides, and napalm) in the Vietnam War and problems in the U.S. chemical weapons program aroused substantial controversy. During the late 1960's, growing international political controversy about chemical and biological weapons led to proposals at the UN General Assembly and in the multilateral disarmament conference in Geneva to prohibit both chemical and biological weapons. Initially, these discussions made little headway, since the United States was staunchly opposed, arguing that a ban could not be verified.

THE BIOLOGICAL WEAPONS CONVENTION OF 1975 (BWC)

The situation changed radically, however, when the United States announced in late 1969 that it would unilaterally renounce biological weapons and destroy its stockpiles. The renunciation was expanded in early 1970 to include toxins (i.e., toxic chemicals produced by living organisms). This remarkable about-face led quickly to a U.S.-Soviet initiative, based on an initial UK proposal, for an agreement banning biological and toxin weapons. Multilateral agreement was reached at the Geneva disarmament conference in 1971 and the agreement, dubbed the "Biological Weapons Convention" entered into force in April 1975. Although the United States was unwilling to negotiate a ban on chemical weapons as proposed in 1969 by the Soviet Union and its allies as part of an agreement outlawing both chemical and biological weapons, it did agree to a provision in the Biological Weapons Convention (Article IX) that stipulated that effective prohibition of chemical weapons was a "recognized objective" and committed BWC States Parties "to continue negotiations in good faith with a view to reaching early agreement on banning chemical weapons."

Provisions

The Biological Weapons Convention provides for a comprehensive prohibition of biological and toxin weapons and requires the destruction of existing stockpiles. It does not, however, require declaration of stockpiles or facilities or provide for international verification. Instead, monitoring is left to the individual member states and their national technical capabilities. Since the United States had unilaterally renounced biological and toxin weapons, it did not consider that verification was essential to protect its security interests and furthermore assessed that the treaty was unverifiable.

The Convention also requires that each State Party take measures to ensure that the provisions of the Convention are effective on its territory. In practice, this means that a State Party must adopt criminal legislation against development, production, stockpiling, etc. of biological or toxin weapons. This requirement has taken on added importance with the rise of concern about possible bioterrorism.

Status Report

The Biological Weapons Convention has achieved broad international acceptance. As of May 2017, the BWC had 178 States Parties and nine signatory states. All of the world's major powers have joined. The most important non-Parties are Egypt, Israel, and Syria.

A serious violation of the BWC by the Soviet Union was uncovered in the 1980's. Under the Soviet program, ton quantities of anthrax, smallpox and other disease-causing organisms were secretly and illegally produced in very large production facilities. An early indication of this program was provided by the April 1979 accidental airborne release of anthrax spores from a production facility in Sverdlovsk (now Ekaterinburg), which led to the death of approximately 100 people from inhalation anthrax. Russia has only partially acknowledged this program and their official statements are very unclear about measures taken to end it.

After the 1992 completion of the Chemical Weapons Convention, which contains detailed provisions for monitoring, as well as international assistance and cooperation, an effort was made between 1994 and 2001 to develop a protocol to the BWC containing similar provisions. These negotiations came to an abrupt end in 2001, when the United States decided that the draft protocol would not be effective and would adversely affect U.S. security and commercial interests. Although U.S. opposition was most apparent, dissatisfaction was more widespread.

After 2001, the States Parties established a pattern of two meetings per year - one week in late summer for discussions among experts of technical issues related to strengthening security against biological threats and one week in late autumn to discuss political issues related to the Convention. They also set up a small staff ("Implementation Support Unit"), currently consisting of three people, to prepare for the two weeks of meetings and to assist member states with their implementation efforts.

Every five years a special meeting of the States Parties is held to review the overall operation of the Convention. The most recent Review Conference was held in November 2016. Unfortunately, the 2016 Review Conference was unable, due largely to obstruction by Iran, to reach consensus on continuing and strengthening the intersessional discussions that had been the pattern for the previous 15 years.

In addition to these political difficulties, the BWC - along with several other multilateral arms control agreements - is experiencing serious financial problems. These problems stem in part from the failure of a major State Party - Brazil - to pay its contributions to meeting and staff costs for almost a decade. This problem has been complicated by the UN's adoption of more rigorous financial management practices and by the vague financial rules followed for financing of BWC activities. Member States and the UN are now discussing how to put the financing of the BWC (and several other agreements) on a sound long-term footing.

The 2017 meeting of the States Parties, to be held in Geneva in early December 2017, will again attempt to reach consensus on the form and content of intersessional discussions, including establishment of working groups to discuss scientific and technological advances, national implementation, protection and response against biological attacks, and international cooperation. In addition, measures to stabilize the long-term finances will be on the agenda.

CHEMICAL WEAPONS CONVENTION OF 1997 (CWC)

As noted above, international interest in a ban on chemical weapons was dormant for more than thirty years, between the early 1930's and the late 1960's, and even then strong opposition arose. Only after entry into force of the Biological Weapons Convention in 1975 – with its commitment to further negotiations on a chemical weapons prohibition – was discussion on banning chemical weapons revived in Geneva in the 40-nation Conference of the Committee on Disarmament, although the talks were relatively general and slow-moving.

Negotiations

Serious negotiations on a Chemical Weapons Convention only got underway in 1984, when then Vice President George H.W. Bush presented a draft treaty proposal in the renamed Conference on Disarmament. The negotiations gradually grew in political prominence as the United States and the Soviet Union reactivated their bilateral discussions on a chemical weapons ban (June 1984) and the use of chemical weapons in the Iran-Iraq War, and particularly the Iraqi use against its own Kurdish population (Halabja attack; March 1988), aroused international concerns and pressure for completion of the negotiations. High-level international conferences in Paris (March 1989) and Canberra (September 1989) stressed the importance of outlawing chemical weapons. The multilateral negotiations were completed in September 1992 and the Chemical Weapons Convention was opened for signature in January 1993 in Paris, more than twenty years after the Biological Weapons Convention had been signed.

Main Provisions

The Chemical Weapons Convention provides for a comprehensive prohibition of chemical weapons and requires declaration and destruction of chemical weapons and chemical weapons production facilities. In addition, it establishes a complex international inspection system for confirming the declarations and the destruction of weapons and production facilities, as well as for monitoring certain chemical industry facilities to ensure that legitimate commercial activities are not used as a cover for chemical weapons production. The treaty also establishes an international organization, the Organization for the Prohibition of Chemical Weapons, headquartered in The Hague, The Netherlands, to oversee implementation. In addition, the CWC provides mechanisms for resolving concerns about compliance, including short-notice challenge inspections, for assisting states that are attacked with chemical weapons, and for promoting international cooperation in the field of chemistry.

Status Report

In a relatively short period, the Chemical Weapons Convention has achieved near-universal international support. As of May 2017, 192 countries had become States Parties, leaving only Israel (which has signed, but not ratified), Egypt, North Korea, and South Sudan as non-parties. The CWC thus has been more broadly accepted than any other international arms control agreement.

Unquestionably, the CWC has very substantially reduced the threat of the use of chemical weapons. The United States and Russia have each destroyed tens of thousands of tons of chemical agents and numerous production facilities. As of May 2017, more than 90 per cent of their declared stocks had been destroyed and the end of destruction clearly in sight. A number of countries have disclosed their previously secret chemical weapons activities, including India, France, South Korea, Albania, Libya, Syria, Iraq, Iran, and China, and destroyed prohibited stocks and/or facilities.

Over the two decades since entry into force, the OPCW has built a reputation as a lean and effective international organization of more than 400 staff members. Remarkably, despite having a large number of member states, its policy-making bodies have almost always been able to take decisions by consensus. It has firmly established itself as a model for international security cooperation. The contribution of the OPCW to strengthening international peace and security was recognized by the award in 2013 of the Nobel Peace Prize.

ELIMINATING SYRIA'S CHEMICAL WEAPONS PROGRAM

Since late summer 2013, the OPCW has been charged with eliminating the Syrian chemical weapons program. This is without doubt the most important challenge the organization has faced in its twenty years of existence. OPCW involvement began in spring 2013 when the Executive Council, reacting to reports of the use of chemical weapons by government forces in the Syrian civil war, condemned the use of chemical weapons by anyone and expressed its support for a UN investigation. Since Syria was not a CWC State Party, the task of investigating fell to the UN Secretary-General under a mechanism established in the 1980's. A nerve gas attack on August 21, 2013 in Ghouta, a Damascus suburb, which killed roughly 1500 people galvanized the international community into action. Under strong international pressure, Syria accepted a framework for elimination of its chemical weapons developed by the United States and Russia in early September and agreed to join the Chemical Weapons Convention. In late September 2013, the OPCW Executive Council, quickly followed by the UN Security Council, endorsed the U.S.-Russian framework and established a timetable for Syrian destruction activities and procedures for strict oversight over Syria's compliance.

When Syria became bound by the CWC in mid-October 2013, the OPCW quickly deployed inspectors to confirm the stocks that had been declared and that Syria was destroying empty munitions and chemical weapons production equipment. All of the declared chemical weapons stocks (i.e., mustard agent and nerve agent precursors) were destroyed; most had to be removed from Syria for destruction. A joint OPCW-UN mission was established to oversee this difficult process. Mustard agent and the nerve agent sarin precursor DF were destroyed on a specially-equipped U.S. cargo vessel at sea, while precursor chemicals for the nerve agent VX and for mustard agent were shipped to commercial disposal facilities in Europe and the United States. The removal and destruction activities were extremely complex, costly, and demanding and were accomplished only through an extraordinary international cooperative effort.

Although destruction of chemical weapons production equipment was accomplished without difficulty, the destruction of the facilities housing the equipment proved much more difficult. Underground tunnels had been used to house some production equipment, while mixing and filling equipment for munitions was sited in reinforced concrete aircraft hangars designed to withstand aerial attack. Lengthy negotiations were required to reach agreement on destruction methods and destruction activities stretched into late 2015. Significantly, several facilities remained inaccessible through 2016; including two in rebel-controlled territory.

Use of Chemical Weapons by Syria

Although Syria's declared stockpile was under international supervision, reports in early 2014 alleged that Syrian government aircraft had attacked rebel-controlled towns with improvised chlorine bombs. At the urging of concerned OPCW member states, the Director-General decided in late April 2014 to send OPCW experts as a "Fact-Finding Mission" to establish the facts and report back to him and the Executive Council. The first detailed report, in September 2014, was based largely on interviews of victims, witnesses and medical personnel, as well as examination of open-source videos, and indicated that chlorine had been used as a weapon. While the Fact-Finding Mission had no authority to determine who was responsible for the attacks, the interviews pointed invariably toward Syrian government forces. In response, the OPCW Executive Council stressed that the individuals responsible for the use of chemical weapons should be held accountable.

Holding Syria Accountable

The Executive Council, as well as the UN Security Council, have stressed the importance of accountability for those involved in the use of chemical weapons. The OPCW Fact-Finding Mission, however, only has a mandate to determine whether or not chemical weapons were used, not who used them. That task falls to the UN Security

Council, which on 7 August 2015 (Resolution 2235) established the OPCW-UN Joint Investigative Mechanism (JIM) to identify those involved in those specific incidents in which the OPCW Fact-Finding Mission determined that chemical weapons were used in Syria.

The three-person JIM panel, headed by UN official Virginia Gamba, reported to the Security Council in late August and October 2016 its finding that chlorine attacks in three separate incidents in 2014 and 2015 had been carried out by the Syrian military. They concluded that insufficient information was available on several other attacks to determine responsibility, although they also noted that opposition forces did not possess aircraft, which had been the means of delivery. The United States and several other countries pressed for the Council to impose sanctions on individuals and organizations likely to have been involved, but the resolution was vetoed by Russia and China. Subsequently, the United States and a number of European countries imposed national sanctions on a number of Syrian officials.

The JIM's mandate was renewed on 17 November 2016 (Resolution 2319) and a new three-person panel established; the head is Guatemalan diplomat Edmond Mulet.

Continuing Uncertainty about Syria's Disclosures

Since the initial declaration and additional disclosures by the Syrian government in the fall of 2013, the OPCW has been seeking to clarify and resolve gaps, discrepancies and omissions in the information provided. A group of technical experts, the Declaration Assessment Team (DAT), has held numerous meetings with Syrian officials and made a number of visits to Syrian facilities. While the Syrian government has claimed it is cooperating, few issues have been fully resolved and in fact the list of issues has increased. As a result, the OPCW Director-General reported to member states in July 2016 that he is unable to fully verify the Syrian declaration of its chemical weapons program as accurate and complete. These unresolved issues range across the full spectrum of activities in a chemical weapons program. The task has been complicated by the Syrian insistence that no original documents exist. As of July 2017, a stalemate appears to have been reached. Emphasizing the unresolved gaps and discrepancies in Syria's disclosures, as well as the chemical weapons attack on April 4, 2017 at Khan Sheikhoun, the United States and other countries believe that Syria failed to declare its entire chemical weapons program and has retained some agent and/or agent precursors. The OPCW Fact-Finding Mission has reported that the nerve agent sarin was used in this attack. The OPCW Director-General has transmitted the report to the OPCW-UN Joint Investigative Mission, which is mandated by the UN Security Council to seek to determine who was responsible for the attack.

USE OF CHEMICAL WEAPONS BY THE SO-CALLED "ISLAMIC STATE IN THE LEVANT" (ISIL)

In mid-2015, reports also appeared that ISIL forces were using mustard agent in Syria and Iraq. The OPCW Fact-Finding Mission has investigated and confirmed that mustard agent was used in August 2015 in Marea, Syria and also in mid-September 2016 at Um-Housh, Syria. An Iraqi government team, assisted by the OPCW, also confirmed that mustard agent was used in Kurdistan in August 2015. Numerous small-scale attacks, using very crude mustard agent delivered by mortar rounds, have been reported since then, continuing into 2017. The agent appears to have been made by ISIL technicians, rather than being obtained from some other source. In response to the attacks, the United States has sought to attack and destroy ISIL's chemical weapons capabilities.

FUTURE CHALLENGES

Chemical Weapons Convention

The most immediate challenge with respect to chemical weapons is to end the continuing use of such weapons in the fighting in Syria and Iraq.

Longer-term political challenges are to adapt the CWC regime for an international security environment in which the task of eliminating declared stocks and facilities has been accomplished and the chemical threats to national and international security no longer arise primarily from national chemical weapons programs. (Russia is close to completing destruction and the United States is well above 90% finished.) In contrast, use of toxic chemical by non-State actors, which was not a great concern during the treaty negotiations, has become a serious security threat. While the CWC, particularly its requirement for national implementation and commitment to protection and assistance, provides relevant tools, the principal responsibility for dealing with non-State actors falls to individual member states. In this environment, maintaining the interest in the CWC regime will require creative and energetic activities to adapt OPCW activities to the needs and interests of member states in dealing with this new chemical security threat.

On the technical side, implementation of the Convention will need take into account far-reaching changes in the way chemicals are made and in the way the chemical industry is organized. Furthermore, the rapidly advancing understanding of the chemistry of biological systems means that new ways to disrupt them are also possible. While the text of the Convention is drafted broadly enough to cover these new potential threats, their significance needs to be assessed and methods of verification adapted. It is also important to recognize that these advances also possess great potential for strengthening protective measures against chemical weapons.

Biological Weapons Convention

For the Biological Weapons Convention, the biggest political challenge is to remain relevant while rapid changes are taking place in the life sciences, both in the knowledge about the functioning of biological systems, but also in the use of this knowledge in the biotechnology industry. In principle, the security implications of these developments could – and should – be discussed among the BWC member states. In recent years, however, a highly-politicized debate about reviving failed negotiations on a protocol to the Convention has overshadowed efforts to establish a productive international discussion on the ways in which member states can assist each other in meeting modern-day biological security threats and in building confidence that treaty provisions are being complied with. On the technical side, the challenge is to understand the security implications of the rapid advance in knowledge of living systems and to adapt to the growing overlap produced by the convergence of the chemical and biological sciences.

CONCLUDING REMARKS

Security threats from chemical and biological weapons have evolved in the last hundred years since the first large-scale use of chemical weapons and remain very real. These are not purely abstract or theoretical concerns.

Chemical and biological weapons have a special political status that has caused them to be singled out for comprehensive prohibition through international treaties. These agreements have almost universal international support and have largely been effective in eliminating national programs.

The biggest current challenge is to end the use of chemical weapons in Syria and Iraq. The longer-term challenge is to ensure that both treaties remain relevant to national security concerns. This will require addressing security challenges as they evolve.

Acknowledgement

The author wishes to thank his colleague Mallory Stewart for making this presentation to the Short Course in his stead, as a last minute replacement, when he experienced a medical emergency.

Implications of the Joint Comprehensive Plan of Action

George Perkovich

Carnegie Endowment for International Peace
1779 Massachusetts Av., NW
Washington, DC 20036 USA

Abstract. This essay describes the background behind the July 2015 Joint Comprehensive Plan of Action that was negotiated to redress the crisis that had developed around Iran's nuclear activities, and summarizes some of the agreement's key features. The essay then highlights political and strategic factors that enabled the diplomatic breakthrough, and draws lessons that could inform approaches to future proliferation challenges. The conclusion suggests how some of the agreement's innovative features could be built upon and applied more broadly to reduce risks that civilian nuclear energy programs could be diverted for military purposes and to inform approaches to nuclear disarmament in the future.

INTRODUCTION

This essay posits some of the more important implications of the Iran nuclear deal, also known as the Joint Comprehensive Plan of Action (JCPOA), which was completed on July 14, 2015 by China, France, Germany, Russia, the United Kingdom, the United States, the European Union, and Iran. Predicting the future is always risky. In the case of the Iran nuclear issue, much of the past is disputed. Scholars and politicians inevitably will disagree about what the Iran nuclear deal means for the future. Iranians will debate this question with each other, as will Americans, Russians, and others who were involved in these thorny matters for the past fifteen years or longer. These internal debates will be mild compared with the international arguments over what should be learned from the Iran experience and what it means for the future of the global nuclear order.

Fortunately, many of the basic facts that created the nuclear crisis with Iran and led to its resolution have been documented by the International Atomic Energy Agency (IAEA). The IAEA is as close to an objective source as we have in these affairs. Thus, in the first section of this essay I rely heavily on the IAEA's record to summarize the technical circumstances that led up to the 2015 agreement, and then describe some of the deal's key features. In the next section I describe some important political and strategic factors that allowed the crisis to be resolved diplomatically and attempt to draw lessons for future nuclear proliferation crises. This is followed by an analysis of how the experience with Iran could or should change the ways that key states may approach the challenges of preventing nuclear proliferation in the future. Then I briefly discuss how the resolution of the Iran case could affect the nuclear aspirations of Iran's neighbors. I conclude with some thoughts on how leading states in the global nuclear order might enhance the fulfillment of the nuclear nonproliferation regime's basic objectives – nonproliferation, the peaceful use of atomic energy, and nuclear disarmament.

The premise throughout this essay is that Iran and the other parties to the Joint Comprehensive Plan of Action will implement it. This means that they will fulfill the terms set out for the first ten years, then fifteen years, and subsequently they will abide by commitments that extend even further into the future. These include the reaffirmation "that under no circumstances will Iran ever seek, develop or acquire any nuclear weapons," and that Iran will not "engage in activities, including at the R&D level, that could contribute to the development of a nuclear explosive device, including uranium or plutonium metallurgy activities."[1] Disputes inevitably will arise along the way, but the assumption for the purpose of analysis here is that all parties ultimately will act in accord with the JCPOA. Of course, this optimistic premise may prove untenable. The agreement could collapse for many reasons with countless imaginable effects. The purpose of this essay is not to speculate on possible failures and their implications, but rather to explore what success might suggest about the future.

Nuclear Weapons and Related Security Issues
AIP Conf. Proc. 1898, 040001-1–040001-15; https://doi.org/10.1063/1.5009224
Published by AIP Publishing. 978-0-7354-1586-7/$30.00

ORIGINS OF THE IRAN CRISIS AND SUMMARY OF ITS RESOLUTION

In August 2002, the National Council of Resistance of Iran publicly revealed the existence of two previously secret nuclear facilities – a uranium enrichment plant near Natanz and a heavy-water research reactor near Arak.[2] Such facilities would be invaluable for the production of fissile materials – specifically enriched uranium and plutonium. These fissile materials could be used for civilian purposes and/or nuclear weapons. The Natanz and Arak facilities had not been declared to the International Atomic Energy Agency and Iran had no apparent civilian need for these facilities. This discovery added to earlier international suspicion that Iran was secretly seeking to develop nuclear weapons, which would violate its obligations under the Nuclear Nonproliferation Treaty (NPT).

The International Atomic Energy Agency duly began an intense investigation. The first year of this investigation proved, in the words of the IAEA's Director General, "that Iran had failed to meet its obligations under its Safeguards Agreement with respect to the reporting of nuclear material imported into Iran and the subsequent processing and use of the material, and the declaration of facilities and other locations where the material was stored and processed."[3]

The IAEA continued to investigate whether Iran's declarations had been "correct" and "complete," and continued to discover that they had not been. Further verification was needed to ascertain whether Iran had diverted nuclear materials, perhaps for military purposes, and whether the character of Iran's nuclear activities was exclusively peaceful, as required under the NPT. This investigatory process continued through 2015. It yielded a long list of instances in which Iran had failed to declare activities, materials, and facilities, including some that plausibly had military applications.

Throughout the course of this investigation, a pattern emerged whereby Iran would first promise full transparency and cooperation with the IAEA and then withhold information and cooperation, or provide new information only after the IAEA provided evidence that disproved earlier Iranian claims. Perhaps the most dramatic example of this pattern was exposed in 2009 when several governments detected that Iran was secretly building a new enrichment facility underground called Fordow near the city of Qum, notwithstanding its various pledges of transparency and full disclosure.[4]

These early developments, of course, occurred in the midst of the Iraq War, intense U.S. pressure for regime change in Iran, threats of Israeli military action against Iran's nuclear program, and political tension within Iran. This context motivated actions by all the parties that were involved in negotiations that ultimately produced the JCPOA. Iran found itself under intense pressure. While it was compelled to admit to having violated its obligations to the IAEA, it also felt that the nuclear issue was being used for broader purposes of threatening its government and its development aspirations. The other states and the IAEA were deeply concerned that Iran could seek to develop nuclear weapons, which would cause grave international security risks and severely undermine the international nonproliferation regime. At the same time, many countries and the leadership of the IAEA feared that Israel and/or the United States would use military force against Iran, causing a war in the region's largest state.

From 2003 through the summer of 2005, the United Kingdom, Germany, and France – on behalf of the European Union – volunteered to negotiate a peaceful resolution of the nuclear issue to avoid the necessity of referring the matter to the UN Security Council (as was called for under the IAEA's statute). Mindful of Iraq, the three European states and the IAEA leadership wanted to avoid U.S. military action in Iran. This negotiation led to a short-lived voluntary arrangement between the EU-3 and Iran in October 2003, whereby Iran volunteered to suspend its enrichment activities. Then, after Iran interpreted the agreement differently than the other parties, a more defined agreement was renegotiated in November 2004. In return, the EU-3 offered Iran a plan for the improvement of broader political, economic, and nuclear cooperation relations. This process broke down in the summer of 2005, before presidential elections in Iran. Iran then declared that it would resume activities to develop its fuel-cycle capabilities. In August 2005, Iran announced that it would resume converting uranium yellowcake into gas at the Esfahan facility with the declared intention of subsequently enriching it at the pilot facility near Natanz.[5] This occurred as the confrontational Ahmadinejad administration took office in Tehran, casting grave doubts on the prospects of diplomatic progress. In September 2005, the Iran "file" was referred for the first time to the UN Security Council. In December 2006, the Security Council imposed (relatively mild) sanctions on Iran. In August 2007, Israeli jets bombed a suspected plutonium production reactor secretly being built by North Koreans in Syria. Among other things, this heightened the sense that military action might be taken against Iran's nuclear program. Unknown at the time, Israel and the United States were preparing a highly secret cyberoperation to sabotage Iranian centrifuges at Natanz.[6]

In 2008, the U.S. presidential campaign resulted in the rather surprising nomination of Barack Obama as the Democratic Party candidate. In his campaign Obama differed from his Democratic opponents and subsequently the

Republican nominee, John McCain, in saying that he would reach out to Iran and seek diplomatic resolution of the nuclear issue and other disputes. This was controversial in the U.S., and welcomed in much of the rest of the world. After Obama took office in January 2009, he quietly pursued diplomatic overtures to Iran. In October 2009, negotiators from the P5+1 (China, France, Russia, the United Kingdom, and the United States, plus Germany) met with Iranian counterparts in Geneva and appeared to reach agreement on a plan whereby Iran would ship 80 percent of its low-enriched uranium stockpile from Natanz to Russia, from which it would then be shipped to France for the fabrication of fuel rods for the Tehran Research Reactor.[7] However, when Iranian negotiators returned to Tehran, opponents of President Ahmadinejad lashed out at the tentative agreement and the deal collapsed.

In 2010, the foreign ministers of Turkey and Brazil met with Iranian counterparts to try to revive the 2009 deal. However, in the meantime Iran had increased its stockpile of low-enriched uranium. This development, and other terms of the bargain agreed to by Turkish and Brazilian representatives, made the 2010 deal unacceptable to the P-5+1. To increase pressure on Iran, which was steadily increasing the number of centrifuges enriching uranium, the UN Security Council adopted more stringent sanctions in June 2010. From 2011-2013, Iran's economy suffered noticeably from restrictions on investment, imports, the financial sector, and particularly oil exports.[8]

In December 2011, President Obama dispatched then-Senator John Kerry to meet secretly with officials of Oman to ask if they would be willing to explore Iran's interest in establishing a direct negotiating channel with Washington.[9] Having received a positive reply from Tehran (via Oman), President Obama, in March 2013, authorized two trusted emissaries, Deputy Secretary of State William Burns and Vice President Biden's National Security Advisor Jake Sullivan, to travel to Oman for secret talks with Iran. These talks involved at least five meetings, the last four of which took place after new Iranian President Hassan Rouhani took office in August 2013. The victory in June 2013 of Hassan Rouhani, who had led Iran's nuclear negotiations from 2003-2005, surprised the Iranian establishment. (The victories of Mohammad Khatami in 1997 and Mahmoud Ahmadinejad in 2005 had also surprised the Iranian establishment). Rouhani and his new foreign minister, Javad Zarif, were eager to pursue nuclear diplomacy. Rouhani and Zarif had been intimately involved in the earlier nuclear diplomacy and Zarif had lived and worked in the U.S. for many years. In November 2013, the two countries revealed the heretofore secret talks and pledged to bring them to a conclusion.[10] This announcement and the subsequent negotiations involving the P5+1 and the EU engendered intense discord in the U.S., Iran, Israel and Washington's allies in the Persian Gulf.

The course of diplomacy between 2013 and the conclusion of the Joint Comprehensive Plan of Action in July 2015 are familiar enough that they do not need to be summarized here. However, it is worth noting that Russia continued to cooperate with the U.S. and European powers to bring the nuclear deal to conclusion, even though Moscow and the Western powers were clashing over developments in Ukraine.

The 159-page JCPOA is remarkable in its detail and several of its innovative features. From the perspective of nuclear nonproliferation its most outstanding features in strategic terms are the following:

- Elimination of 97 percent of Iran's stockpile of low-enriched uranium, and limit of 300 kilograms LEU that Iran may possess for 15 years. Numbers of centrifuges reduced from approximately 19,000, to 6,104 first-generation IR-1s, of which no more than 5,060 can enrich for first 10 years.
- Enrichment level limited to 3.67%, for 15 years, with enrichment to be conducted only at Natanz, and no construction of new facilities allowed.
- Research and Development of more advanced centrifuges is restricted in time, scale, and application.
- Modifications of the Arak research reactor to render it, in essence, not a proliferation threat, and commitment not to reprocess plutonium for 15 years.
- Prohibition on activities, including at the R&D level, that could contribute to the development of a nuclear explosive device, and specification of such activities.
- Verification
    - Iran may procure specified important equipment for its nuclear program only through a monitored channel.
    - Continuous surveillance of centrifuge production for 20 years and uranium mining for 25 years.
    - Time-bound access to undeclared sites.
    - Implementation of the Additional Protocol and the Modified Code 3.1 of the Subsidiary Arrangement to Iran's Safeguards Agreement.

From Iran's perspective, the outstanding features of the JCPOA include:

- Recognition of its "right" to enrich uranium for peaceful purposes and to develop a civilian nuclear program with international cooperation.
- Sanctions relief including access to $100 billion in oil export revenue that had been frozen.
- An agenda for removing its nuclear "file" from the UNSC.
- Greater legitimacy on the regional and world stage.

As noted in the introduction, only time will tell if all of the actors required to implement the JCPOA will do so to each other's satisfaction. The purpose of the remainder of this essay is to assess the implications of this "story" and the JCPOA under the assumption that it will be implemented.

KEY FACTORS BEHIND THE DIPLOMATIC RESOLUTION

It is easy to overlook the importance of the fact that Iran is a party to the Nuclear Nonproliferation Treaty and thereby committed not to seek to acquire nuclear weapons and to abide by its safeguards agreement with the International Atomic Energy Agency. These obligations established the basis for the international campaign that ensued when Iran was observed to have broken its safeguards obligations. Had Iran not undertaken such obligations, the story would have evolved very differently. The international community would have had much less leverage to press Iran to verifiably limit its nuclear activities and demonstrate that its nuclear program is exclusively peaceful.

Cooperation among the permanent members of the UN Security Council – China, France, Russia, the United Kingdom, and the United States – was vital to the success of this diplomatic campaign, including its coercive elements. Internationally-binding sanctions compelled Iran to accept terms it would not have otherwise. These sanctions would not have occurred or have been effective without UN authorization and major power enforcement. So, too, cooperation from the European Union and with the IAEA created a wall of pressure that Iran could not break through or maneuver around. Had the P5 been divided, Iran could have exploited differences and ameliorated the pressure it felt from the international community. Similarly, had the P5 been divided, the IAEA Board of Governors could have become split and could have impeded the valuable cooperation of the IAEA Secretariat in managing the Iran case. In addition to cooperating, the P5, the EU, and the IAEA also generally acted with urgency in pressing for the resolution of this case. This sense of urgency further diminished Iran's opportunities to avoid complying with its obligations under the IAEA, the NPT, and the UN Security Council.

The cooperation of the P5+1 was unusually powerful in part because it was difficult to maintain. France took a more hardline position than the U.S. on some of the demands to limit Iran's nuclear activities, including enrichment. This complicated Iran's negotiating strategy. Russia expressed a keen interest in continuing cooperation with Iran in the nuclear energy sector and in arms supplies, while at the same time firmly insisting that Iran accept significant limits on its fuel-cycle activities. This, too, complicated Iran's calculations. China demonstrated an intention to maintain and even expand trade with Iran, and yet, when Iran proved recalcitrant, Beijing cooperated in strengthening sanctions. Ahmadinejad's government had hoped to split the Western states from Russia, China, India, and other Eastern powers, but found this route for escaping international pressure blocked. Russia and China could not be portrayed as tools of the U.S., so when Beijing and Moscow held firm, Iranian elites concluded that they had no option but to negotiate seriously.

Other contextual factors played an important role. As noted, the U.S.-led intervention in Iraq, beginning in early 2003, motivated Iran and the EU-3 (France, Germany and the United Kingdom) to seek a diplomatic resolution. Russia and China shared this interest in averting war. The interest in pursuing diplomacy was intensified by the fact that the leading justification for the war in Iraq – Iraq's alleged ongoing nuclear, chemical, and biological weapons activities – proved to be erroneous. This created demand for internationally validated evidence regarding Iran's nuclear activities, which in turn allowed space and time for diplomacy. As the IAEA painstakingly acquired evidence of Iran's nuclear misdeeds, the international diplomatic pressure increased on Iran.

At the same time, the threat of force from Israel and/or the United States also compelled Iran and other international players to accept that robust restrictions on Iran would be required in order to resolve the crisis diplomatically. Iran naturally sought a resolution that would restrict it as little as possible and would allow it to retain as much secrecy as possible over its past and ongoing nuclear activities. Other parties to the negotiation could have been inclined to accept terms that would have been less restrictive on Iran so as to allow easier and quicker agreement. The threat of Israeli or American force constituted an alternative that could be avoided only if Iran would agree to extensive, verifiable limitations on its future nuclear operations. While U.S. officials, especially in the Pentagon, doubted the utility of military action to durably stop Iran from acquiring nuclear weapons, and evinced concern over the implications of another war in the region, Obama had repeatedly declared he would act to stop Iran from acquiring nuclear weapons. The president and others did not appear to think that Iran's enrichment activities per se were a sufficient cause for military action, but rather that a combination of indicators that Iran was moving decisively to produce nuclear weapons would trigger the use of force. Meanwhile, Israel's apparently lower threshold for military action added coercive pressure on Iran and the U.S. Israel could attack Iran on its own and/or could draw the U.S. into a conflict with Iran. This complicated the calculations of all players, including those among the P5 that were especially averse to war. If war was not acceptable to Moscow, Beijing, and others, then effective economic sanctions were the only viable alternative.

The sanctions mobilized against Iran were indispensable. From 2011 to 2013, oil exports fell from 2.5 million barrels per day to 1.1 million barrels per day, reducing Iran's oil revenue from $100 billion to $35 billion. Fourteen blacklisted Iranian banks were denied access to Swift (Society of Worldwide Interbank Financial Telecommunications), an information network for financial institutions that facilitates electronic transactions. These two effects of sanctions compounded the Ahmadinejad government's mismanagement of economic policy such that the inflation rate rose to between 50% and 70%, the value on unofficial markets of the national currency, the rial, plummeted 56% percent between January 2012 and January 2014, and major enterprises were non-productive and approaching bankruptcy.[11]

The purpose of economic sanctions was to motivate Iran to negotiate meaningful, verifiable limits on its nuclear activities and resolve outstanding issues with the IAEA and the UN Security Council. The sanctions, in effect, were a tax on Iran for violating nuclear rules and also on Iran's trade partners who sacrificed revenues to serve the larger international purpose. The premise was that if and when Iran agreed to terms that satisfied the international community, the withheld funds would be returned to Iran and, indirectly, to its trade partners. The effectiveness of the sanctions led many opponents of the JCPOA in the U.S. and the Middle East to criticize the return of funds to Iran. However, this criticism overlooks the fundamental fact that the nuclear sanctions would not have been imposed internationally in the first place if there had not been a commitment to relieve them when Iran agreed to terms with the P5+1. That is, the motivating power of sanctions depends also on the positive inducement of relieving them if and when the adversary changes its behavior.

Clandestine coercion also, arguably, helped motivate Iran to negotiate. Beginning in 2007, Israel and the United States mounted an extremely sophisticated cyberattack on Iran's centrifuge program – now known as Stuxnet. This operation damaged a significant number of Iran's machines and slowed its acquisition of enriched uranium. Perhaps more importantly, this operation, along with the discovery of the Fordow facility in 2009, demonstrated to Iranian leaders that their nuclear program had been deeply penetrated. This must have alarmed Iranian decisionmakers that further secret nuclear activities would be discovered and either attacked by the U.S. and/or Israel, or used to intensify sanctions and other international pressure against Iran.

Iranian domestic politics also played an important role in paradoxical ways. The international politics surrounding the Iran nuclear issue have always been in part a contest over whether the world feared U.S. and Israeli exertions more than it feared Iran's. In the period between 2002 and 2005, many feared that the U.S. and Israel were more reckless and dangerous than Iran. The election of Mahmood Ahmadinejad as president in 2005 led to more confrontational Iranian rhetoric and actions. While the new Iranian government advanced Iran's threatening nuclear activities and capabilities, it also made the international community more wary of Iran's intentions – nuclear and otherwise. After Ahmadinejad's election, the balance of fear tilted against Iran. This became more dramatic in 2009 when the Fordow facility was discovered. The international sanctions and backlash against Iran during Ahmadinejad's tenure, and the economic hardship that Iranians experienced, enabled Hassan Rouhani to win the presidential election in 2013.

Similarly, domestic politics in the United States created favorable conditions for diplomatic resolution. Barack Obama's victory in 2008 and subsequent willingness to suffer political attack for pursuing negotiations with Iran demonstrated that the U.S. and its partners were genuine. This development further bolstered international support for the U.S. and its partners in the contest with Iran, and further put Iran on the defensive. If the U.S. was now willing to negotiate genuinely, then failure to reach agreement could be blamed on Iran, creating a basis for further sanctions.

The change in U.S. leadership opened the way to a fundamental strategic development that made the JCPOA possible: the willingness to match coercive pressure with an offer of a major gain for Iran. This factor – like the importance of Iran's membership in the NPT – may be obvious, but it was slow to emerge. Specifically, by late 2009 the U.S. and the other powers negotiating with Iran indicated they would accept as a matter of fact, if not yet of principle, that Iran had won the "right" to enrich uranium. The NPT itself neither grants such a "right" nor denies it.

But for years, the U.S. and some other states, along with many nonproliferation experts, insisted there was no such right and that Iran should agree to forego fuel-cycle activity as part of any arrangement to rectify its non-compliance with its IAEA and post-2005 UN Security Council obligations. Other states, particularly developing countries, rejected this contention. Over time, Iran positioned itself as a leader of the cause for establishing something like a right to enrich. By 2009, it was apparent that the U.S. and others would accept some ongoing enrichment in Iran as part of a verifiable, comprehensive agreement. Then-Senator John Kerry reportedly conveyed this position directly to Iranian interlocutors in December 2011. Iran could count this as a major gain for its exertions, both at home and internationally. However, in accepting this concession, the U.S. and other parties gained leverage to motivate Iran to curtail development of nuclear weapons capabilities and to complete negotiation of what became the JCPOA.[12]

INSIGHTS OTHER STATES MAY HAVE DRAWN FOR FUTURE POLICIES

The previous section highlighted major factors that created conditions for negotiating the JCPOA. No doubt other factors were important, and the ones discussed here can be interpreted differently and debated. In this section I probe deeper into lessons that the Iran experience may offer for future cases of proliferation concern and for contemporary international politics more broadly. I divide the lessons into two categories: high politics and state techniques.

High Politics

As discussed above, cooperation among the P5+1, the European Union, and the IAEA was necessary to make a diplomatic resolution with Iran possible. In future potential cases of proliferation concern, the likelihood of such cooperation will influence how states (or non-state actors) calculate the risks and benefits of seeking nuclear weapons and/or of complying with international rules. If potential proliferators believe that one or more permanent members of the UN Security Council would, in effect, protect them and block international cooperation to enforce nonproliferation norms and rules, then diplomatic resolution of proliferation issues will be extremely difficult to achieve.

Cooperation is inherently difficult to obtain among states whose interests are as diverse as those of the United States, Russia, China, France, and the United Kingdom. It requires hard work through the politics within each state, especially in the U.S. with its fractious system, and in negotiations among these states. The interests of each of these states' allies or partners also complicate their freedom to maneuver. These challenges vividly appear now in Ukraine, Syria, Libya, and the South China Sea, for example. Because no single state is powerful enough to impose its preferred approach and outcome on the others, each must be willing to at least partially accommodate the preferences of the others. This occurred gradually in the international approach to Iran. First, in 2005, the Bush Administration lowered some of its resistance to negotiation. Then, in 2009, President Obama encouraged more energetic diplomacy. Moscow displayed this sort of statesmanship when it stayed aligned with the Western powers in the Iran negotiations even after the Ukraine conflict erupted. In future proliferation cases, when the state of concern may be regarded as a particular friend or adversary of another major power, for example of Russia or China, the key actors may be located in Moscow, Beijing, or another capital. Diplomatic success, then, will depend on whether combinations of coercive pressure and positive inducements can enable the major powers to align their approaches sufficiently to form a common front. This challenge is evident now regarding North Korea, where China is seen as pivotal in determining whether sufficient pressure can be mobilized to make leaders in Pyongyang limit its nuclear and missile-related activities.

The Iran case revealed a deeper matter that may affect how states approach future proliferation situations and the potential for resolving them diplomatically: whether a proliferation dispute is framed as a matter of compliance with rules or, instead, as a matter of bargaining for a fair deal. Compliance and fair bargaining need not be mutually exclusive concepts and processes. If rules are fair, then compliance with them should be fair too. But in international politics, all states do not necessarily perceive all rules to be fair. Nor do states always agree on the fairness of proposed ways of enforcing compliance with rules.

The U.S., France, the United Kingdom and, probably to a lesser extent Russia and China, saw the Iran case primarily as a compliance problem. Iran broke the rules of the nonproliferation regime – particularly IAEA safeguards and disclosure requirements. The IAEA and subsequently the UN Security Council issued binding requirements of transparency and confidence-building, including a demand that Iran "suspend all enrichment-related and reprocessing activities." Iran was not meeting these requirements. Thus, the international community continued to pressure Iran with sanctions until it complied.

Iran could not effectively dismiss the concept of compliance, but Iranian officials sought to portray the dispute as a matter of fairness or justice. They said Iran had no choice but to conduct illicit nuclear activities because the U.S. and Israel were illegally sabotaging Iran's declared activities. They insisted that all countries have a "right" to enrich uranium, and the U.S. and its allies were engaged in neo-colonial nuclear repression. Iran also argued that the nuclear-weapon states have failed to live up to their NPT obligations to eliminate all nuclear arsenals, and that they frequently exhibit double standards in choosing when, where, and how to apply nonproliferation rules.

The P5+1 were correct that compliance was and is at the heart of the diplomatic struggle with Iran. Civilization and international security are advanced by the creation and enforcement of rules to regulate technologies and activities that can threaten international peace and security. The NPT is the foundation of a rule-based global nuclear order. Therefore, it is vital to enforce compliance with rules derived from the NPT.

Yet, the international system generally lacks judicial and enforcement mechanisms that are as legitimate and effective as those in well-governed states. The International Atomic Energy Agency admirably performs some important functions in the rule-based system, but it does not have clear enforcement authority. The actions of its Board of Governors can be subject to the exertions of major powers and resistance by dissident states. The UN Security Council is the entity that is expected to mobilize international power in response to violations of NPT-related rules when the IAEA reports violations to it. The Security Council has authority to enforce mandates related to international peace and security. Enforcement most commonly takes the form of political demands and sanctions. Military action can be the ultimate means of enforcing rules, and in the international system the UN Security Council is supposed to be the authorizer of legitimate uses of force. Yet, the Security Council is a highly politicized body and the five permanent, veto-wielding members enjoy disproportionate influence within it. As a result, the Security Council is often inefficient in catalyzing international action, particularly the use of force. In sum, disputes over questions of nuclear rules and their enforcement are subject to the distribution of power within the international system and the politics that flow from that. This makes it difficult in practice to separate compliance with rules from debates over fairness.

The tensions inherent in this enforcement context are exacerbated by the fact that the NPT is based on "bargains" that have been unevenly and inadequately fulfilled. Some states frequently argue that Article IV's promise of nuclear cooperation in return for nonproliferation is not fulfilled. These states feel that restrictions on access to nuclear technology continue to grow tighter, whereas the "sacrifices" that nuclear-weapon states are supposed to make are unfulfilled. Many states and civil society groups believe, with reason, that Article VI represents a bargain of nonproliferation in return for the eventual elimination of all nuclear arsenals. This was affirmed in the 1995 decision to extend the NPT indefinitely. Many argue that, despite the dramatic reductions in U.S. and Russian nuclear arsenals, the world's nuclear-armed states are not seriously committed to pursuing the abolition of these weapons. Overall, while few if any states argue that the world would be more secure without the NPT, widespread dissatisfaction exists over how its terms have been interpreted and enforced.

The Iran case brought these divergent perceptions of fairness and compliance with the NPT to the center of international politics. The U.S. and the other nuclear-weapon states argue either that they have fulfilled their obligations, or that the obligations are not what Iran and others say they are. While there are merits to the nuclear-weapon states' arguments, it is also the case that the nuclear-weapon states would resist allowing others the authority to judge their compliance, for example with Article VI's call for disarmament. Other examples of double standards, selective enforcement, and perceived unfairness include differences in the ways that major powers treat India, Pakistan, and Israel. None of these three states signed the NPT, so they did not forego the "right" to acquire nuclear weapons. But different major powers have treated each of them differently, and as a group they have been treated more indulgently than Iran, Iraq, and other states that did sign the NPT. There are sound legal and strategic reasons for treating states that have legally agreed not to acquire nuclear weapons differently from those that did not sign the NPT, but this does not alleviate the sense that the nuclear order is unfair.

Iran broadened and politicized the debate by invoking considerations of fairness that appealed not only to the Iranian population but also to leaders and citizens of other states. To rebut claims that the central issue is objective compliance, Iranians argued that the U.S. and its partners were subjective and prejudicial against Iran in their handling of this case as distinct from others.

The competing frames of compliance versus fairness have important implications that are not often analyzed. If fairness is central, the notion of negotiating and bargaining among parties with equal standing informs how one sees the contest. Compromise naturally emerges as an acceptable outcome. However, in the compliance frame, the contest is not among actors with equal standing, but rather is an effort by authorities to compel a deviant actor to comply. Negotiation and compromise are not obviously required. We don't pay a violator of rules to correct its behavior and comply; we compel it. In one frame, fairness involves give and take. In the other, fairness is simply compliance.

Iran created, perhaps, two major precedents through the long campaign that resulted in the JCPOA. The most frequently commented upon is the "right" to enrich uranium. The other, less obvious, precedent is that compliance with nonproliferation rules is a matter of bargaining. To be sure, Iran paid an enormous price for winning these points. It lost hundreds of billions of dollars in revenue and trade as a result of sanctions, and was politically isolated. It had to accept more comprehensive monitoring and verification of its future nuclear activities than other states. But, given the long pattern of Iran's illicit nuclear activities and the evidence to doubt that its nuclear program was exclusively peaceful, Iran was able to bargain for more, arguably, than it was entitled to under the NPT. This occurred in part because of the fundamental reality mentioned earlier: the rule-based international system, unlike domestic governance, does not have predictable recourse to physical power to enforce compliance. War is the final arbiter in the international system. If war is not a welcome option among the countries that would have to wage it in order to enforce compliance – which was the case regarding Iran -- then negotiation and bargaining must be tried. And in a bargaining process, the value of fairness naturally comes into play.

While Russia insisted that the JCPOA should not be regarded as a precedent, and the text itself states this, the realities of international politics make it likely that other states will try to use the Iranian case to their advantage. The concluding section of this essay discusses how some of the innovative features of the JCPOA could be adopted more widely to enhance peaceful nuclear cooperation while strengthening confidence that civilian nuclear programs will not be misused to produce nuclear weapons. More immediately, the de facto acceptance of a "right" to enrich uranium for civilian purposes, and the success of Iran in making compliance with IAEA and UN Security Council demands a subject of bargaining, will be noted by leaders of other states who might be interested in acquiring nuclear weapon capabilities. Saudi Arabia, Egypt, Turkey, and South Korea are states most frequently mentioned in this regard.

Techniques

Beyond its higher-level political implications, the Iran case also offers lessons about modern techniques of statecraft. Three categories are most obvious: sanctions, intelligence collection, and operations.

The economic sanctions imposed on Iran were exceptionally effective. Indeed, they countered a long-standing assertion by many that sanctions are weak tools. Political, legal, and technical factors were vital. Politically and legally, agreement among the five permanent members of the UN Security Council meant that sanctions on Iran would be universal and legally binding. Iran could still seek to encourage states and enterprises to trade with it illicitly, but any potential partner would have to weigh the value of potential trade with Iran against the costs of the world's leading economic and political powers holding them to account and penalizing them. Thus, unlike in other cases where sanctions are imposed by one state or one group of states such as the European Union, the sanctions authorized by the UN Security Council severely limited Iran's options.

Global economic interdependence further strengthened the effectiveness of sanctions on Iran. States and businesses that conduct significant international financial transactions depend on entities subject to the jurisdiction of the United States, the European Union, Japan, and other states whose governmental authorities are highly able to enforce compliance with sanctions. In earlier eras when states and businesses were not so internationally entwined, they found it easier to immunize themselves from the demands of those who would seek to impose sanctions.

Iranian enterprises and entities that would do business with major international counterparts were severely constrained by impediments to finance and insurance. The international financial system has become deeply integrated as banks and other financial institutions such as insurance companies depend on a global network to borrow, lend, exchange, insure, invest, and clear funds. Few major economic enterprises or states can conduct significant transactions without involving entities subject to U.S., European, Japanese, and other allied states' jurisdiction. This gives the U.S. exceptional leverage (as long as Washington does not abuse it and prompt other major financial powers to take countermeasures). The U.S. early on identified and prosecuted major Western banks that failed to comply with U.S. sanctions and others authorized by the EU and/or the UN Security Council. One major case involved BNP Paribas, which in 2015 forfeited $8.83 billion and paid a $140 million fine for violating U.S. sanctions against Iran, as well as Sudan and Cuba. Other banks which have paid smaller penalties for violations include HSBC, Deutsche Bank, ING, and Crédit Agricole.[13] The scale of these penalties made international businesses, and their host states, extremely wary of the risks of doing business with Iran and motivated them to take extra care to monitor their own compliance with sanctions.

Advanced information technologies greatly enhance the capacities of government regulators and intelligence communities to monitor compliance with sanctions, identify suspect activities, and mobilize law enforcement. The technical, intelligence, and enforcement infrastructure developed to counter global terrorism and money-laundering were readily applied to enforcing economic sanctions on Iran.

Technological advances also greatly augmented the capacity of leading states to gather intelligence on Iran's nuclear activities. Much of this capacity remains classified, but the discovery of the secret Fordow enrichment facility and the development and implantation of the Stuxnet "virus" into systems controlling Iran's centrifuges demonstrated the increased difficulty of hiding large-scale nuclear-related activities, facilities, and equipment. So, too, the assassination of at least five key figures in Iran's nuclear and missile programs indicated the capacity of at least one intelligence service to identify, track, and target technical leaders. (Noting the capacity to identify and target these individuals is not to endorse this tactic on moral, legal, or strategic grounds.) Spatial imagery, signals, and cyber intelligence-gathering, remote sensing, and other technologies have advanced dramatically since the late-1980s and early-1990s when Iraq and North Korea managed to secretly develop capabilities to produce fissile materials.

These advances in technology and other capacities to gather intelligence, implement sanctions, and conduct covert operations heightened international leverage to motivate Iran to accommodate international demands. These capabilities also will increase the risks facing other actors that in the future might contemplate the pursuit of nuclear weapons. However, the Iran case also indicates that the use of advanced technology and covert operations is not risk-free. The Stuxnet attack on Iran led to the unintended infection of Programmable Logic Controllers (PLC), a type of industrial control system, in other countries. It also intensified Iran's interest in developing capacities to conduct offensive cyber operations, which was apparent in the 2012 cyberattack on Saudi ARAMCO's computers[14] and attacks on U.S. banks such as PNC, SunTrust, and BB&T.[15] It is impossible to know whether Iranian actors would have undertaken such attacks if Iran had not first been attacked via Stuxnet. It is difficult to know whether the balance of benefits and costs of this series of actions still made the Stuxnet operation advantageous. Similarly, the assassination of Iranian nuclear and missile program personnel may have hardened Iran's resolve and precluded concessions that otherwise could have been possible in the negotiations.

The techniques of statecraft that were developed and applied in the Iran case presumably will continue to be enhanced. This enhancement and policy deliberations on whether and how to use them should be informed by the unfolding process of implementing the JCPOA. There is much here to inspire confidence that actors contemplating the illicit acquisition of nuclear weapons capability in the future will experience severe costs and other impediments. At the same time, it should be expected that prospective proliferators will study the Iran case and search for new ways to avoid detection, political isolation, and coercion.

IRAN'S NEIGHBORS

For years, officials and commentators around the world have speculated and asserted that Iran would cause other states in the Middle East to seek nuclear weapons. In the 1990s and 2000s, the assumption was that Iran would reject diplomacy and march on to acquire nuclear weapons, which would then drive Saudi Arabia, and/or Egypt, and/or Turkey to do the same. Some officials and experts in various countries concluded, therefore, that military strikes were the only way to stop Iran and prevent regional proliferation. When the JCPOA was finalized, the assertion changed. Now the concern is that Iran has gained recognition, acceptance of its ongoing enrichment program, and sanctions relief, all of which make its neighbors fear Iran's growing potential power so much that they will seek nuclear weapons.[16] Indeed, Saudi Prince Turki al-Faisal stated in March 2015, "if Iran has the ability to enrich uranium to whatever level, it's not just Saudi Arabia that's going to ask for that."[17] Prince Turki later stated at a conference in Seoul, "Whatever the Iranians have, we will have, too."[18]

More careful analysis suggests that the threat of nuclear weapons proliferation in the Middle East is exaggerated. The JCPOA, if it is implemented, creates opportunities to further reduce this threat. This can be the case even though the JCPOA may make it more difficult to persuade other countries to forswear acquisition of capabilities to enrich uranium.

Saudi Arabia, Egypt, and Turkey are the states most likely to seek to balance Iran's potential nuclear power. However, they lack basic technical capabilities, physical and social infrastructure, and political-economic space that would allow them to sustain the efforts and risks required to acquire nuclear weapons against the resistance of the international community.

Saudi officials have been the most vocal of Iran's adversaries in expressing interest in acquiring countervailing nuclear capabilities. However, the Kingdom faces major technical limitations in the nuclear domain. It possesses no nuclear reactors of any type. Nor does it have indigenous or imported capabilities to conduct even experimental work to enrich uranium and/or produce and separate plutonium. The country does not appear to have scientific or engineering cadres who could design, develop, manufacture, and test a nuclear explosive device. Acquiring the many capabilities necessary to indigenously produce the fissile material and other components for nuclear weapons would take many years. Most – but not all – countries that have sought to clandestinely develop nuclear weapons did so under the cover of civilian nuclear programs that had operated for decades. Saudi Arabia has the economic resources, in principle, to finance such a nuclear program, but has not yet mobilized to do so. By comparison, for example, the United Arab Emirates over the past decade has dedicated enormous resources to finance and import nuclear power plants and related technical and regulatory expertise, which has been welcomed by the international community because the UAE has taken pains to demonstrate that it will not seek to acquire nuclear weapons. Most importantly in this regard, the UAE has committed not to seek capabilities to enrich uranium or separate plutonium. If Saudi Arabia were interested in acquiring an option to develop nuclear weapons, it would not likely follow the UAE example and forswear enrichment and reprocessing capabilities. This in turn would complicate international cooperation with the Kingdom and slow its route to potential nuclear-weapon capability.

Some observers speculate that Saudi Arabia would seek to import fissile materials and other requisites for developing nuclear weapons, or to import assembled nuclear weapons. No other country has acquired nuclear weapons this way. The challenges this route would pose to Saudi Arabia and to the international community would be new. Pakistan is often mentioned as the most likely source of potential nuclear imports to Saudi Arabia, though no hard evidence has been produced to suggest that this may occur. Pakistan has benefited over the years from Saudi financial support of various kinds, and until recently the two states' relations had been very close.[19] Still, Pakistan would run significant risks to its own interests were it to transfer fissile materials and/or nuclear weapons to Saudi Arabia. Iran borders the economically valuable and rebellious Pakistani province of Balochistan; Tehran could impose severe costs on Pakistan in retaliation for Pakistani nuclear transfers to Saudi Arabia. The U.S., India, and others would also seek to isolate and punish Pakistan. A key unknown is how China would react. China is Pakistan's indispensable ally. It is difficult to imagine Pakistan exposing itself to the consequences of proliferating to Saudi Arabia without obtaining at least tacit assent from China.

One of Saudi Arabia's leading motives for acquiring nuclear weapons capability would be prestige and political power vis-a-vis Iran. Yet, buying nuclear weapons capabilities from Pakistan would underline the fact that Saudi Arabia lacks indigenous means in the nuclear domain and would highlight the Kingdom's inferiority to Iran. (Iran imported centrifuges and a nuclear blueprint from Pakistan, but relied largely on indigenous wherewithal to utilize these assets and develop a significant nuclear program.) In terms of security, if Saudi Arabia acquired nuclear weapons from Pakistan and then found itself in a security crisis with Iran, could Riyadh rely on its capacity to use the nuclear capabilities that Pakistan provided? Could the Kingdom trust the resolve of Pakistan to join it in the possible escalation of such a crisis? In 2015, Islamabad demurred when Saudi Arabia asked it to contribute forces to the military intervention in Yemen. Would Pakistan be more likely to cooperate in potential nuclear brinksmanship with Iran?

The JCPOA increases the impediments to any Saudi effort to acquire nuclear weapons. In terms of motive, if Iran adheres to the JCPOA, the nuclear threat that Iran could pose to Saudi Arabia is significantly reduced. This should ameliorate Saudi interest in nuclear weapons. Iran's adherence to the JCPOA also should be expected to harden international determination to resist any Saudi efforts to acquire nuclear weapons capabilities. Saudi Arabia would be seen as threatening the benefits the world has attributed to the JCPOA. Alternatively, if Iran violates the JCPOA, the U.S. and other international powers should be expected to impose severe constraints and penalties on Tehran, in part to prevent proliferation by Saudi Arabia. Either way, the JCPOA makes it more difficult for Saudi Arabia to take significant steps towards acquiring nuclear weapons as long as Iran is complying with its obligations.

Saudi Arabia's capacity to resist international nonproliferation pressure has decreased in parallel with the decline of oil prices and the rise of the threat of the Islamic State. The Kingdom's revenue has fallen just when it needs to undertake major economic reforms to enhance its development prospects and meet the aspirations of its youth so as to avoid popular challenges to the monarchy. It is spending resources to contest Iran's exertions in Yemen, Iraq, and Syria. Nuclear weapons capability would not improve the prospects of Saudi Arabia's success in these contests, or in preventing discord among the Kingdom's Shia population. Iran's interventions are more subtle than the sorts of threats that nuclear weapons can effectively deter or defeat. If Saudi Arabia nevertheless decided to pursue nuclear weapons, it could readily be exposed to international sanctions because the purchasers of Saudi oil can find alternate supplies, including from Iran. In other words, Saudi Arabia is more vulnerable than ever before to international pressure that could be mobilized if the Kingdom were to seek nuclear weapons.

Egypt has the most advanced indigenous nuclear capabilities of any Arab state. (The United Arab Emirates will have the only advanced nuclear power plants in the Arab world, which are being built by a South Korean consortium.) Egypt possesses two research reactors, a pilot fuel-manufacturing plant, and a pilot-scale plant for converting uranium yellowcake to gas. Egypt also has experience in managing and treating nuclear waste, including hot cells capable of separating plutonium on a very small scale. Still, in recent decades Egypt has been unable to act on plans to develop its nuclear program significantly, including for electricity production. Egypt and Russia in November 2015 signed an agreement whereby Russia would finance and build a nuclear power plant in Egypt, to be completed by 2022 according to press reports.[20] However, recent experience with new nuclear power plant construction in Finland, France, India, Turkey, and the United States suggests that there is no chance that a plant will be built in Egypt by 2022. The political turmoil that has engulfed the country since 2011 has preoccupied elites and society alike. The military dictatorship struggles to maintain political control and satisfy economic demands. Egypt's infrastructure remains underdeveloped. This creates an exceedingly poor environment for international nuclear cooperation which would be necessary for Egypt to develop enough nuclear infrastructure to contemplate acquiring nuclear weapons.

Moreover, the Egyptian regime needs secret cooperation with Israel to help manage security threats posed by internal and external militant groups. The Cairo regime also relies on ongoing financial and military assistance from the United States, Saudi Arabia, and others. This assistance could be jeopardized if Egypt were perceived to be seeking nuclear weapons. As Iran adheres to the JCPOA, the international community will be determined to block any other states in the region from acquiring even latent nuclear weapons capabilities. Saudi Arabia and others want Egypt to remain a bulwark against Sunni extremism (and Iran), but intra-Arab competition means that no Arab state wants another to be the first to acquire nuclear weapons.

Turkey probably possesses the technical capability and economic resources to develop nuclear weapons over the 10-15 year period of the JCPOA if doing so were a top priority of the government. However, Turkish leaders have steadfastly emphasized their membership in the NPT as a non-nuclear-weapon state. Turkey has only recently devoted considerable attention to developing nuclear energy, relying primarily on the import of a nuclear plant from Russia with exceptionally generous Russian financing. This project has been cast into doubt by the tensions in Russo-Turk relations.[21] Turkey's ability to attract other international suppliers of nuclear energy depends on its complete and full adherence to nonproliferation norms and rules. Thus, for Ankara to risk the future availability of nuclear energy in favor of seeking nuclear weapons, its leaders would need an overwhelming security imperative to justify it.

Turkey is a member of NATO. The U.S. continues to base approximately 60-70 air-deliverable tactical nuclear weapons in Turkey at Incirlik Air Base under the auspices of NATO.[22] To the extent that Russia could pose a severe security threat to Turkey in the near to mid-term, NATO's legal obligations to defend Turkey, including if necessary with nuclear weapons, would be more reliable than an uncertain indigenous effort to develop nuclear weapons (against the resistance of NATO states, Russia, Iran, and others). Turkey's most immediate concern is the war in Syria, the re-eruption of conflict with the Kurdish minority, and the Islamic State. Whether or not ongoing membership in NATO greatly augments Turkey's capacity to deal with these threats, a program to develop nuclear weapons would not do so, and yet would risk Turkey's membership in NATO. The international costs of such an effort would further complicate President Recep Erdogan and the ruling Justice and Development Party's struggle to maintain majority power in Turkey. These costs would be all the more severe as long as Iran adheres to the JCPOA. Thus, as the recent book, *Turkey's Nuclear Future* concludes, "for the foreseeable future Turkish leaders will determine that the national interest is best served by eschewing moves to acquire an independent nuclear force."[23]

Taking into account the considerations of Egypt, Saudi Arabia, and Turkey, a highly informed diplomat of a leading Gulf Cooperation Council state concludes that for the next fifteen years, at least, the threat of a regional nuclear arms race is "very unlikely." The most probable candidates, he noted, lack the requisite capabilities. They have much more pressing priorities within their countries and the region. And, he added, if any of these countries sought to pursue acquisition of dual-use nuclear capabilities, the international community would at a minimum press them to adopt modalities for monitoring and verifying nuclear activities similar to those established by the JCPOA in Iran. As long as Iran adheres to the JCPOA and its permanent obligation not to acquire nuclear weapons, any other state in the region that would seek options to break out from the NPT would face at least as many constraints and penalties as were imposed on Iran.

ENHANCING THE NUCLEAR NONPROLIFERATION REGIME

Iran's membership in the Nuclear Nonproliferation Treaty and safeguards agreement with the International Atomic Energy Agency provided the basis for the international campaign that motivated Iran to negotiate the JCPOA. At the same time, the NPT's lack of specificity on some particular matters provided an opportunity for Iran to test its terms in ways that others might be tempted to try – for example, by acquiring fuel-cycle capabilities that are unnecessary for civilian purposes, and/or by conducting research and development related to nuclear explosives. Article IV of the Treaty, as widely noted, affirms the general right of states to develop atomic energy for peaceful purposes, and encourages cooperation among states to this end. Yet these provisions apply only *as long as* such activities are consistent with the obligations not to acquire or facilitate the acquisition of nuclear weapons, under Articles I and II. Unfortunately, the NPT does not define "nuclear weapon" or delineate activities central to developing nuclear explosives that should not be permitted in non-nuclear-weapon states. Nor does the Treaty specify that states have or do not have a "right" to acquire and utilize capabilities to produce fissile materials. Thus, much of the dramatic and potentially escalatory contest between the P5+1, the IAEA, and Iran revolved around whether Iran's nuclear activities were weapons-related or exclusively peaceful, and whether or not Iran should or could be required to abandon efforts to produce fissile materials.

The JCPOA addresses and in some places clarifies these issues in ways that will inevitably inform the future behavior and perspectives of individual states and the broader international community. It contains provisions that could be emulated in the future to enhance confidence that nuclear programs in non-nuclear-weapon states will remain peaceful and not be misused for proliferation. This, in turn, could augment international nuclear cooperation for civilian purposes. The JCPOA also contains provisions that could facilitate an eventual nuclear disarmament process. For example, states that have agreed to eliminate all their nuclear arsenals would probably also need to prohibit research and development activities that are vital to making nuclear explosives, as the JCPOA does. Similarly, the JCPOA's monitoring, verification and inspection procedures could be adapted for a disarmament regime. Much of the debate surrounding the JCPOA ignored these potentially positive systemic features.

The JCPOA, in practice, established that the NPT does not a priori deny states the "right" to acquire and utilize capabilities to produce fissile materials for peaceful purposes. This is a major basis of criticism of the agreement from the standpoint of nonproliferation. But the JCPOA also defines conditions, as well as monitoring and verification mechanisms, that would significantly bolster international confidence that a state's fuel-cycle activities would remain exclusively peaceful.

The agreement provides a model for making a state's fuel-cycle activities commensurate with demonstrable civilian requirements. It limits the level of enrichment to 3.67 percent, which is the typical requirement for nuclear power reactors. (For nuclear weapons, uranium needs to be enriched to 90 percent). The JCPOA limits the volume of enriched uranium that can be accumulated in the state before a requirement for large quantities of nuclear reactor fuel manifestly exists. In the case of Iran, this limit is 300 kilograms for a period of 15 years, with an understanding that after that period Iran's enrichment capability would expand according to a plan shared in advance with the IAEA. The principle of commensurating the production and accumulation of fissile material with demonstrable civilian requirements is also evident in how the JCPOA addressed the issue of plutonium production and separation. Iran has no civilian need for plutonium. Accordingly, it agreed to redesign the Arak heavy water reactor "to minimise the production of plutonium and not to produce weapon-grade plutonium in normal operation."[24] Iran also agreed, for 15 years, not to engage "in any spent fuel reprocessing or construction of a facility capable of spent fuel reprocessing," and added that it does not intend to reprocess thereafter.[25]

Monitoring and verification procedures in the JCPOA significantly enhance confidence that Iran will not be able to divert fuel-cycle facilities and material to weapons or to conduct clandestine fuel-cycle activities. Iran's uranium mining activities will be monitored for 25 years. For 20 years Iran will declare all locations and equipment "that are used for production of centrifuge rotor tubes or bellows, and will permit the IAEA to implement continuous monitoring."[26] For 15 years enrichment is allowed only at one facility, Natanz, to which the IAEA will be given daily access. These provisions, among others, significantly decrease the prospects that Iran could divert the outputs of declared facilities to nuclear weapons. More importantly, these provisions, taken together, increase the probability that the IAEA and international intelligence services would detect clandestine nuclear activities in Iran. There is no legitimate functional reason why these forms of monitoring and verification could or should not be applied in any other non-nuclear-weapon state that seeks to engage in uranium enrichment and spent-fuel reprocessing.

Turning to nuclear weapons acquisition, which is forbidden under Article II of the NPT, the JCPOA establishes that "Iran will not engage in activities, including at the R&D level, that could contribute to the development of a nuclear explosive device"[27] Section T in Annex 1 specifies a number of such activities that are prohibited:

- 82.1. Designing, developing, acquiring, or using computer models to simulate nuclear explosive devices.
- 82.2. Designing, developing, fabricating, acquiring, or using multi-point explosive detonation systems suitable for a nuclear explosive device, unless approved by the Joint Commission for non-nuclear purposes and subject to monitoring.
- 82.3. Designing, developing, fabricating, acquiring, or using explosive diagnostic systems (streak cameras, framing cameras and flash x-ray cameras) suitable for the development of a nuclear explosive device, unless approved by the Joint Commission for non-nuclear purposes and subject to monitoring.

- 82.4. Designing, developing, fabricating, acquiring, or using explosively driven neutron sources or specialized materials for explosively driven neutron sources.

These prohibitions, now codified in a legally binding UN Security Council resolution, have several implications conducive to preventing weapons proliferation, defining nuclear disarmament, and facilitating peaceful nuclear cooperation. They specify some, though not all, activities that should be regarded as indicators of efforts to acquire nuclear explosive devices and therefore unacceptable in non-nuclear-weapon states (unless the host state demonstrates plausible non-explosive purposes for these activities). This specification – for the first time – clarifies activities other than those related to nuclear materials that should be considered inconsistent with purely peaceful programs. As a result, the barrier to proliferation is widened and at the same time detail is provided for defining elements of a future nuclear disarmament regime. Activities deemed essential for developing nuclear explosives should not be conducted in states that purport to be nuclear-disarmed. Conversely, states with peaceful nuclear programs that affirm their commitment not to conduct such activities, and allow procedures to verify adherence to this commitment, could be more readily provided nuclear cooperation.

The JCPOA also affirms the authority of the IAEA to verify compliance with nonproliferation obligations beyond the narrow question whether nuclear material has been diverted or illicitly acquired. The Vienna-based agency is now tasked with verifying that Iran does not conduct activities that could contribute to the development of a nuclear explosive device. Experts have generally assumed that the IAEA is the most appropriate body to verify adherence to nonproliferation commitments and eventual nuclear disarmament, but the latter role had not been specified prior to the JCPOA.

Finally, the JCPOA contains an ingenious enforcement mechanism that could be employed in future cases of grave proliferation concern or, more problematically, of violations of a disarmament agreement. In the case of an unresolved dispute over compliance with the agreement, any of its parties is allowed to call for a vote in the UN Security Council on whether to keep sanctions on Iran lifted. If any one of the permanent members of the Council votes against such a resolution, then the provisions of extant sanctions resolutions would be re-imposed. Essentially, the veto mechanism has been re-engineered so that instead of allowing one permanent member of the Security Council to block imposition of sanctions, which has long undermined enforcement, now one member could re-impose sanctions in response to an unresolved dispute over compliance. Iran stated that if sanctions were reinstated in this manner, "Iran will treat that as grounds to cease performing its commitments under [the] JCPOA in whole or in part." Nevertheless, in a future case when the UN Security Council has found a state to be non-compliant with its nonproliferation obligations and imposed sanctions, the "snap-back" mechanism of the JCPOA could be adopted. The mere prospect of this can strengthen deterrence of nonproliferation rule violations.

Whether and how the innovative provisions of the JCPOA discussed here will be adapted and adopted more broadly will inevitably be the subject of intense debate among states. Russia and other states initially have insisted that the JCPOA is not a precedent that should be applied widely.[28] However, as the JCPOA is implemented over the next 10-15 years, it would be folly for the international community not to utilize its mutually beneficial provisions more broadly. States could include relevant provisions in bilateral nuclear cooperation agreements. The Nuclear Suppliers Group could require acceptance of relevant provisions as a guideline for trade with states that seek to acquire enrichment and/or plutonium extraction capabilities. Future NPT review conferences could debate and recommend adoption of the prohibition on specified activities that could contribute to the development of a nuclear explosive device, as both a nonproliferation measure and a requisite of nuclear disarmament. The IAEA Board of Governors could update safeguards to include monitoring and verification procedures like those in the JCPOA in states with fuel-cycle programs.

Discussion and debate over these matters will take years to resolve, but the key provisions of the JCPOA itself apply for 10-15 years and beyond, allowing plenty of time to apply its welcome features more broadly. Doing so would have the additional benefit of encouraging Iran to abide by the letter and spirit of the agreement indefinitely. Indeed, failure to apply the salutary lessons of the Iran experience and provisions of the JCPOA would squander the prodigious international effort and sacrifices that led to the peaceful resolution of this major nuclear crisis.

REFERENCES

[1] "Joint Comprehensive Plan of Action," Vienna, July 14, 2015, available at: http://www.state.gov/documents/organization/245317.pdf

[2] National Council of Resistance of Iran, "Remarks by Alireza Jafarzadeh on New Information on Top Secret Projects of the Iranian Regime's Nuclear Program," Iran Watch, August 14, 2002, http://www.iranwatch.org/library/ncri-new-information-top-secret-nuclear-projects-8-14-02

[3] Director General Mohamed ElBaradei, "Implementation of the NPT Safeguards Agreement in the Islamic Republic of Iran," IAEA Board of Governors, August 26, 2002, pg. 1.

[4] David E. Sanger and William J. Broad, "U.S. and Allies Warn Iran Over Nuclear 'Deception,'" *New York Times*, September 25, 2009, http://www.nytimes.com/2009/09/26/world/middleeast/26nuke.html?pagewanted=all

[5] CITATION FROM IAEA.

[6] This secret operation, now known as Operation Olympic Games, began in 2006. David E. Sanger, "Obama Order Sped Up Wave of Cyberattacks Against Iran," *New York Times*, June 1, 2012, http://www.nytimes.com/2012/06/01/world/middleeast/obama-ordered-wave-of-cyberattacks-against-iran.html?pagewanted=all&_r=0

[7] Semira N. Nikou, "Timeline of Iran's Nuclear Activities," U.S. Institute of Peace, http://iranprimer.usip.org/resource/timeline-irans-nuclear-activities

[8] Kenneth Katzman, "Iran Sanctions," Congressional Research Service Report, January 21, 2016, http://www.fas.org/sgp/crs/mideast/RS20871.pdf

[9] David Ignatius, "How the Iran Deal Became the Most Strategic Success of Obama's Presidency," *Washington Post*, September 15, 2015, https://www.washingtonpost.com/opinions/the-obama-bet-that-paid-off/2015/09/15/e46b80f6-5be6-11e5-8e9e-dce8a2a2a679_story.html .

[10] "Report claims secret US-Iran talks laid groundwork for nuclear deal," *Associated Press*, November 24, 2013, http://www.foxnews.com/politics/2013/11/24/report-claims-secret-us-iran-talks-laid-groundwork-for-nuclear-deal.html

[11] Kenneth Katzman, "Iran Sanctions," Congressional Research Service Report, January 21, 2016, http://www.fas.org/sgp/crs/mideast/RS20871.pdf

[12] This important point is not fully appreciated, especially in the U.S. For example, in a December 2015 article, the eminent *New York Times* reporters David Sanger and William Broad reported that the IAEA had concluded that Iran in 2009 ceased conducting activities to design nuclear weapons – http://www.nytimes.com/2015/12/03/world/middleeast/Iran-nuclear-report-atomic-agency.html. Sanger and Broad wrote that Iran's efforts "ended after 2009, or just as Mr. Obama was taking office and accelerating the sanctions and cybersabotage program against Iran's nuclear facilities that ultimately brought Iranian officials to the negotiating table." Sanger and Broad did not mention the possibility that Obama's willingness to negotiate forthcomingly with Iran, including the acceptance of ongoing enrichment in Iran, also played a role in Iran's decision. Nor did they mention that Iran had negotiated with the EU-3 between 2003 and2005, and that those negotiations collapsed in part because the EU-3 (and the U.S. in the background) would not agree to accept that Iran had a right to enrich uranium in the future. The point here is not that the Iran's position should have been accepted then or in 2009 and thereafter. Rather, the point is that many reporters, commentators, and officials fail to recognize that coercion alone did not cause Iran to curtail activities related to acquiring nuclear weapons and negotiate the JCPOA. Positive inducements were necessary too.

[13] These cases involved violations of a range of sanctions against Iran, not just those related to its nuclear activities.

[14] Jose Pagliery, "The inside story of the biggest hack in history," CNN, August 5, 2015, http://money.cnn.com/2015/08/05/technology/aramco-hack/

[15] Siobhan Gorman and Danny Yadron, "Banks Seek U.S. Help on Iran Cyberattacks," *Wall Street Journal*, January 16, 2013, http://www.wsj.com/articles/SB10001424127887324734904578244302923178548

[16] See, for example, the debate between Robert Einhorn and Olli Heinonen, "Beyond the Vote (Part 2): Implications for Proliferation in the Middle East," Washington Institute for Near East Policy, September 21, 2015, http://www.washingtoninstitute.org/policy-analysis/view/beyond-the-vote-part-2-implications-for-proliferation-in-the-middle-east

[17] Barbara Plett Usher, "Iran deal could start nuclear fuel race – Saudi Arabia," *BBC News*, March 16, 2015, http://www.bbc.com/news/world-middle-east-31901961

[18] David E. Sanger, "Saudi Arabia Promises to Match Iran in Nuclear Capability," *New York Times*, May 13, 2015, http://www.nytimes.com/2015/05/14/world/middleeast/saudi-arabia-promises-to-match-iran-in-nuclear-capability.html?_r=0

[19] Saudi-Pakistani relations soured in 2015 when Pakistan declined Riyadh's request to send ground forces to augment the Saudi military campaign in Yemen.

[20] "Egypt, Russia sign deal to build a nuclear power plant," *Reuters,* November 19, 2015, http://www.reuters.com/article/us-nuclear-russia-egypt-idUSKCN0T81YY20151119

[21] Sinan Ulgen, "Is this the end of Moscow-Ankara nuclear cooperation?" *Bulletin of the Atomic Scientists*, January 13, 2016, http://thebulletin.org/end-moscow-ankara-nuclear-cooperation9059.

[22] "Turkey," Nuclear Threat Initiative, last updated March 2016, http://www.nti.org/learn/countries/turkey/

[23] Sinan Ulgen and George Perkovich, eds., *Turkey's Nuclear Future* (Washington, DC: Carnegie Endowment for International Peace, 2015), 214.

[24] JCPOA, ANNEX 1, para 2
[25] JCPOA, para 12.
[26] JCPOA, para 80.2
[27] JCPOA, para 16.
[28] Paragraph xi of the JCPOA states: "All provisions and measures contained in this JCPOA are only for the purpose of its implementation between E3/EU+3 and Iran and should not be considered as setting precedents for any other state or for fundamental principles of international law and the rights and obligations under the NPT and other relevant instruments, as well as for internationally recognised principles and practices."

The Future of the NPT and Measures to Reduce Nuclear Dangers in the Age of Trump

Daryl G. Kimball

Executive Director
Arms Control Association
1200 18th Street, NW (Suite 1175)
Washington, DC 20036, USA

Abstract. Through the decades, the international nuclear disarmament and nonproliferation enterprise, though imperfect, has curbed nuclear proliferation and limited the number of nuclear-armed states to nine, forced reductions in major-power nuclear arsenals, ended nuclear testing by all but one state, and created an informal taboo against nuclear weapons use.

The cornerstone of global nuclear risk reduction and elimination efforts, the 1968 Nuclear Nonproliferation Treaty (NPT), will be 50 years old in 2018. The treaty has, overall, been a success, but is under increasing stress.

The NPT system has made it possible to pursue and negotiate agreements to enforce the rules and guard against new nuclear-armed states. The 2015 EU-3+3 nuclear deal with Iran was a vital and historic agreement designed to address Iran safeguards violations and severely curtail its ability to produce material that could be used for nuclear weapons purposes.

But today, there are some very tough challenges that pose a serious threat to the international nuclear order. Relations between Washington and Moscow are worse than ever. Although U.S. and Russian nuclear stockpiles are down from their Cold War peaks, the risk of conflict and unconstrained nuclear competition remains. At present, there is no plan to replace or extend the New START agreement, and both countries are on track to replace their excessive nuclear arsenals at enormous cost.

The United States and Russia are each estimated to have more than 4,000 warheads stockpiled and assigned for military use. Under the 2010 New Strategic Arms Reduction Treaty (START), each is be limited to 1,550 deployed strategic nuclear weapons on as many as 700 nuclear delivery vehicles until Feb. 5, 2021. (Figure 1)

Deteriorating U.S./NATO relations with Russia over Moscow's violation of Ukraine's territorial integrity and other issues increase the possibility of a direct confrontation that could escalate into the nuclear realm. With hundreds of long-range thermonuclear weapons on each side poised for "launch under attack," the fate of the world still depends on the good judgment and restraint of the U.S. and Russian presidents in a crisis. If these weapons were used even in a "limited" way, the result would be catastrophic.

Worse yet, the United States and Russia are planning to spend hundreds of billions in the next decade to replace their Cold War-era nuclear strike systems. Washington has recently determined that Moscow is deploying ground-based cruise missiles in violation of the 1987 Intermediate Nuclear Forces (INF) Treaty, putting that treaty and future arms control agreements in doubt. Some U.S. lawmakers want the United States to pursue its own INF Treaty-noncompliant missiles in response.

The next largest nuclear arsenals are those of France, with approximately 300, and China, with 280 warheads of all types. With the conclusion of the 1996 Comprehensive Test Ban Treaty, all nuclear-armed states except for North Korea have halted nuclear testing. Though the CTBT has helped reinforce a global norm against nuclear weapons testing, which can be used to develop new and more dangerous nuclear weapons, the treaty has not yet entered into force because eight key states, including the United States and China, have not yet ratified.

Nuclear Weapons and Related Security Issues
AIP Conf. Proc. 1898, 040002-1–040002-10; https://doi.org/10.1063/1.5009225
Published by AIP Publishing. 978-0-7354-1586-7/$30.00

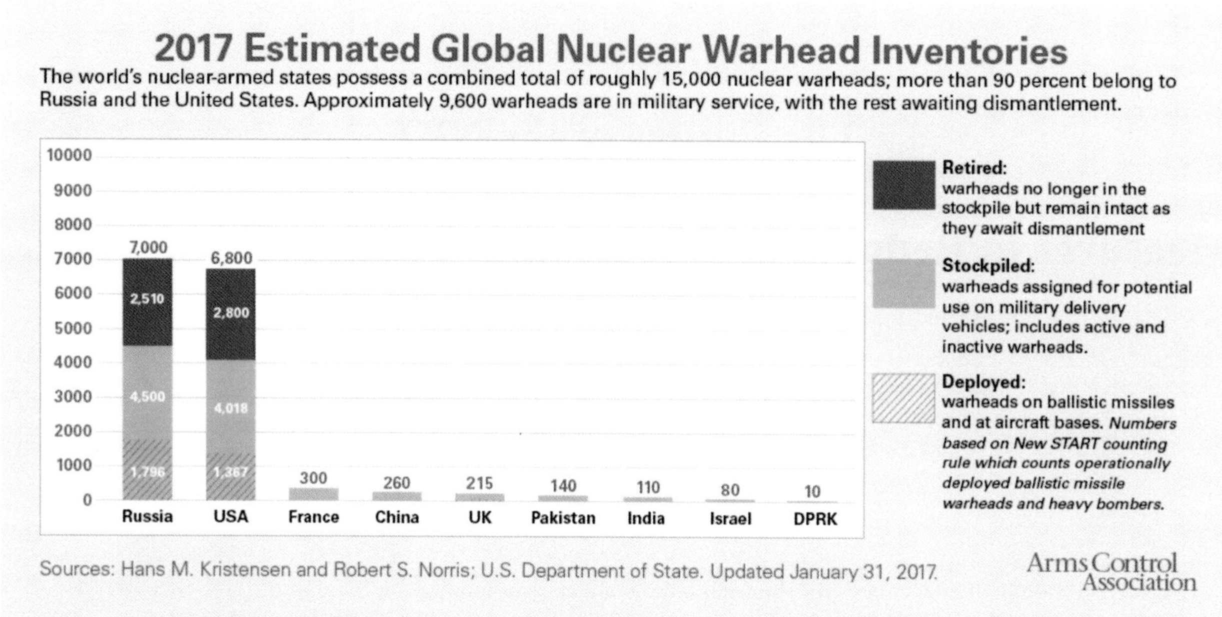

FIGURE 1. 2017 Estimated Global Nuclear Warhead Inventories (Arms Control Association.)

The nuclear geometry in Asia is also becoming more complex and far more dangerous. North Korea's unconstrained nuclear and missile programs pose a long-term threat to the global security and the nonproliferation and disarmament enterprise. Without a diplomatic agreement to halt and reverse North Korea's efforts and build toward a peace regime on the Korean peninsula, North Korea may soon have an operational arsenal of nuclear-armed ballistic missiles that can hit all of East Asia, and eventually the United States.

India and Pakistan continue to amass more fissile material and both are deploying new and more destabilizing nuclear delivery systems. China has begun to put multiple warheads on its arsenal of some 100 long-range missiles and may increase the launch readiness of its nuclear forces.

"Today, the danger of some sort of a nuclear catastrophe is greater than it was during the Cold War and most people are blissfully unaware of this danger," warned William J. Perry, the former U.S. Secretary of Defense, in his 2016 memoir, *My Nuclear Journey*.

With the election of Donald J. Trump, the situation is even more precarious. In his rambling first news conference as president on Feb. 17, Donald Trump made it clear he is aware, in general terms, of the danger. Speaking about Russia, Trump said: "…we're a very powerful nuclear country and so are they. There's no upside … nuclear holocaust would be like no other."

So far, however, he has not made it clear whether he knows how to avoid it. Trump made a number of controversial, seemingly impulsive, and sometimes contradictory comments on nuclear weapons, before and after Election Day. Some Trump appointees have made statements that put into question the United States commitment to meet its nuclear disarmament and nonproliferation obligations.

In March, Christopher Ford, the National Security Council's senior director for weapons of mass destruction and counter-proliferation, said an examination of whether global nuclear disarmament "is a realistic goal," would be conducted as part of a wider nuclear weapons policy assessment called the Nuclear Policy Review.

In reality, the pursuit of nuclear disarmament is not negotiable. It is a legally binding commitment in Article VI of the NPT. Progress on the next steps on nuclear disarmament as outlined in the 2010 NPT Review Conference Action Plan remains unfulfilled.

These trends have driven the non-nuclear weapon state majority to conclude negotiations on a treaty to prohibit nuclear weapons on July 7.

Trump has also criticized the 2015 nuclear deal between six world powers and Iran and his administration is pressing for tougher, non-nuclear sanctions against Iran. Even though Iran is complying with the stringent nuclear restrictions mandated by the agreement, the Trump administration has hinted that it may seek to blow-up the agreement.

What and how the new president and his fledgling administration will do on these and other nuclear weapons matters—and how U.S. allies and rivals respond—will have long-lasting ramifications for the NPT and the global nuclear nonproliferation and disarmament enterprise.

The bottom line is that the pillars of the global nuclear order, including the NPT, cannot be taken for granted.

The following is a brief examination of how U.S. nuclear weapons policy could change under Trump and how the United States and its allies and partners might help to reduce rather than increase global nuclear weapons dangers.

TRUMP ON NUCLEAR WEAPONS

Unlike Barack Obama, who campaigned for the presidency in 2008 on the basis of a detailed nuclear risk reduction strategy that he first outlined in legislation he introduced in 2007, Mr. Trump has not yet, in any formal way, outlined his thinking on nuclear weapons.

Instead, Trump has made a number of poorly thought out statements on nuclear weapons during the campaign and since.

If translated into formal U.S. policy, some of Trump's pronouncements on nuclear weapons would represent a radical break from long-standing U.S. policies to reduce the role and number of nuclear weapons, thereby increasing tensions with other nuclear-armed states.

Nuclear Use Doctrine

On Jan. 3, 2016, early in the campaign for the presidency, Trump was asked if he would rule out the use of nuclear weapons against terrorist groups. He said: "I'm never going to rule anything out ... because, at a minimum, I want them to think maybe we would use it, OK?"

This contrasts sharply with the Obama administration's conclusions about the threat of use of nuclear weapons. In a January 15, 2017 speech, Vice-President Joe Biden said: "Given our non-nuclear capabilities and the nature of today's threats—it's hard to envision a plausible scenario in which the first use of nuclear weapons by the United States would be necessary, or make sense. President Obama and I are confident we can deter—and defend ourselves and our Allies against—non-nuclear threats through other means."

Preventing Proliferation

In a March 26, 2016 interview, with *The New York Times* Trump suggested "it wouldn't be a bad thing for us if Japan, South Korea, Saudi Arabia acquired nuclear weapons."

> "It's going to happen anyway. It's only a question of time," he said about nuclear proliferation in another interview on May 4, 2016.

When asked about Trump's comments at his Senate confirmation hearing Jan. 11, Secretary of State Rex Tillerson said he did not agree. "I don't think anyone advocates for more nuclear weapons on the planet," Tillerson said.

> "One of the vital roles of the State Department ... has to be the pursuit of nuclear nonproliferation. We just simply cannot back away from our commitment to see a reduction in the number of these weapons on the planet," he said.

Nuclear Disarmament

Trump has contradicted himself on whether he wants to increase or reduce the number of nuclear weapons.

As President-elect, Trump reportedly told Kazakhstan's President Nursultan Nazarbayev, "There is no more important issue than nuclear disarmament and nonproliferation to be addressed in a global context," according to a Kazakhstan-issued statement on their Nov. 30 phone call.

In a pre-inauguration interview in January 2017 with the *Times of London* Trump said "nuclear weapons should be way down and reduced very substantially," and he suggested that such a deal might be linked to the easing of sanctions against Russia for its annexation of Ukrainian territory.

But Trump has also pledged to "greatly strengthen and expand" U.S. nuclear weapons capabilities, which are already substantial, and he has criticized the New START agreement with Russia, suggesting he may be looking to change nuclear policy in significant ways.

On Dec. 22, he launched a tweet declaring, "The United States must greatly strengthen and expand its nuclear capability until such time as the world comes to its senses regarding nukes." When asked by MSNBC to clarify, Trump reportedly said, "Let it be an arms race. We will outmatch them…and outlast them all."

Shortly after Inauguration Day, Trump denounced the agreement in a phone call with Russian President Vladimir Putin, suggesting the two countries might agree to extend the treaty for another five years (to 2026), according to a Reuters account published Feb. 9.

In a February 23 interview with Reuters, Trump vowed "…if countries are going to have nukes, we're going to be at the top of the pack" and called New START "one-sided."

However, in his Jan. 11 confirmation hearing, Trump's Secretary of State Rex Tillerson articulated a different approach. "We have to stay engaged with Russia, hold them accountable to commitments made under the New START and also ensure that we are in a position to meet our accountability as well," he said. He's right. New START has increased stability and predictability in the U.S.-Russian relationship and put verifiable caps on both sides nuclear arsenals.

In response to a question from Senator Jeanne Shaheen (D-N.H.), Tillerson also said he supports the long-standing bipartisan policy of engaging with Russia and other nuclear armed states to verifiably reduce nuclear stockpiles.

Trump's Nuclear Posture Review

Sorting out what the actual Trump administration policies on nuclear weapons actually are will take some time. In January, Trump ordered his Defense Secretary James Mattis to lead a new Nuclear Posture Review (NPR), the fourth since the end of the Cold War. There will be a parallel review of U.S. missile defense policy.

The nuclear policy review, which is scheduled for completion before the end of 2017, could potentially set into motion significant changes regarding

- the role of nuclear weapons in U.S. strategy;
- targeting requirements for U.S. forces;
- the plans for maintaining and upgrading nuclear forces, nuclear force structure requirements, and costs;
- whether new types of nuclear weapons are required, and
- the overall U.S. approach to nuclear arms control and nonproliferation.

The review will be completed just as the Trump administration is making decisions on key nuclear policy matters with long-term implications.

Before the end his term Trump, along with Russian president Vladimir Putin will need to decide whether to extend the 2010 New Strategic Arms Reduction Treaty (New START) and its monitoring regime past its February 2021 expiration date for another five years, negotiate a follow-on agreement, or go forward without legally-binding, verifiable limits on the world's two largest nuclear arsenals.

The NPR will also affect the course of the ongoing programs to sustain and upgrade each of the elements of the U.S. strategic arsenal, including: a new, long-range, stealthy strategic bomber; a fleet of 1,000 new nuclear-capable air-launched cruise missiles; a new fleet of 14 strategic submarines; 400 new ground-based intercontinental ballistic missiles; and upgraded nuclear command and control systems.

The latest estimates from the Congressional Budget Office put the cost in excess of $400 billion between fiscal years 2017-2026, which is six percent of the projected total costs of the fiscal 2017 national defense budget request. CBO estimates that over the next 30 years, the cost of nuclear weapons will exceed $1 trillion. (Figure 2)

If the forthcoming NPR accelerates or expands upon these plans, it would worsen global nuclear competition and undermine support for the 1968 nuclear Nonproliferation Treaty, which requires the nuclear-armed states and other NPT parties to "pursue negotiations in good faith on effective measures relating to cessation of the nuclear arms race at an early date and to nuclear disarmament …."

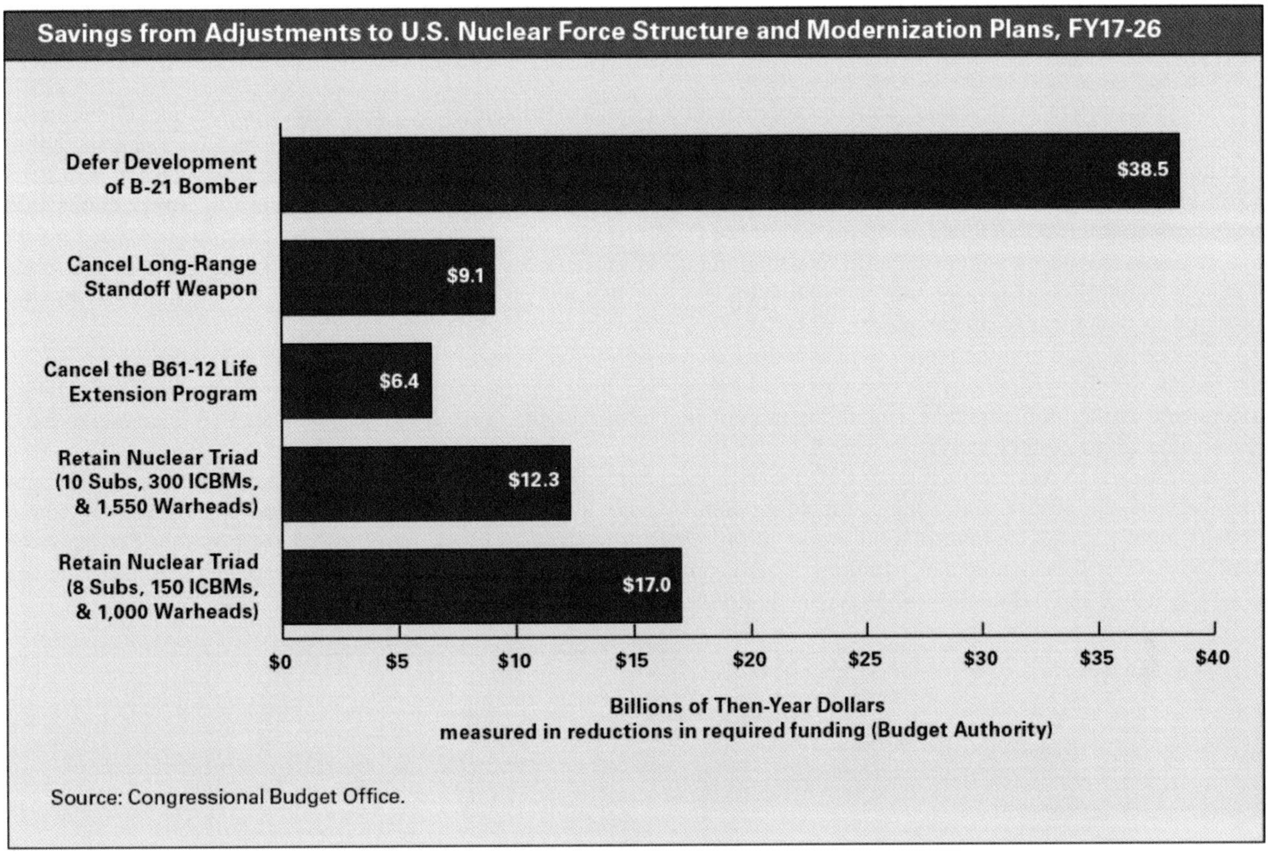

FIGURE 2. Savings from Adjustments to US Nuclear Force Structure and Modernization Plans, FY17–26. (Congressional Budget Office).

Though he appears unlikely to do so, Trump also has the option of scaling back the nuclear weapons recapitalization plan. According to estimates from CBO, tens of billions of dollars could be saved by scaling back or delaying key nuclear weapons programs.

The Potential Impact of Congress on U.S. Nuclear Policy

U.S. nuclear policy is never entirely determined by the executive branch. The Republican-led Congress may have a significant influence—positive or negative.

Trump's suggestion that the United States must increase the "capacity" of its nuclear stockpile could encourage some hawkish members of Congress to seek to overturn the Obama-era policy of "no new nuclear warhead designs" and approve funding for the development of new types of "more usable" nuclear warheads.

A December 2016 Defense Science Board report prepared for the new administration encourages the Departments of Defense and Energy to build an entirely new nuclear warhead and questions their ability to maintain the warheads in the absence of explosive testing, which is prohibited by the 1996 Comprehensive Test Ban Treaty, which the United States has signed but not ratified. The report recommends "a more flexible nuclear enterprise that could produce, if needed, a rapid, tailored nuclear option for limited use," ostensibly for a conflict in Europe with Russia.

There is already opposition to the concept from several Democratic Senators who were part of earlier, successful efforts to thwart George W. Bush administration proposals to enhance low-yield warhead capabilities. Senator Dianne Feinstein (D-CA) wrote on March 3 in *The Washington Post*: "There is no such thing as "limited use" nuclear weapons …This is even more problematic given President Trump's comments in support of a nuclear arms race."

Feinstein cited the 2015 testimony of Deputy Defense Secretary Robert Work who said, "Anyone who thinks they can control escalation through the use of nuclear weapons is literally playing with fire. Escalation is escalation, and nuclear use would be the ultimate escalation."

Other members of the House and Senate, including Rep. Joe Wilson (R-SC) and Sen. Tom Cotton (R-AR) have introduced legislation to restrict funding for the Comprehensive Test Ban Treaty Organization, which is responsible for monitoring global compliance with the CTBT. The bill also calls on Congress to declare that a Sept. 23, 2016 UN Security Council resolution does not "impose an obligation on the United States to refrain from actions that would run counter to the object and purpose" of the CTBT, which bans nuclear test explosions.

If Congress were to adopt this bill, it would signal to other states that that the United States is seeking to back out of its commitment to a global, verifiable nuclear test ban and is considering, after a 25 year moratorium, the resumption of nuclear weapons testing.

Trump has said nothing about nuclear testing so far, but his Secretary of State Rex Tillerson said in responses to the Senate Foreign Relations Committee in January: "So long as the reliability of our nuclear deterrent can be guaranteed through other means, I think the moratorium has served us well. It would not serve U.S. interests to have Russia and China resume nuclear testing."

In response to reports of Russian INF Treaty non-compliant ground-launched cruise missiles, several influential members of the House and Senate, including Sen. Marco Rubio (R-FL) have introduced legislation to pursue development of a U.S. ground-based cruise missile in Europe and to block funding for the possible extension of New START if Russia does not withdraw its INF non-compliant cruise missiles. Such a response would divide the NATO alliance and hasten the end of verifiable limits on U.S. and Russian nuclear forces. (Figure 3)

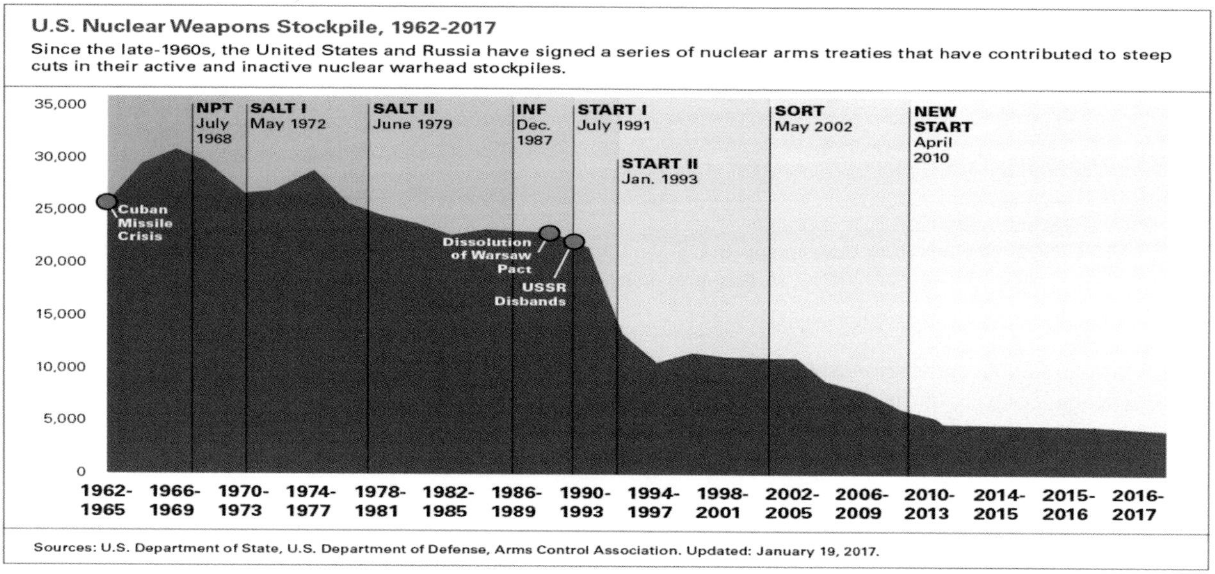

FIGURE 3. U.S. Nuclear Weapons Stockpile, 1962–2017. (Arms Control Association.)

Key Decisions and Options

For decades, U.S. presidents from both parties have sought to reduce the risk of a nuclear confrontation, cut bloated nuclear stockpiles, and prevent proliferation. Some have been more successful than others, but all have tried. Maintaining progress on U.S. and Russian nuclear disarmament helps to lower tensions and maintain strategic stability and is vital to helping reinforce efforts to prevent the spread of nuclear weapons.

Trump's early statements on nuclear weapons and nuclear arms control are deeply troubling. Key U.S. allies in Europe and Asia have an important role to play in encouraging the Trump administration to respect and uphold past U.S. nuclear risk reduction commitments and to seek ways to achieve further maintain progress in several key areas:

Reduce Nuclear Tensions. When Trump and Putin meet later this year, the two leaders could reduce worries about nuclear missteps by reaffirming the statement by U.S. President Ronald Reagan and Soviet leader Mikhail Gorbachev from 1985 that "a nuclear war cannot be won and must never be fought." They should also be encouraged reaffirm their commitment to the quarter-century-long U.S. and Russian moratoria on nuclear weapons test explosions and the prompt entry into force of the 1996 Comprehensive Test Ban Treaty, which both have signed but only Russia has ratified.

Extend New START and Seek Deeper Cuts. As President Barack Obama noted in his final press conference, "[T]here remains a lot of room for both countries to reduce our nuclear stockpiles." With up to 1,550 deployed strategic nuclear weapons allowed under New START, Russia and the United States can safely cut their bloated nuclear stockpiles further without negotiating a new treaty.

By agreeing to extend New START and its verification provisions by five years, to 2026, Trump and Putin could confidently pursue further, significant parallel reductions of warhead and delivery system inventories by one-third or more and still meet their respective nuclear deterrence requirements. This step would ease tensions and reduce fears of a new nuclear arms race, plus it would reduce the skyrocketing price of nuclear weapons.

Address INF Treaty Violations. Russia's deployment of ground-based cruise missiles prohibited by the landmark 1987 Intermediate-Range Nuclear Forces (INF) Treaty is a serious matter. Trump said on Feb. 23 he would take up the issue with Putin when they meet. He should, but he should remember that Russia's missile deployments do not significantly alter the military balance. The two sides should agree to discuss the United States evidence of the violation at another meeting of the treaty's Special Verification Commission and to work to resolve all outstanding compliance issues.

If Moscow continues to deploy the banned ground-launched cruise missiles, U.S. and NATO leaders should insist that the weapons would need to be counted under the limits set in the next round of nuclear arms reductions. Washington should also continue to support ongoing NATO efforts to bolster the conventional defenses of those allies that would be potential targets of Russian aggression or intimidation.

Adjust U.S. Missile Defense Plans. The United States has followed through on its phased adaptive approach for limited missile defenses in Europe to counter Iran's medium-range missile arsenal. With the successful implementation of the 2015 Iran nuclear deal, Washington can suspend the deployment of more advanced Aegis missile interceptors in Poland, as well as a possible ground-based, strategic interceptor site on the U.S. East Coast—both of which are oriented to counter a long-range, nuclear-tipped Iranian missile threat that has not materialized. Failing to adapt missile defenses will only deepen Russian suspicion the system is also directed at them and increase the likelihood of dangerous Russian countermoves.

Reaffirm the Commitment to the CTBT. For more than two decades, the United States has led global efforts to verifiably halt nuclear testing — an activity that is not necessary for maintaining the U.S. nuclear arsenal but that can be used to advance other states' nuclear weapons capabilities and that fuels nuclear proliferation. As U.S. Secretary of State Rex Tillerson wrote in his responses to questions from his January 2017 confirmation hearing, "I think the [nuclear test] moratorium has served us well."

The United States and China have signed but not yet ratified the 1996 Comprehensive Test Ban Treaty (CTBT). Six other CTBT "hold-out" states must also ratify to trigger its formal entry into force.

In the meantime, it is essential that all states help to reinforce the global taboo against nuclear testing and provide the financial and technical support to complete and maintain the Comprehensive Test Ban Treaty Organization's (CTBTO) International Monitoring System, its data center, and on-site inspection capabilities designed to detect and deter clandestine testing.

In September 2016, on the 20th anniversary of the opening for signature of the CTBT, 42 states, including Israel, adopted UN Security Council Resolution 2310, which encourages states to "provide the support required" to the CTBTO and the IMS, and urges states to refrain from nuclear testing and urges those states that have not ratified to do so; and takes note of a September 15 joint statement by the five permanent Security Council members formally "recognizing" that a nuclear-weapon test explosion would "defeat the object and purpose of the CTBT."

Building on UNSC Resolution 2310, the U.S. and Chinese leaders should reaffirm their commitment to the nearly quarter-century-long global moratoria on nuclear weapons test explosions and move forward on consideration of the ratification of the treaty.

Enforce and Preserve the Iran Nuclear Deal. Two years ago on July 14, 2015, six world powers and Iran finalized an historic nuclear agreement with Tehran that removed the threat of an Iranian nuclear weapon. The nuclear deal, known as the Joint Comprehensive Plan of Action, has proven to be a nonproliferation success. The agreement has significantly restricted Iran's nuclear activities and imposed an intrusive monitoring and verification regime. The threat of a nuclear-armed Iran no longer looms over the international community.

Despite the success of the deal, some policymakers in Washington are recklessly urging the Trump administration to abandon the agreement on the basis of inaccurate claims that Tehran is violating the accord. Six reports from the International Atomic Energy Agency demonstrate that Iran is meeting its commitments and the Trump administration certified in April that Iran is living up to its end of the deal.

Before taking unilateral steps that risk the success of the agreement, Washington should carefully consider the consequences. Abandoning an agreement that is verifiably blocking Iran's pathways to nuclear weapons is irresponsible and dangerous. It would further destabilize the region, and could open the door to a nuclear-armed Iran or increase the prospect of a costly war in the Middle East. Additionally, pulling out of a multilateral agreement that benefits international security sends the message to U.S. allies and partners that Washington cannot be trusted to follow through on its commitments.

Given the current instability in the Middle East, it is now more vital than ever that Washington continue to support the nuclear deal with Iran and look for options to build on the agreement. Full implementation of the accord benefits global security and international efforts to prevent the spread of nuclear weapons.

Diplomacy to Halt North Korea's Nuclear Expansion. Pyongyang's advancing nuclear and ballistic missile programs are the most urgent nuclear proliferation challenges today.

North Korea already has the capability to deliver a warhead on a short- or medium-range ballistic missile. With more nuclear tests, North Korea can further refine its warhead designs to increase the explosive yield and develop a lighter, more compact warhead. North Korea's next nuclear test explosion, which would be its sixth, could be imminent.

North Korea is estimated to have some 50 kilograms of plutonium, enough for 10 nuclear explosive devices. International Atomic Energy Agency Director-General Yukiya Amano said on March 20, 2016 that North Korea has doubled its capacity to produce highly enriched uranium. By 2020, it may have enough fissile material for 100 warheads. Worse yet, it could soon begin flight-testing an intercontinental ballistic missile.

If the current action-reaction cycle continues, it will not only diminish the prospect of denuclearization of the Korean peninsula, but it will increase the risk of a devastating nuclear war.

The new administration of President Donald Trump, along with U.S. allies and partners in Northeast Asia, have a rapidly closing window of opportunity to pull back from potential escalation and war and pursue a diplomatic course that halts and eventually reverses Pyongyang's nuclear weapons program.

The Trump administration has completed its review of North Korea policy and is pursuing a variation on past policy that they describe as "maximum pressure and engagement." The question is what kind of pressure? What are the terms of engagement and when? Are China and our allies in the region on board?

Ideally, engagement should begin soon. Trump should offer direct "talks about talks." To move forward, Trump should take note of Pyongyang's statement from July 2016, which called for denuclearization of the Korean peninsula, to make clear that the United States remains committed to denuclearization as the end state of the process. If North Korea is willing to negotiate, initial talks should focus on obtaining a moratorium to prevent additional nuclear and ballistic missile tests in order to buy time for a more far-reaching and lasting solution.

For that, the United States will need to be prepared in return to put something on the table. In consultation with the incoming government in Seoul, Washington should consider scaling back or delaying the next round of joint military exercises with South Korea. The exercises could be scaled up if North Korea breaks off talks or conducts further nuclear or ballistic missile tests.

If the initial moratorium holds, North Korea and the United States could proceed to steps that would roll back Pyongyang's nuclear activities, including a verifiable halt to fissile material production that would be monitored by international inspectors at North Korea's nuclear sites. In return, the United States might extend to North Korea limited sanctions relief, offer to adjust or postpone U.S.-RoK military exercises, issue negative security assurances against military attack under certain conditions, and begin formal talks on a formal end to the Korean War.

To maintain leverage, the United States and its partners must strengthen implementation of existing UN Security Council sanctions. China's decision to halt coal imports from North Korea is a good start. The latest report of the UN panel of experts on sanctions against North Korea said that the Security Council must compel better enforcement by imposing penalties against non-North Koreans involved in evasion.

Other policy approaches pose very high risks and have lower chances of success. New sanctions without an offer of talks would be opposed by China, and North Korea would, as leader Kim Jong Un said Jan. 1, "continue to build up" its nuclear forces, which he sees as a guarantor of regime survival. Pre-emptive U.S. military strikes would be operationally difficult and provoke a strong, likely military response from Pyongyang that could trigger a second Korean War.

The Treaty on the Prohibition of Nuclear Weapons. In the midst of all of this is a new initiative that could further reduce the salience of nuclear weapon: the new Treaty on the Prohibition of Nuclear Weapons.

Fundamentally, the initiative aims to spur action on nuclear disarmament and risk reduction and to further delegitimize their possession.

Although the world's nuclear-armed states have boycotted the negotiations, the process and the final product could help strengthen the legal and political norm against their use—a worthy goal, especially in light of the uncertainty surrounding U.S. nuclear policy under Trump's leadership.

Contrary to some skeptics, this process is not a "distraction," nor will it undermine the NPT, as some fear – so long as ban treaty advocates recognize its value and its limitations and so long as the nuclear weapon states do not continue to suggest that the ban treaty is the source of the nuclear nonproliferation regime's problems.

In reality, the stresses and strains on the NPT are due to the actions of North Korea, the inability of the major nuclear armed states to make progress on disarmament commitments, the technological arms race by the nuclear weapon states, and the failure of key states in the Middle East to agree on the agenda for a conference on a WMD-free zone in their region – not the ban treaty negotiations.

In order to attain a world free of nuclear weapons, it ***will*** be necessary, at some point, to establish a legally-binding norm to prohibit such weapons. As such, the pursuit of a treaty banning the development, production, possession and use of nuclear weapons is a key step along the way. This new treaty has the potential to further delegitimize nuclear weapons and strengthen the legal and political norm against their use—a worthy goal.

CONCLUSION

In the seven decades since the attacks on Hiroshima and Nagasaki, nuclear weapons have become less and less relevant to the security of possessor states and their allies and more harmful to international security and human survival.

Today, the world's nuclear-armed states still face significant security threats, but none can be effectively resolved with nuclear weapons or the buildup of nuclear capabilities. Rather, it is more essential than ever that the leaders of all of the world's states must redouble efforts to strengthen the NPT and head-off renewed nuclear competition.

APPENDIX: U.S. Nuclear Policy Before Trump

Since the end of the Cold War every administration has conducted a comprehensive review of U.S. nuclear weapons strategy and policy. The most recent review by the Barack Obama administration Nuclear Posture Review (NPR) was completed in April 2010.

The 2010 NPR posited that the top priority of the U.S. nuclear agenda should be to move toward a world free of nuclear weapons through U.S. leadership to bolster the nuclear nonproliferation regime, including "reversing the nuclear ambitions of Iran and North Korea" and impeding illicit nuclear trade, accelerating efforts to secure vulnerable nuclear weapons-usable materials worldwide. Obama's NPR also emphasized the importance of further efforts to reduce the number of nuclear weapons, including New START, while at the same time reinforcing "strategic stability" with Russia and China.

The 2010 NPR and follow-on *Nuclear Weapons Employment Guidance Report* issued by the Pentagon in 2013 recognized that a disarming surprise nuclear attack against U.S. forces has diminished significantly since the Cold War. However, it still directs the DoD to maintain a significant portion of the U.S. nuclear force in a "launch under attack" mode in the event that Russia initiates a major nuclear attack.

The 2010 NPR emphasized that the "fundamental role of U.S. nuclear weapons is to deter nuclear attack" on the United States and its allies and it declared that "the United States will not use or threaten to use nuclear weapons against non-nuclear weapon states that are party to the NPT and in compliance with their nuclear non-proliferation obligations."

The 2010 NPR, like its 2002 predecessor, argued that the United States "extended deterrence" to allies has relied increasingly on non-nuclear elements including forward U.S. conventional presence and effective theater missile defenses."

The Obama NPR also underscored that "as long as nuclear weapons exist, the United States must sustain a safe, secure, and effective arsenal." It called for sustained investments in recapitalizing the "triad" of nuclear delivery systems and the ongoing warhead life extension programs, but determined that the United States "will not conduct nuclear [explosive] testing" and "will not develop new nuclear warheads."

The 2010 NPR said warhead life-extensions will "use only nuclear components based on previously testing designs, and will not support new military missions or provide for new military capabilities."

Under the Obama administration, however, the United States made improvements to other parts of some of its weapons. For instance, the B-61-12 nuclear gravity bomb life extension program incorporates a new guidance package and tail kit that makes it more accurate.

International Atomic Energy Agency Safeguards: Challenge and Response

Leonard S. Spector

Executive Director
Washington, DC, Office
James Martin Center for Nonproliferation Studies
1400 K Street, Washington, DC, 20005

Abstract: This article provides a critical review of the nuclear accounting and inspection system of the International Atomic Energy Agency (IAEA), known as "IAEA safeguards." The article focuses on the multiple challenges the Agency confronts in verifying that all nuclear activities in the countries under its safeguards system are being pursued for exclusively peaceful purposes. The principal challenges noted are those posed by: undeclared facilities, the development of enrichment and reprocessing capabilities, illicit procurement activities, denial of inspector access, difficulties in verifying absence of weaponization activities, and difficulties in establishing that all nuclear-relevant activities in a state are peaceful. The article is in the form of annotated PowerPoint briefing slides.

INTRODUCTION – THE BASICS OF IAEA SAFEGUARDS

This article provides a brief review of the role of the nuclear accounting and inspection system of the International Atomic Energy Agency (IAEA), known as "IAEA safeguards," and examines current challenges to their effectiveness.

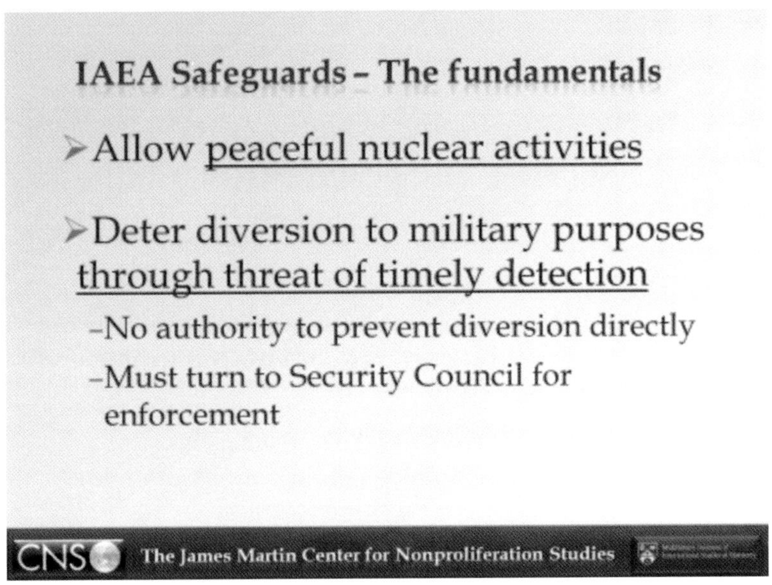

FIGURE 1.

Nuclear Weapons and Related Security Issues
AIP Conf. Proc. 1898, 040003-1–040003-12; https://doi.org/10.1063/1.5009226
Published by AIP Publishing. 978-0-7354-1586-7/$30.00

The IAEA is an intergovernmental organization, based in Vienna, Austria, with 168 member states. One of its primary responsibilities is operating a system of nuclear material accounting and inspections, known as "safeguards."[1] Safeguards seek to deter the diversion of nuclear materials from peaceful uses through the threat of timely detection. The Agency has no authority to intervene to prevent such diversions, but must refer cases of suspected diversion or other noncompliance with its safeguards system to the UN Security Council. In the case of Iran, for example, in 2006, the Agency determined that the country was not in compliance with its safeguards obligations and referred the case to the Security Council, which subsequently imposed sanctions against Iran.[2]

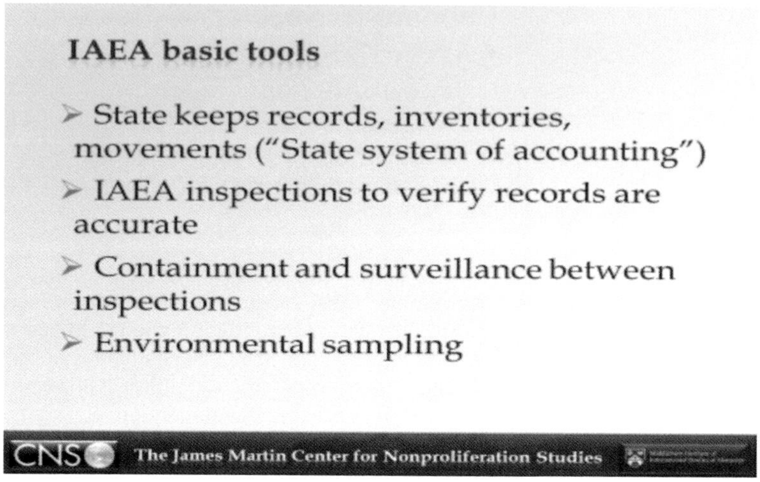

FIGURE 2.

The Agency's safeguards system requires a state to keep records of all domestic production, imports, and exports of nuclear materials, as well as the location of such materials within the state. The Agency then undertakes periodic inspections to verify the state records. The frequency of inspections varies according to the nature of the nuclear material at issue, with materials directly usable as the core of nuclear weapons, namely highly enriched uranium and plutonium, receiving the most frequent monitoring. Between inspections the Agency uses containment and surveillance measures – video cameras and seals – to ensure materials have not been moved surreptitiously. It also has authority to take environmental samples at inspected sites to check for the movement of declared materials and the presence of undeclared materials.

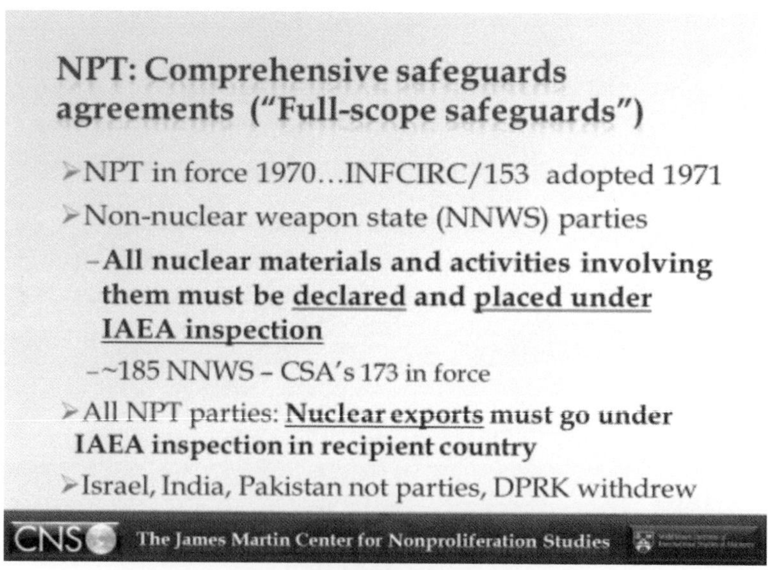

FIGURE 3

The nuclear Nonproliferation Treaty (NPT) entered into force in 1970 and now has 189 parties. Under the treaty, states that detonated a nuclear explosive before January 1, 1969 – China, France, the Soviet Union/Russia, the United Kingdom, and the United States – are considered "nuclear weapon states." They are not required to renounce nuclear weapons, but pledge to work in good faith to end the nuclear arms race and achieve nuclear disarmament. The other parties to the treaty are considered non-nuclear-weapon states. They are obligated to renounce nuclear weapons and, to verify their compliance with this pledge, must declare all their nuclear materials and accept IAEA safeguards on *all* of these (in effect, on all of their nuclear activities), an arrangement known as "full-scope" or "comprehensive" safeguards. Under the treaty, all parties agree that transfers of nuclear materials to non-nuclear-weapon states must be placed under IAEA safeguards upon arrival.

India, Israel, and Pakistan are not parties to the treaty, and North Korea announced its withdrawal from the pact in 2003. All have announced their possession of nuclear weapons or, in the case of Israel, are widely believed to possess them. These states can submit individual facilities containing nuclear materials (usually imported ones) to safeguards, but do not allow inspections at facilities associated with their nuclear weapon programs.

The document setting out the framework for implementing comprehensive safeguards is known as Information Circular 153 (INFCIRC/153), originally adopted in 1971. Under that framework, states sign "comprehensive safeguards agreements" (CSA's) with the Agency, adapting INFCIRC/153 to the specific circumstances in that state. There are 185 NPT non-nuclear-weapon states of which 173 have CSA's. (The others have insignificant amounts of nuclear material in their territories, making this unnecessary.)

CHALLENGES TO THE IAEA SAFEGUARDS SYSTEM

Undeclared Facilities

INFCIRC/153 depends for its effectiveness on states' providing complete information on their nuclear materials and the facilities processing or using them. It was recognized that states might cheat and not declare all such information, which could allow the state to build a clandestine nuclear weapon production capability. Using these secret facilities, a state could sneak out of the NPT and declare itself a nuclear power, presenting the world with a nuclear fait accompli.

FIGURE 4

It was not until 1991, however, that the issue was recognized as a serious challenge to the IAEA system. The trigger was the exposure of Iraq's secret uranium enrichment research and development program after the first Persian Gulf War through Security Council-mandated inspections that went well beyond those of INFCIRC/153.[3] During the years the Iraqi program was under way, Iraq was a party to the NPT and subject to INFCIRC/153 inspections – but only at the handful of nuclear facilities it had declared to the Agency.

Some six years later, in 1997, to address the issue of undeclared nuclear activities in NPT non-nuclear weapon states, the IAEA issued an enhancement of its safeguards regime known as the Model Additional Protocol.[4] The major additions to the IAEA safeguards system set out in that document are noted in Figure 4. Because many IAEA members believed that some of these additional measures went beyond the authority conferred on the Agency by its statute, it was agreed that NPT non-nuclear-weapon states would be asked to adopt the additional protocol voluntarily as an addition to their Comprehensive Safeguards Agreements with the Agency. As of mid-2017, 129 states had done so, and seventeen others had taken preliminary steps towards bringing Additional Protocols into force.[5]

At their 2016 Lubeck Summit, the Group of Seven (Canada, France, Germany, Italy, Japan, the United Kingdom, and the United States) agreed to make a state's adoption of the Additional Protocol a condition for the supply of nuclear goods, such as power reactors and fuel, by G-7 members.[6] These states constitute the majority of the suppliers of such goods. Russia and China have not adopted this condition formally, but as a practical matter, all NPT non-nuclear-weapon states to which they are currently supplying or are planning to supply nuclear power reactors have Additional Protocols in force.[7]

Several additional episodes underscore the challenge posed by clandestine facilities. In August 2002 Iran's partially completed efforts to construct a secret uranium enrichment facility at Natanz were exposed. Iran was a non-nuclear-weapon states party to the NPT, but at the time, it had not brought an Additional Protocol into force. In the wake of the controversy triggered by revelation of the Natanz plant, in 2003, Iran agreed to accept the protocol's enhanced IAEA inspections on a de facto basis, but it ceased adhering to that instrument in 2006. Thereafter, in October 2009, it was learned that it had been constructing a second enrichment plant, at Fordow, again accentuating the threat posed from clandestine facilities.

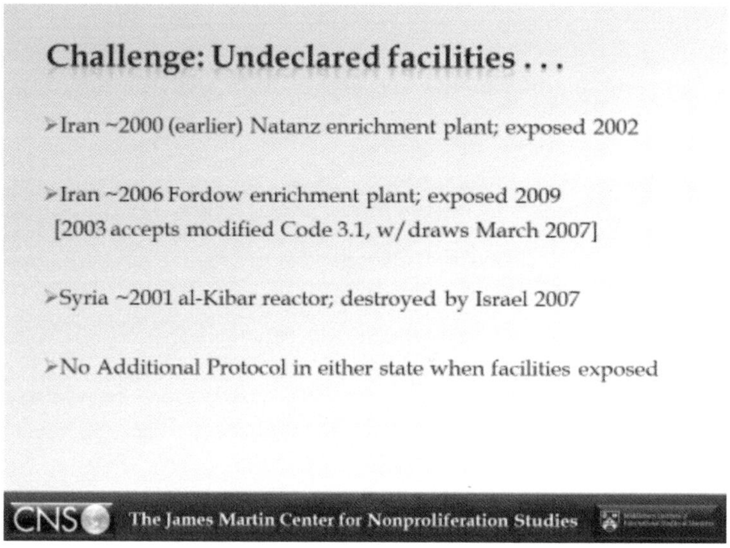

FIGURE 5.

Meanwhile, in 2007, a clandestine plutonium production reactor at al-Kibar in Syria was destroyed in an Israeli air strike. The facility had been under construction since 2001.[8] Like Iran, Syria was a non-nuclear-weapon state party to the NPT but had not brought its Additional Protocol into force. The IAEA later conducted an inspection of the reactor site, which Syria had cleared to obscure any evidence that it had housed a nuclear facility. The Agency was, nonetheless, able to detect traces of processed uranium at the site. This and satellite images of the facility during the clean-up efforts, which revealed the contours of the reactor, left little doubt as the nature of the facility.

Returning to Iran, to date, it has not formally adopted an Additional Protocol. Tehran has, however, agreed voluntarily to adhere to the instrument's provisions for enhanced IAEA monitoring under its July 2015 nuclear agreement with China, France, Germany, Russia, the United Kingdom and the United States, known as the Joint Comprehensive Plan of Action (JCPOA).[9] As part of that agreement, it has pledged that it will formally bring its Additional Protocol into force at a later stage of the agreement's implementation.

Importantly, the inspection provisions of the JCPOA go beyond those of the Additional Protocol to include use of the most advanced monitoring techniques; declaration and monitoring of all of Iran's stocks of uranium ore concentrate, which could be converted into uranium hexafluoride for enrichment; and declaration and monitoring of Iran's stocks of key components for the manufacture of uranium enrichment centrifuges, along with key equipment for centrifuge production.[10] In effect, Iran's negotiating partners believed that given the history of Iranian disregard of IAEA requirements, measures beyond those of the Additional Protocol were essential. Still more draconian monitoring measures were imposed on Iraq in 1991 by the Security Council. Thus, in a series of special cases, at least three echelons of monitoring have been deployed at various times that go beyond INFCIRC/153.

Challenge: Enrichment and Reprocessing

The IAEA system must also contend with another difficult scenario under which an NPT non-nuclear-weapon state, while fully adhering to its Comprehensive Safeguards Agreement and its Additional Protocol, constructs facilities that are capable of producing nuclear-weapon-usable highly enriched uranium or plutonium, claiming these are part of a peaceful nuclear program. Once it is operating facilities of this kind, the state in question can then produce stocks of one or both weapons-usable materials and, by allowing IAEA monitoring, remain in compliance with its nonproliferation obligations. However, with weapon-usable material in hand, the state could suddenly terminate the monitoring and, if it had secretly fabricated the non-nuclear components for nuclear arms, could possess a small arsenal of nuclear weapons in a matter of weeks, in effect, breaking out of the nuclear restraints it had previously accepted before the international community could mount an effective response.

FIGURE 6

In the case of Iran, this scenario so troubled the international community that in the JCPOA, Iran's negotiating partners insisted not only that Iran implement the Additional Protocol and certain further enhancements to IAEA monitoring noted earlier, but also required that Iran (1) greatly reduce its uranium enrichment operations; (2) produce only non-weapons-grade low enriched uranium (LEU); and (3) limit stocks of this material, which could potentially be further enriched to weapon-grade, so that the stocks, if further enriched, would not produce enough highly enriched uranium for a single nuclear weapon.[11]

The 48-member Nuclear Suppliers Group has attempted to limit transfers of enrichment and reprocessing technology to NPT non-nuclear-weapon states with considerable success. In addition, the United States in its agreements establishing the terms for nuclear trade with NPT non-nuclear-weapon states has sought in recent years to obtain pledges that they will not construct enrichment or reprocessing facilities. The United Arab Emirates and Taiwan have agreed to such restrictions, while Viet Nam has declared it has no current intention of doing so. Russia and France seek to curb enrichment and reprocessing through a different mechanism, arranging to supply all enriched uranium fuel needed for reactors they sell and arranging to take back plutonium-bearing used (spent) nuclear fuel to be reprocessed in Russia and France, respectively, with the plutonium retained there and only non-weapon-usable high-level radioactive waste returned to the partner country.

These various restraints obviously go beyond the requirements of IAEA monitoring, in effect, recognizing the inherent limitations of the IAEA safeguards system and attempting to compensate for them.

Challenge: Illicit Procurement of Nuclear-Relevant Dual-Use Goods

Contributing both to the sneak out and break out challenges to IAEA safeguards is the challenge of the illicit procurement of nuclear dual-use goods to support nuclear programs of concern Although, as noted, the nuclear supplier states have refrained from transferring enrichment and reprocessing technology to NPT non-nuclear-weapon states, both Iran and North Korea (previously a party to the NPT) have been able to construct enrichment plants using indigenous capabilities augmented by the illicit procurement from abroad of essential high-tech nuclear-relevant goods. Important illegal acquisitions have included high-strength steel, aluminium, and carbon fiber, various process controls and measuring devices, and specialized magnets and lubricants.

FIGURE 7

Figure 7, based on information in the 2014 report of the UN Panel of Experts that supported Security Council implementation of sanctions against Iran, highlights Iran's use of illicitly acquired, export controlled materials for its IR-1 centrifuges, the type used at the Natanz facility.[12] Both commodities mentioned, aluminium-7000 and maraging steel, were subject to a Security Council embargo on transfers of nuclear-relevant dual-use goods to Iran.

Supplier countries have extensive systems to try to thwart such illegal procurement activities, including export control lists, export licensing requirements, customs inspections of outgoing cargoes, and post-export interdiction efforts. These appear to have slowed the progress of Iran and North Korea in developing sensitive nuclear facilities, but were unable to halt these efforts.

Trade in nuclear-relevant goods involves two categories of commodities: goods especially designed or prepared for nuclear use (in effect, nuclear-specific goods having no other uses) and dual-use goods, items or materials that are used in both nuclear and non-nuclear settings. An example of the former are control-rod drives used in nuclear power plants, and an example of the latter are triggered spark gaps, used in kidney-stone smashing lithotripters and as triggers for nuclear weapons. Controlled especially designed or prepared items are listed under the guidelines of the Nuclear Suppliers Group under Part I of its control list, and nuclear-relevant dual-use goods, under Part II.[13]

The NPT requires that transfers of especially designed or prepared goods to non-nuclear-weapon states be placed under IAEA safeguards (or in safeguarded facilities) in the recipient country to ensure that they are not used for nuclear explosive purposes. But the NPT is silent on the issue of transfers of nuclear-relevant dual-use goods. The Nuclear Suppliers Group requires members to control such goods through an export licensing process, but does not require that the goods be monitored by the IAEA. Nor has the IAEA sought to monitor such goods and, indeed, the IAEA Board of Governors has rejected this possibility as beyond the Agency's authority.

The result, as seen at the Iranian Natanz enrichment plant, can be disturbing. A number of months after the Natanz facility was exposed in August 2002, Iran agreed to place the facility under IAEA monitoring, which has continued ever since, with no violations reported. From the time safeguards were first applied until the July 2015 JCPOA, the number of centrifuges at the facility increased from zero to 19,000. All of these were first generation IR-1 models.

As shown by the UN Panel of Experts finding above, however, Iran used controlled materials for key components in all of these centrifuges. Since none of the limited number of suppliers of maraging steel, 7000-level aluminum, or high-strength carbon fiber is known to have deliberately sold such embargoed goods to Iran, it must be assumed that the materials were acquired illegally. Unfortunately, although the Natanz facility was thus contaminated with illegally procured material – and there were other illegally obtained items observed in photos of the facility published by the Iranian government – the IAEA's mandate was limited to determining whether nuclear material processed at the facility had been diverted. Since no evidence of diversion was found, the Agency declared the facility to meet the IAEA's requirements, never addressing the tainted origins of its equipment.

The JCPOA seeks to address this problem in the case of Iran by establishing a "white" channel through which Iran can acquire materials and equipment needed for its permitted nuclear activities – and proscribing all other acquisition of nuclear goods under the threat of renewed sanctions, which, under the JCPOA are triggered automatically for any material violation of the agreement, including illicit procurement activities. To date, Iran appears to be complying with these terms.

Challenge: Denial of Access

The IAEA is granted the authority under its statute to undertake "special inspections" in order to locate undeclared nuclear activities in states under its safeguards system, authority reinforced in states that have accepted Additional Protocols to their Comprehensive Safeguards Agreements. In practice, however, the Agency has had difficulty in gaining access to suspect sites in a number of important cases.

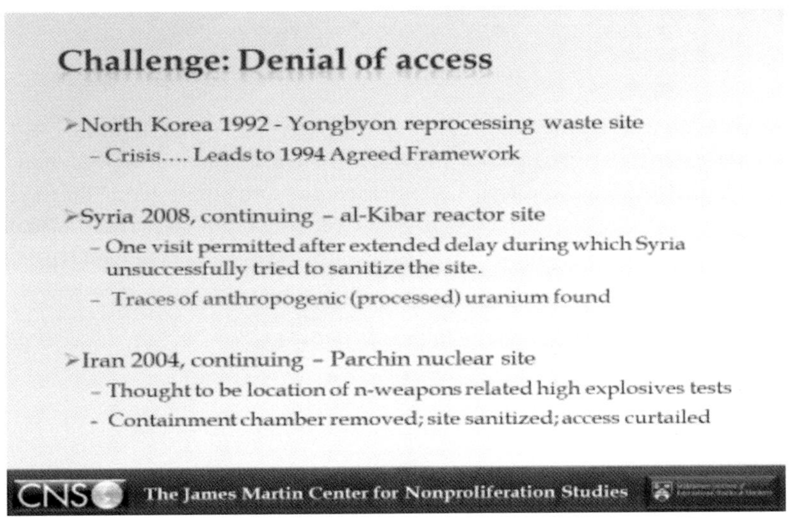

FIGURE 8

The first such episode involving states of current concern took place in 1992, when North Korea after some seven years of delay, allowed the IAEA access to the country to begin implementing safeguards under the NPT. The Agency found traces of separated plutonium in a processing facility at North Korea's Yongbyon nuclear complex. As North Korea had not declared past reprocessing of spent fuel to separate plutonium, the Agency sought to probe further and sought access to a nuclear waste storage facility at the site, hoping to find residues from reprocessing activities. Pyongyang refused to acknowledge the existence of such a waste facility, which satellite imagery indicated North Korea had attempted to disguise by covering it with earth. An impasse was reached, leading the Security Council to threaten to impose sanctions if North Korea did not accede to the Agency's request, an action Pyongyang declared it would consider an act of war. As tensions mounted, a deal was struck known as the Agreed Framework, under which North Korea froze its plutonium production activities, promising to permit access to the contested location at a later time after key milestones in other aspects of the deal concerning the supply to North Korea of a nuclear power plant had been reached. In 2002, however, before this milestone was reached, the Agreed Framework collapsed, after it was discovered that North Korea had been secretly constructing a uranium enrichment facility at Yongbyon. Thus, the IAEA never gained access to the waste site.

In 2007, following Israel's destruction of Syria's al-Kibar reactor, after considerable delay, the IAEA, as noted earlier, was allowed to visit the site of the facility, by which point remnants of the reactor had been razed and a new building erected in the same location. The Agency sought access to the debris from the site as a means for confirming what activities had taken place there, but Syria refused to comply, nor did it allow subsequent Agency visits to the site. The IAEA Board of Governors referred the Syria matter to the Security Council in June 2011 for further action. The Council has not addressed the matter, however. Russia, for one, argued that the Council was already embroiled in two difficult cases, Iran and North Korea, and there was little to be gained from its involvement in the Syrian case, since any danger posed by the reactor had been eliminated by its destruction.

In another episode, in 2004, the IAEA sought access to the Parchin site in Iran, where it was believed Iranian scientists had conducted experiments with conventional high explosives and natural (non-weapon grade) uranium to test the design of a nuclear weapon detonation package. During an early visit to the site, the Agency was not able to observe the building at issue, and satellite imagery showed that extensive clean-up activities were later conducted at the site, leaving in doubt whether full access to the building would yield meaningful information.

The controversy over Parchin was a component of a broader clash with the Agency over whether Iran engaged in nuclear weapon development activities, referred to in Agency documents as the "possible military dimension" of the Iranian nuclear program. Although Iran provided some additional information on aspects of this issue, the matter was never fully resolved and, in effect, was put aside as agreement was reached on the JCPOA.[14] U.S. officials have argued that Washington had sufficient knowledge, based on intelligence, regarding the activities at the site and regarding the overall Iranian nuclear weapon development effort, making further investigation by the Agency unnecessary.[15]

Under the JCPOA the Agency will have added authority to demand access to suspect locations. The agreement sets up an elaborate mechanism for enforcing these rights, and if Iran persists in refusing the access that the Agency is seeking sanctions against Iran that were lifted under the accord will be reimposed.

Challenge: Weaponization Activities

As the discussion of Parchin suggests, the Agency has had difficulty in obtaining a full picture of weaponization activities in Iran, despite extensive investigation that identified a number of components of the Iranian program. Figure 9 lists many of the areas where the IAEA gained insight into Iran's work on nuclear weapons.[16] However, because many of these activities can take place in laboratories, which are difficult to monitor, and have few tell-tale signatures – unlike the operation of nuclear facilities – exposing such work is inherently challenging.

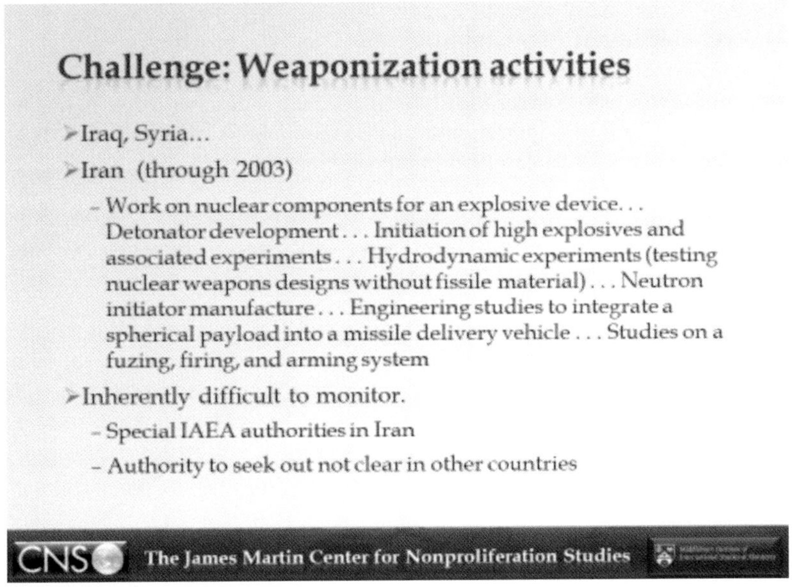

FIGURE 9.

The IAEA will have added authority under the JCPOA to monitor Iranian weaponization activities and certain activities in this realm are expressly prohibited under the accord. As noted a congressional hearing on the deal, the prohibited activities include work on, "explosively driven neutron sources and multiple-point detonation systems. These commitments are indefinite. In addition, Iran will not pursue plutonium or uranium or uranium alloy metallurgy for 15 years."[17] But these special restraints will not be routinely available in other cases, leaving uncertainty as to how this subject will be managed in the future.

Challenge: The "Broader Conclusion"

As safeguards have evolved to manage the above issues, it became clear that to provide confidence that an NPT non-nuclear-weapon state was not pursuing nuclear weapons, the IAEA would need not only to determine that declared nuclear materials and facilities in that state were not being misused – the IAEA's classic conclusion – but that all nuclear-relevant activities the country were exclusively peaceful – now known as the "broader conclusion." Since 1999, the Agency has attempted to reach the latter conclusion in as many states as possible. This is a painstaking effort that relies on assiduous tracking of a wide range of industrial and research and development activities. Nonetheless, in its annual summary of its safeguards activities for 2016, the Agency declared:

> One hundred and twenty-four States had both comprehensive safeguards agreements and additional protocols in force:
>
> (a) For 69 of these States, the Secretariat found no indication of the diversion of declared nuclear material from peaceful nuclear activities and no indication of undeclared nuclear material or activities. On this basis, the Secretariat concluded that, for these States, all nuclear material remained in peaceful activities.
>
> (b) For 55 of these States, the Secretariat found no indication of the diversion of declared nuclear material from peaceful nuclear activities. Evaluations regarding the absence of undeclared nuclear material and activities for each of these States remained ongoing. On this basis, the Secretariat concluded that, for these States, declared nuclear material remained in peaceful activities.[18]

Other NPT non-nuclear-weapon states lacking Additional Protocols had positive conclusions regarding declared materials (only) or were not under traditional safeguards because of the small quantity of nuclear material on their territory.

Under the JCPOA, the IAEA will attempt to reach a broader conclusion for Iran, which may prove difficult, given the past secrecy surrounding many aspects of the Iranian program. However, Iran appears to have been cooperating with the Agency under the JCPOA and given the Agency's broad powers under that agreement, it may find that reaching the broader conclusion is possible. Certain restrictions in the JCPOA, concerning embargoes on arms sales and missile technology transfers to Iran will be lifted once that conclusion is reached or, at the latest, five or eight years, respectively, from the initial implementation of the JCPOA. The ability of the Agency to make this finding will be a key test of the JCPOA's monitoring provisions and an important benchmark of the IAEA's capabilities.

CONCLUSIONS AND A LOOK AHEAD

In 2016, the Agency applied safeguards at 709 installations in 181 states, providing confidence that nuclear materials under Agency monitoring were not being diverted to non-peaceful purposes. In 69 of these states, as noted, it was able to conclude further that no undeclared nuclear materials were present. It is difficult to overstate the importance of this effort to international stability. Moreover, although the Agency has suffered setbacks from the challenges discussed above, actions at the IAEA or related international efforts have repeatedly strengthened the safeguards system or introduced other measures to reduce the risk of future setbacks in most of these spheres.

The JCPOA incorporated many of these measures and establishes de facto precedents for the future.

- It employs the Additional Protocol, as well as further monitoring techniques that may be made future components of the IAEA safeguards system;
- It restrains enrichment and reprocessing, shining a spotlight on the threat these technologies pose and on the suspicions they arouse that will likely make other states cautious in pursuing these capabilities;
- It bans weaponization efforts – a matter not expressly banned under the NPT or IAEA system – relying on the augmented monitoring above to ensure compliance;
- It reinforces the access rights of the IAEA and establishes mechanisms that will trigger significant sanctions for violations;
- It establishes strict rules for the procurement of nuclear goods and has mechanisms that will trigger significant sanctions for violations; and
- It highlights the importance of the IAEA broader conclusion as an imprimatur of a state's compliance with its renunciation of nuclear weapons under the NPT.

It is also important to recognize the repeated contributions of the Security Council in supporting the Agency and expanding its authority in specific cases. As noted, the Council, in Resolution 687 (1991), tasked the IAEA to uncover and eliminate Iraq's nuclear activities after the 1991 Persian Gulf War, giving it expanded authority and resources as an arm of the UN Special Commission on Iraq (UNSCOM). Similarly, the Council validated and reinforced the Agency's pursuit of Iran's weaponization activities in Resolution 1737 (2006) and in endorsing the JCPOA through Resolutions 2231 (2015). The Council has not acted on the Syria al-Kibar reactor dossier forwarded by the IAEA, however.

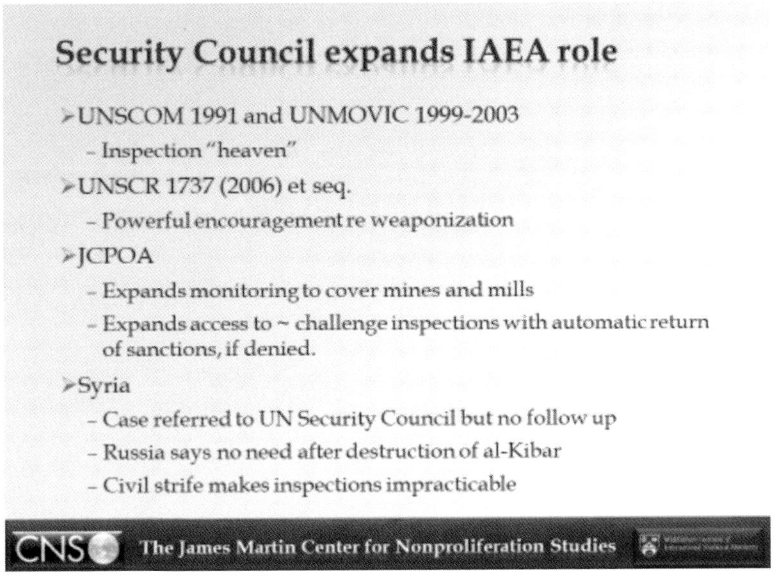

FIGURE 10

Looking ahead, as noted in Figure 10, it is likely that the Additional Protocol will be adopted as a condition of supply for transfers of nuclear reactors and fuel whether through a formal decision of the Nuclear Suppliers Group or, as seen in the decision at the G-7 Lubeck summit, through informal action by the handful of Nuclear Supplier Group states actually able to supply such items. As the Additional Protocol becomes more widely adopted, it is also likely that the Agency will gradually increase its monitoring of uranium mines and mills, as well as its focus on weaponization efforts, as it is doing under the JCPOA.

The Agency may hold to its current reluctance to take major initiatives to address illicit procurement activities, but it might take partial measures, such as seeking enhanced transparency measures for facilities that are contaminated with illegally obtained technology and equipment. De facto limits on enrichment and reprocessing in additional states are also likely to gain momentum, limiting this burden on the safeguards system. Finally, if present trends continue, in coming years, the IAEA is likely to reach its broader conclusion in an increasing number of states, possibly including Iran.

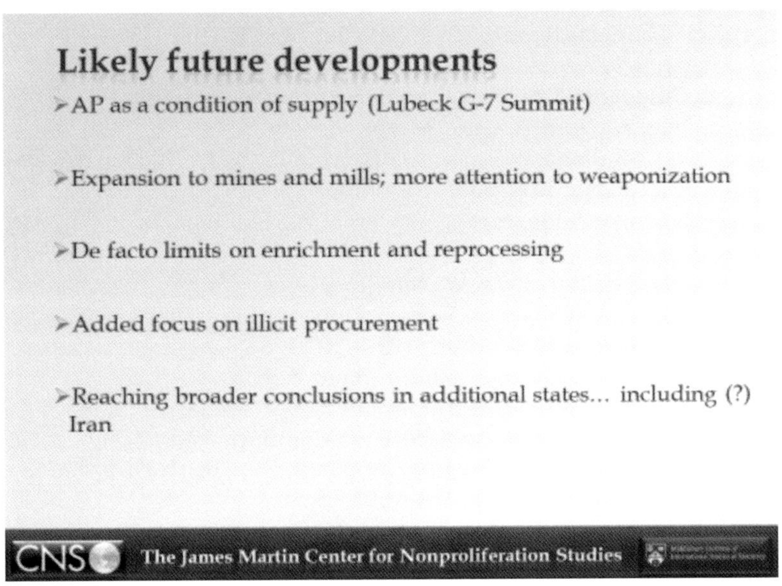

FIGURE 11.

In sum, the Agency plays a unique role on the international stage, in a dynamic environment. The Agency has adapted with a fair measure of success to the challenges it confronts, through internal reforms, grants of new authority from the Security Council to meet special contingencies, and complementary actions of the major powers.

REFERENCES

[1] "IAEA Safeguards – An Overview," International Atomic Energy Agency website, last updated August 31, 2016; https://www.iaea.org/newscenter/multimedia/videos/iaea-safeguards-overviewhttps://www.iaea.org/newscenter/multimedia/videos/iaea-safeguards-overview.

[2] "IAEA Overwhelmingly Votes to Send Iran to UN Security Council," Deutsch Welle, April 2, 2006, http://www.dw.com/en/iaea-overwhelmingly-votes-to-send-iran-to-un-security-council/a-1892077; "Statement by the President of the Security Council," March 29, 2006, S/PRST/2006/15, https://www.iaea.org/sites/default/files/unsc_sprst2006-15.pdf; Security Council Imposes Sanctions on Iran for Failure to Halt Uranium Enrichment, Unanimously Adopting Resolution 1737 (2006)," United Nations Press Release, December 23, 2006, https://www.un.org/press/en/2006/sc8928.doc.htm.

[3] Security Council Resolution 687 (1991), http://www.un.org/Depts/unmovic/documents/687.pdf.

[4] International Atomic Energy Agency website, "Additional Protocol." https://www.iaea.org/topics/additional-protocol.

[5] International Atomic Energy Agency website, "Status of Additional Protocol, as of May 19, 2017." https://www.iaea.org/topics/additional-protocol/status.

[6] "G7 Statement on Non-Proliferation and Disarmament," Lübeck, April 15, 2015.

[7] Iran is an exception to this rule. Russia did not insist that Iran ratify an Additional Protocol to its CSA when Moscow agreed to construct a nuclear power plant for Iran at Bushehr. See discussion of Iran and the Additional Protocol, below.

[8] "Full Text of CIA Briefing," ArmsControlWonk blog, April 25, 2008, www.armscontrolwonk.com/archive/201865/full-text-of-the-syria-briefing/.

[9] Joint Comprehensive Plan of Action (JCPOA), Vienna, July 14, 2015.

[10] *The Iran Nuclear Deal: A Definitive Guide,* Belfer Center for Science and International Affairs, August 15, 2015, Section 4.1.1, http://www.belfercenter.org/sites/default/files/legacy/files/IranDealDefinitiveGuide.pdf.

[11] JCPOA, op. cit.

[12] UN Security Council, Panel of Experts Report 2014, February 13, 2015, , http://www.un.org/ga/search/view_doc.asp?symbol=S/2015/131.

[13] "Nuclear Suppliers Group Guidelines," Nuclear Suppliers Group website, http://www.nuclearsuppliersgroup.org/en/guidelines.

[14] See, David Albright and Serena Kelleher-Vergantini, "Parchin in the IAEA's Final Assessment on the Possible Military Dimensions to Iran's Nuclear Program," December 3, 2015. http://isis-online.org/uploads/isisreports/documents/Parchin_Final_Assessment_on_PMD_Issues_3Dec2015-Final.pdf.

[15] Author interview with U.S. official, June 2015, Washington, DC.

[16] For details, see "Final Assessment on Past and Present Outstanding Issues regarding Iran's Nuclear Programme," IAEA Board of Governors document, GOV/2015/68, December 2, 2015, https://www.iaea.org/sites/default/files/gov-2015-68.pdf.

[17] "Statement of the Honorable Ernest Moniz, Secretary of Energy, U.S. Department of Energy, *"Iran Nuclear Agreement: The Administration's Case,* Hearing Before the Committee on Foreign Affairs, July 28, 2015, http://docs.house.gov/meetings/FA/FA00/20150728/103823/HHRG-114-FA00-Transcript-20150728.pdf.

[18] International Atomic Energy Agency, "Safeguards Implementation Report - Statement for 2016," GOV/2017/23, June 2017, https://www.iaea.org/sites/default/files/statement_sir_2016.pdf.

Nuclear Safeguards in Brazil and Argentina: 25 Years of ABACC[1]

Togzhan Kassenova

Fellow, Nuclear Policy Program
Carnegie Endowment for International Peace
1779 Massachusetts Ave., NW
Washington, DC 20016, USA

Abstract. As possessors of advanced nuclear technology, Brazil and Argentina bear special responsibility for helping the international community and neighbors in their region feel confident that their nuclear programs are peaceful, secure, and safe. Over the past 25 years, the Brazilian-Argentine Agency for Accounting and Control of Nuclear Materials (ABACC) has played an indispensable role in strengthening such confidence by implementing nuclear safeguards in the two countries. Today, ABACC carries out safeguards inspections at a total of 76 nuclear facilities in Brazil and Argentina. This article describes how Brazil and Argentina view trends in the global nonproliferation regime and international nuclear safeguards, and explains how these trends relate to unique challenges and opportunities facing Brazil, Argentina, and ABACC.

POLITICS OF NUCLEAR NONPROLIFERATION: DIFFERENCES IN ASSESSING THE THREAT

Historically, nuclear proliferation has been the preoccupation of advanced nuclear countries. Several concerns feed the fear of the spread of sensitive technology. In a strategic sense, acquisition of nuclear weapons technologies by new countries can disrupt the established nuclear order and the balance of power. The dual-use nature of nuclear technology calls for caution—the spread of nuclear technology for peaceful purposes comes with increased risk that some of that technology and associated material might be diverted for military use by states. Non-state actors can also represent a risk, as well-funded and well-organized groups might obtain nuclear or radioactive material and use it in their attacks. The threat of nuclear terrorism, like nuclear proliferation, has remained predominantly a concern of Western countries.

On the opposite side of the spectrum, the majority of non-Western, developing countries do not view nuclear proliferation as a fundamental threat to their security for a number of reasons. Due to the absence of strategic conflicts and existential threats, the use of nuclear weapons remains unlikely in many parts of the world. More apparent security threats that these countries face usually emerge from other sources—drug and arms smuggling, crime, economic and political instability, public health issues, and others. The risk of nuclear proliferation and its potential consequences (new nuclear-armed states or acquisition of nuclear material or weapons by non-state actors) remain abstract when compared with the more immediate risks facing these nations. Furthermore, the majority of non-Western developing countries believe that nuclear terrorism does not present an immediate or probable threat to their security. Western cities are still perceived as primary targets for global terrorism, despite the more recent history of terror attacks in non-Western countries.

Countries in South America belong to the latter group. For the last several decades the region has enjoyed peace. No major inter-state wars or deep ideological conflicts have occurred. Unlike the animosity between India and Pakistan that drives the nuclear arms race in South Asia, the regional competition between Brazil and Argentina

Nuclear Weapons and Related Security Issues
AIP Conf. Proc. 1898, 040004-1–040004-6; https://doi.org/10.1063/1.5009227
Published by AIP Publishing. 978-0-7354-1586-7/$30.00

never grew into an arms race. Moreover, the two countries have cooperated in the nuclear field for decades. Drivers for any country in South America to seek military nuclear programs are close to non-existent.

Many South American countries present the lack of progress on disarmament by nuclear-weapon states as one of the main qualms and sources of hesitation by non-nuclear-weapon states to embracing additional nonproliferation commitments. Since the United States leads the efforts to promote nuclear nonproliferation, countries such as Brazil pay special attention to what the United States does or does not do on the nuclear disarmament front.

Yet, at closer examination, lack of progress on the disarmament front does not appear to serve as the primary concern for either Brazil or Argentina. Instead both countries appear to be more concerned that developing countries' access to nuclear technology might become hostage to the nonproliferation goals of developed countries.[2] Past experience feeds these concerns. Particularly in the 1970s-1980s, the United States exerted pressure on suppliers of nuclear technology to limit cooperation with other countries in order to curb potential proliferation.

Arguments about fairness and justice in the global nuclear order also feature prominently in the discourse. Concerns about double standards in the global nuclear order are not new. The Nuclear Nonproliferation Treaty (NPT), which provides the foundation for the nuclear order, has long been considered to promote the division of countries into nuclear "haves" and nuclear "have nots."

In recent years, however, voices expressing inherent unfairness codified in the NPT resounded more widely for two reasons. First, the Humanitarian Initiative, a coalition of states and civil society actors, which focuses on the humanitarian impact of nuclear weapons, gained impressive momentum. The 2010 NPT Review Conference final document included the language on the catastrophic consequences of use of nuclear weapons. At the 2015 NPT Review Conference more than 150 countries supported the most recent statement by the Humanitarian Initiative. The Humanitarian Initiative, and the widespread support it received, amplified the voices of non-nuclear weapon states within the NPT forum.

Second, several emerging powers, by virtue of their growing standing in the international system and possession of advanced nuclear programs, became stronger participants in the discourse on global nuclear issues. These states include Brazil and Argentina, which have nuclear power infrastructure, are important players in the Nuclear Suppliers Group and in the IAEA, and hold leverage in the NPT forum.

A favorable security environment, concerns about access to nuclear technology for development and economic needs, and a sense of injustice about the existing global nuclear order all explain why Brazil and Argentina are less concerned about nuclear proliferation threats than major Western powers.

However, the regional context should serve as an impetus for Brazil and Argentina to adhere to high nonproliferation standards. It is in the interest of both nations that their neighbors in the region do not have any serious concerns about their advanced nuclear programs. Such concerns might include questions about power projection (such as in the case of Brazil's nuclear-powered submarine program), nuclear material security, and nuclear safety (in relation to nuclear power plants in Brazil and Argentina).

Confidence in the peaceful nature of Brazil's and Argentina's nuclear programs remains by far the most critical component for regional confidence-building and stability. Existing arrangements, such as the nuclear safeguards regime implemented by ABACC, serve a very important role in that sense. But the existing arrangements have limitations, as will be discussed below, and the two countries might want to consider strengthening them.

TRENDS IN NUCLEAR SAFEGUARDS

Evolving trends in the field of international safeguards hold relevance to Brazil and Argentina in multiple ways.

Nuclear Suppliers Group (NSG) Guidelines for Sensitive Trade

Until 2011, the NSG called for restraint with transfers of sensitive fuel technology – uranium enrichment and reprocessing (ENR). In 2011, the NSG adopted more specific guidelines, urging NSG members to limit transfer of ENR technology only to countries who signed Additional Protocols with the IAEA.

Brazil and Argentina insisted that ABACC safeguards arrangements be recognized as sufficient when considering ENR transfers. As a result, the language of the NSG guidance includes a notable acknowledgement of ABACC. According to adopted guidance, ENR technology can be transferred to countries with IAEA Additional Protocols in place and those "implementing appropriate safeguards agreements in cooperation with the IAEA, including a regional accounting and control arrangement for nuclear materials, as approved by the IAEA Board of Governors."[3]

Different perceptions have emerged, however, regarding what the language on regional safeguards arrangements implies. It appears that Brazil and Argentina did not see it as a requirement to eventually conclude Additional Protocols. The Brazilian ministry of external affairs, for example, stated that the NSG started recognizing ABACC's safeguards as "an alternative criterion to the Additional Protocol."[4]

But the exact wording of the relevant passage on ENR transfer criteria includes an important caveat: it recognizes regional safeguards arrangements *pending* adoption of an Additional Protocol. Nonproliferation analysts suggest that this implies an expectation that Brazil and Argentina should sign Additional Protocols.

Experts closely familiar with the matter commented: "The use of the phrase 'pending this' in the NSG decision links the concepts of the Model Additional Protocol and ABACC as an expression of hope that comfort with the ABACC arrangement will be bolstered in due course by the conclusion of additional protocols by both countries or at least the adoption of the key elements of the protocol that provide the highest level of confidence."[5] Then NSG chair Piet de Klerk described it the following way: "we have formulated that the Additional Protocol is the norm when it comes to the export of sensitive facilities and equipment and technology."[6]

The IAEA and the Future of Safeguards

Among developments within the IAEA, two stand out as relevant to Brazil and Argentina. The first concerns the future of safeguards as it relates to the State-Level Concept (SLC). While the State-Level Concept is not new, in 2012 the discussion about it grew into a heated debate involving the IAEA and the member states. The concept promotes the idea that any given country should be evaluated as a whole on the basis of all available information instead of on the basis of disparate information about individual facilities. The rationale driving this concept was a desire to utilize the IAEA's resources more efficiently by tailoring safeguards to each state.

Several countries raised serious concerns about the concept, including but not limited to issues such as use of intelligence, potential subjective judgments about countries, and lack of clear information on how implementation of the State-Level Concept will affect them in practice.[7]

Brazil and Argentina were among the countries that raised concerns about the State-Level Concept both in terms of the substance of the concept and the way the IAEA Secretariat initially communicated with the states.

In 2013 after the IAEA Director General reported to IAEA member states on State-Level Concept (The Conceptualization and Development of Safeguards Implementation at the State Level (GOV/2013/38)), Brazil then voiced concerns on the substantive issues related to SLC and lack of clarity on what it entailed. Fundamentally, Brazil called for clearer answers on how the IAEA would draw safeguards conclusions about states. For example, Brazil sought answers on how member states could be assured that SLC would be in strict adherence to the different categories of safeguards arrangements that countries have with the IAEA (countries with Comprehensive Safeguards Arrangements (CSA), countries with CSA and Additional Protocol, countries with voluntary offer agreements, and countries under INFCIRC 66 provisions). Brazil voiced concerns about subjectivity of some factors that would feed into safeguards conclusions, such as "level of cooperation with the Agency." Brazil noted that the GOV/2013/38 did not elaborate on what "safeguard-related information" entailed and on how open-source and third-party information would be used.[8]

Similarly, Argentina had concerns with the lack of clarity of how SLC was presented and the contradictions in terms of its contents and objectives.

Brazil and Argentina played an instrumental role in encouraging the IAEA Secretariat and the IAEA Director General to provide more information and to engage with member states in a productive dialogue.

The consultations that ensued and further information provided by the IAEA Secretariat, including in the Supplementary Document to the Report on the Conceptualization and Development of Safeguards Implementation at the State Level (GOV/2013/38) and its Corrigenda on State-Level Concept (GOV/2014/41 and GOV/2014/41/Corr1), alleviated Brazil's main concerns.[9] Of critical importance to Brazil was the affirmation from the IAEA Secretariat that SLC would not place additional rights or obligations on the states and would not substitute the Additional Protocol, and that acquisition path analysis would be focused exclusively on the acquisition of nuclear material in a nuclear weapon or nuclear device and would rely on technical methods and not on judgements about intentions.[10] Brazil's diplomats described as unprecedented the consultations and technical meetings between the Secretariat and member states and acknowledged that "for Brazil, this process introduced a new and encouraging dynamic in how safeguards issues should be dealt with within the IAEA."[11] Yet the country continues to monitor the process to verify it remains in conformity with the states' respective safeguards agreements and does not shy away from pointing out remaining issues surrounding SLC that require clarification.[12]

In an interview with the author, a representative of Argentina acknowledged that the commitment of the IAEA Secretariat to consult fully and openly with countries provides a good basis for bilateral discussions and may help clarify issues. He noted an improvement in the IAEA Secretariat's communication with states but added that more can be done in terms of its engagement.[13]

The second on-going issue within the IAEA that draws attention is the IAEA budget and funding. With a steady increase in the amount of nuclear activities around the world, the safeguards burden on the IAEA continues to increase because of the growing number of nuclear facilities and activities it must verify. According to a U.S. National Nuclear Security Administration report from 2008, the number of facilities under safeguards more than tripled over a 25 year period.[14] In addition, the Iran Deal placed a considerable new burden on the IAEA to verify Iran's nuclear activities. IAEA Director General Yukiya Amano stated that monitoring implementation of Iran will cost the IAEA $10 million per year.[15]

The implementation of strengthened safeguards since 1991 has increased the Agency's costs. To meet the requirements of the State-Level Concept, the IAEA will also need to invest in its infrastructure, facilities, and expertise.[16]

Zero real growth in its budget means that the IAEA depends on voluntary and extra-budgetary funding. Although the IAEA's regular budget includes expenses to carry out nuclear safeguards, even safeguards work depends on voluntary funding due to a limited budget.[17] Brazil believes that all safeguards activities foreseen in the safeguards agreements between the states and the IAEA should be covered from the IAEA's regular budget. In the case of the Iran Deal implementation, Brazil's position is that the activities under the safeguards agreement and the Additional Protocol signed by Iran must be covered from the regular budget and the confidence-building measures beyond the comprehensive safeguards agreement and the Additional Protocol must be paid by extra-budgetary voluntary contributions.

Argentina similarly believes that, in principle, verification should not be subject to voluntary contributions and as a rule statutory activities should be covered from the IAEA regular funds.[18]

Late payments by member states of their contributions to regular budget further complicate the IAEA's funding problems. Brazil and Argentina find themselves in the company of the United States and several other countries, who owe significant contributions to the IAEA regular budget. According to the IAEA financial documents, as of September 2015, Brazil owed 16 million Euros and 2 million US dollars and Argentina owed 2.2 million Euros.[19] At the beginning of 2016, Brazil paid a small part of its contribution to the 2014 regular budget and, according to officials, aims to pay off its remaining debt as soon as possible. As of August 2016, Argentina has cleared most of its arrears.[20]

Currently Brazil and Argentina belong to a category of developing states "shielded" from paying their full share of the IAEA's safeguards costs. Starting in 2017 they will no longer remain in this category and will have to pay the full share.[21]

ABACC AT 25: CHALLENGES AND OPPORTUNITIES

At twenty-five, ABACC's main challenge is deciding whether and how the organization should evolve. Responding to this challenge creates opportunities for ABACC to strengthen its standing and make an even greater contribution to international and regional security.

The global trend towards higher standards in nuclear safeguards implies that sooner or later Brazil and Argentina might feel compelled to reconsider their positions on the Additional Protocol.

The arrival of the new administration in Buenos Aires in late 2015 rekindled a conversation in Argentina on the question of an Additional Protocol. Key advisors of President Mauricio Macri believe that signing an Additional Protocol will serve Argentina's interests.[22] In fact, Argentina might have more compelling reasons than Brazil to eventually sign an Additional Protocol. Given Argentina's more export-oriented nuclear industry, signing an Additional Protocol might deliver political benefits and greater international confidence in its nuclear program.

Without a doubt, Argentina would prefer to make a joint decision with Brazil on whether both countries should sign an Additional Protocol. It remains in the interests of both Brazil and Argentina to move in tandem on this issue since the existing safeguards arrangements rely on bilateral cooperation within the ABACC framework. However, a scenario of Argentina moving first cannot be excluded.

Strong concerns remain in Brazil and Argentina among some stakeholders that signing an Additional Protocol would undermine the existing bilateral system of nuclear safeguards. In Argentina, in what appears to be a response to a renewed discussion on an Additional Protocol within the government, Grupo Convergencia, a group of prominent academics and specialists in international relations, defense, and security, released a statement asserting strong opposition to the possibility that Argentina might sign the Additional Protocol. In the document entitled "An Inconvenient Initiative in Nuclear Politics" the authors state the following: "To advance unilaterally in that direction would imply putting an end to a system of guarantees and safeguards constructed 25 years ago by Argentina and Brazil that has made it possible that the use of nuclear energy in both countries could be carried out with exclusively peaceful ends."[23]

Yet what is seen as a challenge can be turned into an opportunity. A decision by Brazil and Argentina to sign an Additional Protocol or move in a direction aligning more closely with its provisions could allow ABACC to evolve beyond its original mandate. One potential area of growth might involve granting ABACC authority to verify the absence of undeclared nuclear activities and undeclared nuclear material.

Another long-term challenge for ABACC will be safeguarding Brazil's nuclear-powered submarine program. No precedent exists for safeguarding such a program in a non-nuclear-weapon state, and developing safeguards for naval fuel might be interpreted as a good policy and technical challenge for Brazil, Argentina, and ABACC. The uncertainty surrounding the future of the nuclear submarine program, however, may push this particular challenge into the future. The on-going large-scale anti-corruption investigation (Lavo Jato) that implicated several major contractors will likely delay construction.

Looking forward, a growing safeguards burden for the IAEA may open a window of opportunity for ABACC. ABACC could eventually take on more safeguards responsibilities, perhaps with the possibility that its mandate currently limited to implementing safeguards in Brazil and Argentina could evolve into a regional one. If that goal is pursued, important questions relating to the legal foundation, procedures, budget, recruitment of inspectors beyond Brazil and Argentina, and others will need to be addressed.

Nuclear safeguards, and the nonproliferation regime as a whole, are not static. By thinking ahead and responding to global trends, ABACC can become a more influential and consequential actor in this regime.

ACKNOWLEDGMENTS

The author gratefully acknowledges invaluable research assistance provided by Scoville Fellow Chelsea Green and Junior Fellow William Ossoff, formerly with the Carnegie Endowment for International Peace. A visiting fellowship from the Center of International Relations of FGV (FGV CPDOC) offered the author an opportunity to spend two months in Brazil in 2015 and to learn about the evolution of Brazil's nuclear policy. The author thanks Laercio Antonio Vinhas, Ambassador, Permanent Mission of Brazil to the IAEA and CTBTO and Rafael Grossi, Ambassador and Permanent Representative of the Argentine Republic to International Organizations in Vienna for answering questions related to their countries' policies and Mark Hibbs of the Carnegie Endowment for International Peace for comments on an earlier draft. The contents of this paper are the sole responsibility of the author.

REFERENCES

1 This is an abridged version of a previously published work "Brazil, Argentina, and the Politics of Global Nonproliferation and Nuclear Safeguards" in Odilon Antonio Marcuzzo do Canto, ed., *O Modelo ABACC: Um Marco no Desenvolvimento das Relações Entre Brasil e Argentina (The ABACC Model: A Milestone in the Development of the Relationship Between Brazil and Argentina)*, Federal University of Santa Maria, Editora UFSM, Brazil, 2016.

2 For relevant discussion on the importance of access to technology see the essays on Brazil and Argentina in Toby Dalton, Togzhan Kassenova, Lauryn Williams, eds., *Perspectives on the Evolving Nuclear Order* (Washington, D.C.: Carnegie Endowment for International Peace, 2016), http://carnegieendowment.org/2016/06/06/perspectives-on-evolving-nuclear-order-pub-6371.

3 Mark Hibbs, "New Global Rules for Sensitive Nuclear Trade," Nuclear Energy Brief, Carnegie Endowment for International Peace, July 28, 2011, http://www.carnegieendowment.org/2011/07/28/new-global-rules-for-sensitive-nuclear-trade/4avp.

4 "Nuclear Suppliers Group (NSG) Recognizes the Quadripartite Agreement as an Alternative Criterion to the Additional Protocol," ABACC, June 28, 2011, http://abacc.org.br/?p=3846&lang=en.

5 David S. Jonas, John Carlson, and Richard S. Goorevich, "The NSG Decision on Sensitive Nuclear Transfers: ABACC and the Additional Protocol," *Arms Control Today*, November 5, 2012.

[6] "The NSG in a Time of Change: An Interview with NSG Chairman Piet de Klerk," *Arms Control Today*, September 20, 2011.

[7] Laura Rockwood, "The IAEA's State-Level Concept and the Law of Unintended Consequences," *Arms Control Today*, August 28, 2014; Mark Hibbs, "Intel Inside," *Foreign Policy*, December 10, 2012; "IAEA Safeguards Development and Russia's National Interest," 2014 Moscow Nonproliferation Conference, November 22, 2014.

[8] Brazil's Statement at the IAEA Board of Governors on Agenda Item 6 (B) The Conceptualization and Development of Safeguards Implementation at the State Level, delivered by H.E. Ambassador Laercio Antonio Vinhas, September 2013; Brazil's Statement at the IAEA Board of Governors on Agenda Item (7b) Safeguards Implementation Report, delivered by Ambassador Laercio Vinhas, June 2013.

[9] Supplementary Document to the Report on the Conceptualization and Development of Safeguards Implementation at the State Level (GOV/2013/38), GOV/2014/41, report by the Director General, https://armscontrollaw.files.wordpress.com/2014/09/iaea-state-level-safeguards-document-august-2014.pdf.

[10] Brazil's Statement on the State-Level Approach at the IAEA Board of Governors, delivered by H.E. Ambassador Laercio Antonio Vinhas, September 2014.

[11] Statement by H.E. Antonio de Aguiar Patriota Ambassador, Permanent Representative of Brazil to the United Nations, 2015 Review Conference of the Parties to the Treaty on the Non-Proliferation of Nuclear Weapons, New York, April 28, 2015, http://www.un.org/en/conf/npt/2015/statements/pdf/BR_en.pdf.

[12] Brazil's Statement on the State-Level Approach at the IAEA Board of Governors, delivered by H.E. Ambassador Laercio Antonio Vinhas, September 2014.

[13] Author's interview with Rafael Grossi, Ambassador of the Argentine Republic and Permanent Representative to International Organizations in Vienna, August 19, 2016.

[14] "International Safeguards: Challenges and Opportunities for the 21st Century," NNSA, https://nnsa.energy.gov/sites/default/files/nnsa/inlinefiles/NGSI_Report.pdf.

[15] Laura Smith-Spark and Marilia Brocchetto, "U.N. Watchdog: We Need Money to Monitor Iran Nuclear Deal," CNN, August 25, 2015, http://www.cnn.com/2015/08/25/world/iaea-iran-nuclear-deal/.

[16] Trevor Findlay, What Price Nuclear Governance? Funding the International Atomic Energy Agency, Belfer Center, Harvard University, March 2016, pp. 19-20.

[17] Ibid, p. 30.

[18] Author's interview with Rafael Grossi, Ambassador of the Argentine Republic and Permanent Representative to International Organizations in Vienna, August 19, 2016.

[19] Statement of Financial Contributions to the IAEA: Report by the Director General, GC(59)/INF/6, Vienna, September 11, 2015 quoted in Trevor Findlay, What Price Nuclear Governance? Funding the International Atomic Energy Agency, Belfer Center, Harvard University, March 2016, p. 39.

[20] Author's interview with Rafael Grossi, Ambassador of the Argentine Republic and Permanent Representative to International Organizations in Vienna, August 19, 2016.

[21] Trevor Findlay, What Price Nuclear Governance? Funding the International Atomic Energy Agency, Belfer Center, Harvard University, March 2016, p. 37.

[22] Eduardo Diez, "National Development and Argentina's Nuclear Policy," in Toby Dalton, Togzhan Kassenova, and Lauryn Williams, eds., *Perspectives on the Evolving Nuclear Order* (Washington, D.C.: Carnegie Endowment for International Peace, 2016), http://carnegieendowment.org/2016/06/06/perspectives-on-evolving-nuclear-order-pub-6371.

[23] Grupo Convergencia, "Una Iniciative Inconveniente en Política Nuclear," in Spanish, translation by author, March 28, 2016, https://grupoconvergenciablog.wordpress.com/.

Proliferation Risks From Nuclear Power Infrastructure

Sharon Squassoni

Senior Fellow and Director, Proliferation Prevention Program
Center for Strategic & International Studies
1616 Rhode Island Avenue, NW, Washington, D.C.
June 27, 2017

Abstract. Certain elements of nuclear energy infrastructure are inherently dual-use, which makes the promotion of nuclear energy fraught with uncertainty. Are current restraints on the materials, equipment, and technology that can be used either to produce fuel for nuclear electricity generation or material for nuclear explosive devices adequate? Technology controls, supply side restrictions, and fuel market assurances have been used to dissuade countries from developing sensitive technologies but the lack of legal restrictions is a continued barrier to permanent reduction of nuclear proliferation risks.

INTRODUCTION

It has been more than seventy years since humankind learned how to unleash the tremendous energy that binds atoms together. Since then, we have pursued both military and civilian uses of atomic energy, with abundant risks and rewards. A quarter-century after the first atomic bomb explosion, countries signed a treaty that recognized the dangers inherent in widespread proliferation of nuclear weapons (the Nuclear Nonproliferation Treaty, or NPT) and established supply controls as bulwarks against further proliferation. Only four countries (India, Pakistan, Israel, and North Korea) since have sustained nuclear weapons programs outside of the NPT but several others abandoned their programs, including South Africa, which had six assembled weapons. Non-nuclear weapon state parties to the NPT agreed to forego the development of nuclear weapons for a price: a guarantee that the peaceful uses of nuclear energy would not be restricted and that nuclear weapons would eventually be eliminated.

The three-part bargain – nonproliferation, peaceful uses of nuclear energy, and nuclear disarmament – has endured for almost fifty years but not without setbacks. The unfortunate technical fact is that certain inputs for peaceful nuclear energy are inherently dual-use. Specifically, uranium enrichment and spent fuel reprocessing can be used to make benign fuel or nuclear explosives. More generally, scientists, engineers, equipment and technologies applied to peaceful nuclear power could potentially be used to assist a nuclear weapons program, or the whole peaceful nuclear program could provide cover for a country acquiring capabilities. For countries with advanced nuclear energy capabilities, nuclear power infrastructure might include fissile material production capabilities that provide a latent nuclear weapons option. States could misuse avowedly peaceful nuclear facilities to make undeclared material for use in nuclear weapons or could build secret facilities to produce fissile material for nuclear weapons.

These are exactly the scenarios that experts worried about regarding Iran's nuclear program, and why the Joint Comprehensive Plan of Action signed in 2015 contains significant restrictions on Iran enriching uranium and reprocessing nuclear spent fuel. Iran is a special case because it was formally found in non-compliance with its safeguards obligations under the NPT. But as long as there are no legal restrictions on uranium enrichment or spent fuel reprocessing within the treaty itself, the nonproliferation regime will continue fighting a rearguard action against the risks that these capabilities pose.

NUCLEAR POWER INFRASTRUCTURE

The dominant fuel cycle in commercial operation today is known as the Light Water Reactor fuel cycle, which uses lightly enriched uranium (called low-enriched uranium or LEU) fuel and light water (as opposed to heavy water, or D_2O) as a moderator to control neutron flux. About 85 percent of all commercial nuclear power reactors utilize low-enriched uranium for fuel.

Nuclear Weapons and Related Security Issues
AIP Conf. Proc. 1898, 040005-1–040005-14; https://doi.org/10.1063/1.5009228
Published by AIP Publishing. 978-0-7354-1586-7/$30.00

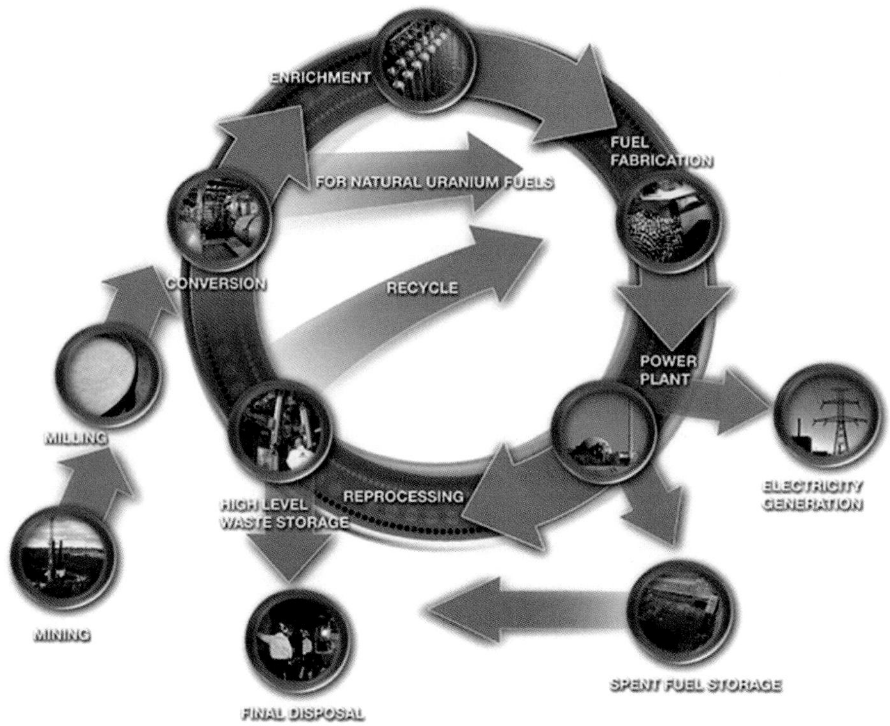

FIGURE 1. The Light Water Nuclear Reactor Fuel Cycle[1]

The nuclear fuel cycle (Fig. 1) is commonly divided into two stages: the front end, in which uranium is processed through conversion and enrichment plants and then fabricated into fuel for reactors; and the back end, in which fuel is handled after irradiation in a reactor. Reactors are generally considered the operations process between the front and back ends.

Components of the front end include:

- Uranium exploration, mining, and milling: Uranium ore is mined and processed to produce uranium ore concentrate, commonly called U_3O_8 or yellowcake.
- Conversion: The concentrate is converted to uranium hexafluoride gas (UF_6).
- Enrichment: The UF_6 is enriched to contain up to 5 percent of the isotope U-235 in the enriched uranium product.
- Fuel fabrication: The enriched uranium product is processed into uranium dioxide (UO_2) powder and pressed into pellets. The pellets are then encased in metal tubes to form fuel rods, which are arranged into a fuel assembly ready for introduction into a reactor.

Components of the back end include:

- Spent fuel interim storage: Spent fuel removed from the reactor is temporarily stored in spent fuel pools, which helps to dissipate heat and radioactivity. It is then sometimes removed to massive air-cooled dry casks for further storage.
- Reprocessing: Fuel can eventually be taken to a reprocessing facility, where it is chemically separated into its three components: uranium, plutonium, and waste, which contains fission products. The uranium from reprocessing, which typically contains a slightly higher concentration of U-235 than occurs in nature, can be reused (recycled) as fuel after conversion and enrichment. The plutonium can be directly made into mixed oxide fuel, in which uranium and plutonium oxides are combined. In fast neutron spectrum reactors, both plutonium and fission products can be consumed in the reactor.
- High-level waste disposal: Spent fuel and other high-level wastes are expected to be placed in a permanent geological repository (none are yet built).

In the past, countries opted to "close" their fuel cycles with the eventual aim of endlessly recycling nuclear fuel to take advantage of the production of plutonium (Pu-239) in reactors from the transmutation of natural uranium (U-238). The few countries that have chosen a closed fuel cycle – France, Japan, Russia, and the UK (and in the future, possibly China) – have only been able to reuse a relatively small amount of fuel in conventional or advanced reactors, but continue their research and development of multiple recycling of fuel for fast reactors. Most countries have chosen to use the nuclear fuel only once, storing spent fuel at reactor pools, or in interim storage facilities, and eventually storing it in a permanent geological repository. No matter which choice of fuel cycle, a permanent disposal site for high level waste is required.

Uranium Production

Uranium mining and milling are concentrated in the countries with the largest uranium resources. Three countries accounted for about two-thirds of global uranium production in 2014: Kazakhstan (with 41 percent of primary production); Canada (with 16 percent); and Australia (with 9 percent). Niger and Namibia produce 7 and 6 percent of global uranium, respectively, and Russia and Uzbekistan each produce 5 percent.[2] Major uranium producers include private and government-owned entities, and there are many partnerships, joint ventures, and other investment or ownership arrangements. Some of the larger uranium producers such as Cameco, AREVA, and Rosatom are also active in other parts of the fuel cycle.

Uranium is not a publicly traded good, unlike other commodities and precious metals. About 90 percent or more of the trade in uranium has historically been through long-term contracts (from 5 to 15 years) which require the mining company to meet production specifications (such as purity and isotopic concentration) and transport milled uranium to a conversion facility by a pre-agreed date. Utilities often enter into several of these contracts with various suppliers to ensure security of supply. Utilities also purchase uranium on the spot market to take advantage of lower purchase prices.

Conversion

To prepare the uranium for commercial enrichment, milled uranium is converted to a gaseous form (uranium hexafluoride). There are five principal conversion suppliers – AREVA, Rosatom, Cameco, Westinghouse, and ConverDyn. Nuclear power plant operators contract with these major firms using long-term agreements. Suppliers like AREVA and Rosatom are vertically integrated companies and encourage their customers to buy a range of services, including uranium, conversion, enrichment, and fabrication.

The availability of secondary uranium sources, including not only spot market material, but also Russian HEU downblended under the US–Russian Megatons to Megawatts program, has dampened need for conversion services and current capacity exceeds demand. For companies like AREVA, a capability in conversion may simply be the necessary cost of providing full cycle fuel services and supporting more profitable areas of the cycle.

Enrichment

Six companies commercially enrich uranium—Urenco, AREVA, Eurodif (59 percent owned by AREVA), CENTRUS (formerly US Enrichment Corporation, or USEC), Russia's Tenex, and the China National Nuclear Corporation. The top three – Urenco, AREVA and Tenex – dominate 90 percent of the enrichment market.[3] Most uranium is enriched under long-term contracts (five years or more), with some enrichment traded in a spot market. The movement toward centrifuge technology is likely to reinforce this trend, since centrifuge capacity is generally added only upon securing long-term enrichment contracts.

Centrifuge technology has transformed the enrichment landscape over the last few decades. The United States, once the dominant supplier, is in danger of disappearing from this market altogether. The private U.S. enrichment corporation CENTRUS estimates it has about 4 percent of the global enrichment market. The prospects for commercial laser enrichment in the United States are dim, given that GE-Hitachi announced in 2016 that it was seeking to divest its share in the Global Laser Enrichment Services consortium (together with Cameco) after initial technology demonstration. Australian SILEX, the developer of the laser enrichment process, has the option of purchasing 76 percent of the shares and is actively seeking investors.

Fuel Fabrication

Fuel fabrication is a more specialized engineering process than mining, conversion, and enrichment, because fuel rods must be tailored to the specific needs of each reactor. The dimensions of the fuel pellets and other components of the fuel assembly are precisely controlled to ensure consistency in the fuel characteristics.

Reactor vendors and their affiliates tend to be the major fuel fabricators. Most fuel contracts require the fuel fabricator to provide sufficient information to develop a detailed fuel design for the reactor, a process that takes about five years and involves a conceptual design followed by years of testing and at least a year's effort to acquire a license. There are now roughly 11 nuclear fuel fabricators in the world, including AREVA in France and the United States, the China National Nuclear Corporation, Korea Nuclear Fuel, Rosatom's TVEL in Russia, and General Electric and Westinghouse in the United States.

Reprocessing

Only a few countries worldwide reprocess nuclear spent fuel, which involves chopping up the fuel assemblies and then chemically dissolving them to extract both plutonium and uranium (known as the PUREX process, but there are other variations).[4] About a third (90,000 metric tons out of 290,000 metric tons) of spent fuel discharged from commercial power reactors has been reprocessed. Annual reprocessing capacity is about 4,000 metric tons, while the annual production of spent nuclear fuel is more than double that amount.

The United Kingdom, France, and Russia currently separate the largest amounts of plutonium, and both the United Kingdom and France temporarily store plutonium for other countries. As of December 2015, the UK, France, Russia and Japan declared 275 metric tons of separated plutonium in their reports under IAEA INFCIRC-549. An estimated 2 metric tons of additional material is stored in Germany as MOX fuel. The United Kingdom accumulates the largest amounts of civilian plutonium each year, although that rate is decreasing because of problems at its Thermal Oxide Reprocessing Plant at Sellafield in Cumbria, which is scheduled to close in 2018 when all existing reprocessing contracts are completed. France's stockpile has decreased slightly since 2012 primarily because AREVA has been shipping it back to foreign customers that have terminated their reprocessing contracts. Russia continues to separate plutonium at a rate of about one to two metric tons per year, and plans to use its excess plutonium for a commercial breeder program starting in 2020. Japan is mulling its options for the use of MOX fuel in its light water reactors, given the present difficulty it finds itself in with only a handful of nuclear power plants running.

AREVA, which previously led the international market for reprocessing, has lost most of its foreign customers, forcing its facility at La Hague to operate at half-capacity. Significant restructuring in 2016 led to the spinoff of its mining, enrichment, and reprocessing activities, which are now conducted under the newly incorporated NewCo. Though several international investors expressed interest in taking a stake in the new entity, only Mitsubishi Heavy Industries and Japan Nuclear Fuel Limited (JNFL) have made final arrangements, purchasing a combined 10 percent stake in NewCo for roughly $538 million. In the United Kingdom, the future of reprocessing is unclear. In 2011 the UK Nuclear Decommissioning Authority (NDA) closed the Sellafield MOX plant, Britain's only commercial MOX fuel fabrication facility. In 2014, the NDA issued a preliminary report outlining three options for the plutonium stockpile—reuse as MOX in light water reactors, reuse in CANDU (EC6) reactors, or reuse in PRISM fast reactors— but no final decision has emerged.[5]

Storage of Spent Nuclear Fuel

Decisions about storing and disposing of spent nuclear fuel and/or high-level radioactive waste can affect fuel cycle choices. Geological disposal is the internationally accepted strategy for permanently isolating waste from the accessible environment, but no country has yet succeeded in opening a repository for commercial waste.

As of the end of 2015, about 265,000 metric tons of spent fuel was in storage worldwide, most of it at reactor sites. About 80 percent of all spent fuel is submerged in fuel ponds and the balance is stored in dry casks. Every year, nuclear power reactors across the world generate about 10,500 metric tons of heavy metal. Of that, roughly 8,500 metric tons goes into long-term storage and about 2,000 metric tons is allocated for reprocessing, but much of that remains in interim storage. The 10 largest nuclear power generators produce about 87 percent of these totals. None of these has yet opened a commercial nuclear waste repository.

Spent fuel pools were never designed to be a long-term storage solution. Many sites have had to install high-density racks and/or add dry storage casks on site. Storage is limited by the size of the pool and criticality-control geometry, so the addition of dry casks sometimes incites controversy among local populations, who may not have consciously agreed to long term storage of nuclear waste. Taiwan, South Korea, and Japan face particularly acute shortages of storage space at plants. At the same time, just like dry cask storage, permanent repositories can become lightning rods for political controversy in many countries. Finland and Sweden have made the most progress of any country because they have selected repository sites and begun licensing and other activities.

RISKS FROM NUCLEAR POWER INFRASTRUCTURE[6]

Like most aspects of nuclear energy, views on proliferation risks can be deeply divided. Some industry advocates claim no country has developed nuclear weapons from a civilian nuclear program and therefore there are no proliferation risks while some industry opponents claim that the proliferation risks are too great to justify any nuclear energy at all. The truth lies somewhere between the two.

Proliferation concerns about the operation of nuclear facilities focus on two elements: fissile material and the facilities that make them. These risks are usually described as diversion of declared material, undeclared production at a declared facility and undeclared production at undeclared facilities. Simply put, operators might divert monitored material, produce more than they declare at a monitored facility, build a separate, clandestine production line at a declared facility, or produce things clandestinely at a different site. A secret parallel program might need to duplicate significant portions of both the front and back ends of the fuel cycle, rather than just enrichment and/or reprocessing facilities in order to avoid detection.

The facilities therefore that pose the highest proliferation risks in the nuclear fuel cycle are uranium enrichment and reprocessing facilities.[7] Although plutonium is technically "produced" in a reactor, it is not useful in a weapon unless the highly radioactive fission products have been removed through reprocessing described above. Civilian facilities rarely produce weapons-grade material, but for a country that has production facilities, this obstacle is not insurmountable. A state that has produced enough LEU enriched to roughly four percent in U-235 has already completed about three quarters of the enrichment effort required to produce weapons grade uranium and can finish the effort through either clandestine batch processing or through a clandestine enrichment line. In the case of reprocessing, reactor-grade plutonium has been and could be used in nuclear explosive devices, although it would probably not be the first material of choice for proliferators. A country with a reprocessing plant almost certainly has reactors that could be used to lightly irradiate fuel to produce better, bomb-grade plutonium. Even in the absence of specific diversion scenarios, reprocessing plants and associated storage and fuel fabrication facilities pose proliferation and security risks by virtue of their substantial stocks of separated plutonium. A modestly sized reprocessing plant (e.g., 800-ton throughput per year) can have a "working stock" of ten tons of separated plutonium, or enough plutonium for thousands of bombs.

The landscape of proliferation risk has shifted over time, both in terms of the technologies that are perceived as posing significant risks and the countries (or non-state actors) that may have clandestine intentions. For example, when the NPT was negotiated, experts assumed that the tremendous costs, energy requirements, and physical footprint of uranium enrichment plants (based on gaseous diffusion technology) would make clandestine enrichment very difficult, if not impossible. This is certainly not the case today, as we have discovered with Iran and North Korea. In fact, detecting such clandestine enrichment has become a major challenge and could become even more difficult if laser enrichment can be commercialized.

Ultimately, negotiators of the NPT opted to limit restrictions to the manufacture, acquisition, or transfer of nuclear weapons, as well as their control or the seeking or receiving of any assistance in manufacturing nuclear weapons. Some diplomats questioned whether drawing the line that high up in a long chain of decisions was adequate. The Soviets, for example, wanted to include "preparation for manufacture" as well as, at the other end of the spectrum, the testing of nuclear weapons. Swedish Ambassador Alva Myrdal, in a statement to the Eighteen Nation Disarmament Conference, asked whether a middle ground could be found that prohibited acquisition and enjoined states to refrain from preparation of nuclear weapons. No such middle ground was found, and as a result, states that are in full compliance with the Nuclear Non-Proliferation Treaty are able to engage in the following activities that could ultimately aid development of a nuclear weapons program:

- producing highly enriched uranium (HEU)
- withdrawing HEU from safeguards for non-explosive military purposes
- accumulating stocks of separated plutonium or low-enriched uranium
- acquiring sensitive technology under safeguards and then withdrawing from the NPT
- developing the technical know-how for a nuclear-weapons program and an indigenous manufacturing capability

This has led to a growing concern that a state within the treaty could slowly and stealthily acquire capabilities to make nuclear weapons without actually violating the NPT. That state could teeter on the brink of a nuclear weapons capability – sometimes referred to as having a "latent nuclear weapon capability" – without actually stepping over the line. Such "virtual" weapons capabilities could strain the international monitoring system that the International Atomic Energy Agency (IAEA) runs and complicate detection of diversion to military uses.

Looking into the future, a handful of advanced countries has been researching other fuel cycles that could potentially utilize plutonium-based fuels, adding to proliferation risks. Fast neutron reactors fuel cycles, if widely adopted, could result in hundreds of tons of separated plutonium put into commercial circulation, increasing the possibility of diversion for illicit weapons programs. If those fast reactors are configured to breed more plutonium than they burn, the risk is greater. Critics argue there is no economic justification for a closed fuel cycle because the once-through cycle is cheaper and uranium is abundant. Efforts to reduce the proliferation risks of closed fuel cycles through either avoiding the use of HEU or Pu in the start-up fuel or by designing "burner" reactors that can use fuel that has been conditioned rather than reprocessed (i.e., where the plutonium is not fully separated from transuranics) are in development, but may take decades to yield results. There is also no consensus that states should close off technical options to reduce proliferation risks. In the meantime, promotion of a closed fuel cycle, even if not yet widely adopted, exerts pressure on an already over-taxed international safeguards system.

RESPONDING TO PROLIFERATION RISKS

The nonproliferation regime has evolved over time and has been surprisingly successful. Even though presidential candidate John F. Kennedy famously predicted in 1960 that 10 to 20 countries could have a nuclear capability by 1964, this has been avoided (only 9 countries have nuclear weapons). Of course, fears of proliferation motivated Presidents Kennedy, Johnson, and their successors to engage in arms control and nonproliferation efforts, enacting treaties, trade controls, sanctions, and cooperation agreements that have helped to dampen proliferation.

Over time, many of these elements have adapted response to changing technical and/or political circumstances. The NPT has remained relatively constant, although IAEA safeguards were strengthened in response to Iraq's nuclear weapons program in the early 1990s (i.e., with the adoption of the Additional Protocol in 1998). Interpretations of the NPT have shifted over time, however. For example, no one believes that peaceful nuclear explosions, which are allowed in Article V, are a good idea. It is also likely that more states now favor stronger withdrawal penalties (particularly following North Korea's exit from the treaty in 2003) and/or stronger penalties for non-compliance (following Iraq's and Iran's noncompliance episodes of the past decades) but the tension among its states parties makes amendment of the NPT a rather risky enterprise. Instead, states have left the treaty alone, but sought to plug gaps through unilateral, bilateral and multilateral trade controls as well as sanctions. Some efforts have been targeted at technology development, although these have been relatively anemic.

Tools to Mitigate Proliferation Risks

As evident from the earlier description of the fuel cycle and its suppliers, most nuclear power programs rely on a handful of outside suppliers for most of their nuclear technology. A successful nuclear power program certainly requires nuclear power plants, but these are often foreign-built and rely on foreign fuel services (whether conversion, enrichment or reprocessing). While enriched uranium is required to fuel light water reactors, there is no need for indigenous enrichment capabilities, and the LWR fuel cycle does not require reprocessing at all.

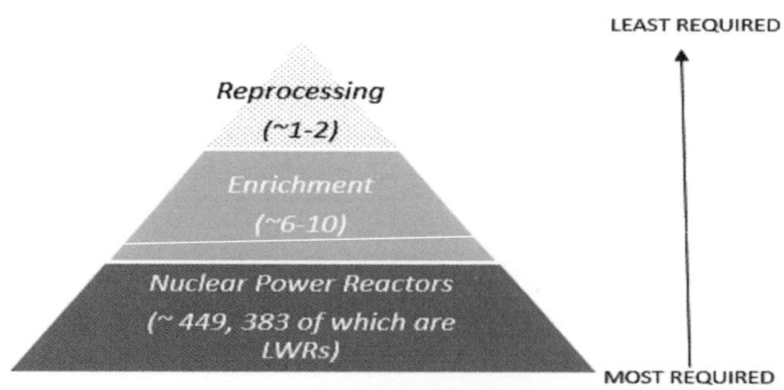

FIGURE 2. Essential Elements of LWR Fuel Cycle (Most to Least Required)

The facilities that present the highest risks – uranium enrichment and spent fuel reprocessing – are the least essential for successful commercial nuclear power programs. This should make them easy to restrict, but the opposite is true. Because the NPT has no specific restrictions on sensitive fuel cycle technologies, states that are unhappy with progress in nuclear disarmament resist efforts to strengthen nonproliferation and fall back on arguments based on state sovereignty to underpin their resistance.

Figure 3 shows the durability of risks from uranium enrichment and spent fuel reprocessing capabilities from the least fluctuating (that is, structural risks), the most fluctuating (changeable supply relations).

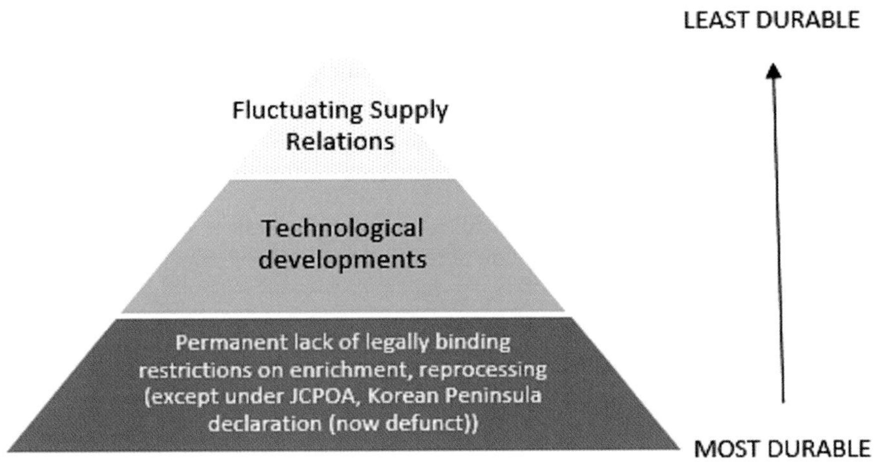

Figure 3. Durability of Proliferation Risks

The tools to address these different sources of proliferation risk work in a complementary fashion (Fig. 3). To counter the most enduring risk – the fact that uranium enrichment and spent fuel reprocessing are not restricted in any way by the NPT – the IAEA applies safeguards to all fissile material in non-nuclear weapon states. The IAEA has developed particular approaches for enrichment and reprocessing plants, but these have been applied at relatively few operating sites (Rokkasho in Japan for enrichment and reprocessing, Resende in Brazil, and some monitoring at Yongbyon for a period of time in North Korea). More recently, Iran, the European Union, the P-5 (US, UK, France, Russia, China) and Germany signed the Joint Comprehensive Plan of Action (JCPOA) for Iran that includes additional monitoring of and restrictions for enrichment and reprocessing.[8] Most efforts at nonproliferation, however, try to dampen both supply and demand for sensitive fuel technologies before they are deployed. To target the demand side, initiatives like the Global Nuclear Energy Partnership and now the International Framework for Nuclear Energy Cooperation (IFNEC) have sought to encourage the development of proliferation-resistant technologies. To target the supply side, countries enact unilateral trade controls (export controls), bilateral trade agreements, often called nuclear cooperation agreements, and peaceful nuclear cooperation agreements with built-in restrictions. Trade relations are loosely coordinated or harmonized within the Nuclear Suppliers Group. The sources of proliferation risk from nuclear power infrastructure and the tools to address them are depicted in Figure 4 and discussed in greater depth below.

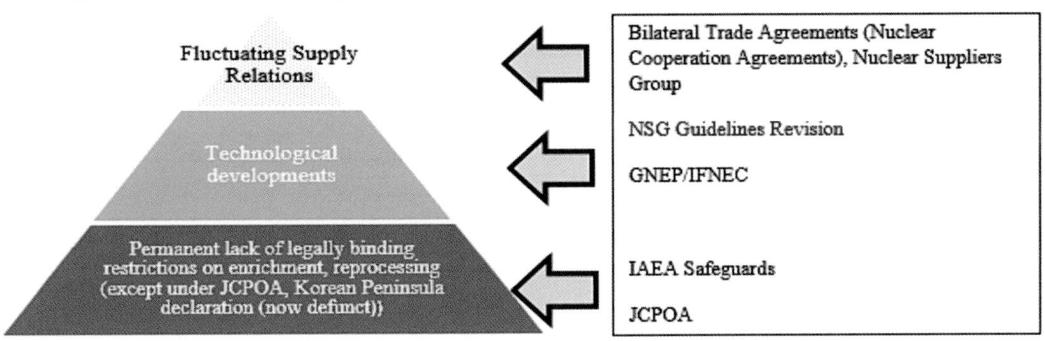

FIGURE 4. Sources of Proliferation Risks and Tools to Address Them.

IAEA Safeguards

Nuclear trade would be nearly impossible without the monitoring system set up by the International Atomic Energy Agency (IAEA). Before its creation in 1957, states supplying nuclear items conducted their own bilateral inspections to ensure peaceful uses of nuclear material and equipment, technology and facilities. After 1957, the IAEA took on some of those inspections and once the NPT entered into force in 1970, the IAEA monitored all nuclear material in the states that joined the NPT as non-nuclear weapon states. The IAEA still monitors material and equipment upon request as a result of bilateral supply arrangements with states that are not NPT members (such as India, Pakistan, Israel, and North Korea) and may monitor some items in the five nuclear weapon states (US, UK, France, China and Russia) that have limited, voluntary safeguards agreements with the IAEA.

Comprehensive safeguards agreements under the NPT focused Agency efforts on material control and accounting – in other words, safeguards follow the nuclear material. A comprehensive safeguards agreement requires countries to "provide the Agency with information concerning nuclear material subject to safeguards under this Agreement and the features of facilities relevant to safeguarding such material." Although the Agency always had the mandate to verify the absence of undeclared activities, materials, and sites, its ability to detect undeclared nuclear activities and access to suspicious sites for many years was limited.[9]

Accounting for material inventory starts at conversion plants and extends through the rest of the fuel cycle.[10] Because reactors produce plutonium as uranium is irradiated, IAEA safeguards follow fresh fuel into the reactor and are maintained through irradiation to storage and reprocessing, if applicable.

In 1991, the IAEA's focus on nuclear material and an arguably passive, accounting-oriented style of inspection, failed to uncover Iraq's nuclear weapons program despite ongoing inspections. Following the discovery in Iraq in 1991 of a massive, Manhattan Project-like nuclear weapons program and North Korean deceptions in 1993 and 1994, the IAEA adopted a more investigative approach. Before the end of that decade, member states voted to adopt the Additional Protocol, which strengthened safeguards (INFCIRC/540) to address those shortcomings.

Five elements encompassed the Additional Protocol program:

- provision of intelligence information to the IAEA by member states about suspect nuclear activities;
- access for inspectors to any location on a timely basis;
- new safeguards technology, including environmental sampling;
- measures to promote complete transparency and reporting of all nuclear commerce (far beyond the traditional starting point of safeguards at conversion plants);
- sufficient financial resources to carry out the IAEA's expanded responsibilities.

There is no doubt that the Additional Protocol has strengthened the safeguards system. Nonetheless, there are relatively few sites where safeguards are applied to enrichment and reprocessing plants. At Urenco facilities in the UK, Germany, and the Netherlands, monitoring is primarily conducted by EURATOM inspectors and the results are verified by the IAEA. At Brazil's enrichment plant, the IAEA conducts inspections, but Brazil does not have an Additional Protocol in place. Japan is the only non-nuclear weapon state that has both uranium enrichment and spent fuel reprocessing plants and they are under comprehensive (INFCIRC/153) safeguards. In nuclear weapon states, there are some limited safeguards (for example, in China as a result of bilateral trade requirements) but few inspections are conducted in Russia or the United States.[11] This largely means that if, in the future, there is a significant spread of enrichment and reprocessing facilities outside of the nuclear weapon states, additional measures might be needed to provide better assurances that material is not diverted and that facilities are not misused.

A fundamental concern is that enrichment and reprocessing production processes themselves pose technical risks. First, these facilities handle material in bulk (versus discrete items that can be counted) and often process so much material that the statistical errors in accounting could accommodate several bombs' worth of material. Second, newer technologies for enrichment might make it easier for diversion (e.g., changing valve switches in centrifuge cascades instead of diverting piping). And, a national nuclear program that includes manufacturing of enrichment components could provide the opportunity to establish a parallel clandestine enrichment program.

Systemic weaknesses in the International Atomic Energy Agency (IAEA) safeguards also pose technical risks. The primary objective of IAEA safeguards, per Article III.1 of the NPT, is to prevent "diversion of nuclear energy from peaceful uses to nuclear weapons or other nuclear explosive devices." The IAEA has set goals for timely detection to enable the international community to intervene before a state attempting a diversion is able to produce nuclear weapons, but this system has two problems. First, at bulk-handling facilities, such as uranium enrichment and spent fuel reprocessing plants, the quantities of material that statistically fall into the norms of acceptable "material unaccounted for" (MUF) exceed the levels that the IAEA has set for "significant quantities" (the IAEA's fissile material threshold at which a nuclear weapon could be made—though it is generally acknowledged that it is possible to make a nuclear weapon with much less material). Verification of non-diversion must be based on additional information provided by containment and surveillance measures (seals and cameras); given material accountancy alone does not reduce uncertainties to an acceptable level. This problem is multiplied in facilities with large throughputs, which have even higher acceptable amounts of MUF. Second, although detection of anomalies may be timely, delays in reporting from the IAEA Secretariat and in action by the Board of Governors can reduce warning time dangerously. The IAEA Board of Governors and/or the IAEA Secretariat can conclude that a state is in non-compliance with its safeguards agreement. In the case of the Board, politics can muddle the picture. In the case of the Secretariat, the Agency has worked hard to overcome institutional biases against finding noncompliance, particularly through the adoption of strengthening measures like the Additional Protocol. Few would argue that findings of noncompliance, because they are rare, are strictly technical in nature. Lastly, the United Nations Security Council has limited ability to enforce resolutions and affect behavior, even when the IAEA does report non-compliance.

Joint Comprehensive Plan of Action (JCPOA)

Signed in July 2015, the JCPOA is an example of an "upgraded" approach to solving the conundrum of enrichment and reprocessing. When it became clear that Iran would not foreswear enrichment (it did foreswear reprocessing) during negotiations with other parties to bring Iran back into compliance with the NPT, it became necessary to negotiate additional monitoring and reporting requirements that would provide confidence in the peaceful uses of its nuclear program, even as limited enrichment capabilities remained.[12] For the purposes of this discussion, the most important aspect of the Joint Comprehensive Plan of Action (JCPOA) is its recognition that IAEA safeguards alone are not sufficient to provide assurances of the peaceful nature of safeguarded enrichment. The second most important aspect of the JCPOA is that it is temporary, suggesting that states interested in reducing risks from enrichment and reprocessing have a limited window to universalize some of the JCPOA tools.

Under the JCPOA, the IAEA will monitor Iran under three separate nonproliferation agreements: the Iran Safeguards Agreement, the Additional Protocol, and the Joint Comprehensive Plan of Action itself. The starting point for the Iran deal was to ensure that Iran maintained its obligations under the NPT. Iran agreed to modernize its safeguards agreement to provide for early notification of facility design changes (so-called Code 3.1 of subsidiary arrangements), but the JCPOA adds explicit procedures for its version of "challenge inspections". What is more, the IAEA, pursuant to the JCPOA, laid out specific methods for resolving previously unresolved issues regarding Iran's program.

The JCPOA requires Iran to implement its Additional Protocol, which Iran had implemented between 2003 and 2005, provisionally until ratification. Iran is not committed to seeking ratification until 2023, a turning point for the JCPOA. Implementing the Additional Protocol helps by providing much more information to the IAEA about a broader range of nuclear activities, including research and development, intermediate uses of nuclear material, and activities involving nuclear waste. "Complementary access" provisions in the Additional Protocol grant IAEA inspectors access to all sites necessary to resolve concerns, and the ability to carry out "visual observation; collection of environmental samples; utilization of radiation detection and measurement devices; application of seals and other identifying and other identifying and tamper indicating devices," as well as "other objective measures."

The JCPOA also contains specific obligations designed to limit Iran's sensitive capabilities (e.g., in the number and sophistication of centrifuges, the amount of nuclear material it can stockpile, the level of enrichment of that material, and the ability of the Arak reactor to produce bomb-grade plutonium) that are not in the NPT. It also monitors procurement and manufacturing and only allows Iran to procure nuclear-related equipment with UN approval. Annex I, Section Q of the JCPOA allows the IAEA to request access "if the IAEA has concerns regarding undeclared nuclear materials or activities, or activities inconsistent with the JCPOA at locations that have not been declared under the comprehensive safeguards agreement or Additional Protocol." The resolution of disputes is not left solely to Iran and the IAEA but can be taken to the Joint Commission if necessary.

The JCPOA specifically states that its measures are not applicable to other states, but there are many experts who contend that some of the innovations of the JCPOA are improvements to the existing nonproliferation regime. It may be possible to approach these questions at the 2020 NPT Review Conference, although the prospects seem slim that additional nonproliferation restrictions will be palatable to the many states that are still unhappy with the pace of nuclear disarmament thus far.

Global Nuclear Energy Partnership (GNEP) and International Framework for Nuclear Energy Cooperation (IFNEC)

One approach to reducing the risks that technology developments may hold for nuclear proliferation is to collaboratively develop technologies. President Bush announced the Global Nuclear Energy Partnership (GNEP) in his 2006 State of the Union address, stating "Through GNEP, the United States will work with other nations possessing advanced nuclear technologies to develop new proliferation-resistant recycling technologies in order to produce more energy, reduce waste and minimize proliferation concerns. Additionally, [the] partner nations will develop a fuel services program to provide nuclear fuel to developing nations, allowing them to enjoy the benefits of abundant sources of clean, safe nuclear energy in a cost-effective manner in exchange for their commitment to forgo enrichment and reprocessing activities, also alleviating proliferation concerns." Although more than 25 countries joined the program, little was accomplished beyond agreeing to a mission statement. GNEP also had a domestic component to reinvigorate fuel recycling in the United States that was cancelled for lack of significant political and financial support.

The reanimation of GNEP as the International Framework for Nuclear Energy Cooperation (IFNEC) in 2010 reflected little support for the restrictive elements of the program. In Accra, Ghana, members adopted the following mission statement: "The International Framework for Nuclear Energy Cooperation provides a forum for cooperation among participating states to explore mutually beneficial approaches to ensure the use of nuclear energy for peaceful purposes proceeds in a manner that is efficient and meets the highest standards of safety, security and non-proliferation. Participating states would not give up any rights and voluntarily engage to share the effort and gain the benefits of economical, peaceful nuclear energy." There are few indications that IFNEC broke any new ground in creating viable and more proliferation-resistant approaches to the spread of sensitive fuel cycle technology.

Nuclear Suppliers Group

The lack of supply controls within the text of the NPT meant that nuclear supplier countries needed to devise them separately. In 1971, seven NPT nuclear supplier nations formed the Nuclear Exporters Committee, known as the Zangger Committee, to assist in restricting nuclear trade as called for in Article III of the treaty. The group identified materials and equipment other than source or special fissionable materials that would "trigger" the application of IAEA safeguards to ensure that the items were not used to make nuclear explosive devices. The Zangger Committee completed the list by 1974, just two months after the Indian nuclear test. The Zangger list included reactors, reactor components, and certain nuclear materials. Although the Zangger Committee still meets, its role has largely been overtaken by the Nuclear Suppliers Group.

The Nuclear Suppliers Group guidelines, which were made public in 1978 (as INFCIRC/254) called from the start for suppliers to exercise restraint regarding transfers of enrichment and reprocessing technology. At the time, sales of enrichment plants to Brazil and of reprocessing plants to Pakistan, South Korea, and Brazil were pending, amid considerable fear about controlling such technology once transferred. All of those sales were cancelled.

Although NSG guidelines have been strengthened over the years, modifications to its paragraphs on enrichment and reprocessing transfers have only recently been addressed. In 2004, President Bush proposed an outright ban on the acquisition of enrichment and reprocessing capabilities by new states following the scandal over Pakistani scientist A.Q. Khan's sales of enrichment technology to several states. Although the G-8 implemented a moratorium on sales, this finally fell apart under pressure from Canada in 2008. From 2004 to 2011, NSG members debated new criteria for transfers of enrichment and reprocessing technology. These discussions were undoubtedly complicated by the explicit application, for the first time, of restrictions to NSG members. In previous years, most NSG controls were applied to states outside the NSG. But with 48 members today, the NSG is more diverse than ever. Restrictions in the early days applied to countries like South Korea, Argentina, and Brazil, all of which are now NSG members.

At the June 2011 plenary, members agreed on new language in paragraphs 6 and 7 of the guidelines. Specific criteria for transfers of enrichment and reprocessing now take the place of formerly vague language that was predicated on an assumption of denial rather than approval. The guidelines state that suppliers should exercise a policy of restraint and that in the context of the restraint policy, they should not authorize transfers to states that are not NPT members, are not currently in breach of IAEA safeguards, adhere to NSG guidelines and UNSCR 1540, have a nuclear cooperation agreement with assurances of non-explosive use, effective safeguards in perpetuity and retransfer, agreed to physical protection and international safety conventions. More restrictive criteria that addressed implementation of an Additional Protocol and the exclusion of states that had former agreements not to enrich or reprocess (Iran in the future? South Korea now?) were dismissed. In addition, there is a loophole for transfers of such technology for research and development purposes. Since no requests for enrichment and/or reprocessing transfers have yet been made, it is unclear how countries might approach these new guidelines. Do the criteria facilitate approval for a country that meets the guidelines (like South Korea)? Or are they supposed to act as incentives for countries to meet the criteria? It is abundantly clear, however, that the NSG missed an important opportunity to put in place significant restrictions on enrichment and reprocessing. It may be that the discord sown by the United States within the NSG in 2008 over an exception for India from the NSG ban on trade with non-NPT member states made further refinements impossible.

Nuclear Trade Controls/Nuclear Cooperation Agreements

Whether or not states harmonize their nuclear trade controls within the Nuclear Suppliers Group, individual countries have adopted requirements for nuclear trade in general, and in particular, for sensitive nuclear technology. For almost a decade after the first atomic explosion, the United States discouraged the spread of any nuclear technology, advocating international control of nuclear materials and technology to deter or prevent their military use. The 1946 Atomic Energy Act expressly prohibited even exchanges of information until "effective and enforceable international safeguards against the use of atomic energy for destructive purposes" were in place. A few years later, the Soviet and British nuclear tests, as well as nascent nuclear weapons programs in other countries, underscored the futility of trying to keep the lid on this Pandora's box of nuclear energy, and a new approach was born: the Atoms for Peace program. As nuclear technology spread, new mechanisms for control were established also – the IAEA and the NPT.

A focus on restricting sensitive nuclear technologies, however, did not develop until the 1970s. United States provided nuclear training and technical assistance in reprocessing for twenty years before it adopted a policy of restraint. Alarmed by India's nuclear test in 1974, the Ford administration prohibited the export of reprocessing and other technologies and firmly opposed France's and Germany's proposals to sell reprocessing plants to Brazil, South Korea, Pakistan and Taiwan. The Ford administration also called upon all states to join the U.S. in "exercising maximum restraint in the transfer of reprocessing and enrichment technology and facilities by avoiding such sensitive exports or commitments for a period of at least three years." The Carter Administration followed up by directing U.S. nonproliferation policy towards "preventing the development and use of sensitive nuclear power technologies which involve direct access to plutonium, HEU or other weapons-usable material and non-nuclear weapons states, and at minimizing the global accumulation of these materials." (PD/NSC-8, March 24, 1977). Both Presidents Ford and Carter recognized the difficulty in preventing the spread of reprocessing while supporting plutonium recycle at home.

President George W. Bush, on the other hand, promoted reprocessing at home and emphasized denial policies abroad. Both ultimately fell flat. The Obama administration advocated creating incentives for states not to pursue sensitive fuel cycle technologies but acknowledged frustration about the build up of plutonium. President Obama told an audience at Hankuk University in 2012 that "We simply can't go on accumulating huge amounts of the very material, like separated plutonium, that we're trying to keep away from terrorists." His Administration's solution included creating fuel banks, a new international framework for civil nuclear cooperation, and U.S. investment in research and development on new fuel cycles. Fuel banks, i.e. stores of low-enriched uranium (LEU) available to countries in good standing with the IAEA that are cut off from the global market, now exist in the United States, Russia, and Kazakhstan (the last under IAEA control), but there is no evidence that they have had an impact on any state's interest in enrichment.

FUTURE POLICY PROSPECTS

There are seventy years of proposed solutions to the proliferation risks of enrichment and reprocessing, but none with truly lasting success. International ownership of all enrichment and reprocessing plants was proposed in the 1940's but not implemented because of the Cold War. Some multilateral ownership of facilities (Urenco's uranium enrichment and Eurochemie's reprocessing facility in Belgium) has been attempted but not on a global scale. An assortment of other proposals such as fuel banks, enrichment bonds, and LEU banks have also been proposed to tepid response. Some of the most far-reaching efforts have responded to proliferation crises – India's 1974 nuclear test and Iraq's 1991 crash nuclear program – but it is not particularly wise or prudent to wait for a disaster to occur before trying to prevent it. Looking ahead, challenges to the spread of enrichment and reprocessing may emerge in the Middle East if Iran's enrichment program expands significantly after the JCPOA ends or if other states in the region seek to match Iran's enrichment program by seeking one of their own. Challenges may also emerge in Asia if South Korea seeks to enrich uranium for a new nuclear–fueled submarine that the new President Moon Jae-In has expressed interest in pursuing or if China strongly embraces commercial reprocessing for its nuclear power program. These challenges need to be met in advance with principled, rather than regionally focused approaches. The discussion below focuses on what the United States could possibly do to strengthen controls on nuclear fuel cycle capabilities.

Strengthen Current Tools

An incremental approach would strengthen the tools currently in use by the United States: nuclear cooperation agreements, bilateral diplomacy, export controls (national and within the NSG), and international fuel cycle collaboration:

1. Nuclear cooperation agreements. The United States recently has included provisions in its agreements for countries like Taiwan and South Korea to send their spent fuel off to Europe for reprocessing. Instead, future agreements and plans should emphasize assistance on putting spent nuclear fuel in dry-cask storage.
2. Diplomacy. The uranium enrichment market has been depressed as a result of oversupply since the Fukushima accident in 2011 took 50 reactors off the market. The U.S. should argue that, given the current overcapacity of the existing suppliers and the resulting historically low price of enrichment services with no end in sight, countries don't need a national enrichment capability. Further, it should argue that building small-capacity enrichment plants of the kind that Argentina, Brazil, Iran and Japan have planned have no economic justification. A second effort should focus on a global ban on enriching beyond 4-6 percent in U-235. This would build on U.S. nuclear security efforts to take HEU out of civilian commerce. Most enrichment plants do not enrich above 6 percent and there is an abundance of HEU in Russia and the United States from the Cold War that should be blended down. The NSG could also adopt a guideline calling for an NSG right of prior approval for enrichment.
3. Export controls. The United States should urge the NSG to develop policy options under its guidelines to require rather than "encourage" multinational involvement in any new enrichment or reprocessing plant built with transferred technology.
4. International fuel cycle collaboration. With regard to enrichment, selected governments and industry representatives should at least harmonize approaches and preferably develop international standards for "black-boxing" centrifuge enrichment technology. With regard to reprocessing, the U.S. should discourage reprocessing as an unnecessary, costly, and risky way to condition spent nuclear fuel for disposal. The United States could also champion deep borehole disposal of spent fuel, a potentially politically more acceptable alternative to mined repository disposal of spent fuel, as a research priority in international collaboration.

Approach the Problem Holistically

The United States alternatively could adopt a broader approach to convince other countries to help limit the spread of enrichment and reprocessing and the follow-on challenges of material stockpiles (à la Iran). This would require a major investment of time and political capital and would need to address the concerns of countries that would be giving up what they see as their right to engage in peaceful enrichment and reprocessing. It would be an effort to negotiate a new grand bargain with developing countries but also would be less discriminatory by applying new standards to all existing civilian enrichment and reprocessing facilities, including in the nuclear-weapon states. It would include seeking to multilateralize any new enrichment facility and convert existing enrichment facilities eventually to multilateral control and oversight. It would also entail substantial assistance in dealing with spent fuel, and as well as legal guarantees for the supply of enriched fuel. It would target production, use, and disposal of fissile material and create specific objectives for each of them.

1. Production: The objective would be to have no new national enrichment or reprocessing plants. The U.S. could continue to oppose any expansion of reprocessing and encourage the retirement of existing reprocessing plants with the argument that reprocessing is uneconomic, unnecessary for nuclear power, and poses unacceptable proliferation risks. Although currently there is no foreseeable need for new enrichment capacity, in the future it may be necessary. The objective would be that any future enrichment plant would be limited to enrich to less than 6 percent, have diversified ownership, and be situated in a country that has an Additional Protocol. Furthermore, the facility must be subject to IAEA safeguards even if it is in a

nuclear-weapon state. Existing enrichment plants in the weapon states would have safeguards applied to them, as would existing reprocessing plants (with no exceptions).

2. Use: The fundamental objective would be to reduce the amount of directly nuclear weapons-usable material. Critical elements would include a phase-out of HEU in the civilian sector and limits on enrichment levels for naval fuel (e.g., 5-20 percent). States would have to politically commit to no HEU production, no new facilities licensed for HEU production, no new civil HEU-fueled facilities, and political commitments and regulatory assurances of no plutonium separation unless existing stocks have been reduced to near-minimum working levels.
3. Disposal: The objective would be to create incentives for states to agree to new restrictions on sensitive technology by including assistance for the back end of the fuel cycle, i.e. long-term surface storage of spent fuel and/or deep underground disposal, including a multinational repository.

Any such grand new bargain would require consequences for noncompliance with whatever bargain emerges, which could include targeted limits on uranium supply to the country and on access to services such as conversion, fuel fabrication, or spent fuel storage. The ability to purchase or use the product of fuel cycle facilities, or the ability to send high-level waste or spent fuel to a multinational repository, could also be affected. Taken together, such limits could constitute a general ban on nuclear-related international commerce for and with the country in question.

There is a window of opportunity to strengthen the barriers against the spread of enrichment and reprocessing that could help to prevent future proliferation cases like Iran. Within the next ten to fifteen years, countries need to develop a framework on enrichment and reprocessing that could prevent potential proliferation of these capabilities in the Middle East and Asia in particular. An approach that shifts direction from technologies to the broader picture (nuclear security writ large) could help minimize concerns of non-nuclear weapons states about discriminatory treatment, address virtually all of the current NPT challenges and at the same time increase the barriers to the spread enrichment and reprocessing.

ACKNOWLEDGMENTS

This paper has been drawn from previously published work, including the CSIS-NTI Report, *A New Approach to the Nuclear Fuel Cycle*, by Squassoni, Hinderstein, Newman, Hartigan, (Washington, DC: CSIS, 2014), congressional testimony by Sharon Squassoni and an as-yet unpublished book chapter entitled, "The Nuclear Nonproliferation Regime," in an edited volume by Bruce MacDonald. Amelia Armitage helped proofread and edit the copy.

REFERENCES

1. "Nuclear Fuel Cycle," *IAEA*, undated, www.iaea.org/OurWork/ST/NE/NEFW/Technical-Areas/NFC/images/nfc-image-big.jpg.
2. The Nuclear Energy Agency and the International Atomic Energy Agency, "Uranium 2016: Resources, Production and Demand," OECD, 2016: pp. 62-63.
3. Centrus Energy Corp., "Form 10-K: Annual Report Pursuant to Section 13 or 15(d) of the Securities Exchange Act of 1934," (31 March 2017), pp. 9.
4. See fact sheet by the United Kingdom's Nuclear Decommissioning Authority: "Fact Sheet: Spent fuel reprocessing," *Nuclear Decommissioning Authority*, undated, http://ukinventory.nda.gov.uk/wp-content/uploads/sites/2/2014/01/Fact-sheet-spent-fuel-reprocessing.pdf.
5. "Progress on approaches to the management of separated plutonium: Position Paper," Nuclear Decommissioning Authority, January 2014.
6. Proliferation risks used to be described in terms of what states might accomplish, but assessments now also look at the proliferation risk from non-state actors acquiring material. There is an underlying assumption that non-state actors such as terrorist organizations would find it too difficult to put together an entire nuclear weapons acquisition program, but groups like Aum Shinrikyo attempted this in the past. Conventional wisdom is that terrorists might seek ready-made material rather than seeking to produce it on their own. This means, however, that stockpiles of material, if not controlled properly, could become targets of opportunity for terrorist groups. This paper addresses primarily the risks of state proliferation.
7. While separated plutonium can be diverted while stored (for example, awaiting fabrication into fuel), it is generally more "countable" in discrete items in storage than in many parts of the reprocessing plant where it is being processed, sometimes contained in solutions that can be difficult to measure.
8. The JCPOA prohibits Iran from reprocessing spent nuclear fuel but allows Iran to produce radioisotopes for medical and peaceful industrial purposes. Iran agreed to foreswear reprocessing for the duration of the agreement, stating it

"does not intend to engage in reprocessing after the expiration of the JCPOA." Iran conducted some reprocessing-related experiments and was building a heavy water reactor at Arak that could have produced weapons-grade plutonium, but did not have a reprocessing plant before the JCPOA was signed.

9. See early criticisms of the IAEA safeguards system in S. Moglewer, "IAEA Safeguards and non-proliferation," in *Bulletin of Atomic Scientists*, **37**:8, pp. 24 ff (October 1981), and Senate testimony by R. Richter, also reprinted in *Bulletin of Atomic Scientists*, **37**:8, pp. 30 ff (October 1981).
10. The starting point of safeguards is a technical term, defined in the IAEA Safeguards Glossary for states with comprehensive safeguards agreements (INFCIRC/153) as the point "when any nuclear material of a composition and purity suitable for fuel fabrication or for being isotopically enriched leaves the plant or the process stage in which it has been produced, or when such nuclear material, or any other nuclear material produced at a later stage in the nuclear fuel cycle, is imported into a State." See: "IAEA Safeguards Glossary," *International Nuclear Verification Series* **3** (2001), https://www.iaea.org/sites/default/files/iaea_safeguards_glossary.pdf.
11. See, for example, John Carlson's paper on expanding IAEA safeguards in the nuclear weapons states: J. Carlson, "Expanding Safeguards in Nuclear Weapons States," *NTI* (2012), http://www.nti.org/media/pdfs/NWS_safeguards_carlson_fin.pdf?_=1337718775.
12. Iran's safeguards failures have been recorded meticulously by the IAEA in its regular reports to the Board of Governors from about 2003 to 2015. See: "Monitoring and Verification in Iran," *IAEA*, https://www.iaea.org/newscenter/focus/Iran for official IAEA documentation on Iran.

Nuclear Proliferation and Testing: A Tale of Two Treaties*

Pierce Corden[a] and David Hafemeister[b]

[a]*American Association for the Advancement of Science*
Washington, DC 20005

[b]*Physics Department*
California Polytechnic State University
San Luis Obispo, CA 93407

Abstract. Despite progress in reducing stockpiles after the end of the Cold War, the disturbing actions of some nations could spread nuclear weapon capabilities and enlarge existing arsenals.

The first nuclear weapon explosion, code-named Trinity, took place at Alamogordo, New Mexico, on 16 July 1945. Since then eight countries have carried out more than 2000 nuclear tests with yields from a fraction of a kiloton to some 50 megatons. (Yield is the amount of TNT needed for an equivalent explosion.) The public anxiety over nuclear testing began shortly after World War II, as the powerful but polluting atmospheric explosions exposed entire communities, known as downwinders, to the health hazards of radioactive fallout. Governmental anxiety followed in the late 1950s, and the US, USSR, and UK entered a moratorium on nuclear testing, only to have it fall apart three years later when the French began testing and the three countries resumed testing ever more powerful weapons with a vengeance.

In 1963, three years after President John F. Kennedy warned the public that as many as 20 countries might obtain nuclear capabilities by the end of the decade, the Limited Test Ban Treaty (LTBT) was negotiated to help curb the spread. That treaty prohibits testing in the atmosphere, under water, and in outer space, but not underground. Satellite surveillance and seismology were both very new at the time, and effective verification of a ban on underground explosions was widely thought impossible without onsite inspections, which were not part of the LTBT. Nonetheless, in 1968 the international commitment to comprehensively end nuclear testing was reaffirmed in another agreement, the Nuclear Non-Proliferation Treaty.

The NPT was built on a strategic bargain: The overwhelming majority of countries agreed to join the treaty as non-nuclear-weapon states, to not acquire nuclear weapons, and to allow inspections of their nuclear facilities. In return, the five nuclear weapon states—the US, UK, and USSR (now Russia), later joined by China and France and collectively known as the P5—pledged to work toward the elimination of their nuclear arsenals and to allow the sharing of peaceful nuclear technology with the non-nuclear-weapon states.[1]

At the time, the NPT signatories understood that the total elimination of all nuclear weapons was likely to be far in the future. Nearly half a century later, that's still the case. But India, Pakistan, Israel, and North Korea are not parties to the NPT, and dealing with their capabilities and those of the P5 proceeds in parallel with efforts involving the non-nuclear NPT parties. Figure 1 outlines the history of nuclear weapon proliferation since the first test in 1945. The US and USSR–Russia stockpiles have each declined by almost an order of magnitude from their peaks; the UK and French stockpiles have modestly declined. For others, either the picture is uncertain or the numbers are increasing.

---

* Reproduced from *Physics Today*, Vol. 67, April 2014, p. 41–46 with the permission of AIP Publishing.

Nuclear Weapons and Related Security Issues
AIP Conf. Proc. 1898, 040006-1–040006-8; https://doi.org/10.1063/1.5009229
Published by AIP Publishing. 978-0-7354-1586-7/$30.00

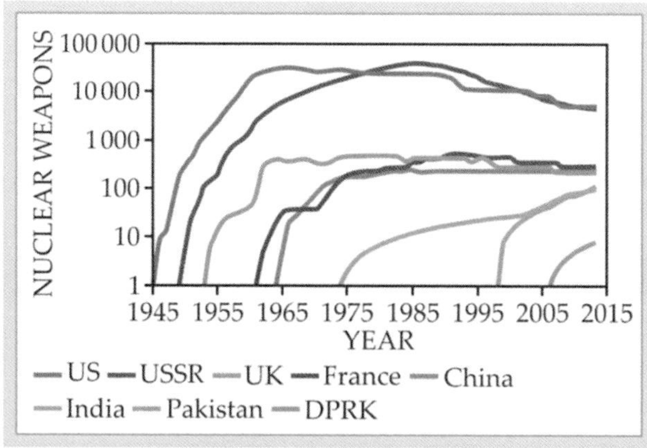

FIGURE 1. Nuclear weapons, counted. The approximate number of stockpiled weapons in each country is keyed to the date of its first nuclear test. (Adapted from data of Hans Kristensen, http://www.fas.org.) The following are 2013 totals: US, 4650 stockpiled, 2120 operational; Russia, 4300 stockpiled, 1600 operational; UK, 225; France, 300; China, 250; India, 90–110; Pakistan, 100–120; and North Korea (DPRK), 5–10. [P. Corden]

If Iran were to build the bomb, in violation of its NPT obligations, how would NPT non-nuclear-weapon states, such as Saudi Arabia and Egypt, react? If the US, China, or Russia were to resume nuclear testing, would that lead to testing by others and to new deployments of nuclear weapons? The NPT "regime" extends beyond the treaty in numerous ways to help counter those and other risks. Tools such as diplomacy between nations, arms control, economic sanctions, safeguards and protocols of the International Atomic Energy Agency (IAEA), physical security for nuclear materials and weapons, counterproliferation military strikes, port and border security, limits on the production of weapons-useable uranium and plutonium, and the Proliferation Security Initiative—a global effort to stop trafficking in weapons of mass destruction—are all part of the broader regime. That interconnected system must be resilient to unpredictable changes. Further nuclear testing or withdrawal by a nation from the NPT would pose serious challenges to the regime.

In November 2013 two dozen experts spoke at a short course, cosponsored by the American Physical Society's Forum on Physics and Society and the George Washington University's Elliott School of International Affairs, to discuss those and other current threats to international security posed by nuclear weapons and related technologies.[2] This article focuses on the NPT and national security—particularly as both are affected by the legal status of the closely linked Comprehensive Nuclear-Test-Ban Treaty (CTBT) and the ongoing developments in Iran and other nations.

THE NPT–CTBT INTERSECTION

International law is not domestic law. Enforcement of international nonproliferation agreements includes measures decided by the UN Security Council. But those measures are effective only if most of the global community of nations adhere to them. Unfortunately, confidence in the NPT has been complicated by the fact that the CTBT, opened for signature in 1996, has yet to enter into force—that is, be ratified into law by China, Egypt, Iran, Israel, India, Pakistan, North Korea, and the US. (See the article by Jeremiah Sullivan, *Physics Today*, March 1998, page 24.)

The holdouts are troubling. For one thing, there is broad international support for the CTBT: The treaty has been signed by 183 states, has been ratified by 162, and bans nuclear tests at any yield, at any place, and for all time. For another, during negotiations to secure a permanent extension of the NPT, France, Russia, the UK, and the US sent a letter, dated 19 April 1995 (China later agreed), to the NPT Review and Extension Conference promising a CTBT:[1]

> We reaffirm our determination to continue to negotiate intensively, as a high priority, a universal and multilaterally and effectively verifiable comprehensive nuclear test-ban treaty, and we pledge our support for its conclusion without delay. . . . We call upon all States parties to the [NPT] to make the treaty provisions permanent. This will be crucial for the full realization of the goals set out in Article VI.

In 1999, when the US Senate considered giving its advice and consent to ratification, it declined to do so. In the hearings and floor debate, the verifiability of the treaty and the effectiveness of the US nuclear weapons stockpile in the absence of nuclear-explosive testing were principal issues cited as problematic. The contention caused by the failure of the US and the other seven states to ratify the CTBT does not help develop the global consensus needed to enforce nonproliferation undertakings. Without the CTBT, the NPT is perceived as a treaty of unbalanced obligations.

A ban on nuclear tests is a nonproliferation and arms-control measure that raises the barrier for states to move toward nuclear weapons and blocks existing nuclear powers from substantial new technical advances. Although such a barrier would not prevent the development of simple gun-type weapons, which typically do not require testing to work, it prevents development of thermonuclear weapons and the miniaturization of nuclear weapons for missiles.

At the 2010 NPT Review Conference, all 189 NPT members viewed the CTBT's entry into force as essential for a strengthened nonproliferation regime. In December 2013 the United Nations General Assembly urged all nations to sign and ratify the CTBT on a vote of 181 yes, 1 no (North Korea), and 3 abstentions (India, Mauritius, Syria). The Obama administration has declared its support for Senate action to ratify the treaty.

SEISMIC MONITORING

Since 1963, when the LTBT was adopted, the capabilities of seismology to detect underground nuclear explosions have steadily improved. By the time the CTBT was negotiated at the Geneva Conference on Disarmament in the mid 1990s, seismic instrumentation had matured so dramatically that it and other test-detection technologies provided a very effective means to verify states' compliance with the new treaty.

The CTBT establishes the International Monitoring System (IMS), networks comprising 321 seismic, radionuclide, hydroacoustic, and infrasound sensors around the world—84% of them certified, 5% installed, 6% in construction, and 5% planned. Thanks to the relatively high density of the 170-station seismic network, states have access to a richer set of data than was possible when the LTBT was negotiated, when there was no regular access to territories of states of interest, particularly the USSR. The seismic network can detect explosions at more reduced yield levels than previously. "National technical means"—the nationally owned and operated seismic and other sensors, reconnaissance satellites, ships, and aircraft—provide an additional and powerful source of verification that exceeds the capability of the IMS networks.

Regional and long-range seismic waves are used to distinguish between explosions and earthquakes, estimate the magnitude of an event, and determine its location so that follow-up onsite inspections can be carried out in ambiguous cases. Consider Figure 2, which compares seismograms from a low-magnitude earthquake, a chemical explosion, and the 9 October 2006 nuclear explosion in North Korea. The nuclear test was detected widely by the IMS and other stations in South Korea, China, and Japan. A later North Korean nuclear explosion, on 12 February 2013, was detected by 96 IMS stations, which were able to locate the test site to within 8 km.

Regional waves predominantly travel through Earth's crust and upper mantle and are strongly affected by local geology, which can complicate interpretation, but they offer better, more detailed results than long-range waves. In 2001 seismologists Vitaly Khalturin, Tanya Rautian, and Paul Richards analyzed the regional signals collected between 1961 and 1989 at distances 500–1500 km from underground explosions at the former Soviet test site at Semipalatinsk in Kazakhstan. Usefully detailed seismic information was found in the old Soviet data from all but two of 340 tests with yields over a ton.[4]

Newer broadband seismographs are sufficiently sensitive that, according to a 2002 National Academy of Sciences (NAS) study,[5] underground nuclear explosions conducted in hard rock anywhere in Europe, Asia, North Africa, and North America can be reliably detected and identified as explosions down to a yield of 0.1 kt using IMS data. The denser the seismic network in a particular region, the greater the detection probability. A 2012 National Research Council (NRC) report has mapped the magnitudes and yields that a nuclear test anywhere in the world would have to exceed in order to be detected.[6] As shown in figure 3, explosions well below a kiloton can be detected with 90% confidence.

The probability that a state seeking to violate the CTBT can repeatedly evade detection drops rapidly with an increased number of tests. Additionally, geophysicists Ola Dahlman, Svein Mykkeltveit, and Hein Haak point out[7] that the violator would have to reduce its test explosion's yield by a factor of three to reduce the seismic detection probability from 90% to 10%. Further compounding the risk of being detected, violators must take into account the synergy provided by other IMS networks, possible on-site inspections, possible excess yields, and information available from the thousands of seismic stations operated by academic and disaster-preparedness organizations to monitor earthquakes.

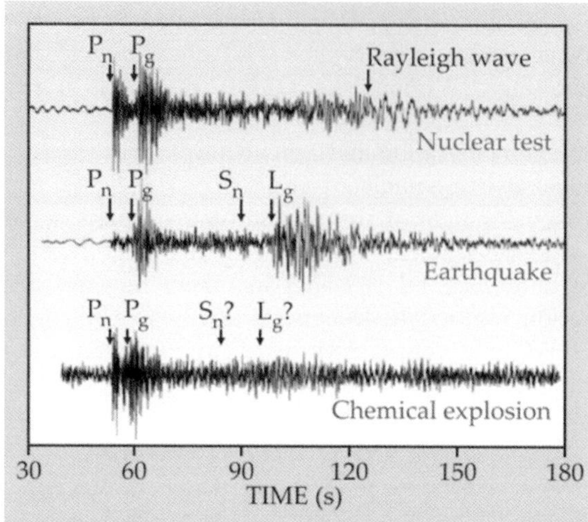

FIGURE 2. Seismographs from a station in Mudanjiang, China, record the ground-velocity amplitudes as a function of time of three events a few hundred kilometers away in North Korea: a 9 October 2006 nuclear test (top) of magnitude 4 (0.7 kt equivalent yield); a 16 December 2004 magnitude 4 earthquake (middle); and a 19 August 1998 underground chemical explosion (bottom; 0.002 kt equivalent yield). The nuclear and chemical explosions cause the sudden onset of compressive waves that travel either through the crust and mantle (Pn) or through the crust alone (Pg) and the later onset of weak shear waves (Sn and Lg) and Rayleigh waves, a type of surface wave. Compressive waves in earthquakes emerge more slowly and their shear waves are stronger than in the explosions; indeed, the shear waves are so weak in both nuclear and chemical explosions that labeling their emergence in spectra is difficult. (P. Richards via RightsLink, Adapted from ref. 11.)

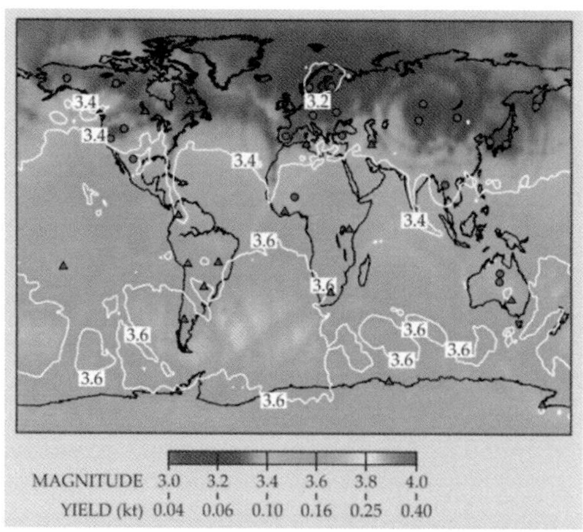

FIGURE 3. The detection capability of the International Monitoring System's seismic network is correlated to the magnitude of Earth's seismic events. In 2007, when this map was made, the primary network consisted of 38 stations (dots and triangles) around the world. The map's colors correlate the minimum seismic magnitude to the explosive yield in kilotons in hard rock required for 90%-confidence-level detection. For example, one can be 90% sure of detecting any tests above 0.09 kt in Asia, Europe, and North Africa for tamped explosions in hard rock with better propagation, and 0.22 kt in rock with poorer propagation. If the confidence level is relaxed to 10% detection, the yields are reduced by a factor of three. (Courtesy of NAS, Adapted from ref. 6, Tormod Kvaerna and Frode Ringdal.)

One evasion scenario, envisioned in the 1950s, is "cavity decoupling"—the use of a cave or underground cavity intended to muffle the seismic waves from a nuclear explosion and thus reduce the radiated signal. Even in that case, according to the 2012 NRC report, "an evasive tester in Asia, Europe, North Africa, or North America would need to restrict device yield to levels below 1 kiloton (even if the explosion were fully decoupled) to ensure no more than a 10 percent probability of detection for IMS and open monitoring networks" (page 10).[6] And that doesn't account for the deterrent provided by states' national technical means and from the IMS network of radionuclide sensors.

IRAN'S CENTRIFUGES

To build a nuclear weapon, one must first produce the fissile material that powers it. Historically, the most common path to that end was to extract plutonium from the spent fuel rods of heavy-water reactors that used natural uranium. That method was India's path to its 1974 test explosion. A less common approach was to enrich uranium to increase the abundance of its fissile isotope ^{235}U. Uranium enrichment was difficult with bulky gaseous diffusion technology, and active diplomacy convinced several nations in the 1970s to abandon plans for reprocessing spent fuel.

Thirty years later, centrifuge technology has advanced sufficiently that it is now the chosen path for acquiring highly enriched, bomb-grade uranium. Because gas centrifuges require much less power and space than reactors and reprocessing plants, they are much easier to hide. And their growing use represents a challenge: how to prevent the development of nuclear weapons and still allow the peaceful use of nuclear energy. Both goals are part of the NPT. (See the article by Houston Wood, Alexander Glaser, and Scott Kemp, *Physics Today*, September 2008, page 40.)

The enrichment of uranium is measured in separative work units (SWUs) and is a nonlinear process. A common light water reactor (LWR) producing about a gigawatt of electric power consumes each year approximately 25 tons of 3.75%-enriched fuel, obtained from 210 tons of natural uranium, and requires 120 000 SWU. Figure 4 offers an example of the extent to which the separative work diminishes as uranium is enriched from its natural form of 0.7% ^{235}U, to its highly (90%) enriched form. It requires about 800 SWU per ton of uranium to make 130 kg of 4%-enriched reactor fuel. An additional 300 SWU yields 26 kg of 20% highly enriched uranium, and a further 200 SWU yields 5.6 kg of 90% HEU. Put another way, it takes about 60% of the total work to produce low-enriched LWR fuel, 25% of the total to enrich the fuel to 20%, and another 15% of the total to raise that amount to 90%. Thus, a stockpile of 20% enriched material can significantly reduce the time a country requires to make a nuclear weapon.

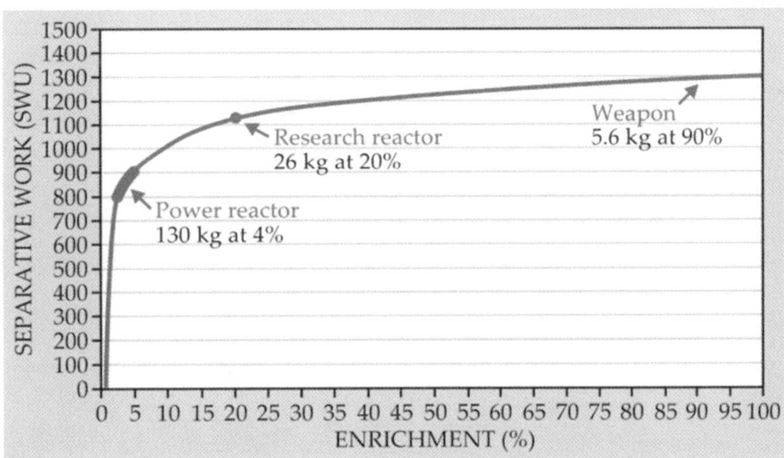

FIGURE 4. The effort required to enrich uranium and increase the concentration of its fissile isotope ^{235}U diminishes exponentially as the concentration rises. The effort is measured in separative work units. Starting with 1 ton of natural uranium, it takes 800 SWU to produce 4%-enriched reactor fuel, an additional 300 SWU to produce 20%-enriched material for research reactors, and 200 more SWU to produce 90%-enriched material. (Courtesy of Olli Heinonen.)

Argentina, Brazil, China, France, Germany, India, Iran, Japan, the Netherlands, North Korea, Pakistan, Russia, the UK, and the US currently enrich uranium. Commercially enriched stock is primarily supplied by Europe's URENCO (10 million SWU/year), Russia's Rosatom (26 million), and France's AREVA (10 million). Iran's centrifuges produce 0.8–4 SWU/year,[8] and the country has claimed the right to pursue uranium enrichment in line with the NPT. But its claim has been a subject of dispute, considering that Iran had not reported its past enrichment activities to the IAEA. The developments have prompted condemnation in UN Security Council resolutions. Other nonweapon states that enrich uranium have negotiated acceptable reporting and inspection agreements with the

IAEA.

Figure 5 shows the growing numbers of centrifuges Iran has installed and is operating at its large facility in Natanz since 2007. (To date, Natanz and the other main centrifuge facility, at Fordow, together house 19 000 centrifuges.)

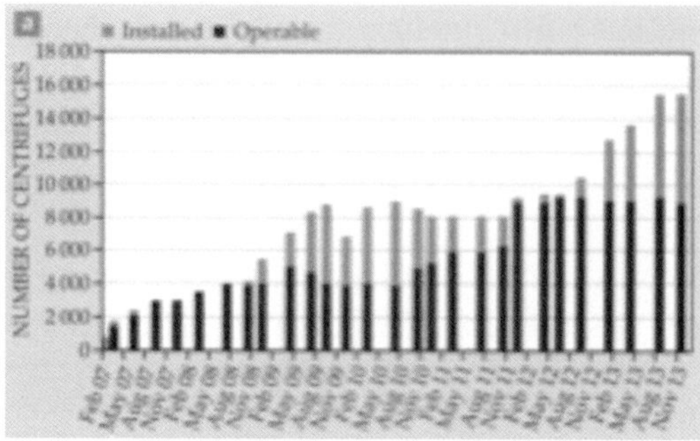

FIGURE 5. The number of Iran's centrifuges at Natanz (a) in 2013 grew to nearly 16 000, a roughly 60% increase from the previous year. Combined with the number at a smaller facility in Fordow, the country currently has a total of 19 000 centrifuges. Operable centrifuges (black) are enriching uranium hexafluoride. Inoperable centrifuges (green) include installed machines not yet fed with UF_6 and a large fraction damaged by sabotage in 2009 and 2010 when the computer virus Stuxnet infected the operating system. The plateau in late 2013 may be related to the negotiations between Iran and the P5-plus-1 (the US, UK, Russia, China, France, and Germany). (Plot courtesy of David Albright.)

P5 PLUS 1

An agreement known as the Joint Plan of Action between Iran and the P5 (plus Germany) took effect on 20 January 2014. In the agreement, which will last six months, Iran allows constraints to be placed on its nuclear program in exchange for a reduction of $7 billion in economic sanctions. The agreement approximately doubles the "breakout time" needed to obtain sufficient HEU to build a nuclear weapon; the IAEA counts 25 kg of 90% HEU as a "significant quantity," sufficient for a weapon.

According to the plan,[9] Iran agrees to halt uranium enrichment above 5% and cap the production of 3.5%-enriched uranium; dilute or oxidize the country's near-20%-enriched stock; stop the installation of additional centrifuges at Natanz, Fordow, and elsewhere; freeze advances at the country's heavy-water reactor in Arak; allow monthly verification inspections of the reactor and daily inspections of the Natanz and Fordow facilities; provide access to centrifuges, production facilities, uranium mines, and mills; and notify the IAEA in a timely manner about any new nuclear facilities.

FIGURE 6. The P5–PLUS–1 discuss the terms of the Joint Plan of Action. [US State Department]

The primary goal of the constraints is to maintain at least a six-month breakout time so that the UN can respond well before Iran can make sufficient weapons-grade uranium if the country reneges on the agreement. Detection of the production of uranium enriched over 5% would be a violation of the joint agreement. David Albright, president of the nonprofit Institute for Science and International Security, and colleagues calculate a breakout time of six months if Iran kept 150 kg of 20%-enriched material and 2000 IR-1 centrifuges, 250 kg of 20%-enriched material and 1000 centrifuges, or no amount of 20%-enriched material and 6000 centrifuges.[10]

An inspection regime can weigh and examine cylinders containing feed, product, and wastes to determine enrichment levels. Radio-frequency identification tags can track the cylinders. Similarly, tags, seals, and surveillance can be used to keep track of nuclear activity. Uranium stocks, records, and receipts can be audited. Environmental samples of soil and water can be obtained in and outside a plant. Unannounced inspections are useful to detect undeclared material and operations.

The IAEA and the US have had considerable experience in nuclear monitoring. The technology used to monitor the "blend-down" of 500 tons of Russian HEU to reactor fuel is instructive. The density of ^{235}U in the uranium hexafluoride gas can be determined by measuring the emission of 186-keV gamma rays from the uranium. And the transmission of 122-keV gamma rays from cobalt-57 through the UF_6 gas determines total uranium density. The ratio of the two values yields the enrichment level. Twenty years of monitoring by Oak Ridge and Los Alamos (LANL) National Laboratories have verified the dilution of Russian HEU in the equivalent of 20 000 HEU warheads.

CONFIDENCE IN UNTESTED NUCLEAR WEAPONS

The enduring US stockpile consists of several types of warheads; they include tactical and strategic bombs and land- and submarine-launched ballistic missiles. The US has not tested any of them since 1992. The directors of LANL, Lawrence Livermore (LLNL), and Sandia National Laboratories have concluded that nuclear-explosive testing is not needed to resolve technical issues concerning the US nuclear stockpile.

At the November 2013 short course,[2] Texas A&M University's Marvin Adams surveyed the US Stockpile Stewardship Program and agreed with a central conclusion from the NRC's 2012 report (reference 6, page 1):

> Provided that sufficient resources and a national commitment to stockpile stewardship are in place, the committee judges that the United States has the technical capabilities to maintain a safe, secure, and reliable stockpile of nuclear weapons into the foreseeable future without nuclear-explosion testing. . . . Sustaining those technical capabilities will require . . . :
>
> - a strong science and engineering base. . . ;
> - a vigorous surveillance program;
> - adequate ratio of performance margins to uncertainties;
> - modernized production facilities; and
> - a competent and capable work force.

Stockpile stewardship assesses the status of warheads, including the impact of changes made to them, and takes any needed actions to maintain near certainty that weapons would meet requirements if used. (See the article by Raymond Jeanloz, Physics Today, December 2000, page 44.) That assessment, in practice, translates into monitoring the warheads, repairing any observed degradation, dismantling retired warheads (now about 250 per year), and extending the life of the remaining stockpiled weapons.

Scientists at LANL and LLNL have determined a lower limit of 85 to 100 years for the lifetime of a plutonium "pit," the part associated with the fission stage of a nuclear explosive. The JASONs, a group of technical experts who advise the Defense Department, agreed with that assessment. One may simply reuse the plutonium-based pit when a weapon's life is extended. However, long-term stockpile stewardship requires the ability to manufacture new pits. Currently the Los Alamos Pit Production Facility fulfills that role and can produce 6–10 pits per year; further investment could boost its production to as many as 80 pits per year.

Modifications in the warheads are being considered to enhance their safety, security, and performance margins as part of upcoming life-extension programs (LEPs); see Physics Today, December 2013, page 26. According to Adams, the actions being considered in LEPs can include refurbishing existing warheads, reusing the nuclear components from different warheads, and replacing nuclear components with designs not previously in the stockpile. All LEPs require some manufacturing, and any change in a weapon requires the assessment of experts familiar with the latest advances in experiments, theory, and computation. Opportunities to introduce changes are relatively rare because LEPs are carried out on a multidecade time scale.

The following points summarize developments in stockpile stewardship since the NAS's 2002 CTBT study was carried out:

- LEPs are a reality, successfully carried out for land- and submarine-based warheads.
- Peta-scale computation is now available to aid design, stewardship, and understanding of a weapon's behavior.
- The National Ignition Facility and Dual-Axis Radiographic Hydrodynamic Test Facility have been completed for studying relevant physics.
- The future of the Chemistry and Metallurgy Research Replacement Facility at LANL and the Uranium Production Facility at the Department of Energy's Y-12 National Security Complex at Oak Ridge are in doubt because of cost.

Although the NAS and NRC panels concluded that the US can maintain a reliable stockpile without nuclear-explosive testing, the stewardship program's future success isn't guaranteed. The biggest concern is the ability to attract, develop, and retain a first-class work force. Adams urges several steps: Professionals in the US should be involved with related nonproliferation, counterterrorism, and science issues to maintain their expertise and skills. The weapons labs should continue to maintain first-rate facilities, some for unclassified work. Experts outside the weapons facilities can provide helpful assessments, as it is difficult to assess quality "from the inside." The gaps in understanding of weapons' behavior can be addressed with ongoing experiment, theory, simulation, and analysis of past data.

FINAL THOUGHTS

There's a highly compelling reason the US and other nuclear-weapon-possessing states don't test nuclear weapons: To do so would not only invite a new arms race, it would make addressing the nuclear programs in Iran and North Korea much harder and put the nonproliferation regime at risk.

India and Pakistan are examples of what happens in the absence of an arms-control agreement. Both countries have roughly the same nuclear capability and are continuing to build larger and more sophisticated arsenals. International efforts to deal with their competition have not led to a halt, let alone a reversal. A 2006 agreement between India and the US gives India more flexibility on nuclear imports. And the US war in Afghanistan loosened international pressure on Pakistan's nuclear programs. One can hope that at least for the near term, the development of confidence- and security-building measures will give diplomacy time to prevent a nuclear crisis.

REFERENCES

1. T. Graham Jr, D. Hafemeister, *Disarmament Diplomacy* 91, 16 (2009).
2. P. Corden, D. Hafemeister, P. Zimmerman, eds., *Nuclear Weapons Issues in the 21st Century*, AIP Conf. Pro., Vol. 1596.
3. For a description of the technical objectives of testing, see P. Corden, in *CTBT at 15: Status and Prospects*, Arms Control Association, Washington, DC (October 2012), p. 10; P. S. Corden, *Phys. Soc.* 42 (2), 11 (2013).
4. V. Khalturin, T. Rautian, P. Richards, *Pure Appl. Geophys.* 158, 143 (2001).
5. National Academy of Sciences, Committee on Technical Issues Related to Ratification of the Comprehensive Nuclear Test Ban Treaty, *Technical Issues Related to the Comprehensive Nuclear Test Ban Treaty*, National Academies Press, Washington, DC (2002).
6. National Research Council, Committee on Reviewing and Updating Technical Issues Related to the Comprehensive Nuclear Test Ban Treaty, *The Comprehensive Nuclear Test Ban Treaty: Technical Issues for the United States*, National Academies Press, Washington, DC (2012).
7. O. Dahlman, S. Mykkeltveit, H. Haak, *Nuclear Test Ban: Converting Political Visions to Reality*, Springer, Dordrecht, the Netherlands (2009), p. 167.
8. D. Hafemeister, *Physics of Societal Issues: Calculations on National Security, Environment, and Energy*, 2nd ed., Springer, New York (2014), p. 167.
9. D. Albright, in ref. 2; *Department of State, Background Briefing on the Implementation Plan of the P5+1 and Iran's First Step Nuclear Agreement*, 13 January 2014, http://www.state.gov/r/pa/prs/ps/2014/01/219571.htm.
10. D. Albright et al., *Maintaining at Least a Six-Month Breakout Timeline: Further Reducing Iran's Near 20 Percent Stock of LEU*, http://isis-online.org. (2014)
11. P. Richards, W.Y. Kim, *Nat. Phys.* 3, 4 (2007).

Nuclear Terrorism – Threat or Not?

Miles A. Pomper[a] and Gabrielle Tarini[b]

[a]*Senior Fellow, James Martin Center for Nonproliferation Studies, Monterey, CA 93940*
[b]*Research Associate, James Martin Center for Nonproliferation Studies, Monterey, CA 93940*

Abstract. A terrorist attack using nuclear or radiological materials is a low-probability event, but if executed, would lead to unprecedented socio-economic, material, and psychological disruption and damage. This chapter seeks to provide a sound assessment of the scope and nature of the threat by examining the different types of nuclear terrorism, each of which poses different risks, involves different barriers to success, and requires different terrorist capabilities. In addition, the chapter aims to provide an overview of the sources and nature of terrorists' motivations to employ a nuclear attack.

INTRODUCTION

Nuclear terrorism is a low probability, high consequence threat. That is, the probability that a terrorist organization would utilize a nuclear strategy is far lower than the probability of many other types of terrorist attacks, but the risk posed by nuclear terrorism – the probability multiplied by the immense consequences of such an event – is unacceptably high. Indeed, former US President Barack Obama has stated that the danger of a terrorist group obtaining and using a nuclear weapon is "one of the greatest threats to global security."[1] Given the potentially catastrophic consequences of nuclear terrorism, scholars and policymakers have justifiably focused substantial efforts and resources on understanding and attempting to prevent the threat.

Although nuclear terrorism as a threat entered the international discourse in the mid-1970s with the negotiation of the Convention on the Physical Protection of Nuclear Materials (CPPNM) and the development by the IAEA of the first set of international guidelines for physical protection, contemporary concerns about nuclear terrorism came to the fore after the collapse of the Soviet Union left 35,000 nuclear warheads and thousands of tons of nuclear material spread across Russia, Ukraine, Kazakhstan and Belarus.[2] The events of September 11, 2001 caused purported restraints on terrorist violence to be re-examined, and gave credence to the idea that a capable terrorist organization motivated by the desire to cause mass casualties might succeed in mounting a nuclear attack. Thus, in the years following 9/11, there was an explosion of US and international initiatives designed to combat the threat of nuclear terrorism. In recent years, the rise and global reach of powerful and well-funded terrorist groups such as the Islamic State of Iraq and Syria (ISIS), combined with the spread of civilian nuclear energy to new countries across the globe, has yet again placed nuclear terrorism high on the political agenda of many states.

But what exactly is the nature of the threat of nuclear terrorism? Fortunately, there have been no instances in which non-state actors have used nuclear weapons, and the historical record of pursuit by terrorists of intact nuclear weapons, compared to other weapons and tactics, is sparse. However, as this chapter will demonstrate, nuclear terrorism can occur in a variety of ways, and may be achieved by terrorist groups through numerous pathways. At the same time, the execution of nuclear terrorism involves significant technical, financial, political, and security-related hurdles that organizations must overcome to break the nuclear threshold. Moreover, it is not sufficient to assume that all terrorists who have the opportunity or capability to carry out an act of nuclear terrorism will automatically do so. It is necessary to consider how well nuclear terrorism fits in with an organization's ideology, goals, and strategic and organizational objectives.

This chapter will proceed in three main sections. First, it will take stock of some of the challenges and limitations inherent in scholarly attempts to assess the threat of nuclear terrorism. Second, it will review in detail the three scenarios of nuclear terrorism that experts consider most likely: radiological terrorism, the attack or sabotage of a nuclear facility, and the detonation of a nuclear device, and also provide insights from the historical record. Finally, this chapter will review the oft-neglected "demand side" of nuclear terrorism, that is, the sources and nature of terrorists' motivation to engage in nuclear terrorism.

Nuclear Weapons and Related Security Issues
AIP Conf. Proc. 1898, 050001-1–050001-16; https://doi.org/10.1063/1.5009230
Published by AIP Publishing. 978-0-7354-1586-7/$30.00

THE LIMITS AND CHALLENGES OF ASSESSMENT

Before reviewing the substance of nuclear terrorism in the proceeding sections, it is necessary to acknowledge several challenges in assessing the scope and nature of the threat. First, the small sample size of prior nuclear and radiological events and the (fortunate) absence of catastrophic events makes a sound analysis of nuclear terrorism extremely difficult. Secondly, empirical evidence of nuclear terrorism is often limited to proxy evidence, or analyses of frequently-examined cases such as the Japanese doomsday cult Aum Shinrikyo, Al Qaeda, or rebel factions in Chechnya.[3] Moreover, the details of even these cases are often sketchy and ambiguous, complicating a rigorous assessment of the threat.[4]

Compounding these problems is a general lack of information on the subject due to secrecy on the part of both the terrorist organizations and national governments. Secrecy and stealth complicate valid threat assessments.[5] Terrorist organizations, in an effort to evade national counterterrorism efforts, rely heavily on secrecy. Thus, there is a notable lack of reliable evidence in the public domain on the capabilities and intentions of the terrorists themselves. Another barrier that scholars face in their analysis is the fact that, as Micah Zenko notes, "Nuclear terrorism lies at the nexus of three areas the US government treats with supreme secrecy – intelligence, nuclear weapons, and terrorism." [6] As a result, it is more than likely that all of the relevant information is not available for scholarly consumption. Finally, military and security officials often downplay the seriousness of, for example, incidents at nuclear energy or weapons facilities, or do not report breaches or vulnerabilities. Even in the US, which has been much more open about its security vulnerabilities, many historical cases have only recently become known.[7] In the absence of the above information, it is difficult for scholars to provide accurate and complete threat assessments. With these challenges in mind, the next section of this chapter will provide a detailed overview of what experts see as the three key threats involving nuclear or radiological weapons and non-state actors.

SCENARIOS OF NUCLEAR TERRORISM

Experts have typically divided nuclear terrorism dangers into three categories: (1) radioactive dispersal devices (RDDs, which include conventional explosives that disperse radioactive materials, known as dirty bombs) or radioactive emission devices (REDs, or fixed radiological sources that expose potential victims to radiation); (2) attacks or sabotage of nuclear facilities; and (3) terrorist acquisition and detonation of a nuclear device. Each type of nuclear terrorism poses different risks and requires different capabilities from terrorist organizations. The following sections will analyze each type of nuclear terrorism in turn while providing examples from the empirical record.

Radiological Terrorism

Of the three types of nuclear terrorism, an act of radiological terrorism would likely cause the lowest scale of destruction, and would not lead to catastrophic levels of death and injury on the scale of a nuclear weapons detonation. The largest impact of most dirty bomb events would be economic, for example, the substantial costs involved in the evacuation, relocation, and cleanup of areas contaminated by radiation.[8] Radiological terrorism could also cause significant short- and long-term health problems for people in the area exposed to the radiation, and would undoubtedly engender widespread distress and panic. This mode of nuclear terrorism would be the easiest for a terrorist organization to execute due to the ubiquitous nature of radiological materials and the relatively less complex technical demands of such an undertaking.

Radiological devices as weapons of mass disruption

Radiological terrorism could take several forms. The two most commonly cited scenarios include the use of a radiological dispersal device (RDD), or a radiological exposure device (RED). The use of an RDD would involve the intentional dispersal of radioactive material, either by using conventional explosives (a.k.a. a "dirty bomb") or spreading the material actively (e.g. using a sprayer) or passively (i.e. without the assistance of a powered mechanical device).[9] Despite media and other reports to the contrary, a dirty bomb is in no way similar to a nuclear bomb. A nuclear bomb creates an explosion that is millions of times more powerful than that of a dirty bomb. The cloud of radiation from a nuclear bomb could spread tens to hundreds of square miles, whereas a dirty bomb's radiation would be dispersed only within a few blocks or miles of the explosion.[10] RDDs are considered "weapons of mass disruption" because they are not expected to kill large amounts of people but instead destabilize the economy and functioning of a society through panic, the destructive force of the blast, and the spread of radiation and radioactive materials. On the other hand, a RED emits radiation without spreading radioactive material itself. This distinguishes a RED from a passive RDD because the passive RDD still requires the dispersal of radioactive material over an area.[11] For example, a terrorist organization could place a concealed radioactive source in a highly-trafficked location such as a train station, sports arena, or concert hall.

However, as James Acton, Brook Rogers, and Peter Zimmerman note in a 2007 study, exposure to radioactive materials by the "inhalation, ingestion, or immersion" is another, largely overlooked, mechanism of radiological terrorism that may be attractive to terrorist organizations.[12] If administered in this manner, radioactive materials can be lethal in tiny quantities. For example, the murder of Alexander Litvinenko, a former Russian spy, in November 2006 by polonium-210 ingestion was likely the first provable act of radiological terror.[13] Relatively simpler versions of these attacks are likely within the skillset of reasonably technically competent terrorists.

Incidents such as the inadvertent dispersal of cesium-chloride in Goiânia, Brazil in 1987 illustrate the scale of damage that could be caused by radiological terrorism. On September 13, 1987 scavengers broke into an abandoned cancer clinic and took a metal canister from a radioteletherapy machine, which contained 1,375 curies of cesium-137. Without realizing what it was, they broke open that canister and gave the material to friends and family, eventually giving it to a junkyard. The junk dealer opened the canister, causing the Cs-137 powder to disperse, which spread further as a result of runoff and wind. The Brazilian National Nuclear Energy Commission (CNEN) found that 200 people had been contaminated, of those 200 people four died, one individual had an arm amputated, and 28 suffered radiation burns. Other direct consequences of the incident included a massive clean-up of contaminated land, at a cost of US$20 million, while indirect costs included an estimated millions of dollars of lost tourism.[14] This incident, in addition to the attempted theft of cobalt-60 in Chechnya in 1999 which led to three deaths, and the hijacking of a truck carrying cobalt-60 in Mexico City, Mexico in 2013, illustrate the potential dangers that might occur should terrorists obtain and use radiological materials to physically or economically damage or intimidate a population.[15]

Radiological materials are widespread and often vulnerable

It would not be too difficult for terrorist organizations to obtain radioactive materials for use in a plot. According to the International Atomic Energy Agency (IAEA), millions of radioactive sources have been distributed worldwide over the past fifty years, with hundreds of thousands currently being used, stored, and produced in familiar locations like hospitals and universities.[16] According to one estimate, radiological sources large enough to pose a serious danger exist in over 13,000 buildings in more than 100 countries.[17] Radioactive materials are used widely for numerous applications covering various sectors such as medicine (e.g., teletherapy, radiography, and blood-irradiation), industry (e.g. level, dredger, conveyor, and well-logging gauges), and sterilization (e.g., medical instruments and food). Some of these sources use high-risk radiological materials – radioactive materials that are extremely dangerous (Category 1), very dangerous (Category 2), or dangerous (Category 3) to people not shielded properly – that could result in serious injury or death if misused.

Moreover, the quality of both regulatory and physical security arrangements for radiological sources is highly variable by country, state, or even institution. Many medical, commercial and industrial groups that handle radiological materials are ill-equipped to secure them, and a lack of national regulatory controls in many countries has led to thousands of missing or stolen radiological sources. Indeed, even in the US, a country with stricter regulatory guidelines, it is estimated that every day one radioactive source is "orphaned," in other words, it falls outside of an institution's accounting system.[18] In the developing world, such controls are generally far weaker, particularly in Central Asia and the Caucasus area, unstable regions that inherited many radioactive sources following the collapse of the Soviet Union.

Attack or Sabotage of a Nuclear Facility

Nuclear power plants have long been recognized by experts as potential targets for terrorists. As Bennett Ramberg noted in his 1980 book, *Nuclear Power Plants as Weapons for the Enemy*, "Nuclear power plants turn conventionally armed enemies into nuclear enemies."[19] Indeed, one need not look further than recent events in Belgium to understand the potential dangers posed by the confluence of a violent non-state actor and nuclear facilities. In 2012, two employees at Belgium's Doel-4 nuclear power station left their jobs to become fighters in Syria.[20] In August 2014, an employee at the same nuclear reactor intentionally drained the lubricant for the reactor turbine, causing it to overheat and shut down for five months.[21] The perpetrator and the motive still remain unknown to Belgian authorities. And finally, in 2015, Belgian police discovered that the perpetrators of the Paris terror attacks monitored and surveilled an official at multiple Belgian nuclear research sites, which housed a wide range of nuclear and radiological materials, including highly-enriched uranium.[22]

Terrorists are indeed interested in nuclear facilities, which exist in numerous locations worldwide. According to the IAEA, there are 440 power reactors in operation (with 60 under construction) around the globe, along with 218 operational research reactors, which are used to produce medical isotopes and train nuclear scientists.[23] The global nuclear industry also includes hundreds of plants that enrich uranium and fabricate fuel for reactors.[24] The security measures at these facilities can vary widely, as there are no binding global standards for physical protection at civilian nuclear facilities. The IAEA recommends general provisions to protect reactors against attack or sabotage, for example, INFCIRC 225, but IAEA inspectors do not verify whether these measures have actually been implemented at sites. Rather, each country adopts its own domestic laws and regulations which operators are supposed to implement at their facilities. The following sections will describe several possible scenarios in which an attack or sabotage of a nuclear facility could occur, and provide examples from the empirical record.

Attacking a nuclear facility to steal nuclear or radiological material

Since non-state actors cannot currently manufacture their own nuclear material, attacking a facility where highly enriched uranium (HEU) or plutonium (Pu) is housed is a potential means to obtain the materials necessary to build a nuclear device. Nuclear facilities also often store large amounts of radioactive material, spent fuel, and other nuclear waste products that terrorists could use in a dirty bomb. As Gary Ackerman correctly notes, although non-state actors may be the ultimate recipient of the nuclear material, the perpetrators of the theft at the facility could be experienced, profit-driven criminal groups who would then sell the material to a violent non-state customer.[25]

The disappearance of nuclear material from a reactor in Kinshasa is one of the few cases where a perpetrator was successful in actually stealing substantial amounts of nuclear material. During the late 1970s, unknown perpetrators stole two fuel rods, which contained small amounts of uranium enriched to twenty percent, from a small research reactor at the University of Kinshasa's Kinshasa Nuclear Research Center in what is now the Democratic Republic of the Congo (then, Zaire).[26] While twenty-percent low-enriched uranium isn't particularly useful for either an actual nuclear weapon or a dirty bomb, this case symbolizes potential intent or terrorist to acquire more dangerous material. At the time, the theft raised no alarms at the facility, and the rods were only noticed to be missing during a routine inventory check.[27] The reactor was said to be poorly guarded by a token fence and small group of security guards.[28] It is worth noting that the fuel rods used in the reactor were provided to Zaire by the US with almost no security stipulations as part of the Atoms for Peace Program. The fuel rods remained unaccounted for over the next two decades until the late 1990s, when an Italian smuggling ring obtained one of the rods and sold it to an Italian police sting operation whom they believed to be a Middle Eastern buyer. While the Kinshasa case is considered to be uniquely significant in that the perpetrators managed to steal a significant amount of nuclear material, according to the Nuclear Facilities Attack Database (NuFAD) at the National Consortium for the Study of Terrorism and Responses to Terrorism (START), there have been seventeen successful incidents of nuclear or radiological theft (along with two unsuccessful attempts) at nuclear facilities from 1961 to 2014.[29]

Sabotaging a nuclear facility to disperse radioactive material

A second way in which terrorists might target a nuclear facility for malicious purposes is by causing harm to the facility itself, in other words, by sabotaging the normal operations of the facility with the intention of causing a meltdown of the reactor core and releasing radioactivity into the environment. This scenario would be attractive to terrorists seeking to cause considerable damage. As John Holdren notes, a successful attack on a nuclear power reactor could, "destroy the facility itself, worth hundreds of millions to billions of dollars; produce tens to hundreds or even thousands of early fatalities and tens of thousands of delayed cancer deaths; and severely contaminate hundreds to thousands of square miles of land, requiring removal of much of it from habitation, commerce and agriculture for periods ranging from months to many decades."[30] The disaster at the Fukushima Daiichi nuclear plant in 2011 underscores the vulnerability of nuclear power plants and demonstrates the terror, disruption, and massive costs that could be caused by a major nuclear accident.[31] Although it took an extraordinary natural disaster to take out both the normal and emergency cooling at the plant, this type of event could also be caused by the intentional sabotage of a power reactor by a motivated terrorist group. An attack on a nuclear power plant would also play on the public's fear of radiation exposure if the attack released radioactive material into the environment.

There are several kinds of nuclear sabotage that are discussed in the public domain. First, as the attacks of September 11, 2001 underscored, terrorists could use an airliner or plane packed with explosives as a guided missile to crash into a nuclear power plant, specifically a number of potential targets: the reactor core or the reactor's spent-fuel storage pool would be the most dangerous targets, but a mixed-oxide fuel fabrication plant, a dry-cask spent-fuel storage facility, or a nuclear waste-repository also represent targets.[32] It is worth noting that only as recently as 2009 has the US Nuclear Regulatory Commission (NRC) required all new nuclear power plants to incorporate design features that would ensure that, in the event of a crash by a commercial airliner, the plant's reactor core would remain cooled or the reactor containment would remain intact, and radioactive releases would not occur from spent fuel storage pools.[33] The fact that these guidelines are only beginning to take shape at nuclear power plants in the US, whose nuclear power plants are widely considered to be the best guarded in the world, does not bode well for the security of plants in other nuclear power states, particularly those just beginning to develop nuclear power.

In addition to the threat posed by aircraft, more recently, government officials have also been forced to take into account the emerging threat of small drones. In 2014, thirteen nuclear power plants in France were plagued by around twenty unexplained drone overflights, while a similar unmanned aircraft was spotted flying over Belgium's Doel-4 nuclear site that same year. [34] Terrorist groups could harness drones as an intelligence-gathering tool in order to capture precise images of nuclear plants for use in a future breach attempt, or could be packed with explosives to cause immediate damage to the plant.

Second, terrorists could attack a nuclear facility by land using mortars, rockets, or other emplaced explosives.[35] One of the most recent manifestations of this type of attack occurred in 2014 during the conflict between Hamas and Israel. On one day of the conflict, Hamas fired seventy-four rockets from the Gaza Strip into Israel's territory, at least three of which targeted the Negev Nuclear Research Facility, which houses an operational nuclear reactor.[36] Israel's Iron Dome anti-missile system intercepted one of the missiles before it could land, while the other two landed in open areas. While Hamas has been thought to target the reactor in the past, this incident represents the first time that the group explicitly stated that it had been attempting to strike the reactor.[37] According to Gary Ackerman and James Halverson, "Although the attack caused no damage, this incident represents perhaps the most significant instance of a non-state actor attacking a nuclear facility in either a genuinely airborne or "standoff" capacity."[38] This scenario is particularly concerning given its feasibility: Mortars and rockets are relatively low cost, unsophisticated weapons already possessed by many non-state actors. Moreover, most nuclear facilities around the world are not protected by expensive missile defense systems like Iron Dome. One could imagine such an attack being carried out on one of South Korea's twenty-five nuclear power plants by North Korea, which is rapidly developing and expanding its ballistic missile arsenal. This scenario is all the more concerning as a recent report by South Korea's operator, the Korea Hydro and Nuclear Power Company (KHNP), revealed that the outer protective walls of South Korean reactors were never meant to withstand a missile strike or other forms of concerted attack.[39] Again, while the IAEA recommends in its nuclear security guidance that countries "protect targets against stand-off attacks consistent with their design basis threat (DBT)," this recommendation is not required and few states have encoded it in their domestic legal framework.[40]

A third mode of sabotage terrorists could carry out on a nuclear power plant is a ground assault by a commando team. This commando-style attack could be mounted using a small armed force, whose members are potentially familiar with the plant layout or possibly aided by accomplices inside the plant, to gain entry to a facility in order to use explosives or other means in an attempt to release radioactivity into the area.[41] In 2007, eight men, split into two four-man teams, simultaneously attacked the outer security perimeter of the Pelindaba Nuclear Facility in South Africa, where hundreds of kilograms of weapons-grade uranium is stored. One of the teams breached the outer perimeter and exchanged gunfire with a guard until they were eventually chased away, but the other team disabled the electrified outer fence and proceeded to disable the alarm while raising no alarms.[42] Once inside, the intruders disabled the security cameras and entered the emergency control center where they shot a guard in the chest.[43] The four intruders spent a total of forty-five minutes within the facility's security perimeter without being apprehended by security forces, and then escaped the same way through which they arrived. Although the perpetrators did not set off explosives inside or steal material, it is a reminder of the highly capable adversaries that nuclear security measures must be designed to prevent, detain, and deter. In the US, prior to the September 11, 2001 attacks, the NRC included in the national DBT the possibility of a small commando-style attack. However, governmental and nongovernmental experts raised concerns that the commando force in the DBT was too small, as compared to the nineteen attackers during the 9/11 attacks.[44] The revised US DBT may require defense against larger commando groups, but the NRC has not published the new DBT due to security sensitivities.

Attackers could also potentially use trucks, vans, or other vehicles filled with explosives against a nuclear power plant. This type of weapon produced devastating effects on military bases, embassies, other government buildings, and commercial buildings throughout the 1990s.[45] Alternatively, terrorists could mount a waterborne attack on a nuclear facility, as many power plants are accessible by sea, lake, or river.[46]

Finally, cyber-attacks are an emerging vulnerability for nuclear power plants. Several notable cyberattacks have already occurred at nuclear facilities, including the Stuxnet virus that damaged Iran's Natanz centrifuge facility, the placement of a virus into the computers of the Ignalia nuclear power plant in Lithuania in 1992, and the hacking of the computer systems of KHNP, South Korea's nuclear operator by presumably North Korea.[47] Attackers could use cyber techniques to undermine security at nuclear reactors to facilitate, on one end of the spectrum, the theft of confidential or proprietary information, or the release of radiation on the other. According to a 2015 report published by Chatham House, "An adversary with sufficient technical knowledge and adequate resources could mount an attack on a nuclear power plant that could trigger the release of ionizing radiation. All nuclear power plants need offsite power to operate safely and all have a standby generator which is designed to be activated when a loss of main power occurs. Attacks on the offsite power supply and on the on-site backup system could create some of the effects that occurred following the 2011 earthquake and tsunami at Fukushima Daiichi, although multiple failures of the many safety features at modern nuclear power plants could also need to occur at the same time as that loss to offsite power and the disruption of standby generators."[48] While the concern of radiation release is remote, even a small-scale cyber incident at a nuclear facility would likely have a disproportionate effect on the public's opinion of nuclear energy and nuclear industry. Although terrorists are not currently believed to possess an advanced cyber capability, groups like the Islamic State of Iraq and Syria (ISIS) have developed a sophisticated strategy of online recruitment, which utilizes not only Facebook and Twitter but also encrypted platforms on mobile devices.[49] Moreover, it is not implausible that a group's desire to cause damage at a nuclear facility could lead them to develop the necessary skills or employ a profit-motivated cyber-criminal group to do this.[50] Cyber defense of nuclear power plants requires a significant financial and intellectual investment on the part of states. According to the NTI Nuclear Security Index, many countries require virtually no cyber security measures at nuclear facilities and do not have a legal framework for these requirements.[51]

Sabotaging a nuclear facility to disrupt the operation of the facility

A third category of non-state motivation for attacking a nuclear facility may be to disrupt the operations of the facility or shut the facility down. For example, non-state actors might wish to disrupt or eliminate the energy source for a particular area. There are a number of vulnerable components at power reactors whose destruction could disrupt energy production for weeks or even months including the main turbine generator, the electric generator, conductors, transformers, circuit breakers and switches.[52] This action might be linked to a broader political struggle, for example, an ethno-nationalist group shutting down a civil power reactor to disrupt electricity supplies to a local population in an attempt to extract loyalty or other concessions.[53] This mode of attack could also be carried out by anti-nuclear groups who oppose the use of nuclear energy and, ironically, hope to stop the operations of a reactor in order to prevent the harm caused by nuclear materials.

Intrusion as a form of protest

The intrusion of a nuclear power plant as a form of protest, while not necessarily an act of attack or sabotage on a nuclear facility, is nonetheless included in this discussion because the intrusion of protestors or activists into a nuclear facility highlights the vulnerabilities that could be exploited by individuals or groups with violent or malicious intentions. In these scenarios, the perpetrators are usually anti-nuclear or environmental activists and do not intend to acquire material, shut down a facility, or even cause serious harm to the facility. Rather, they seek to draw attention to their cause, which often times includes pointing out the deficiencies in security at nuclear plants.[54] Greenpeace activists have broken into nuclear facilities around the globe numerous times. In October 2012, Greenpeace activists entered two nuclear power plants in Sweden by breaking open a gate and scaling fences without being interdicted by security guards. Four of the activists hid on the roof of one reactor overnight before being discovered by guards and surrendering the next morning.[55] In 2014, Greenpeace activists broke into the Fessenheim nuclear plant near the German border and hung a banner from the reactor building.[56]

One of the most serious and well-known cases of protestor intrusion at a nuclear power plant was the 2012 break-in at the Y-12 National Security complex in the US. This incident is of particular interest because Y-12 is ostensibly one of the most stringently guarded nuclear facilities in the world, as it houses the main US repository of weapons-grade uranium. Indeed, according to Matthew Bunn, before the intrusion, nuclear security managers at the US Department of Energy (DOE) would have pointed to the security program at Y-12 as one of the strongest at DOE.[57] Nevertheless, an 82-year-old nun along with two other activists breached the outer layer of security at the complex, with little more than a pair of bolt-cutters, and succeeded in getting to the building where thousands of bombs' worth of HEU was stored.[58] The activists evaded multiple cameras, alarms, and other sensors. The activists proceeded to vandalize the building, using their own blood and sledgehammers, for two hours before being detected and arrested by a single security guard.[59] Stunningly, because the facility had experienced hundreds of false alarms in months leading up to the break-in, security guards heard the alarms but did not act upon them because they thought they were not real.[60] Similarly, the guards heard the protestors hammering on the building's wall but assumed that the sounds were being made by maintenance workers.[61] While their intent was not to cause serious harm to the facility, the protestors revealed major weaknesses in the complex's nuclear security and highlighted the damage that could easily be done by adversaries with more nefarious intents.

High-level observations

The Nuclear Facilities Attack Database (NuFAD), a database complied by the University of Maryland's National Consortium for the Study of Terrorism and Responses to Terrorism (START), examines all incidents of assaults, sabotage, and breaches of nuclear facilities from 1961 to 2014. The database points to some interesting high-level observations about the scenarios discussed above. First, there are eighty cases of attacks, sabotage, or breaches against nuclear facilities in the empiric record, a sizeable number compared to the general perception of the frequency of these events. While many of the notable and reported cases were discussed above, there are many more cases that have not received the same level of media attention. Second, according to the database, at least sixty percent of the attempted attacks on nuclear facilities were deemed "successful," that is they met the defined or ostensible goals of the perpetrator.[62] Third, in one-third of all cases examined, the perpetrators penetrated structures where nuclear or radiological materials were actually housed. Finally, in more than seventy-five percent of the cases examined, the perpetrators managed to breach the outer defenses of the facility. Fortunately, for now, the most successful types of perpetrators appear to be criminals and anti-nuclear or environmental activists. However, the empirical record demonstrates that successful attacks, sabotage, and breaches of nuclear facilities are indeed possible. Combined with a historic interest by violent non-state groups in nuclear weapons, which will be discussed in a later section of this chapter, it may simply be luck that these two forces have not yet coincided.

Detonation of a Nuclear Device

The final scenario of nuclear terrorism that will be discussed in this chapter is the detonation of a nuclear device. Although this is the most devastating type of nuclear terrorism, it is the least likely to occur, given the significant technical and security-related challenges involved. To accomplish the detonation of a nuclear device, terrorists could either acquire a ready-made nuclear weapon from a state's arsenal through theft, purchase, or gift, or they could construct a crude or improvised nuclear device using weapons-grade nuclear material acquired through theft or diversion. These two scenarios, along with their associated challenges and limitations, will be discussed in the below sections.

Terrorists could steal or buy a nuclear weapon

The existing arsenals of the nine nuclear-weapons states represent the most obvious means by which terrorist groups could acquire an intact nuclear weapon. The global stock of nuclear weapons is estimated to be 14,900 warheads.[63] Precise information about the security of nuclear weapons is highly classified in nuclear-weapons states, but experts agree that stealing an intact nuclear weapon, and then detonating that weapon, could be extremely difficult, if not nearly impossible. A terrorist group would need considerable financial and technical resources, as well as a high degree of organizational, military, and technical competency, to mount an attack on a weapons storage facility, transfer the weapon to a target without being discovered by the authorities, and finally overcome safeguard measures on the weapons designed to prevent unauthorized detonation.[64] Indeed, because nuclear weapons are considered to be the "crown jewels" of many countries, they are necessarily protected with every means at the state's disposal.[65] Moreover, as alluded to above, even if the terrorist group managed to steal a fully assembled nuclear weapon, they might not be able to detonate it. Many modern nuclear weapons are equipped with safety features, what are known as "permissive action links," or PALs, which require an authorized code before detonation. Many nuclear weapons are also equipped with features that are designed to prevent the weapon from detonating unless it has completed the expected flight to its target.[66] As Charles Ferguson explains, "You'd have to run it [the nuclear weapon] through a specific sequence of events, including changes in temperature, pressure and environmental conditions before the weapon would allow itself to be armed, for the fuses to fall into place and then for it to allow itself to be fired. You don't get it off the shelf, enter a code and have it go off."[67]

The "new" nuclear-armed states perhaps represent the most alarming possibilities for the theft of a nuclear weapon by a terrorist group. History demonstrates that political turmoil, government instability and crisis situations put the security and control of nuclear weapons at risk.[68] Elena Sokova notes that these incidents include the 1961 coup in Algeria when a French nuclear site and a nuclear device were at the center of a battle for competing loyalties; the internal power struggle within the Chinese nuclear research and development program and within the Chinese strategic missile forces during the Cultural Revolution in 1966; and the storming by anti-Moscow rebels of an army base with nuclear weapons in Azerbaijan in 1990.[69] More recently, during an attempted coup against Turkish President Recep Erdogan in 2016, the US temporarily lost access to the Incirlik Air Base in Turkey, located sixty-eight miles from the Syrian border. Incirlik is home to fifty American B61 gravity bombs. During the coup, Turkish forces loyal to Erdogan surrounded the base and cut off power for days.[70]

In particular, the regional instability and ongoing terrorist activity in South Asia, as well as the potential for collapse in North Korea, present concerns for the security of nuclear weapons. For example, Pakistan is the most oft-cited example where a "loose nuke" scenario could occur. Public statements from the US and Pakistani officials express confidence in the security of Pakistan's nuclear arsenal, but there is evidence that these statements may have less merit than it appears. In 2013, the *Washington Post* reported that secret documents provided by National Security Agency whistle blower Edward Snowden revealed that then Director of National Intelligence James Clapper had warned that "knowledge of the security of Pakistan's nuclear weapons and associated material encompasses one of the most critical sets of...intelligence gaps," which was concerning "given the political instability terrorist threat, and expanding inventory [of nuclear weapons] in that country."[71] Other experts note that the Pakistani authorities have a dismal record thwarting insider threats – the AQ Khan network operated under the noses of the Pakistani establishment for years – and that the greatest risk of a loose nuke stems from insiders in the Pakistani nuclear establishment working with outside Islamist militants.[72] Moreover, the divide within the Pakistan's military, particularly the fact that senior military officials have been found to have proven ties to religious extremists, make safeguarding its nuclear weapons much more difficult.[73]

Terrorists could construct a crude nuclear bomb

Instead of acquiring an intact nuclear weapon from another state's arsenal, terrorists could construct what is known as an "improvised nuclear device." The most difficult part of making a nuclear bomb is acquiring the nuclear material. The production of fissile material, whether by enriching uranium or producing plutonium in a reactor and separating it from spent fuel, is a technically complex and expensive feat for even advanced states, and as such is well beyond the plausible capabilities of terrorist groups. [74] Thus, obtaining fissile material for a bomb through purchase or theft, though still exceedingly difficult, is the most realistic option for terrorists. However, as mentioned in this chapter's section on the attack or sabotage of nuclear facilities, the essential ingredients for nuclear weapons are housed in facilities whose security measures range from excellent to dismal. According to Matthew Bunn, the highest risks of nuclear theft by non-state actors today appear to be in Pakistan, in Russia and at HEU-fueled research reactors (many of which are located on relatively insecure university campuses) around the world.[75] Moreover, since the end of the Cold War, there have been approximately twenty documented cases of theft and smuggling of plutonium or HEU.[76] Fortunately, none of the known incidents have come close to the quantity of material necessary for a nuclear device. For example, in an incident reported in 2007, 79.5 grams of uranium enriched to eighty-nine percent were seized in Georgia.[77] While the enrichment level of this material is alarmingly high, the amount is still far short of what would be required for a nuclear device.

Should terrorists get hold of the fissile material necessary, they would then need to fashion a nuclear explosive device that could be delivered to a target. While constructing such a working device from these materials is by no means a trivial task, numerous and repeated assessments by the US government, other governments, and nongovernmental organizations confirm that it is indeed plausible that an advanced terrorist group, with technical assistance and sufficient resources, could make a primitive device if it acquired enough HEU or Pu.[78] Terrorists could plausibly fashion two types of crude nuclear devices. The simplest type of nuclear bomb terrorists could make, a gun-type bomb, is similar to the *Little Boy* bomb the US dropped on Hiroshima during World War Two. A gun-type bomb essentially involves slamming two pieces of HEU together at an extremely high speed. Building such a bomb would require "some ability to cast and machine uranium, a reasonable knowledge of the nuclear physics involved, and a good understanding of cannons and ballistics."[79] Knowledge of chemical processing may also be needed, but often times these processes are no more advanced that those undertaken by certain criminal groups in the illegal drug industry.[80] A second albeit more difficult device for terrorists to construct is an "implosion-type" bomb, in which "precisely arranged explosives crush nuclear material to a much higher density, setting off a chain reaction."[81]

The chances for success of these efforts would certainly be improved by the addition of experts with direct experience in a national nuclear weapons program. Nevertheless, a study from the now-defunct congressional Office of Technology Assessment found: "A small group of people, none of whom have ever had access to the classified literature, could possibly design and build a crude nuclear explosive device...Only modest machine-shop facilities that could be contracted for without arousing suspicion would be required."[82] Critically, this assessment was reached in 1977, before the age of the internet, which made troves of relevant information on nuclear weapons more widely accessible. Moreover, this crude bomb would not need to be a safe, secure, reliable weapon that could be delivered by a ballistic missile, like those nuclear weapons found in states' arsenals. The bomb, likely of an uncertain and unreliable yield, could be transported to its target by truck, boat, or land. Particularly with regard to land transportation, many countries have poorly-guarded borders, so this feat would not necessarily be very difficult to accomplish.

IDENTIFYING THE ADVERSARY

The above sections reviewed the various scenarios in which terrorists could theoretically deploy a nuclear strategy, in other words, the supply side of the equation. However, to materialize in practice, these scenarios would require a perpetrator with the intent to engage in this particular type of violence. As Gary Ackerman notes, "One all too often comes across the unquestioned assumption that any terrorist capable of employing a nuclear weapon would automatically want to do so...the history of terrorism reveals, however, that not every terrorist group has sought to maximize its killing power; indeed, the vast majority of terrorists have purposefully curtailed the scale of their violence."[83] Terrorist interest in nuclear weapons is typically seen as lacking because of (1) the technical and material difficulties associated with executing a nuclear attack (2) fears of reprisal, and (3) political constraints on overwhelming violence from a terrorist organization's base. Nevertheless, with the increase in the number of terrorist groups who have expressed the desire to inflict violence on a large scale, the traditional restraints on terrorists might be breaking down. The following sections will consider the other side of the equation, the demand side, that is, the motivational dynamics and decision-making processes of terrorist organizations contemplating one of these nuclear options, and review the universe of non-state actors that might wish to acquire and use nuclear weapons.

Why might terrorist organizations pursue a nuclear strategy?

There are a variety of factors that might induce a terrorist organization to pursue a nuclear strategy. A nuclear strategy might appear attractive to terrorist organizations for a number of reasons. For example, the most obvious reason for terrorists to pursue this strategy would be to inflict massive casualties upon their perceived enemies. Other reasons include the disproportionate psychological impact that nuclear weapons cause, the prestige of nuclear weapons, the ideology of a particular terrorist group, atomic fetishism, and more.[84]

However, in general, if one is to look at terrorist organizations as rational actors, the use of nuclear weapons must serve specific objectives for which conventional or other weapons are less suited.[85] An organization's leaders would ask themselves, "Will a nuclear weapon help us achieve our objectives faster or more effectively than other means? Will the possession or use of such a weapon bring us a considerable amount of prestige or power to intimidate our enemies?" As noted above, the strategic deliberations of most terrorist organizations have steered them away from nuclear weapons. Especially before 9/11, many terrorism scholars observed that there were few strategic benefits a group could derive from the detonation of a nuclear device or the sabotage of a nuclear facility compared to other less extreme measures. Brian Jenkins' well-known maxim that "Terrorists want a lot of people watching, not a lot of people dead" from the 1970s sums up this point.[86] Importantly, a terrorist group might consider whether a nuclear attack would be counter-productive by generating, for example, condemnation among the group's base and their potential supporters. By crossing the nuclear threshold, the group could undermine its popular support, "encouraging a perception of the group as deranged mass murderers, rather than righteous vanguards of a movement."[87] Finally, a nuclear attack may be counterproductive for a group seeking to influence the political process by provoking a government to expand its efforts to destroy the group.

Terrorist organizations, to be sure, are not always driven by strictly strategic motivations.[88] Groups may be motivated by organizational factors, such as expanding their size, influence, or wealth, fending off intervention by authorities, and destroying competing groups. Groups whose behavior is influenced by organizational politics might find that employing a nuclear strategy furthers (or hinders) their organizational goals. For example, the group "Avenging Israel's Blood," a group of Holocaust survivors determined to avenge the murder of six million Jews, was determined—against all political and strategic considerations— to poison Nazi SS storm troopers with Zyklon

B.[89] In this case, a leader's obsession with a particular class of weapons led to the employment of that weapon becoming the overriding group objective.

One of the most prominent explanations for why some terrorist organizations may choose to utilize a nuclear strategy while others will not, focuses on the "type" of organization in question. When examining the spectrum of terrorist groups, scholars have differentiated between types of organizations that are supposedly more constrained from committing acts of mass destruction and those that might be less inhibited, and indeed, might find incentives to commit such acts. According to this theory, each terrorist type has a different ideology, psychology, and motivation that can explain its propensity to use or avoid nuclear weapons. Jerrold Post identifies a diverse typing of sub-state terrorist groups—social-revolutionary terrorism, nationalist-separatist terrorism, religious extremist terrorism, right-wing terrorism, and single-issue terrorism—and analyzes the motivations and goals of each type in order to determine the constraints against or incentives to use a nuclear strategy.[90] For example, Post argues that the goal of social-revolutionary and nationalist-separatist terrorist groups is to establish a new socio-political order or state. In so far as these groups are seeking to influence and change their societies, they are significantly constrained from indiscriminate acts of violence that cause mass causalities among their own countrymen or cause negative reaction in international audiences. Thus, they would be less likely to use nuclear weapons.

In contrast, scholars have asserted that terrorist group types that are the most conducive to the pursuit of a nuclear strategy are those that reflect either an apocalyptic millenarian or religious fundamentalist character, a threat that has come to the fore since 9/11. For example, religious fundamentalist groups act to attain the greatest possible benefits for themselves and their coreligionists only. Violence is a "sacramental act or divine duty executed in response to some theological demand or imperative," which is said to eliminate the group's ambivalence about indiscriminate killing.[91] Bruce Hoffman takes up a similar line of argument in his analysis of the "traditional" terrorist versus the "new" terrorist. Hoffman argues that in contrast to traditional terrorism, which is practiced by a collection of individuals belonging to an organization with a defined set of political, social or economic objectives, new terrorism is motivated in whole or in part by religious imperatives and has led to more intense acts of violence that have produced considerably higher levels of fatalities.[92] In sum, certain groups but not others may seek to use nuclear weapons due to the nature of their particular ideology, psychology, motivations, and goals.

Who are the most likely candidates?

Aum Shinrikyo was the first terrorist organization to make genuine efforts to acquire a nuclear capability. The group ultimately settled for the pursuit of chemical weapons, culminating in its successful release of sarin gas in the Tokyo subway system in 1995. Aum, a well-resourced apocalyptic millenarian organization that was highly active in the 1990s in Japan, attempted on numerous occasions to purchase, produce, or otherwise acquire a nuclear capability. Aum tried to buy nuclear weapons from Russia and purchased land in Australia that contained uranium. A combination of the group's apocalyptic ideology, vast resources, impressive persistence in the face of failure, and lax Japanese law enforcement enabled it to undertake a concerted, if unsuccessful, effort to acquire nuclear weapons.[93] While analysts at the time believed that this confluence of factors was unlikely to be repeated, the rise of Al Qaeda and other transnationalist jihadi movements challenged this conventional wisdom.

At the present, the universe of non-state actors that might seek to acquire and use nuclear weapons is largely confined to jihadists and far-right groups. These two groups of terrorist organizations have the longest and most documented interest in nuclear and radiological, as well as chemical and biological, weapons. Each group will be discussed in turn below.

Efforts by Al Qaeda, the diffuse network of jihadists responsible for many of the deadliest, mass-casualty terrorist attacks in the past decade, to acquire nuclear weapons have been well-documented. Indeed, Al Qaeda's willingness to inflict mass casualties, its repeated efforts to attain weapons of mass destruction (WMD), and its stated intention to use WMD have been enough to convince many policymakers and academics that it is only a matter of time before Al Qaeda will indeed employ WMD. In various statements and videos, Al Qaeda has asserted its "right" to kill four million Americans in retaliation for the deaths of innocent civilians in the Muslim world, for which it holds the US and Israel primarily responsible.[94] Al Qaeda's former leader, Obama bin Laden, stated in 1998 that it is a "religious duty" to acquire and use WMD, and repeated this promise to use "massive weapons" in a 2007 video.[95] Bin Laden also sought and was granted a fatwa, or religious edict, from a Saudi cleric in 2003 authorizing the use of nuclear weapons.[96] Al Qaeda has sought nuclear material in Sudan and southern Africa, technical documentation on radiological and nuclear weapons have been found in Al Qaeda's possession, and on at least one occasion (the Jose Padilla case) a plot to detonate a dirty bomb in the US was actively contemplated but scrapped on feasibility grounds.[97] Al Qaeda in the past also reportedly contacted the AQ Khan network for assistance with its weapons program, although the Khan network rejected them for unknown reasons.[98] Finally, Al Qaeda's attempts to obtain nuclear weapons included a well-known case in which two extremist Pakistani nuclear weapons scientists met with bin Laden and his deputy, Ayman al-Zawahiri, in 2001 to discuss nuclear weapons, which reportedly included the scientists handing over a design for a nuclear bomb.[99]

While the evidence with respect to Chechen jihadists is much sparser compared to that of Al Qaeda, this variety of terrorist group has also been shown to have an interest in nuclear weapons. Chechen terrorists have repeatedly threatened to sabotage nuclear facilities or to carry out dirty bomb attacks.[100] Moreover, Russian officials have confirmed two cases of Chechen groups conducting reconnaissance activities at two nuclear weapons storage sites, and the Russian state newspaper reported two additional cases of terrorist reconnaissance of trains transporting nuclear weapons.[101]

The Islamic State of Iraq and Syria (ISIS), a global jihadist group which rose to prominence in 2014, has not been as outspoken or as clear about their intentions with regards to nuclear weapons as Al Qaeda. Only one article in the group's English-language magazine, *Dabiq*, mentions a hypothetical scenario in which ISIS acquires a nuclear weapon and smuggles it into the US for detonation.[102] Nevertheless, the idea that ISIS could take advantage of chaos and weak governance in Iraq and Syria to acquire dangerous materials has prompted officials to take precautions to protect against the threat of a nuclear or radiological ISIS. For example, the US is providing Iraq with training and equipment that will enhance Iraq's capability to "locate, identify, characterize, and recover orphaned or disused radioactive sources in Iraq thereby reducing the risk of terrorists acquiring these dangerous materials."[103] Moreover, many scholars have assumed that due to the similarity in the groups' ideologies, and the fact that ISIS continues to hold Osama bin Laden in high esteem, ISIS can be assumed to share Al Qaeda's general views on nuclear and other weapons of mass destruction.

However, as noted by Thomas Hegghammer and Andreas Hoelstad Daehli in their comprehensive survey of jihadi writings, chemical, biological, radiological and nuclear (CBRN)-related texts make up a relatively small proportion of jihadi texts on tactics and weapons. Moreover, nuclear and radiological weapons make up the minority of the CBRN texts.[104] In fact, bin Laden never stated that nuclear or radiological weapons were necessarily his WMD of choice, and all of his statements speak about nuclear weapons alongside chemical and biological weapons.[105] In sum, the doctrine and writings of jihadi groups point to a clear interest in mass casualty terrorism, with nuclear and radiological weapons being only one means to this end, with Al Qaeda writing more on radiological and nuclear terrorism than most other terrorist organizations.

In addition to jihadi groups, far-right militant groups have repeatedly demonstrated a desire and willingness to inflict mass casualties and an interest in nuclear weapons and other WMD. Interestingly, according to Peter Bergen, between 2001 and 2013, individuals with connections to far-right ideologies or groups were responsible for thirteen incidents involving either chemical, biological, or radiological weapons, while during the same timeframe in the US jihadists were responsible for zero.[106] The worldview of far-right individuals and groups points to nuclear intent. For example, the manifesto of Anders Behring Breivik, the Norwegian terrorist who bombed a government building and carried out a shooting spree on a Labour party youth summer camp, included a one hundred-page section on musings about CBRN weapons. Breivik endorses the use of these weapons, calling small nuclear weapons "ideal for annihilating concentrations of category A and B traitors" (though he does not advocate for nuclear weapons for surface detonation above .2 kilotons as they would cause too many civilian casualties), and recommends sabotaging nuclear reactors in order to cause "Chernobyl-style damage" to the enemy."[107] Interestingly Breivik's manifesto also contains detailed instructions for how to sabotage a nuclear reactor, and included among other things, six different assault strategies that would-be terrorists could utilize. Breivik apparently also called for taking radioactive sources from hospitals, which led Norway to decide to convert all its blood irradiators from cesium chloride (radioactive) to x-rays (non-isotopic).[108] Other far-right groups have documented an interest in nuclear weapons as well; for example, in the mid-1990s an American group called the Aryan Republic Army produced a video in which it claimed to have a bomb containing "yellowcake plutonium residue." While this was probably an empty threat, it shows that far right groups are indeed interested in using nuclear weapons.

Beyond jihadi and far-right groups, scholars have noted that the Basque ETA, the South African ANC, and Northern Ireland's IRA have been linked to actual attempts at radiological or nuclear terrorism.[109] This subset of groups does not include groups that have only considered this option.

CONCLUSION

This chapter has attempted to provide a sound assessment of the threat of nuclear terrorism. The chapter detailed the various ways in which a terrorist organization could carry out an act of nuclear terrorism – through the use of an RDD or RED, an attack or sabotage against a nuclear facility, or the detonation of a stolen intact nuclear weapon or terrorist-made crude nuclear device – while carefully considering the substantial hurdles any terrorist organization would face in this pursuit. In addition, this chapter has sought to take stock of terrorist motivations for carrying out an act of nuclear terrorism, as well as identify how well nuclear terrorism fits in with various organizations' ideologies, goals, and strategic and organizational objectives.

Without exaggerating the threat, or signaling that a large-scale nuclear terror attack by non-state actors is likely or imminent, this chapter suggests that the international community should take the threat of nuclear terrorism seriously. Fortunately, while nuclear terrorism remains a low-probability event and a significant act of nuclear or radiological terrorism has yet to occur, scholars and the policy community should not minimize the threat due to the disastrous consequences of such an event, and a clear, well-documented interest by non-state groups in nuclear matters.

REFERENCES

[1] Remarks by President Obama and Prime Minister Rutte at Opening Session of the Nuclear Security Summit, 2016, available at https://obamawhitehouse.archives.gov/the-press-office/2016/04/01/remarks-president-obama-and-prime-minister-rutte-opening-session-nuclear.

[2] G. Allison, "What Happened to the Soviet Superpower's Nuclear Arsenal? Clues for the Nuclear Security Summit." Discussion Paper 2012-04, Belfer Center for Science and International Affairs, Harvard Kennedy School, 2012, p. 2.

[3] B. Volders and T. Sauer, "Introduction" in *Nuclear Terrorism: Countering the Threat*, edited by B. Volders and T. Sauer, New York: Routledge, 2016, p. 6.

[4] J. Parachini, "Putting WMD Terrorism in Perspective," The Washington Quarterly 26, 2003, p. 38.

[5] B. Volders and T. Sauer, "Introduction" in *Nuclear Terrorism: Countering the Threat*, edited by B. Volders and T. Sauer, New York: Routledge, 2016, p. 4

[6] M. Zenko, "Intelligence Estimates of Nuclear Terrorism," The Annals of the American Academy of Political and Social Science 607, 2006, 87.

[7] E. Sokova, "Non-state Actors and Nuclear Weapons," in Understanding Nuclear Weapon Risks, edited by J. Borrie, T. Caughley, and W. Wan, UNIDIR Report, 2017, 87.

[8] The Nuclear Threat Initiative, "The Radiological Threat," 2015, available at http://www.nti.org/learn/radiological/.

[9] M. Pomper and A. Gluck, "Promoting Alternatives to High-Risk Radiological Sources," in *Nuclear Terrorism: Countering the Threat*, edited by B. Volders and T. Sauer, New York: Routledge, 2016, p. 57.

[10] The Nuclear Regulatory Commission, "Fact Sheet on Dirty Bombs," 2014, available at https://www.nrc.gov/reading-rm/doc-collections/fact-sheets/fs-dirty-bombs.html.

[11] The Nuclear Regulatory Commission, "Fact Sheet on Dirty Bombs," 2014, available at https://www.nrc.gov/reading-rm/doc-collections/fact-sheets/fs-dirty-bombs.html.

[12] J. Acton, M.B. Rogers, and P. Zimmerman, "Beyond the Dirty Bomb: Re-Thinking Radiological Terror," in Survival 3, 2007, p. 154.

[13] J. Acton, M.B. Rogers, and P. Zimmerman, "Beyond the Dirty Bomb: Re-Thinking Radiological Terror," in Survival 3, 2007, p. 151.

[14] M. Pomper and A. Gluck, "Promoting Alternatives to High-Risk Radiological Sources," in *Nuclear Terrorism: Countering the Threat*, edited by B. Volders and T. Sauer, New York: Routledge, 2016, p. 58

[15] T. Bielefeld, "Mexico's Stolen Radiation Source: It Could Happen Here," in The Bulletin of the Atomic Scientists, 2014.

[16] The International Atomic Energy Agency, "Q & A: Safety and Security of Radioactive Sources," 2003, available at https://www.iaea.org/PrinterFriendly/NewsCenter/Features/RadSources/radsrc_faq.html#qa3

[17] M. Bunn, M. Malin, N. Roth, W. Tobey, "Preventing Nuclear Terrorism: Continuous Improvement or Dangerous Decline?" Project on Managing the Atom, Belfer Center for Science and International Affairs, Harvard Kennedy School, 2016, p. 98.

[18] J. Acton, M.B Rogers, and P. Zimmerman, "Beyond the Dirty Bomb: Re-Thinking Radiological Terror," in Survival 3, 2007, p. 154.

[19] B. Ramberg, *Nuclear Power Plants as Weapons for the Enemy: An Unrecognized Military Peril,* Berkeley: University of California Press.

[20] A. MacFarlane, "How to Protect Nuclear Plants from Terrorists," The Conversation, 2016, available at http://theconversation.com/how-to-protect-nuclear-plants-from-terrorists-57094

[21] M. Bunn, "Belgium Highlights the Nuclear Terrorism Threat and Security Measures to Stop It," The Huffington Post, 2016, available at http://www.huffingtonpost.com/matthew-bunn/belgium-nuclear-terrorism_b_9559006.html

[22] M. Bunn, "Belgium Highlights the Nuclear Terrorism Threat and Security Measures to Stop It," The Huffington Post, 2016, available at http://www.huffingtonpost.com/matthew-bunn/belgium-nuclear-terrorism_b_9559006.html

[23] The IAEA Power Reactor Information System Database on Nuclear Power Reactors, available at https://www.iaea.org/pris/ and the IAEA Research Reactor Database, available at https://nucleus.iaea.org/RRDB/RR/ReactorSearch.aspx.

[24] The IAEA Nuclear Fuel Cycle Information System List of Nuclear Fuel Cycle Facilities, available at https://infcis.iaea.org/NFCIS/Facilities.

[25] G. Ackerman and J. Halverson, "Attacking Nuclear Facilities: Hype or Threat," in *Nuclear Terrorism: Countering the Threat*, edited by B. Volders and T. Sauer, New York: Routledge, 2016, p. 112.

[26] S. Daly, J. Parachini, W. Rosenau, "Aum Shinrikyo, Al Qaeda, and the Kinshasa Reactor: Implications of Three Case Studies for Combating Nuclear Terrorism, report prepared for the United States Air Force by RAND, 2005, p. 58.

[27] G. Ackerman and J. Halverson, "Attacking Nuclear Facilities: Hype or Threat," in *Nuclear Terrorism: Countering the Threat*, edited by B. Volders and T. Sauer, New York: Routledge, 2016, p. 119.

[28] G. Ackerman and J. Halverson, "Attacking Nuclear Facilities: Hype or Threat," in *Nuclear Terrorism: Countering the Threat*, edited by B. Volders and T. Sauer, New York: Routledge, 2016, p. 119

[29] University of Maryland National Consortium for the Study of Terrorism and Responses to Terrorism, Nuclear Facilities Attack Database, available at http://www.start.umd.edu/nuclear-facilities-attack-database-nufad.

[30] J. Holdren, "Threats to Civil Nuclear-energy Facilities," in *Science and Technology to Counter Terrorism: Proceedings of an Indo-U.S. Workshop*, edited by S. Cohen, R. Guenther, R. Narasimha, and A. Kumar, Washington, DC: The National Academies Press, 2007, p. 63.

[31] M. Bunn, M. Malin, N. Roth, W. Tobey, "Preventing Nuclear Terrorism: Continuous Improvement or Dangerous Decline?" Project on Managing the Atom, Belfer Center for Science and International Affairs, Harvard Kennedy School, 2016, p. 114.

[32] J. Holdren, "Threats to Civil Nuclear-energy Facilities," in *Science and Technology to Counter Terrorism: Proceedings of an Indo-U.S. Workshop*, edited by S. Cohen, R. Guenther, R. Narasimha, and A. Kumar, Washington, DC: The National Academies Press, 2007, p. 63.

[33] M. Holt and A. Andrews, *Nuclear Power Plant Security and Vulnerabilities* (Congressional Research Service, Washington, DC, 2014).

[34] M. de la Baume, "Unidentified Drones Are Seen Above French Nuclear Plants," New York Times, 2014, available at https://www.nytimes.com/2014/11/04/world/europe/unidentified-drones-are-spotted-above-french-nuclear-plants.html?_r=0

[35] J. Holdren, "Threats to Civil Nuclear-energy Facilities," in *Science and Technology to Counter Terrorism: Proceedings of an Indo-U.S. Workshop*, edited by S. Cohen, R. Guenther, R. Narasimha, and A. Kumar, Washington, DC: The National Academies Press, 2007, p. 63.

[36] G. Ackerman and J. Halverson, "Attacking Nuclear Facilities: Hype or Threat," in *Nuclear Terrorism: Countering the Threat*, edited by B. Volders and T. Sauer, New York: Routledge, 2016, p. 124.

[37] Y. Lappin, "Hamas: We attempted to hit the nuclear reactor in Dimona," JPost, 2014, available at http://www.jpost.com/Operation-Protective-Edge/Rocket-alert-sirens-sound-in-Zichron-Yaakov-120-km-north-of-Gaza-362087.

[38] G. Ackerman and J. Halverson, "Attacking Nuclear Facilities: Hype or Threat," in *Nuclear Terrorism: Countering the Threat*, edited by B. Volders and T. Sauer, New York: Routledge, 2016, p. 124.

[39] "S. Korea's Nuclear Power Reactors Not Designed to Deal with Military Attacks," Yonhap News Agency, 2016, available at http://english.yonhapnews.co.kr/news/2017/04/16/0200000000AEN20170416002800320.html.

[40] IAEA Nuclear Security Series No. 13 Nuclear Security Recommendations on Physical Protection of Nuclear Material and Nuclear Facilities (INFCIRC/225/Revision 5).

[41] J. Holdren, "Threats to Civil Nuclear-energy Facilities," in *Science and Technology to Counter Terrorism: Proceedings of an Indo-U.S. Workshop*, edited by S. Cohen, R. Guenther, R. Narasimha, and A. Kumar, Washington, DC: The National Academies Press, 2007, p. 64.

[42] M. Bunn, "Is Nuclear Terrorism a Real Threat," in *Debating Terrorism and Counterterrorism* edited by Suart Gottlieb, Los Angeles: CQ Press, 2014, p. 174.

[43] University of Maryland National Consortium for the Study of Terrorism and Responses to Terrorism, Nuclear Facilities Attack Database, available at http://www.start.umd.edu/nuclear-facilities-attack-database-nufad.

[44] C. Ferguson, *Nuclear Energy: What Everyone Needs to Know*, New York: Oxford University Press, 2011, p. 182.

[45] C. Ferguson, *Nuclear Energy: What Everyone Needs to Know*, New York: Oxford University Press, 2011, p. 182.

[46] C. Ferguson and W. Potter, *The Four Faces of Nuclear Terrorism*, New York: Routledge, 2005, p. 218.

[47] M. Bunn, M. Malin, N. Roth, W. Tobey, "Preventing Nuclear Terrorism: Continuous Improvement or Dangerous Decline?" Project on Managing the Atom, Belfer Center for Science and International Affairs, Harvard Kennedy School, 2016, p. 20.

[48] C. Baylon, R. Brunt, and D. Livingstone, "Cyber Security at Civil Nuclear Facilities Understanding the Risks," Chatham House, 2015, p. 6.

[49] M. Bloom, "Should We Regulate the Internet to Prevent Terror," On Point, WBUR National Public Radio, 2017, available at http://www.wbur.org/onpoint/2017/06/05/regulate-internet-prevent-terror

[50]C. Baylon, R. Brunt, and D. Livingstone, "Cyber Security at Civil Nuclear Facilities Understanding the Risks," Chatham House, 2015, p. 5

[51] 2016 NTI Nuclear Security Index, available at http://www.ntiindex.org.

[52] B. Ramberg, *Nuclear Power Plants as Weapons for the Enemy: An Unrecognized Military Peril,* Berkley: University of California Press, p.18.

[53] G. Ackerman and J. Halverson, "Attacking Nuclear Facilities: Hype or Threat," in *Nuclear Terrorism: Countering the Threat*, edited by B. Volders and T. Sauer, New York: Routledge, 2016, p. 112.

[54] G. Ackerman and J. Halverson, "Attacking Nuclear Facilities: Hype or Threat," in *Nuclear Terrorism: Countering the Threat*, edited by B. Volders and T. Sauer, New York: Routledge, 2016, p. 112.

[55] A. MacFarlane, "How to Protect Nuclear Plants from Terrorists," The Conversation, 2016, available at http://theconversation.com/how-to-protect-nuclear-plants-from-terrorists-57094

[56] A. MacFarlane, "How to Protect Nuclear Plants from Terrorists," The Conversation, 2016, available at http://theconversation.com/how-to-protect-nuclear-plants-from-terrorists-57094

[57] [57] M. Bunn, M. Malin, N. Roth, W. Tobey, "Preventing Nuclear Terrorism: Continuous Improvement or Dangerous Decline?" Project on Managing the Atom, Belfer Center for Science and International Affairs, Harvard Kennedy School, 2016, p. 27.

[58] [58] M. Bunn, M. Malin, N. Roth, W. Tobey, "Preventing Nuclear Terrorism: Continuous Improvement or Dangerous Decline?" Project on Managing the Atom, Belfer Center for Science and International Affairs, Harvard Kennedy School, 2016, p. 89.

[59] G. Friedman, *Special Report on Inquiry into the Security Breach at the National Nuclear Security Administration's Y-12 National Security Complex* (U.S. Department of Energy, Washington, DC, 2012)

[60] E. Schlosser, "Break-in at Y-12," The New Yorker, 2015, available at http://www.newyorker.com/magazine/2015/03/09/break-in-at-y-12.

[61] E. Schlosser, "Break-in at Y-12," The New Yorker, 2015, available at http://www.newyorker.com/magazine/2015/03/09/break-in-at-y-12.

[62] G. Ackerman and J. Halverson, "Attacking Nuclear Facilities: Hype or Threat," in *Nuclear Terrorism: Countering the Threat*, edited by B. Volders and T. Sauer, New York: Routledge, 2016, p. 118.

[63] H. Kristensen and R. Norris, "Status of World Nuclear Forces," The Federation of American Scientists, 2017.

[64] C. Ferguson and W. Potter, *The Four Faces of Nuclear Terrorism*, New York: Routledge, 2005, p. 54.

[65] N. Salik, "Nuclear Terrorism: Assessing the Danger," *Strategic Analysis* 38, 2014, p. 177.

[66] M. Bunn, "Securing the Bomb 2010: Securing All Nuclear Materials in Four Years," Project on Managing the Atom, Belfer Center for Science and International Affairs, Harvard Kennedy School, Commissioned by the Nuclear Threat Initiative, 2010, p. 16.

[67] N. Salik, "Nuclear Terrorism: Assessing the Danger," *Strategic Analysis* 38, 2014, p. 178.

[68] E. Sokova, "Non-state Actors and Nuclear Weapons," in Understanding Nuclear Weapon Risks, edited by J. Borrie, T. Caughley, and W. Wan, UNIDIR Report, 2017, p. 85.

[69] E. Sokova, "Non-state Actors and Nuclear Weapons," in Understanding Nuclear Weapon Risks, edited by J. Borrie, T. Caughley, and W. Wan, UNIDIR Report, 2017, p. 85.

[70] J. Cirincione, "Hey, NATO, Let's Move Those 50 US Thermonuclear Weapons Out of Turkey," Defense One, 2017, available at http://www.defenseone.com/ideas/2017/05/dear-nato-get-those-50-us-thermonuclear-weapons-out-turkey/138113/

[71] P. Hoodbhoy and Z. Mian, "Securing Pakistan's Nuclear Arsenal: The Threat From Within" in *Nuclear Terrorism: Countering the Threat*, edited by B. Volders and T. Sauer, New York: Routledge, 2016, p. 183.

[72] R. Mowatt-Larssen, "Nuclear Security in Pakistan: Reducing the Risks of Nuclear Terrorism," Arms Control Today, 2009, available at https://www.armscontrol.org/act/2009_07-08/Mowatt-Larssen

[73] For a full discussion of the obstacles in securing Pakistan's nuclear arsenal, see P. Hoodbhoy and Z. Mian, "Securing Pakistan's Nuclear Arsenal: The Threat From Within" in *Nuclear Terrorism: Countering the Threat*, edited by B. Volders and T. Sauer, New York: Routledge, 2016.

[74] "Module 2: Nuclear Terrorism," The Nuclear Threat Initiative, available at http://www.nti.org/media/pdfs/Securing_The_Bomb_2010.pdf?_=1317159794

[75] M. Bunn, "Is Nuclear Terrorism a Real Threat," in *Debating Terrorism and Counterterrorism* edited by Stuart Gottlieb, Los Angeles: CQ Press, 2014, p. 177.

[76] M. Bunn, "Is Nuclear Terrorism a Real Threat," in *Debating Terrorism and Counterterrorism* edited by Stuart Gottlieb, Los Angeles: CQ Press, 2014, p. 181.

[77] S. Martin, Is Nuclear Terrorism a Real Threat," in *Debating Terrorism and Counterterrorism* edited by Stuart Gottlieb, Los Angeles: CQ Press, 2014, p. 192.

[78] "Module 2: Nuclear Terrorism," The Nuclear Threat Initiative, available at http://www.nti.org/media/pdfs/Securing_The_Bomb_2010.pdf?_=1317159794

[79] M. Bunn, "Is Nuclear Terrorism a Real Threat," in *Debating Terrorism and Counterterrorism* edited by Stuart Gottlieb, Los Angeles: CQ Press, 2014, p. 183.

[80] M. Bunn, "Is Nuclear Terrorism a Real Threat," in *Debating Terrorism and Counterterrorism* edited by Stuart Gottlieb, Los Angeles: CQ Press, 2014, p. 183.

[81] M. Bunn, " Securing the Bomb 2010: Securing All Nuclear Materials in Four Years," Project on Managing the Atom, Belfer Center for Science and International Affairs, Harvard Kennedy School, Commissioned by the Nuclear Threat Initiative, 2010, p. 16.

[82] U.S. Congress, Office of Technology Assessment, "Nuclear Proliferation and Safeguards," Washington, DC: U.S. Government Printing Office, June 1977.

[83] G. Ackerman, "Motivations for Engaging in Nuclear Terrorism," Expert Series, The Fund for Peace, 2006, p. 1.

[84] For further discussion of these motivations, see G. Ackerman, "Motivations for Engaging in Nuclear Terrorism," Expert Series, The Fund for Peace, 2006.

[85] M. Crenshaw, "Theories of Terrorism: Instrumental and Organizational Approaches," in *Inside Terrorist Organizations*, edited by D. Rappaport, London: Frank Cass Publishers, 2001, p. 13.

[86] B. Jenkins, "International Terrorism: A New Mode of Conflict," in *International Terrorism and World Security* edited by D. Carlton and C, Schaerf, London: Croom Helm, 1975, p. 15.

[87] J. Forest, "Framework for Analyzing the Future Threat," Journal of Strategic Security 5, 2012, p. 57.

[88] M. Crenshaw, "Theories of Terrorism: Instrumental and Organizational Approaches," in *Inside Terrorist Organizations*, edited by D. Rappaport, London: Frank Cass Publishers, 2001, p. 18.

[89] E. Sprinzak and I. Zertal, "Avenging Israel's Blood," in *Toxic Terror* edited by J. Tucker, Cambridge: The MIT Press, 2000, pp. 17-42.

[90] L. Denny, P. Kozak, and J. Post, "Weapons of Mass Destruction Terrorism," in *Avoiding the Abyss* edited by B. Schneider and J. Davis, London: Praeger Security International, 2006, pp. 59-64.

[91] B. Hoffman, "Terrorism and Weapons of Mass Destruction: An Analysis of Trends and Motivations," RAND 1999, p. 27.

[92] B. Hoffman, "Terrorism and Weapons of Mass Destruction: An Analysis of Trends and Motivations," RAND 1999, pp. 8-24.

[93] R. Danzig et. al, "Aum Shinrikyo: Insights Into How Terrorists Develop Biological and Chemical Weapons," Center for a New American Security, 2011, p. 4.

[94] G. Ackerman, "Motivations for Engaging in Nuclear Terrorism," Expert Series, The Fund for Peace, 2006, p. 9.

[95] R. Mowatt-Larssen, "Al Qaeda's Pursuit of Weapons of Mass Destruction," Foreign Policy, 2010, available at http://foreignpolicy.com/2010/01/25/al-qaedas-pursuit-of-weapons-of-mass-destruction/

[96] T. Hegghammer and A. Hoelstad Daehli, "Insiders and Outsiders: A Survey of Terrorist Threats to Nuclear Facilities," in *Insider Threats* edited by M. Bunn and S. Sagan, New York: Cornell University Press, 2016, p. 13.

[97] T. Hegghammer and A. Hoelstad Daehli, "Insiders and Outsiders: A Survey of Terrorist Threats to Nuclear Facilities," in *Insider Threats* edited by M. Bunn and S. Sagan, New York: Cornell University Press, 2016, p. 14.

[98] R. Mowatt-Larssen, "Al Qaeda's Pursuit of Weapons of Mass Destruction," Foreign Policy, 2010, available at http://foreignpolicy.com/2010/01/25/al-qaedas-pursuit-of-weapons-of-mass-destruction/

[99] M. Bunn, "Is Nuclear Terrorism a Real Threat," in *Debating Terrorism and Counterterrorism* edited by Stuart Gottlieb, Los Angeles: CQ Press, 2014, p. 176.

[100] M. Bunn, "Is Nuclear Terrorism a Real Threat," in *Debating Terrorism and Counterterrorism* edited by Stuart Gottlieb, Los Angeles: CQ Press, 2014, p. 176.

[101] M. Bunn, "Is Nuclear Terrorism a Real Threat," in *Debating Terrorism and Counterterrorism* edited by Stuart Gottlieb, Los Angeles: CQ Press, 2014, p. 177

[102] T. Hegghammer and A. Hoelstad Daehli, "Insiders and Outsiders: A Survey of Terrorist Threats to Nuclear Facilities," in *Insider Threats* edited by M. Bunn and S. Sagan, New York: Cornell University Press, 2016, p. 14.

[103] G. Moore, Is ISIL a Radioactive Threat," The Federation of American Scientists, Public Interest Report 67, 2014.

[104] T. Hegghammer and A. Hoelstad Daehli, "Insiders and Outsiders: A Survey of Terrorist Threats to Nuclear Facilities," in *Insider Threats* edited by M. Bunn and S. Sagan, New York: Cornell University Press, 2016, p. 14.

[105] T. Hegghammer and A. Hoelstad Daehli, "Insiders and Outsiders: A Survey of Terrorist Threats to Nuclear Facilities," in *Insider Threats* edited by M. Bunn and S. Sagan, New York: Cornell University Press, 2016, p. 14.

[106] Peter Bergen et. al. Jihadist Terrorism: A threat Assessment. Bipartisan Policy Center, 2013, 6.

[107] T. Hegghammer and A. Hoelstad Daehli, "Insiders and Outsiders: A Survey of Terrorist Threats to Nuclear Facilities," in *Insider Threats* edited by M. Bunn and S. Sagan, New York: Cornell University Press, 2016, p. 19.

[108] The Nuclear Threat Imitative, Preventing A Dirty Bomb, 2016, available at http://www.nti.org/media/documents/Preventing_a_Dirty_Bomb_FINAL.pdf

[109] T. Hegghammer and A. Hoelstad Daehli, "Insiders and Outsiders: A Survey of Terrorist Threats to Nuclear Facilities," in *Insider Threats* edited by M. Bunn and S. Sagan, New York: Cornell University Press, 2016, p. 21.

Technologies to Counter Aviation Security Threats

Steve Karoly

Acting Assistant Administrator
Office of Requirements and Capabilities Analysis
Transportation Security Administration
701 S. 12th Street
Arlington, VA 20598, USA

Abstract. The Aviation and Transportation Security Act (ATSA) makes TSA responsible for security in all modes of transportation, and requires that TSA assess threats to transportation, enforce security-related regulations and requirements, and ensure the adequacy of security measures at airports and other transportation facilities. Today, TSA faces a significant challenge and must address a wide range of commercial, military grade, and homemade explosives and these can be presented in an infinite number of configurations and from multiple vectors. TSA screens 2 million passengers and crew, and screens almost 5 million carry-on items and 1.2 million checked bags daily. As TSA explores new technologies for improving efficiency and security, those on the forefront of research and development can help identify unique and advanced methods to combat terrorism. Research and Development (R&D) drives the development of future technology investments that can address an evolving adversary and aviation threat. The goal is to rethink the aviation security regime in its entirety, and rather than focusing security at particular points in the enterprise, distribute security from the time a reservation is made to the time a passenger boards the aircraft. The ultimate objective is to reengineer aviation security from top to bottom with a continued focus on increasing security throughout the system.

INTRODUCTION

The United States Congress created TSA just two months after the tragedy of September 11, 2001. The Aviation and Transportation Security Act (ATSA) makes TSA responsible for security in all modes of transportation, and requires that TSA assess threats to transportation, enforce security-related regulations and requirements, and ensure the adequacy of security measures at airports and other transportation facilities. TSA's scope includes commercial and general aviation; mass transit systems; freight and passenger rail; highways, pipelines, and ports. While TSA continues to evolve and mature, the underlying rationale for its existence is as compelling today as it was a decade and a half ago.

Recent world events are a vivid reminder of the challenging, dynamic threat environment TSA operates in, with a diffuse and dispersed enemy that masterfully manipulates communications and media to inspire attacks. Credible intelligence indicates that terrorist groups will continue to develop plans that target transportation. In 2016, terrorist groups carried out more than 200 attacks on transportation assets worldwide. Although this represents a decrease in the number of overall transportation attacks since 2015, attacks on aviation and freight rail increased. TSA's mission to protect the security of the nation's transportation system remains a national imperative.

Sadly, there are very recent reminders of terrorist attacks against aviation: the destruction of MetroJet Flight 9268 attributed to ISIL in October 2015, and the attack on Daallo Airlines Flight 159 in February 2016 attributed to al Shabaab. In both attacks, the devices were allegedly concealed to circumvent security measures. In addition, Al Qaeda in the Arabian Peninsula (AQAP), another terror group that has attacked aviation in the past, has specifically mentioned aviation and TSA in recent issues of their recurring English language publication "Inspire."

More recently, we have witnessed destruction in the softer targets of the public area of airports. This was true in the Brussels airport bombing in February 2016 and the attack on Ataturk Airport in Istanbul in June 2016. It is prima facie evidence of evolving tactics by terrorists to strike vulnerable areas in spectacular fashion. Securing commercial

Nuclear Weapons and Related Security Issues
AIP Conf. Proc. 1898, 050002-1–050002-7; https://doi.org/10.1063/1.5009231
Published by AIP Publishing. 978-0-7354-1586-7/$30.00

aviation and protecting the traveling public from threats, including hijacking, smuggling of weapons or explosives on board aircraft, and ground-based attacks, are key TSA and DHS priorities.

Terrorists are creative in how they make and control bombs and take full advantage of the openness of Western democracies to strike at civilian targets; they are not shy. Terrorists today routinely broadcast their desire to attack aviation and aviation infrastructure in social media and other propaganda outlets. One only needs to read AQAP's "Inspire" magazine to see the level of understanding these terrorist groups have of aviation security measures.

It isn't just their ability to communicate that has significantly changed. Thanks to the internet and the ease with which information and knowledge can be shared, today's terrorists can readily share technical knowledge and inspire would-be terrorists.

Fifteen years ago, certification standards for explosives detection systems included a few chemicals, all of which were commercial and military grade explosives. Liquid explosives were just beginning to present a threat, and Homemade Explosives (HME) were not stock and trade.

Today, TSA faces a significant challenge and must address a wide range of commercial, military grade, and homemade explosives and these can be presented in an infinite number of configurations and from multiple vectors. As the enemy is willing to continue to take extreme measures for their attacks, person-borne improvised explosive devices (PBIEDs) are becoming more frequent and more dangerous. These explosives can be well disguised. While TSA has been successful in catching many PBIED explosives, more commonly known are the cases that succeed in threatening our global security, particularly for aviation. The shoe bomber in 2001, the underwear bomber in 2009, the suicide vests worn by the attackers in Paris and Brussels and in Istanbul's airport are examples.

Most recently, a Somali man boarded a Daallo Airlines flight in Mogadishu, Somalia's capital. Twenty minutes after the flight took off, his unassuming carry-on bag detonated, blowing a hole in the side of the plane. The bomber was killed, and two others were injured, but if the aircraft had reached cruising altitude, the bomb could have ignited the plane's fuel tank and caused a second, potentially catastrophic blast.

The groups who threaten the transportation system are ruthless and remorseless, but not inevitable, and definitely not invincible.

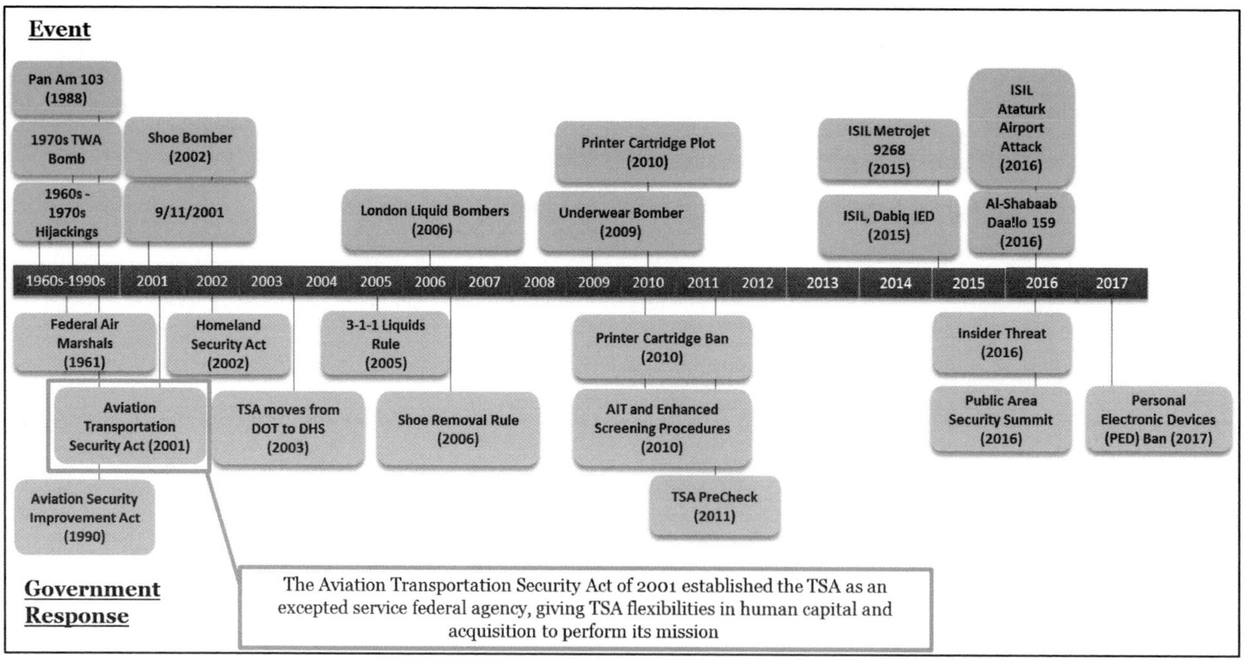

FIGURE 1. Timeline of key aviation security events and responses.

OPERATIONAL CHALLENGES

Before going into the ways the threat can be overcome, it is important to understand TSA's operational challenge. TSA, uniquely in the federal government, is a retail public-facing operation and, for most Americans, is the face of the federal government. After all, TSA screens millions of travelers every day. It is responsible for securing the United States transportation network, and its mission extends to every corner of the United States and its territories, from Puerto Rico to the Pacific Islands and at airports around the world with flights to the U.S.

While TSA works with partners in the public and private sectors to address roadways, railroad tracks, bridges, tunnels, ports and millions of miles of pipeline, its presence is most notable at 440 federal airports. Commercial aviation is vital to our economy. In 2012, commercial aviation generated \$1.5 trillion in economic activity, accounting for 5.4 percent of U.S. GDP.

In today's aviation environment, on a daily basis, TSA screens 2 million passengers and crew, up to 2.5 million in summer months and screens almost 5 million carry-on items and 1.2 million checked bags. All these people and bags go on 22,000 commercial flights operated by 200 U.S. and foreign air carriers and approximately 2,000 flights arrive in the U.S. from international locations. With 233 foreign passenger and cargo air carriers servicing the United States, as well as the 120 U.S. passenger and cargo air carriers with international operations, TSA must maintain a steady presence overseas, accomplished through airport assessment and air carrier inspection programs. The sheer breadth and scope of the aviation operating environment coupled with the dynamic, persistent threat, present significant national security challenges. These, in turn, demand highly effective security measures.

THE TSA APPROACH FOR OPERATIONAL AND SECURITY CHALLENGES

TSA's security measures begin with vetting travelers against the government watch lists to ensure passengers, accessible property, and checked baggage are screened at the appropriate level. With more information about the traveling public, security measures can be more tailored to the specific individual. TSA is partnering with industry to expand Department of Homeland Security Trusted Traveler Programs, such as TSA Pre✓® and U.S. Customs and Border Protection's Global Entry program. These programs benefit security and increase checkpoint efficiency. People can be screened through TSA Pre✓® lanes at a rate of about 240 per hour, compared to standard lanes that can only move about 140-150 travelers per hour.

Since its start in December 2013, over 5 million travelers have enrolled in TSA Pre✓®. Enrollments have increased over time, with approximately 275,000 enrolled in May 2016 versus 91,000 in May 2015. TSA Pre✓® also has a security impact. Large queues are an attractive target. Shorter wait times means shorter queues, and therefore less attractive targets for attacks. Across the nation, wait times in TSA Pre✓® lanes average less than 5 minutes.

While TSA Pre✓® lanes have helped to limit queue lengths, TSA has still had some serious issues with security lines. These problems are often the result of a culmination of three significant factors. First, and most importantly, TSA refocused on security by ending programs that allowed passengers without prescreening to go through expedited security. Secondly, traveler volume has increased in recent years, and as an indicator of the American economy, this is a good thing. Finally, TSA staffing has been on steady decline for a very long time.

To address these challenges in summer 2016, TSA made some significant changes in how it conducts daily operations:

- Attention is focused on seven critical airports that account for 25 percent of originating travelers. By ensuring passengers stayed on schedule at these airports, ripple effects across the system could be avoided (Seven airports: JFK, EWR, ATL, ORD, MIA, DFW, LAX).
- A national Airport Operations Center (AOC) was established. From this command center, TSA monitors checkpoint trends at airports in real-time and helps those in the field to address issues as or before they arise.
- Additional canine teams were deployed to screen passengers at security checkpoints. After passenger screening canines have screened passengers, they can be sent through the faster-moving TSA Pre✓® lanes and not have to remove belts, shoes, laptops that allow lines to move more quickly.

Following increasingly long lines in summer 2016, these efforts were successful and TSA plans to continue to work along this path to keep up with anticipated volume growth of more than 4 percent compared to summer 2016. This means that from Memorial Day to Labor Day, TSA may screen more than 228 million passengers and crew. To prepare for the high passenger volume TSA is adding additional staff and canine teams. By July 2017, TSA will have approximately 3,175 additional frontline staffing full time employees compared to July 2016, and approximately 70 additional canine teams.

TSA is also working to improve coordination with Federal Security Directors (FSD) at critical, high-volume airports to ensure necessary resources are in place. In coordination with FSDs, and the airport and airline community, the AOC continues to monitor operations at critical airports on an hourly basis and deploy resources in real time. Additionally, TSA continues to collaborate with vendors and airlines to open more automated screening lanes and other innovative technologies. These lanes are designed to improve efficiency at the checkpoint, and the TSA Innovation Task Force is continuing to seek out and identify emerging technologies that can improve security, passenger experience, and efficiency.

While TSA Pre✓® is a step in the right direction for improving both security effectiveness and operational efficiency, there are also many levels of security that protect us against threats. One of these layers, and one that most of the public will be familiar with, is made of the technologies in airports that screen passengers, carry-on baggage, and checked baggage.

CURRENT TECHNOLOGIES

Since TSA's creation, there have been many changes in the screening technologies over time. Additional threats have evolved over the years, and we've had to rely on our technologies to mitigate these additional threats. One of the first technologies deployed was the walk through metal detector. Since then, TSA has moved to the Advanced Imaging Technology, or AIT, machines that detect more than metallic threats. This change was a major step forward in closing gaps for passenger-based threats. As the name implies, metal detectors only detect metal.

Advanced Imaging Technology (AIT)

FIGURE 2. Illustration of an AIT machine.

The AIT machines deployed by TSA use non-ionizing electromagnetic radiation similar to that used by wireless data transmitters to detect objects concealed on a passenger's body without physical contact. The AIT systems are considered to be "Active" millimeter wave and operate within the frequency range of 24 – 30 gigahertz.

As the passenger steps into the AIT machine and stands still, transmitters produce millimeter waves that are either absorbed, scattered, or reflected as they pass through clothing, bounce off the person's skin and any potential threats—then return to the receivers. The AIT applies the necessary algorithms to the reflected MMW signals to determine the location of possible anomalies on the body. If it detects an anomaly, a bounding box indicates its location on a generic human image. When the system identifies an anomaly, an operator is required to step in and resolve the alarm with a pat-down. The total processing time for this system, from the start of the scan to the automatic decision, is less than six seconds.

In the future, AIT will continue to be a key component to passenger screening. Adaptable algorithms provide us with the flexibility to enhance security without replacing an entire machine. The next generation of AIT will offer enhanced image resolution by using wide band antennas to assist advanced automated threat recognition (ATR) detection algorithms. Additionally, we will focus on 3-D reconstruction algorithms to address threats concealed in obscured body locations.

Advanced Technology X-ray (AT-2)

In addition to passenger screening, TSA is required to screen all personal items that board an aircraft, both carry-on and checked bags. Advanced Technology-2, or AT-2, systems are the primary technology for screening carry-on items.

The AT-2 systems are equipped with multiple fixed X-ray sources that are set at the same potential—currently, around 140 kilovolts (kV). Electromagnetic radiation, or photons, from these X-ray sources penetrate the carry-on items in the tunnel and the intensity is reduced through attenuation, which includes absorption, refraction, and scattering. Detectors on the opposite side of the tunnel measure the level of attenuation. Detection algorithms can then identify specific threat objects based on a trained library that measures attenuation for these threat objects.

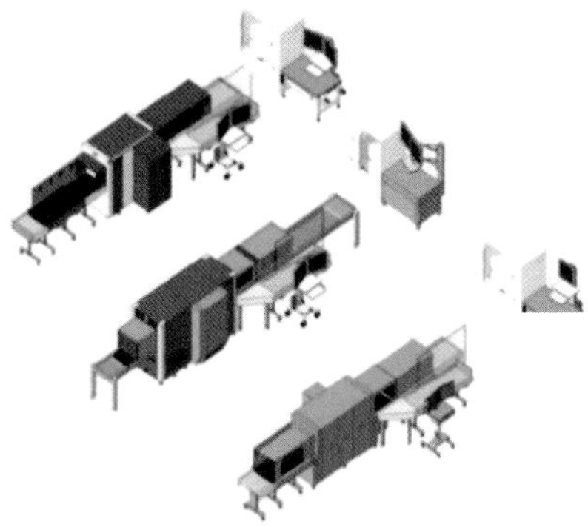

FIGURE 3. Illustration of AT X-ray machines.

TSA-qualified ATs are dual energy systems, which means that the system takes measurements at both high and low level energy levels to characterize the effective atomic number of the objects in the carry-on items. Because the energy from the X-ray source is constant, the AT performs the dual energy measurement using filters and rows of low/high energy detectors. Different materials, including threat objects, will appear in a certain color to the operator to help identify threats. Threat object discrimination has continued to improve through R&D efforts, but comes at a cost of maintaining throughput efficiency and minimizing false alarm rates throughout the system.

Explosives Trace Detection (ETD)

In order to differentiate between benign substances and threats, we utilize secondary screening. Secondary screening is most often performed with Explosives Trace Detection, or ETD, machines.

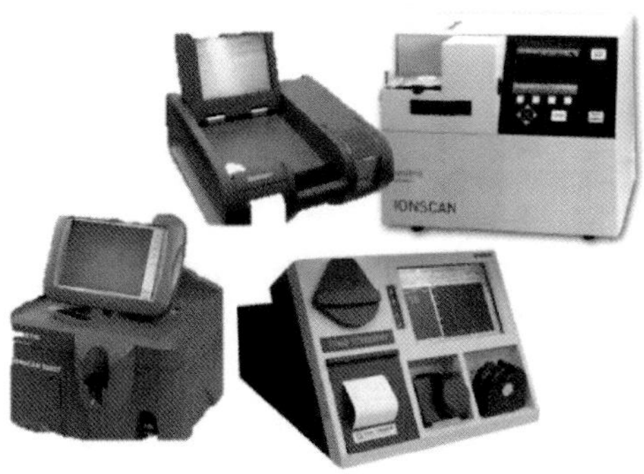

FIGURE 4. Example ETD machines.

TSA ETDs use a technology called Ion Mobility Spectrometry (IMS) to detect nanograms of explosive particles. This technique separates and identifies ionized molecules based on their mobility in the IMS.

When a person or bag goes through secondary screening, a TSA officer will collect samples using a swab, then insert the sample into what is called a desorber. The desorber applies heat to the sample, thereby releasing and converting the harvested explosive materials into a gas form. The gas stream then passes through a region of energetic electrons, which ionize the explosive molecules. These ionized molecules then enter the IMS drift tube, where the time required for them to traverse the chamber and reach a collector plate is measured to determine if explosive materials are present. If the particles are positively charged, they are exposed to a collector plate with a negative electric potential. If the particles are negatively charged, they are exposed to a collector plate with a positively charged electric potential. Some ETDs employ multiple IMS chambers to accommodate both charges, but newer systems have dual polarity IMS chambers that can rapidly switch between the two. The time it takes these particles to reach the collector depends on the particle's molecular weight and mobility through the drift tube, and this information helps the TSA officer determine if the material in question poses a threat. TSA is exploring upgrades to ETDs in the future that use mass spectrometry, which provides significantly better selectivity than traditional IMS solutions.

Explosives Detection System (EDS)

FIGURE 5. Example of an EDS.

Explosives Detection Systems (EDS) are used by TSA to screen checked baggage. EDS use computed tomography (CT) technology to screen checked baggage that has been dropped at the check-in counter before it is loaded onto the plane. CT machines use many X-ray images to produce cross-sectional tomographic images of the bag inside. CT machines use X-ray sources, similar to an AT; but unlike an AT, the CT's X-ray sources rotate around the object across from the sensors. Following each rotation, the object moves forward along the x-ray belt for the next rotational X-ray image scan. Once the whole object has been imaged with the X-ray, the CT reformats the volume of data in various planes and even produces a 3-D representation of the object in the X-ray tunnel.

This data can be used to measure the physical properties (i.e. density, atomic number) of the bag's contents and compare those properties to known explosive threat materials. Compared to an AT, CT offers better material discrimination. TSA is currently exploring the use of CT technology at the checkpoint so that passengers won't be required to remove liquids or electronics from their carry-on bags.

Additional Discriminators

When TSA screens a passenger or his or her baggage, the officers are looking for prohibited items and improvised threat signatures from homemade explosives. Homemade explosives present a challenge for screening because unlike most commercial, and particularly unlike military explosives, they are not manufactured materials with strict quality control and they can have highly variable properties.

To screen for these explosives more effectively, TSA is developing additional capabilities that discriminate between benign objects and potential threats. Broad based detection of improvised explosives threats will benefit from sufficient spatial resolution to enable accurate material characterization, sufficient independent discriminating measurements and, enhanced resolution processes.

TSA is also working with partners to develop advanced algorithms that can upgrade AT and CT machines. These algorithms use, for example, differential phase contrast, X-ray diffraction, and deep learning to discriminate between threats and benign objects.

Differential Phase Contrast (DPC)

Differential Phase Contrast (DPC) enables more in depth determination of the material properties inside a 3-D object to determine whether an object is benign. This kind of imaging provides three signatures: absorption, which is currently used by CT, phase, and dark field or scatter.

DPC provides the screener with additional information to discriminate between threats and benign objects, thereby reducing false alarms and making the screening process more efficient. TSA's goal is to partner with academia and industry to develop reliable, cost-effective system components (both hardware and algorithms) that allow for enhanced baggage scanners with DPC.

X-ray Diffraction (XRD)

X-ray Diffraction (XRD) relies on the dual wave/particle nature of X-rays to obtain information about the structure of crystalline materials of an object. XRD tomography differentiates threat substances that have very similar properties to benign substances, reducing false alarm rates and providing additional discriminators to determine whether an item contains explosive compounds. Any time we reduce the false alarm rate, we can increase throughput at airport checkpoints.

Deep Learning

Deep learning offers a way to automatically screen for prohibited items including explosives, firearms, sharp objects, etc. Deep (or machine) learning uses advanced algorithms that pair shape identification with material properties such as Z_{eff} (effective atomic number) and density measurements, then compares that against a very large library of scanned items. This results in either a "bad" or "good" outcome, with a "bad" outcome indicating that there may be a prohibited item in the individual's bag. As the library expands, the algorithm is able to learn and moves us towards completely automatic identification of weapons, explosives, and prohibited items.

AN EYE TO THE FUTURE

As TSA explores new technologies for improving efficiency and security of the aviation security space, those on the forefront of research and development can help identify unique and advanced methods to combat terrorism. R&D drives the development of future technology investments that can address an evolving adversary and aviation threat. TSA and its international partners actively exchange information to align R&D priorities and use resources in the most efficient manner possible and continues to tap into the vast reservoir of industry and academia expertise and pursue these public/private partnerships with other airports and airlines to install more automated screening lanes at checkpoints across the U.S.

Eventually, the goal is to rethink the aviation security regime in its entirety, and rather than focusing security at particular points in the enterprise, distributing security from the time a reservation is made to the time a passenger boards the aircraft. As passengers get closer and closer to the aircraft, security measures get more and more stringent.

TSA is striving for a more open, secure, and interoperable architecture that will increase automation and integration. This involves designing a true "systems of systems" approach, optimizing open technology platforms to encourage agile capability upgrades, and integrating real-time risk information and data sharing. Such an architecture will allow TSA to continually realign capabilities to maximize effectiveness across the entire aviation security spectrum. The ultimate objective is to reengineer aviation security from top to bottom with a continued focus on increasing security throughout the system.

ACKNOWLEDGMENTS

The author wishes to thank Dr. Eric Houser, Division Director, and Dominic Bianchini, Deputy Division Director, Office of Requirements and Capabilities Analysis, Mission Analysis Division for their assistance and advice.

Terrorist Threats: Technical And Policy Approaches To Countering Them

Anthony Fainberg

Bethesda, MD 20816

Abstract. This paper presents a summary of the history of terrorist threats to the United States and other nations over the last few decades, and will discuss means, both technical and policy-oriented, that can counter such threats. Changing trends in terrorist attacks are discussed, together with the need to develop countermeasures to changes in the types of threats.

INTRODUCTION

For the purpose of discussion, one needs to start with a proposed definition of terrorism. The following is an amalgam of definitions found across different agencies of the U.S. federal government. Terrorism is the use of violence or threats of violence against civilians and other non-combatants to achieve political, social, or religious goals, usually by intimidating the public as well as government officials.

This is suggested as an adequate working definition, capturing the essential elements of terrorism, that is, attacks upon the innocent to achieve political ends. This is meant to include asserted religious ends that may actually be political in nature. Political goals in such cases cover political and often social domination by a given religion of those who do not have the same beliefs. Some commentators will raise the broad objection that one person's terrorist is another's "freedom fighter." Generally though, one might argue that most of us know terrorism when we see it, given this definition, whatever one's political beliefs. This is somewhat analogous to U.S. Supreme Court Justice Potter Stewart's famous 1964 remark concerning pornography: I know it when I see it.

Terrorism is not a new phenomenon. Terrorist acts have occurred for thousands of years. However there has been a resurgence of terrorism, especially by non-state actors, that became wide-spread in the late 1960's and that has since increased in intensity and effect; this has increased public awareness and concern with the phenomenon. The current level has aroused a great level of global interest in the 21st century because of the atrocious behaviors engaged in by two groups, al-Qaeda and its offshoot, Daesh (also known as ISIS).[1] The spectacular mass murders of 9/11 and the frequent videos of atrocities against innocent victims since then, have combined lurid dissemination with widespread murder of innocents in such a way as to embody an apotheosis of the definition given above. A couple of points: first, terrorism has been used by state as well as sub-state actors: Syria's use of poison gas against its own population during the current civil war there is as well as a limited form of auto-genocide. Second, terrorism does not always rely on the mass video shocks provided by Daesh since 2014. In the Syrian case terrorist acts were actually denied by the perpetrators, but still were carried out as an intimidation strategy per the definition, since local victims and their neighbors (and indeed, eventually the world) quickly learned who did what.

The economic, social, and political forces that give rise terrorism are often debated. One view is that terrorism is the last recourse of a failed ideology. This is certainly true on occasion. On a personal level, it may simply be a tool of the depraved. On the other hand, some observers put forth a justification for terrorism: that it is sometimes the only remaining tool of the less powerful to combat opposing, powerful, and oppressive forces, usually of a state:

---

[1] Daesh is the rather hostile name given to this group by its opponents in the Middle East; it derives from an acronym for the Arabic phrase "Islamic State of Iraq and Syria." The name is disliked by the group itself, because it is close to an expression meaning "to trample underfoot." They prefer to be known as Islamic State in Iraq and Syria (or the Islamic State or ISIS), since this validates their assertion that they are already a state that is the rightful inheritor of the Caliphate, as the term was used by Islamic Empires of long ago. See, for example, http://www.bbc.com/news/world-middle-east-27994277, 2015, last accessed June 24, 2017.

Nuclear Weapons and Related Security Issues
AIP Conf. Proc. 1898, 050003-1–050003-18; https://doi.org/10.1063/1.5009232
Published by AIP Publishing. 978-0-7354-1586-7/$30.00

under this ontology, terrorism is just a tool of the weak against the strong. This is often true, but hardly constitutes a moral justification for murdering innocents. There is a danger in using this argument to justify the unjustifiable: violence against the innocent is allegedly moral, because the aggrieved parties have no other resort. I find this line of argument to be substantially dishonest and I will ignore it. Notwithstanding the arguments regarding alleged justifications for terrorist acts, there are many who would like to find ways of countering such behavior and the resulting depredation in order to protect their families, themselves, and their own communities, in addition to protecting societies, writ large. This paper will explore some of the technical, scientific and policy approaches that may help protect civil societies from terrorist attacks. This is not to denigrate efforts to find and counter terrorism's underlying causes: this is a worthy and important field of study, but is beyond the scope of this paper.

To put this in another way, technology and technical approaches furnish only one set of tools, and are of limited effectiveness in themselves. In the first place, the technology must be applied in the framework of systems analyses that set out just how each technological approach should be applied to best effect. But more than that, fighting terrorism effectively will require policy approaches in both the political and social realms. Such approaches include dealing with the underlying reasons that individuals turn to terror, to the degree possible. Examples are: understanding and countering the radicalization process among both individuals and political movements; beyond this, it is vital to develop sources and methods of intelligence that warn of impending terrorist activity. In the absence of the successful advancement of these and other policy processes, technical approaches will be of limited use. In fact, technical approaches are likely far less important overall than policy initiatives.

TERRORIST THREATS

To discover how technology and science may be used to counter terrorism, we need to understand the tactics and tools of terrorism that must be countered. These tools have naturally advanced in conjunction with the advancement of science and technology since the 19th century. In more recent times, terrorist tactics have shifted, thus broadening the types of implements involved. This section will describe specific terrorist tactics and weapons, starting in the distant past, but emphasizing those that have been used relatively recently, that is over the past few decades. At the end of this section, very recent developments will be discussed; there have been significant changes in many tactics used by terrorists. Some of them appear to be almost ad hoc, relying on weapons that are easy to obtain and having goals that are not sophisticated (such as kidnapping, usually for ransom). Such relatively primitive tactics are used by terrorist groups that are not always poised to wage complex attacks, especially when the targets are located far from terrorist safe havens. Some potentially useful countermeasures to these kinds of tactics will be discussed.

While a number of significant terrorist events will be briefly described, this section only provides examples; it is not a detailed summary of terrorist events, even since 9/11. Sources that contain more exhaustive listings of events may be found in the References section below, particularly in the periodic reports produced by the U.S. Department of State and from the archives of START (The National Consortium for the Study of Terrorism and Responses to Terrorism), a Center of Excellence of the Department of Homeland Security led by the University of Maryland.

Ancient terrorism mainly consisted of assassinations, using then common weapons: daggers, swords, and poison. The rise of chemistry and the resulting development of explosives in the late 19th century radically changed the nature of terrorism enabling the fabrication of devastating bombs, which could result in multiple fatalities, often of innocent people in addition to the targets. This trend continues today, with bombings now frequently aimed primarily at civilian targets such as commercial aircraft, religious and political sites, and crowded markets and restaurants. Development of accurate and powerful firearms has furnished important additions to the terrorist tool kit, also primarily aimed at civilian targets. Today, one or a few terrorists can murder scores of people with either bombs or hand weapons. This has occurred frequently over the past few years.

In the past few decades, there have also been sporadic efforts by terrorists to use so-called weapons of mass destruction – usually chemical or biological weapons. These efforts have had limited success to date. And in very recent years, perhaps months, some more primitive tools have occasionally come back in style: not only firearms and knives, but also vehicles. It is possible that this resort to primitivism reflects the difficulties some terrorist groups have experienced in obtaining and training sufficiently effective agents for the purpose of carrying out attacks on civilian targets in those countries where they have limited access. The obvious examples are recent Daesh or Daesh-inspired attacks against civilians in Europe.

Brief Historical Summary

The first examples of activities that could be called terrorist were the tactics of Sicarii Zealots in Judea around the year 70 CE. The Sicarii engaged mainly in assassination of both Roman officials and Jewish opponents using murders of major figures as a tactic to intimidate political opponents. They also engaged in atrocities against the Jewish population in Jerusalem, for example, by destroying the food supply of the Jewish population during a siege of the city by Roman armies; this was aimed at discouraging negotiations with the enemy. A later major example of terrorist tactics, in the 11th-13th centuries CE, was the Assassin political order, a branch of the Ismaili sect of Shi'a Islam that was active mainly in Iran and Syria.[2] The prime terrorist tactic used was the assassination, often in public, of a major official in order to achieve political goals through intimidation. A significant historical murder by the assassins was that of the Grand Vizier of the Seljuk Turk Empire, Nizam al-Mulk in 1092 CE.

Examples of terrorism as a political tool became somewhat common in the 19th century with anarchist bombings and assassinations carried out in both the United States and Europe. Two US presidents and several European leaders, including the President of France and various members of European royalty, were killed by anarchists. The development of explosives, particularly dynamite, provided a tool that went beyond assassination of major figures. Whereas bombs could be fabricated and used as tools for assassination, as was accomplished against Russian Tsar Alexander II in 1881, bombs were progressively used more broadly, sometimes targeting members of the public in addition to political figures. Bombs were used in anti-British incidents in Ireland in the late 19th century; however, in many locations, including Ireland and the United States, terrorist groups such as the Ku Klux Klan still used traditional weapons, primarily firearms, in their terrorist attacks.

In the United States, anarchist groups used bombs in a series of unsuccessful assassination attempts against major political and business figures in 1919, and, more seriously, detonated a bomb on Wall Street in 1920, which killed 38 and wounded hundreds more.

More Recent History

The modern era of terrorism, from the perspective of Western nations – especially the United States – began with the targeting of commercial airlines by Middle Eastern groups as an international tool to weaken and ultimately destroy the state of Israel. An early event of this sort was the bombing of Swissair Flight 330 from Zürich to Tel Aviv in 1970, which resulted in the deaths of all 47 persons aboard. While never positively determined, it is most likely that this was caused by a bomb in the luggage hold. This tactic then became a common terrorist mode of operation against civil aviation for over two decades, until the development and deployment of detectors that could effectively find a bomb in hold luggage were developed. The first such technology was certified in 1994, although not widely deployed until after the 9/11 attacks in 2001.

Although civil aviation has been and continues to be considered a prime target by terrorists, there have also been attempts by some groups to use weapons of mass destruction against the general public on the ground. However, these have often proven somewhat ineffective, at least when tried by sub-state actors. Most recently, the prime terrorist tool has been the use of bombings, often by suicide actors, against civilian targets. Sometimes suicide bombs have been supplemented by automatic weapons, the latter especially in countries such as the United States, where such weapons are readily accessible, even to terrorists. A more recent trend in Europe has been the series of attempts to commit mass murder using more easily available implements, such as knives or motor vehicles. In a fashion, this harkens back to some of the tactics of terrorists of millennia ago.

Civil Aviation as Target

---

[2] U.S. Congress, Office of Technology Assessment, *Technology Against Terrorism: The Federal Effort*, OTA-ISC-481, Washington, DC: U.S. Government Printing Office, July 1991, p. 17.

Terrorist attacks on commercial aircraft have continued since the 1970s, mostly outside the United States, resulting in the deaths of thousands of people. Although there had been few such attacks in the United States, U.S. carriers were often the target of these attempts, which occasionally used hijacking techniques as well as sabotage. The most famous of these is probably Pan American Flight 103 from London to New York, which was bombed and destroyed over Lockerbie, Scotland in 1988. The following year, UTA Flight 772 met a similar fate over Niger, Africa as a result of similar tactics. Over 400 fatalities resulted from these two events. These attacks differed from most others in that the bombing was apparently due to terrorist activities by a state, not a sub-state actor: Libya, under the dictatorship of Muammar Ghaddafi, was considered to bear ultimate responsibility for these attacks, as documented in trials in Scotland and France. In fact, Ghaddafi later agreed to pay billions of dollars in compensation to victims of these attacks, in order to rehabilitate himself in the international community and to deflect the sanctions and political isolation that had been imposed on Libya.

Civil aviation is still a major target for diverse terrorist groups, although tactics have changed somewhat. Rather than placing an explosive charge in the luggage hold, (or occasionally taking the explosive device into the passenger cabin and disembarking at an intermediate stop, which was quite feasible and would allow the terrorist to escape), the trend since 9/11 has been for the terrorist to sacrifice himself and to carry the explosive onto the aircraft, maintaining control over the weapon and deciding when and how to detonate it. One reason for the change may have been the introduction of the effective baggage screening system world-wide. Another is the emergence of the suicide bomber among today's dominant terrorist groups, al-Qaeda and Daesh. Nevertheless, in many countries, increased efforts at security, including the screening of carry-on luggage and of passengers has made the terrorist's task more difficult.

The attacks of 9/11, which applied a novel use of civil aviation – the use of aircraft as the weapon rather than as the target of attack – constituted another major change in tactics, but one that is unlikely to be repeated in the same way, due to fairly obvious countermeasures that have been widely applied, including screening more carefully for weapons carried on one's person, increasing the number of plain-clothed security guards on board many flights, and the hardening of cockpit doors. It is also unlikely, following 9/11, that passengers would remain passive in the midst of a hijacking.

The current trend in attacks on civil aviation may be examined by considering some incidents since 9/11; but first, for comparison, I will note a foiled plot in 1995 that was carried out by some of the same terrorists who later joined al-Qaeda and organized the 9/11 attacks. This was a plan to place bombs simultaneously on eleven United and Northwest Airlines flights on routes from East Asia to the U.S. mainland. It was called Bojinka by the perpetrators. The plan appears to have been to leave the bombs under seats in the passenger compartments and for the terrorists to deplane at an intermediate stop. A preliminary test run was actually attempted on a flight from Manila to Taipei, Taiwan. The bomb detonated and killed the unwitting passenger in the seat above the weapon; however, the pilot heroically managed to land the plane safely and successfully and there were no other fatalities. The plotters were later discovered in Manila, when the explosive materials they were manufacturing in a rented Manila apartment caused a fire, calling attention to the effort. Several were eventually arrested. The ringleader, Ramzi Yousef, the nephew of 9/11 organizer Khalid Sheikh Mohammed, escaped but was later arrested in Pakistan and extradited to the United States, where he was tried and sentenced to life imprisonment. He had previously attacked the World Trade Center with a car bomb in a garage under the building in 1993.

Immediately following 9/11, in December 2001, al-Qaeda operative Richard Reid, now known as the shoe bomber, attempted to detonate a bomb on an American Airlines flight from Paris to Miami. The bomb did not explode. Alert crew members and passengers overpowered him and a tragedy was averted. In 2009, a relatively untrained operative from Nigeria, Umar Farouk Abdulmutallab, unsuccessfully tried to detonate a bomb hidden in his underwear on a Northwest Airlines flight from Amsterdam to Detroit. The results were similar to the previous event, with passengers and crew overpowering the terrorist, who was not able to detonate the weapon.

Since 9/11, there has been also a series of major terrorist attacks on civil aviation against non-American or non-passenger aircraft. There were two attempts to place bombs on cargo aircraft in Yemen (2012), carried out by al-Qaeda on the Arabian Peninsula, and addressed to Jewish institutions in the United States, but with the apparent intention of causing detonations while in fight over the United States. These bombs were cleverly concealed in computer printers. However, these attempts failed; the devices were discovered at intermediate stops: one flight in England and the other in Dubai, mostly because of detailed intelligence that had been obtained.

In another country, Russia, there were two instances of bombs placed on commercial aircraft by Chechen separatist groups in 2004, which resulted in the destruction of both planes with the loss of all on board. These bombs were carried onto the plane by passengers in a suicide attack. Much more recently, in 2015, Russian Metrojet Flight 9268, carrying tourists from Egypt's Sinai back to Russia was destroyed with hundreds of fatalities. The attack was carried out by the Daesh affiliate in the Sinai as a sign of opposition to both the Egyptian and the Russian governments, which had carried out strong anti-Islamist measures in Egypt and the Caucasus, respectively. It was suspected that local airport workers were recruited to facilitate the placement of an explosive device in the passenger cabin. There have also been bomb attacks in Russia against airports and trains, resulting in scores of fatalities, presumably also carried out by Chechen separatists; the most recent one occurred in 2017 in St. Petersburg.

Another example of a recent major attack on civil aviation was an interesting failed attempt on a Somali aircraft in Mogadishu, ascribed to the al-Shabaab movement, which has declared allegiance to al-Qaeda. In 2016, a passenger carried a bomb in a laptop computer aboard a passenger flight. Later video showed how security personnel at the airport wittingly permitted the bomb to pass through security screening. Fortunately, the bomb only killed the perpetrator, blowing his body out of the aircraft through a hole created by the explosion. Since the detonation occurred shortly after take-off, before the build-up of a larger pressure differential between the interior and exterior of the plane, the aircraft survived the blast and was able to return safely to the airport of departure.

In summary, civil aviation is still a major terrorist target, and still continues to be considered by several active groups. Attacks of this sort have been sporadic but persistent over half a century. With minimum expenditure of effort, a major terrorist result can be achieved in terms of publicity and fear. An ultimate goal is to create such fear that civilian passenger traffic will be seriously impacted and reduced around the world.

Other Types of Terrorist Attacks -Weapons of Mass Destruction

While terrorist attacks on the United States, against domestic targets or against its overseas assets, have historically focused on the vulnerabilities of commercial aircraft, other types of terrorism have also taken place. And in other countries, there has been an even greater diversity of attack types. Some efforts have been made by various groups to use weapons of mass destruction (WMD - usually understood to fall into three categories of attacks: chemical, biological, and nuclear/radiological).

There have not been many efforts to use biological or chemical agents as terrorist tools, but there have been a few. Because it seems to require some technical ability to produce and effectively disseminate such weapons, it may first be useful to note three cases in which states, not sub-state actors, have developed and/or used such weapons. First, Iraq, under Saddam Hussein, used a massive chemical attack with nerve gas against Kurdish Iraqi civilians in the town of Halabja in 1988, resulting in over 5000 deaths. Secondly, in the Soviet Union, several large laboratories within a government biological research complex known as Biopreparat were involved in developing enhanced biological agents in violation of its international commitments under the Biological and Toxic Weapons Convention (BTWC) of 1972, which was ratified by the Soviet Union in 1975 and entered into force internationally in the same year. Agents such as anthrax and Marburg virus were investigated and developed at these facilities, not simply for defense, as the Soviet government later claimed, but for offensive purposes as well. A famous incident arose at the facility located in the large city of Sverdlovsk (now again known as Yekaterinburg, its pre-revolutionary name) in 1979. An accidental release of a large amount of anthrax spores eventually killed about 100 civilians, who lived downstream of the escaping plume. For years afterwards, Soviet officials blamed tainted meat in the area for the very large-scale anthrax outbreak, but this explanation was dropped after the fall of the Soviet Union. Worryingly, there is some concern among U.S. government officials that elements of the bioweapons program continue to exist in Russia, as they continued to exist in the Soviet Union after its adherence to the BTWC.[3] If this is the case, the loss of some biological agents from this complex could constitute a danger as a terrorist weapon

The third case of state development of a chemical weapons programs is in Syria. Although it had denied the existence of such a program for decades, in 2014 the Syrian government agreed to destroy its large stockpile of nerve gas, including sarin, following an international outcry over its use against its own civilians during the ongoing

[3] See online report of the Nuclear Threat Institute at http://www.nti.org/learn/countries/russia/biological, last accessed June 28, 2017. Also see Alex Rogers, "Lawmakers Mull Biological Threat from Russia," Time online, May 28, 2004, accessible at http://time.com/91820/russia-putin-congress-biological-weapons, last accessed June 28, 2017.

Syrian Civil War. This use was a clear incidence of state terrorism using WMD, resulting in the deaths of about 1,000 people. The stockpile was destroyed by international inspectors and technicians over the course of two years, at significant expense and effort. Syria also agreed to join the Chemical Weapons Convention (CWC) at this time. However, in 2017, the Syrian military again used sarin against its own population in an air attack against an area of Aleppo that was held by opposition forces, killing nearly 100 civilians. Clearly, either Syria's stockpile was not completely eliminated in 2014, or else its manufacturing capability for this agent continued to be viable. A brief retaliation by US forces caused minimal damage to the air field from which aircraft had launched the chemical attack, but current stockpiles of sarin almost certainly continue to exist in Syria. It should be noted that some opposition forces in Syria, notably Daesh, have used another chemical agent, mustard gas, mainly against Kurds opposing them in another theater of the Syrian Civil War. These efforts caused a small number of injuries and few, if any fatalities, indicating that weaponization of this chemical agent by the terrorist group was not particularly effective. In addition to proscribed (under the CWC) agents, Syrian forces have frequently used chlorine in so-called barrel bombs dropped on areas of the country controlled by opposition forces. Chlorine is not banned as an agent under the CWC because of its vital role in purifying water supplies, but its use as a weapon is proscribed.

In the United States, there have only been sporadic attempts by terrorist groups to use biological or chemical agents. Several such efforts included attempts by individual political extremists to weaponize ricin, a biological toxin that can be refined from castor beans. Because ricin is more effective as an assassination tool rather than as a means to cause mass fatalities, these efforts were not successful. One widely reported and somewhat more successful effort to use biological agents for political ends was the use of salmonella bacteria to infect a salad bar in Antelope, Oregon in 1992. This was an attempt by a bizarre, putatively religious movement, the Bhagwan Shree Rajneesh group. The goal was to sicken enough of the local electorate before an election to enable a takeover of the small town by the cult leaders. This attack, however, failed to achieve victory in the election. There were no fatalities, but hundreds of people were sickened. The cult eventually was forced out of the United States and its remnants exist today in Pune, India.

A better-known biological attack was the case of the anthrax letters, envelopes containing anthrax spores that were sent shortly after 9/11 to members of congress and some television journalists. About five members of the public eventually died from anthrax as a result of infections, and more were sickened. The origin of the letters was attributed by the FBI to a mentally unstable scientist working at the U.S. Army Medical Research Institute of Infectious Diseases at Fort Detrick, Maryland, a biological research facility, but there remains some doubt among experts as to whether the guilty party was correctly determined. In any case, DNA analysis indicated that the anthrax strain used did come from Fort Detrick, in spite of rather crude efforts in the letters to indicate that the attack was part of an Islamic plot.

Internationally, the most effective sub-state chemical terrorist attack was by the Aum Shinrikyo cult in Japan in 1995. Sarin was released in the Tokyo subway, causing 13 fatalities and hundreds of injuries. An earlier release in the city of Matsumoto also caused casualties, including deaths. The cult was eventually broken up and its head is still in prison, possibly awaiting execution. Less widely known is the cult's earlier attempt to disperse anthrax in Kameido, Tokyo.[4] Fortunately, in spite of the fact that the cult membership did include people with advanced degrees in biology and chemistry, the type of anthrax used in this effort happened to be a non-virulent strain that had been isolated by legitimate disease researchers to help investigate the bacterium. Therefore, no injuries or fatalities resulted from this abortive effort.

Al-Qaeda also attempted to isolate and purify anthrax strains when it had access to areas of Afghanistan, accorded them by their Taliban allies, who controlled the country until late 2001. Later analysis of their efforts indicated that they had managed to purify just a small amount but had made little progress towards weaponizing the agent.

Regarding nuclear or radiological weapons, the relatively good news is that it is extremely difficult for sub-state actors to obtain the material to make the former and that the latter are fairly easily detected and not particularly effective at causing mass casualties, although severe disruption and economic costs could well ensue. [5]

---

[4] H. Takahashi et al, "*Bacillus Anthracis* Bioterrorism Incident, Kameido, Tokyo, 1993," *Emerging Infectious Diseases* **10(1)**, January 2004, accessible at https://wwwnc.cdc.gov/eid/article/10/1/03-0238_article, last accessed July 4, 2017.

[5] See C. Ferguson and M. Smith, *Defence Against Terrorism Review* **2(2),** Fall 2009, pp. 15-34, and Peter D. Zimmerman and Cheryl Loeb, "Dirty Bombs: The Threat Revisited," *Defense Horizons*, **38**, January 2004, pp. 4-5

Although a sub-state actor would find it very hard to construct a nuclear weapon by manufacturing the required nuclear fuel – plutonium or enriched uranium – a group could, in principle, purchase or steal a ready-made weapon or nuclear material from an existing nuclear state. This led to worries for many years, regarding the remnants of the former Soviet Union, where an enormous number of weapons and many hundreds of tonnes of material existed. After the collapse of the Soviet Union, societal disruption caused a massive deterioration of security measures protecting existing nuclear weapons and material, which led to world-wide concern. Cooperative efforts involving the United States and other countries helped Russia re-establish its protective measures to more acceptable levels. Recently, Russia has re-asserted its national pride, in part by ending these collaborations; but on the other hand, greater political and economic stability there has resulted in improved protection of nuclear assets, and there is now far less international concern over the security of Russian nuclear material. The other former Soviet states have divested themselves of the weapons and have largely gotten rid of the nuclear material, placing the residue under international safeguards.

Clearly, the less nuclear material that exists in the world, the less likely it would be for some of it to escape national control and end up in the hands of terrorists. And the better such material is accounted for and protected, the lower the possibility that terrorists would somehow be able to obtain this material and weaponize it. International efforts have been underway for decades to reduce the use of highly enriched uranium in many widely-dispersed research reactors, often found at universities. And in those countries that have nuclear material, often produced as a result of nuclear power programs, international safeguards, established and verified by the International Atomic Energy Agency, are employed to account for, verify, and control amounts and locations.[6]

In addition to broad international efforts to counter the possible use of nuclear and radiological weapons, the United States, following 9/11, established the Department of Homeland Security, which was intended to protect the country against major terrorist attacks, including the use of weapons of mass destruction. Many efforts have been underway to improve methods of detecting biological and chemical agents, and to respond rapidly to any sort of attack. Much of the R&D effort in the biological area is also applicable to possible natural outbreaks of epidemics. These are now more feared than in earlier decades, due to the greatly increased amount of international trade and travel, which can aid in the rapid dispersion of new, virulent agents through the concomitant spread of vectors, such as insects. To address the nuclear and radiological areas, a special office was created within the Department in 2005: the Domestic Nuclear Detection Office, which is charged with countering potential terrorism by detection of the unauthorized transportation of nuclear and highly radioactive material.

Some of the efforts to counter WMD terrorism through technical means will be detailed in the rest of this paper.

Other Types of Terrorist Attacks – Reversion to Less Sophisticated Means

In recent terrorist attacks against the United States and elsewhere, some terrorist groups have used less sophisticated weapons, such as small arms and even motor vehicles. Such attacks appear to be currently on the increase, possibly because terrorist groups, particularly Daesh, are under pressure, losing both territory and adherents, and are as a consequence less able than earlier to mount sophisticated attacks. One reaction has been to try to achieve at least a few more sophisticated attacks through their members who have lived in (or, in many cases, having been born in) Western countries, but who have also acquired terrorist techniques through visits to territories controlled by the terrorist group. Another reaction is simply to broadcast, through on-line magazines and through social media messaging, generic calls for attacks by their audiences in home countries,

The following paragraphs describe some selected terrorist attacks in the United States and elsewhere over the past few years. Not all attacks or plots are listed: the ones presented were selected to illustrate the types of attacks and weapons recently used against Western targets.

The major attacks that have occurred in the United States involved a very small number of individuals who had either been encouraged, mostly through communications by emails or through social media, by al-Qaeda affiliates or, in some cases, Daesh members, to engage in terrorist acts of various sorts. The best known of these was at Fort Hood in 2012 when an American-born army officer, Nidal Hasan, used a purchased hand weapon to kill 13 unarmed victims, mostly soldiers, in an office on base. He had apparently been encouraged to do so by members of al-Qaeda

[6] Details of international safeguards applications may be found in many articles and locations. The most authoritative summary should be found at the IAEA website, https://www.iaea.org, last accessed June 28, 2017.

of the Arabian Peninsula through the use of email communications. The San Bernardino massacre of 19 Americans took place in 2015, with the attackers making use of automatic arms that had been easily obtained for them by a friend. This attack was carried out by an American-born individual of Pakistani origin and his Pakistani wife (Syed Farouk and Tashfeen Malik). A last-minute communication by the couple asserted an affiliation with Daesh. A third example, the mass killing of 49 individuals at the Pulse nightclub in Orlando, Florida in 2016 was carried out with automatic weapons by Omar Mateen, an American-born man of Afghan heritage, who asserted his connection with Daesh in the course of his attack.

Many other efforts in the United States were less successful. Several plots to commit mass murder in various ways were intercepted by U.S. authorities and prevented. An example was the abortive effort to explode a car bomb in Times Square, New York in 2010 by Faisal Shahzad, a naturalized American born in Pakistan, who had apparently received some training from al-Qaeda personnel or other terrorists during a visit there some months earlier. The bomb did not detonate, probably due to the perpetrator's insufficient expertise and training.

In Europe, there have been both relatively sophisticated and very crude attacks, reflecting both classes of terrorism stimulated by Daesh, while it is in the process of losing its bases and resources in both Syria and Iraq.

The attack on the satirical political magazine, Charlie Hebdo, in Paris in 2015 was a mass murder using small arms, arranged by two jihadist brothers of North African extraction, benefiting from only minimal direction and guidance by Daesh. A nearly simultaneous set of attacks by another jihadist individual resulted in the killing of a policewoman near Paris and shortly thereafter in the deaths of several persons taken hostage in a Jewish food store in Paris. These primitive attacks, while successful in sowing fear and anger among the French populace, caused the deaths of relatively few and were not at all sophisticated in tactics, beyond the standard criminal ones – the attackers all had criminal histories. These events might fall into the category of attacks generally inspired by Daesh central, but not necessarily organized and controlled by them.

Unlike the 2015 case, the attacks on the Bataclan concert venue in Paris and nearby restaurants and bars, plus the simultaneous attempts to attack a major soccer match nearby, carried out in 2016, appear to be the results of a well-coordinated plot, directed by Daesh and carried out by up to ten individuals operating in a suicide mode. The multi-pronged plot was coordinated by one or two Daesh agents, who succeeded in avoiding capture for several weeks afterwards. About 100 innocent civilians were killed in the attacks at the concert venue and in several restaurants. The aggressors primarily used small arms as weapons. However, interestingly, the attempts to use individuals who functioned as suicide bombers at the soccer match were not successful. In one case, the bomber killed only himself and in the other, only one passerby was killed in addition to the terrorist. In both cases, the terrorists were prevented from entering the soccer stadium by the successful operation of external checkpoints.

By contrast, the attacks on the Brussels airport and on a subway station in the same city, which took place later in 2016, were more successful in constructing and utilizing explosive devices as weapons against a relatively unsuspecting public. About 30 victims were killed in these simultaneous attacks, which were managed and carried out by active Daesh agents. This set of attacks, like the Bataclan atrocity, were not the result of "lone wolves" with minimal direction, but were major projects that required weeks or months of meticulous planning and organization.

Apart from the attack on the Ariana Grande concert in Manchester, England in May 2017, the most recent examples of terrorist attacks against the West, on the other hand, appear to fall into the lone wolf mode, in which perpetrators were single or else very few (2-3) in number. While the Manchester attack required some sophistication in terms of the construction and use of a bomb, and the terrorist may have profited from some coordination with Daesh offshoots in North Africa, other examples give the appearance of almost ad hoc plans, some of which, nevertheless, were quite successful in terms of the number of victims felled by the terrorists. The examples are the Nice, France case on July 14, Bastille Day, 2016, in which one person used a large truck to mow down scores of victims celebrating the national day; a similar attack in Berlin just before Christmas, 2016, had a similar strategy and effect, although the number of victims was somewhat less; and, in 2017, SUV attacks in Stockholm, and London, resulted in fewer fatalities. The last London vehicular attack was continued by the three terrorists, who exited the SUV and then stabbed to death three more individuals encountered on the streets until they were killed in turn by rapidly responding police teams.

Drawing unified conclusions from the diverse nature of terrorist attacks in the United States and Europe over the past few years would be tentative, but some general features are clear. First, those plots in which considerable direction and planning were carried out by a terrorist group were more successful in killing large numbers of people. This sort of terrorist attack is harder to arrange than the mere stimulation of lone wolves by means of propaganda on terrorist websites to just do something. If one examines the results of complicated plots that involve many people and significant coordination, it appears that the terrorist organization benefits considerably from a safe haven in which planning may be undertaken in some calm. It is probably useful to have sufficient time to arrange logistics and to have access to adequate resources including communications – largely using social media – to accomplish as much coordination as possible. The 9/11 attacks benefited greatly from the existence of large safe areas in Afghanistan, in which al-Qaeda could assemble forces and coordinate plans over a period of many months. The Bataclan and Brussels attacks by Daesh appeared to benefit from considerable organization and the deliberate accumulation of resources, which was possible due to their control of large swaths of territory in Syria, particularly around the city of Raqqa. The absence of safe territory for the terrorists would not mean the end of complex, major terrorist attacks, but such high levels of coordinated planning would be constrained, if the terrorist organization had to function in complete secrecy in what for them would be hostile territory. Many terrorist plots, developed both in Europe and in the United States, have actually been prevented over the past 15 years; this includes, for example, one in which a major planned attack on the New York subway system was prevented in 2009. The interruption of the plots occurred in part due to the difficulties the terrorist found in maintaining operational security in hostile territory.

On the other hand, the lone wolf (wolves) attack, as opposed to the centrally-directed plot, is much harder to prevent, particularly if only one person is involved. Intelligence is much more difficult to obtain when communications are minimized, as would usually be the case for one agent operating alone. Such attacks are usually less successful (there are exceptions, of course, such as Hasan or the Orlando attack), but they will likely continue to occur, even if the ultimate instigator – Daesh for example – loses its bases of operations. The utilization of social media and blogs as a means of inspiration of the lone wolf by a terrorist group, can still be accomplished, although probably with less effective dissemination and with less effect, even without territory that functions as a safe haven. For example, using a distributed network of organizers, a terrorist group could still maintain some coordination in its propaganda and still portray itself as a unified organization, worthy of support or emulation.

A final note regarding attacks on civil aviation: attacks on planes have been carried out without the availability of a safe haven by the terrorist group. However, in many cases, the presence of secure territory at a sub-state level has been helpful for terrorist aims. Bomb and hostage attacks on airplanes in the seventies, eighties, and nineties often relied on the availability of state support at various levels: the bombings of PanAm 103 and UTA 772 required sophisticated planning and significant resources. These were obtained using the dedicated assistance of a state: Libya. Several terrorist attacks on aviation in Europe and Asia in this period were carried out by the Abu Nidal Organization, which functioned for years with the support of Libya. Al-Qaeda's attempts to bomb cargo aircraft were carried out through the use of territory that the group controlled in Yemen during earlier phases of the civil war there.

One might conclude from a survey of recent examples that the denial of safe haven territory to a terrorist group can complicate their efforts to carry out a major, sophisticated attack against a target in a distant land. This will not eliminate such threats and indeed, may not have much of a preventive effect against lone wolf attacks. But the frequency and large-scale success of complicated attacks can likely be constrained through the denial of territory to a terrorist group, particularly the denial of large areas that could allow for extensive training and organizing.

TECHNOLOGIES TO COUNTER TERRORIST THREATS

The following discussion of technical means of countering terrorist attacks will be divided according to the type of attack. Each type -- attacks on civil aviation, attacks using WMD, attacks using less sophisticated weapons – will necessarily have its own specific technical (and, of course, policy) responses. However, it should be noted that some policy and operational responses to terrorist threats may be similar and consistent across different types of terrorist tactics. Similar operational responses that would be broadly applicable are, for example, obtaining near-real time intelligence that provides indications or warning of future attacks; cordoning off secure areas as an early defensive measure upon warning (such as secure areas at airports or the deployment of security barriers and checkpoints to cut off paths of terrorist movement); or psychological operations, on social media or elsewhere, to impede recruitment by terrorist groups or to counter their communications and propaganda efforts. The following sections will present some current and potential future uses of technology to counter threats to civil aviation, to the use of WMD and to the recent use of more basic weapons and tactics to attack civilians.

Countering Threats to Aviation Security

Aviation security is perhaps the oldest field in which technology has been developed to defend against criminal and terrorist attacks. Metal detectors for passenger screening and x-ray systems for carry-on screening were first deployed at U.S. airports in 1973, in response to multiple hijackings, primarily to Cuba. A research and development (R&D) program for bomb detection was initiated by the United States Federal Aviation Administration (FAA) in 1975 in response to a bombing at New York's LaGuardia airport. This was expanded following the Air India 182 bombing in 1985. But it was the bombing of Pan American Flight 103 in December 1988 that raised the specter of a serious, unaddressed, and potentially often repeatable threat to the lives of countless aircraft passengers – the emplacement of explosives in the cargo hold within checked baggage – and resulted in the FAA embarking on a greatly expanded research program, aimed at producing equipment that could detect the presence of explosives in checked luggage. The specifications of an explosives detection system had to include a high probability of detection of explosives (in a quantity sufficient to destroy an aircraft in flight), while at the same time, an alarm rate of false positives low enough not to impede commercial air travel. The throughput of any explosives detection system had to be sufficiently high so as not to delay flights significantly. Given an alarm, a secondary inspection of a suspicious item, usually accomplished through a hand search by security personnel, would have to be able to screen baggage adequately for explosives, while at the same time adding only inconsequential delay times in the aircraft boarding process. Specific criteria for the performance of such technical equipment and solutions were set by the FAA with the help of scientific, intelligence, and aviation experts, working through the National Academy of Sciences (NAS) in the early 1990s. There was considerable pessimism about achieving a technical solution for the criteria listed above that would, in addition, not force the introduction of unacceptably large pieces of equipment across the nation's (and, indeed the world's) airports and that would be affordable. A large array of technologies were being investigated at the time of the PanAm103 bombing, including, but not limited to, the use of thermal neutron activation (TNA) (primarily measuring nitrogen content and concentration – nitrogen is an essential element in explosives of interest);[7] dual-energy, backscatter, and computed tomography (CT) x-ray approaches (primarily measuring x-ray attenuation, electron density, and effective atomic number (Z_{eff}) of the material examined within a piece of luggage); and chemical trace measurement approaches (including ion mobility spectrometry, mass spectrometry, gas and liquid chromatography, and thermoluminescence) that identified explosive molecules.

A system based on CT x-ray technology was the first to pass muster and was certified in December 1994 as having satisfied the criteria set by the NAS panel, as tested independently by the aviation security laboratory run by the FAA in Atlantic City, NJ. The first certified Explosives Detection system (EDS), which had been researched and developed by American physicist Fred Roder, was manufactured by Invision, Inc. Later, other companies were able to certify and produce their own EDS. Equipment based on CT technology has since been deployed in most major commercial airports across the globe.[8] The equipment is fairly costly and large, but within the realm of what has been feasible at airports. Unfortunately, there have been difficulties in reducing both size and cost to the degree that this technology would also be considered feasible for screening carry-on baggage (see the next paragraph for a discussion of that problem, which however, is currently being actively researched).

There are other pathways than passenger baggage that could bring explosives aboard aircraft, and these have been used more extensively since 9/11. Until that event, the focus had been on non-suicide attempts to destroy aircraft, in which the terrorist would leave the bomb on the aircraft and deplane before the flight leg on which the bomb would be detonated. This could be done, for example, through the checked baggage route, now made more difficult by the introduction of the successful CT-based EDS, or through bringing an item into the passenger compartment. The latter route could be countered by security measures that involve inspecting the passenger

---

[7] TNA uses a source of thermal neutrons to convert N^{14} into an excited state of N^{15}, which then decays via the emission of a distinctive gamma ray of 10.3 MeV, whose energy puts it well above background and enables detection through the use of scintillators that measure total energy with a reasonable resolution.

[8] One other x-ray technology has also produced a certified system, based on Bragg scattering of crystals of explosive chemicals, but this operates rather slowly and requires multiple units running in parallel to meet the throughput requirement. Systems based on this approach have rarely been deployed.

compartment during intermediate stops. However, the era of the suicide bomber has changed this calculus. If the perpetrator was willing to go down with the aircraft, it would be necessary to screen effectively all items that a passenger might bring aboard the flight in the first place.

To respond to the threat of explosives carried on board by passengers, a number of technologies has been applied, both in the United States and elsewhere. In the first place, there are standard x-ray machines that are used to screen carry-on baggage for suspicious items that could be weapons, explosives, or parts of either. Some items, like weapons, are relatively easy to identify with high-quality x-ray systems, especially those that provide more than one view, but explosive parts are more difficult to identify. A modification of standard x-ray techniques is to measure the transmission of two different x-ray energy spectra, to produce an image that provides both attenuation and atomic number information. These so-called dual-energy measurements can be obtained by switching the kV of the source, inserting and removing a filter from the x-ray beam, a combination of both of these approaches, or by using stacked detector arrays with different energy sensitivities. This last approach is the most commonly used in security x-ray systems. Dual-energy x-ray images allow metallic objects [high atomic number (Z)] to be distinguished from organic objects [low atomic number (Z)], which include explosives. However, because the Z value measured is a complex function of all objects along a beam path, superposition of multiple objects can introduce clutter and make detection more difficult.

Screening of passengers and their carry-on items for explosives can be greatly enhanced by the use of explosive trace detectors (ETDs). The most widely deployed type of ETD utilizes ion mobility spectrometry (IMS). These systems discriminate molecular species on the basis of the speed at which an ion moves between two electrodes in air. Their speed is inverse to mass but also depends on the size and shape of the chemical species, and so is not quite as specific as standard mass spectrometry. However, in airport environments, which are not always tightly controlled, mass spectrometry devices have a greater difficulty in maintaining accuracy and precision for long periods of time. Ion mobility spectrometry has been the technology of choice for this purpose for some 25 years, and is produced by several vendors.

ETDs are extremely sensitive to minute amounts of materials and provide excellent discrimination. The main issue is to be sure that an appropriate sample is taken – usually of bag handles, passengers' hands, and pockets – and to assure that the sample is taken properly. The technology depends on harvesting small particles, usually of dust, that are always present. Many molecules of explosives can adhere to these particles if they are in the neighborhood. The pure vapor pressure of many explosives is far too low to produce enough molecules to be detected, and the presence of these dust particles, to which "sticky" explosive molecules can adhere, essentially amplifies the number of sampled molecules that the detector is able to see. The efficacy of the technique depends strongly on maintaining effectiveness and consistency in the sampling process.

Another familiar and widely-used technique to monitor what passengers bring onto the plane involves screening the passenger him/herself with a millimeter-wave scanner. These are essentially low-powered radars of high frequency that can detect not only metals but other objects hidden in pockets or against the body. Detection is accomplished by measuring dielectric constants of such objects, which differ from those of clothes or skin. Formerly, x-ray backscatter devices had also been used by the Transportation Security Administration (TSA) for this purpose. This technology can similarly find foreign objects hidden in or underneath clothes. However, the x-ray devices were discontinued by TSA in 2012 due to public concerns about being exposed to radiation, even though the level was well below doses characteristically received from cosmic radiation on flights over two hours.

Walk-through metal detectors, which still supplement security screening over much of the world are somewhat problematic in that they cannot detect non-metallic objects. Metal detectors rely on Lenz's Law, which states that an electromagnetic pulse creates counter pulses through eddy currents in a targeted metallic object. It is the secondary radiation from these eddy currents, which is detected. However, there is a concern about objects, such as knives and gun parts made of ceramics or hardened plastic, which could easily be lethal. Walk-through metal detectors by themselves would not be able to unmask such threats to a flight when carried on by a passenger.

In summary, whereas checked baggage threats have been significantly mitigated by CT-based EDS, and whereas even non-metallic threat items carried onto aircraft can be reliably found, there is still room for improvement in detecting dangerous items in carry-on baggage. Work is continuing to develop more effective detectors; it would be ideal if CT-scan devices could fill the gap which exists because of the limitations of standard x-rays and sampling issues with trace chemical detectors. R&D continues to develop, adapt, and test CT-based EDS for screening carry-on items at passenger checkpoints.

Countering Uses of Weapons of Mass Destruction (WMD)

In considering technical means of countering WMD, it is useful to divide the weapons into two classes, those that emit radiation (nuclear and radiological) and those that do not (chemical and biological). Focusing first on those that do not emit radiation, an obvious remark is that they are much harder to detect before release, in most circumstances impossible, without some sort of warning from, for example, intelligence or law enforcement sources that can define fairly precisely where to look. This aspect is made more difficult by the fact that very small amounts of chemical material are sufficient to kill large numbers of people. For biological agents, this situation is even worse. For infectious diseases, which include anthrax, Ebola and Marburg viruses, cholera, smallpox, plague, and many others, a very few initial infectious cells or viral particles can eventually result in a catastrophic epidemic that could affect millions. Even for non-infectious agents, such as botulinum toxin, very small amounts are able to kill (a dose lethal to 50 % of the population is about 2 nanograms per kg of body weight, when injected intramuscularly).

Therefore, detection of chemical or biological agents before release through some kind of technical approach is unlikely to be successful. Lethal quantities are too small to find easily; also, major precautions need to be taken to acquire and examine an unknown sample. On the other hand, if there is some indication of where to look, or about whom to intercept and search, it might be possible at that point to use technical means to examine (quite carefully) a suspicious sample.

The general strategy to deal with chemical or biological weapons is to detect a release as quickly as possible, estimate the direction and coverage of the spreading plume of the agent, and to provide shelter and, where necessary, rapid diagnostics and mitigating treatment for those who may be exposed.

In those cases where an attack location may be anticipated, such as an urban area, where a biological attack in particular could cause up to millions of eventual casualties, one can install permanent monitoring systems that periodically take samples of ambient air and analyze them for the presence of biological agents. Automated biological detectors have been developed that can perform very rapid (30 minutes) analyses. An accurate analysis may be done for only a limited number of agents, so the automated analyzing system tests for agents on a predetermined list. Many of the rapid analytical systems perform a quick DNA analysis using the PCR (polymerase chain reaction) technique,[9] which can identify an agent with high specificity. Detectors have been developed for use in the field, so that the plume carrying the agent can be monitored by technicians in real time. They can be aided by preprogrammed computer simulations that predict the flow of the agent-bearing plume through an urban or other area, given wind and temperature conditions.

Another method to detect the sudden appearance of a biological agent, especially if there has been no specific warning, is to monitor continuously medical information, such as emergency admissions to hospitals and clinics. The information also may extend to observation of major changes in patterns of pharmacological purchases, including over-the-counter medications. These might indicate, for example, a sudden rise in flu-like symptoms among the general population, which could in turn lead to the identification of an unusual disease outbreak.

For chemical weapons, the emphasis is also on rapid detection after release, which will probably be initially manifested through the sudden appearance of poisoning symptoms among people in the vicinity of some unexplained smoke or mist (possibly through an explosion of some sort or the unexpected and unauthorized appearance of equipment spraying unknown substances in populated areas). In both cases, biological and chemical, substantial R&D efforts are underway to develop means of performing rapid diagnostics in the field so that treatment can begin as soon as possible, and so that protection, quarantine, sterilization, and clean-up procedures can be applied. And in both cases, there is extensive development of computer modeling to predict the path of the agent through a region.

The case of radiological and nuclear terrorism is inherently different. Since the material involved has a radioactive signature, it may be possible to find a threat before actuation. This is particularly true in the case of radiological agents, which are inherently highly radioactive, hard to shield, and therefore easier to detect with off-the-shelf radiation detection equipment. The isotopes that are considered to be of interest in manufacturing

[9] PCR is a technique developed over 20 years ago, which amplifies the number of DNA molecules to be sequenced. It has been adapted into several rapid bioagent detectors in a still-developing field. A detailed review of the state of the art eleven years ago is D.R. Christensen et al., *Clinical Chemistry*. **52(1)**, 141-5 (2006), accessible at https://www.ncbi.nlm.nih.gov/pubmed/16391330, last accessed June 28, 2017.

radiological weapons are limited to those in common use for medical and industrial purposes and therefore obtainable in large quantities (hundreds of Curies and more). These include cobalt-60, cesium-137, iridium-192, strontium-90, and a limited number of others. Detecting penetrating gamma radiation is the approach used for discovering the presence of such isotopes. In addition to being an obvious means of detecting the presence of radiation gamma detection can also have the advantage of measuring the energy of the gamma ray; this can provide specific identification of the nuclear isotope in question.

Nuclear material (uranium and plutonium) may be detected through gamma and also through neutron emissions, although the neutron rate from highly enriched (in fissile U-235) uranium is quite small (on the order of a few neutrons per second per kilogram of material – and most neutrons are from the non-fissile isotope, U-238, which constitutes only about one-tenth of the uranium isotopes present in weapons-grade uranium). Of the two elements, plutonium is by far the more radioactive and easier to detect, but even plutonium, if shielded, may not be easily detectable at a range of more than about 10 meters.

The tasks of countering radiological and nuclear terrorism are completely different from each other: radiological terrorism, as mentioned above, is not likely to kill many people, although it can cause immense economic damage by forcing large, economically important areas of cities to be kept off limits for health reasons, pending massive and expensive cleanups. Radiological weapons can, in principle, be readily detected by means of radiation detectors (if one is looking – 100 Curies is equivalent to over 10^{11} radioactive disintegrations per second).

Nuclear weapons, on the other hand, qualitatively different from radiological weapons. First, they are hard to detect if shielded, since the number of particles that they emit is relatively low. Even plutonium in weapon grade form (mostly Pu-239) produces millions of gamma rays and tens of thousands of much more penetrating neutrons per second per kilogram. The neutrons may be detectable, but the gamma rays are mostly of very low energy, many of which are absorbed by the plutonium itself. Uranium is far less radioactive, and, in addition, produces most of its gamma rays at the very low energy of 186 keV, which is easily shielded by centimeters of iron or higher Z (atomic number) metal. Uranium is very difficult to find, especially when some effort is made to shield it.

The problem for gamma detection, even for plutonium, lies in separating radiation from the plutonium from environmental background sources that are naturally-occurring radioactive isotopes in the environment (potassium from soil and many foodstuffs, radium, radon, thorium and uranium, also in the soil in many parts of the country). This separation can be accomplished by using energy-measuring detectors, such as sodium iodide or intrinsic germanium. The energy spectrum of the detected gamma rays can then be compared to those produced by the NORM (naturally-occurring radioactive material) isotopes and to the spectra produced by plutonium isotopes. This can help in separating signal from noise, when one is close to the object in question. These few sentences cover a multitude of issues, however, when one is trying to screen a large number of, say, 40-foot long cargo containers in a short time at rates that do not impede commerce, while at the same time maintaining the capability to detect nuclear material that might be concealed within.

It is obvious that once successfully detonated, a nuclear weapon will instantly produce a large number of deaths on a scale difficult to imagine (tens of thousands immediately, at a minimum, for an explosion with the energy release of the Hiroshima weapon). These results cannot be mitigated, although it will be possible to mitigate, to a degree, later injuries resulting from radioactive fallout and contamination. Therefore, the effort to counter this kind of terrorism focuses in part on developing means of protecting our shores and borders from the illicit introduction of nuclear material, whether in weapons or other forms. The first step is to assure adequate security over the fissionable and fissile material that does exist across the world. One useful part of that step is the ongoing and successful effort by the U.S. Department of Energy, with many international partners, to modify hundreds of nuclear research reactors around the world so that they may function on low enriched uranium, which cannot fuel a nuclear weapon, rather than highly enriched uranium, which can. Another has been to give technical assistance to those states that have nuclear weapons or nuclear material so that they may better protect the material from diversion by sub-state actors.

But, in case these initiatives do not completely succeed, the United States has undertaken to deploy radiation detectors on our borders, in our ports, and in overseas ports from which departing vessels depart for the United States. An office dedicated to this end was created within the Department of Homeland Security in 2005 to address the problem of protecting against the illicit introduction of nuclear material into the United States – the Domestic Nuclear Detection Office (DNDO).

Since the detection range of nuclear material sensors may be as little as a few meters, detection on the borders cannot be the only measure used The goals of DNDO include the scanning of all cargo containers entering the United States, whether by sea, road, rail, or air. The air problem is naturally best addressed at last points of departure of all aircraft on flights into the United States. Less attractively, one could screen for radioactive material when unloading cargo from aircraft at the arriving airports in the United States. This, however, does not protect against the possibility of detonation of a device in the air over the country or at the arriving airport.

Checking cargo containers arriving at U.S. points of entry requires the deployment of large numbers of detectors, both passive and active. This has been done to a large degree. Thousands of passive radiation detectors are deployed at ports of entry to the United States and some active detectors (x-ray imaging systems) are also used to supplement passive detectors for large trucks. Active neutron interrogation detectors are also under development: neutrons can penetrate the large metallic objects often present in cargo containers that would shield gamma rays emitted from nuclear material and make passive detection difficult. The neutrons then can stimulate fission in any fissile material that may be present, causing secondary radiation, which is then detected. One major problem with active neutron interrogation techniques is that the amount of real estate and the shielding required to protect travelers and operators from radiation is large.

At least two parallel R&D efforts are being waged to improve the ability to detect and intercept nuclear material. The first is obvious: to try to improve detection technologies so that the range of detection is significantly increased. However, this approach has natural limits: if there is very little in the way of radioactive emissions, the best detector imaginable may not be adequate. Increasing detector size will improve the likelihood of intercepting a radioactive particle that could be used to identify a threat, but even this will not help at a range of greater than a few meters. So, what can one do to prevent the introduction of nuclear material at points between controlled points of entry? One must bring the detector to the suspected material, and the only way to do this is through law enforcement and/or intelligence work that can detect and identify a person or vehicle that may be bringing nuclear material into US territory. Law enforcement agents can then intercept and hold this target so that radiation detectors may be applied to the vehicle or person in question. This latter strategy is vital to developing a means to protect national territory against nuclear terrorism. It is not clear how to assure success in accomplishing this.

Clearly, having the ability to collect information on terrorist threats through the intelligence community, international partners, and through law enforcement activities (at all levels, federal, state, territorial, local and tribal) is a necessity. To emphasize, technology can be a useful tool in countering terrorism, especially in dealing with WMD, it is only a tool, and probably generally an adjunct tool: the principal method for countering terrorism is to prevent it from occurring, and this involves a wide range of other activities: psychological techniques to convince potential terrorists that an WMD approach will not work for them, the development of predictive warning information that can inform law enforcement as to where to intercept potential WMD and how to look for and verify the presence of such materials. Obviously, social and psychological means of discouraging people from becoming terrorists in the first place must constitute an additional, vital effort.

Countering Low-tech Terrorism

It is vital to bear in mind that there are two general, broad categories of terrorists – not just one – that have been active in the United States in recent years: 1) internationally-oriented Islamist jihadists, some of whom are native-born Americans and others who are naturalized Americans or legal immigrants; and 2) racists/neo-Nazis who are native-born American citizens. The former have been responsible for 68 deaths in the United States in 2015-2016 and the latter have killed 27.[10] The racist terrorists are usually lone wolves; the jihadists have mostly been lone wolves, but some of them also have received significant stimulation and direction from terrorist organizations abroad (notably Daesh, but in some cases, affiliates of al-Qaeda). The discussion below focuses mostly on jihadist terrorism, since in recent years, this ideology has spawned the great majority of major attacks, not only in the United States but also elsewhere, particularly in Europe. However, some discussion of potential countermeasures to

[10] See START infographic dated February 2017; some additional statistics comparing the results of attacks in the United States perpetrated by far-right extremist with those by jihadist extremists, accessible at https://www.start.umd.edu/pubs/START_ECDB_IslamistFarRightHomicidesUS_Infographic_Feb2017.pdf, on the START website, last accessed June 28, 2017.

terrorism, including terrorist attacks by lone wolves, applies both to jihadists and American racists, both of whom make significant use of social media.

Just as in countering WMD, prior indications and warning of a planned attack are essential to a defense. This will usually be attempted through "normal" intelligence operations and police work, in which these professionals use a multitude of sources of information, trying to uncover terrorist plots before they are carried out. Much success has been achieved in the past, both in the United States and Europe. Many plots have been discovered and neutralized, some through information derived from informants and some through monitoring, in various ways, communications between known malefactors and their contacts; the latter sometimes turn out to be newly-appearing terrorist actors. Other options are also possible, such as careful observations of social media communications, monitoring of chatrooms, and attempts to engage in sting operations through the media. Needless to say, there is no guarantee that such methods will always work; indeed, it is clear that they will not always be successful. Therefore, where possible, one also needs to incorporate technological means of defense. The existence of social media sites, such as the Russia-oriented Telegram, are protected by allegedly unbreakable (at least so far) encryption and for this reason such sites have often been used by Daesh and other terrorist groups to contact their agents remotely and provide information, support, guidance, and direction for future attacks.[11] Monitoring social media sites, therefore, is limited in effectiveness; monitoring is most helpful before an individual in communication with a terrorist organization's headquarters goes "underground" by using impenetrable encryption.

However, in dealing with true lone wolves, who appear spontaneously and have little or no communication with directing forces, the situation is even worse: intelligence is often likely to be unavailable. In practice, retrospective analyses of some lone wolf cases have discovered evidences of radicalization and occasionally specific intent to commit terrorist violence openly advertised by the would-be terrorist on, for example, Facebook or Twitter. If observed and acted upon by law enforcement authorities in time, some attacks could have been averted. The case of Omar Mateen and the Orlando Pulse nightclub is one case in which there had been some interactions with law enforcement that indicated that the perpetrator claimed contacts with radical terrorist groups abroad.[12]

The easy availability of automatic and semi-automatic weapons in the United States makes them weapons of choice of terrorists here, both for the racist extreme right and for Islamist jihadists. The use of semi-automatic weapons was demonstrated in two of the deadliest terrorist attacks in the United States since 9/11: the San Bernardino mass murders and the Orlando shootings. A policy (not a technical) countermeasure would clearly be to make it harder to obtain such weapons. One could, for example, prevent the sale of such weapons to persons that have had known contacts or claimed contacts with international terrorist groups. Unfortunately, an effort to pass some more restrictive legislation along these lines failed in the U.S. Congress in 2016.

But it would be even more useful if one could detect signs of a planned attack. How might this be done? The first obvious strategy would be to attempt to gather intelligence of plans to carry out specific terrorist plots aimed at destruction of people or property. Much along these lines has already been done in the United States. Presumably, intelligence agencies try to gather such information from many different sources, using informants, communications intercepts, and many other approaches that include a variety of technologies. There have been many reports of interdicted plots by jihadist terrorists in the United States, and more recently, of plots interdicted by European authorities before they were carried out.[13] But intelligence of this sort can be hit or miss. Many plots can be and have been stopped, but a few have succeeded, both in the United States and in Europe. Intelligence gathering is clearly nothing like a science, even though scientific and technical means are used as working tools to aid standard police and intelligence work. Occasionally, a plot will evade intelligence and be carried out, even in the United States, and more easily in Europe, where it is much less difficult for a terrorist to enter from southwest, south, or central Asia.

---

[11] There are many news articles on Telegram, the use of its encryption by Daesh, and related encryption issues. See for example, David E. Sanger and Nicole Perlroth, "Encrypted Messaging Apps Face New Scrutiny Over Possible Role in Paris Attacks," The New York Times, November 16, 2005, accessible online at https://www.nytimes.com/2015/11/17/world/europe/encrypted-messaging-apps-face-new-scrutiny-over-possible-role-in-paris-attacks.html?_r=0, last accessed June 28, 2017.

[12] E. Perez et al., "Omar Mateen: Angry, Violent 'Bigot' Who Pledged Allegiance to ISIS," CNN, June 14, 2016, http://www.cnn.com/2016/06/13/us/orlando-shooter-omar-mateen/index.html, last accessed June 28, 2017.

[13] One example is S. Smith, "London Terror: 'Active Plot' Foiled in U.K. Police Raid," NBC News online, April 28, 2017, http://www.nbcnews.com/news/world/london-terror-active-plot-foiled-u-k-police-raid-n752336, last accessed June 28, 2017. The following footnote 14 provides a U.S. example. There are many other instances of foiled plots on both sides of the Atlantic.

The potential ability to interdict terrorist plots applies to conspiracies in which there is some directive communication from a terrorist organization not only to stimulate but also to assist in the successful accomplishment of the plot. Two examples are: 1) the attempt to bomb the New York City subway system, planned by a young Afghan, Najibullah Zazi, in 2006[14] together with some assistance from jihadist extremists in New York – a plot that was discovered in time by U.S. authorities and thwarted; and 2) the Paris attacks on the Bataclan concert venue, some restaurants, and a major international soccer game in Paris on the same night – in this case, although some of the planners and perpetrators were known to authorities, warning of the attack was not received and the plot succeeded in most aspects.

Even harder to stop might be plots such as the San Bernardino shootings and the Orlando nightclub atrocity, in which the murderers apparently had minimal contact with a terrorist "central" and were mostly self-directed. If the terrorist's contacts with others are little to none, how might one hope to interrupt the plot? Considering the "easier" problem, are there even any conceivable actions or processes that that could interrupt more substantive terrorist plots in which there are at least some actual communications through social media or other channels?

One might start with the recruiting process, often carried out initially on the internet. Even in the case of lone wolves, the passive observation of terrorist propaganda online has played a role in stimulating an attack.[15] There are also cases in which passive observation converts to active participation as a result of communications with terrorist recruiters; often the potential recruit attempts to contact a well-known terrorist group for which he or she has sympathy. In fact, there have been many cases in which U.S. law enforcement officials have presented themselves on the internet as a member of a terrorist group, and have succeeded in stimulating the recruit to commit an illegal act, upon which he (and it usually has been "he" in the United States) is arrested. For a number of years, such events were fairly frequent; at least it appeared to be so from those incidents reported in the press. This "sting" tactic on the part of police and intelligence agencies may interrupt recruitment of potential terrorists but it cannot not deal with a plot that involves actors who are already agents of the group.

One issue may be that communications can be encrypted to a degree beyond the ability of US intelligence to decode in real time, especially if there are many communications to monitor and survey. One web application that has been used by terrorists to arrange plots has been the site Telegram (see footnote 11), run by a Russian national but located in distributed fashion in a number of European countries. This platform uses encryption technologies that are extremely hard to overcome. There might be technical or policy approaches to solving this problem; given technical difficulties with the high level of encryption used, the latter approach may be more likely. The Russian government has recently expressed serious dissatisfaction with Telegram, which gives some hope that international cooperation may be possible to overcome encryption issues that appear to be vital to terrorist planning for major attacks.[16]

Another option, this one technical, would be to try to invent or discover means of breaking the encryption of this and other such sites used for terrorist planning. It is not certain that the mathematical problems inherent in devising such a solution can be overcome in the near future. A third option, also technical, could be to disrupt the site in such a way that it would no longer be considered secure or reliably accessible by its users or, possibly, no longer be useful to the terrorist group for its communications purposes. This last area of endeavor may comprise

---

[14] D. Amira, "PRISM Stopped Najibullah Zazi From Blowing Up Backpacks in the Subway," New York Magazine online, June 7, 2013, http://nymag.com/daily/intelligencer/2013/06/prism-najibullah-zazi-subway-nsa.html, last accessed June 28, 2017.

[15] The case of Omar Mateen, noted in footnote 12, is one example. Another is the case of Nidal Hasan, the Fort Hood murderer, who communicated through email with Anwar al-Awlaki, an American-born terrorist active in a leadership role with al-Qaeda on the Arabian Peninsula. Al-Awlaki had stimulated and probably helped organize attempts to bomb several commercial U.S. aircraft (the Northwest flight in 2009 and the two cargo flights from Yemen) and had promised to carry out more lethal attacks on the United States. He was killed in a U.S. air raid in Yemen on September 30, 2011, following a presidential decision to treat him as an enemy combatant.

[16] Andrew Roth, "Russian Lawmakers Target Web Anonymity After Deadly Attack," The Washington Post, June 27, 2017, p. A11. Online version accessible at https://www.washingtonpost.com/world/europe/as-russia-targets-internet-anonymity-law-enforcement-says-extremists-planned-a-deadly-bombingusing-an-encrypted-chat-app/2017/06/26/f942e648-5a83-11e7-8e2f-ef443171f6bd_story.html?utm_term=.45bdc969a1bd, last accessed June 28, 2017. The Russian security service, the FSB, announced on June 26, 2017 that Telegram had been used by terrorists in planning the bomb attack on the St. Petersburg metro on April 3, 2017, killing 14 people in addition to the suicide bomber. As a result, legislation is moving through the Russian Duma (parliament) that would ban anonymity on messenger apps. Russian officials also threatened to block Telegram under some circumstances. The article quotes a 2016 study by the Middle East Media Research Institute that claimed that Telegram's "strong encryption and flexibility" have made its app the one of choice for many jihadist elements, including those associated with Daesh. If Telegram became unavailable for terrorists, one might expect them to use other websites offering similar messaging services, but perhaps with less effective encryption capabilities.

several approaches that are worthy of study by computer engineering experts who have a deep knowledge of the web.

There are some terrorist planning scenarios, including those carried out by lone wolf types, in which – perhaps at an early stage of commitment – the actor uses the internet as a platform for venting and for communicating with like-minded individuals and groups and does this more or less in the open, before deciding to "go dark" on a deeply encrypted site. Dealing with such scenarios would require a broad, automated monitoring of platforms such as Facebook and Twitter, as well as many others, using data analytic techniques to look for words, phrases, meanings, or images that may indicate future terrorist behavior. Indeed, speculatively, developing such countermeasures may be in their early stages now. The use of advanced data analysis techniques in near-real time could eventually provide an early filter that would allow intelligence and law enforcement analysts to focus, amidst all the "noise" on the internet, on those individuals, whose expressed anger may more likely indicate an eventual descent into criminal activity and real violence.

An obvious remark needs to be made here: the blanket surveillance and monitoring of a whole population in order to deflect a possible terrorist plot that might not even exist, would meet substantial resistance from much of the public on very rational grounds both in Europe and in the United States. The correct and just balance between the surveillance of possible malefactors (selected for observation by as yet undefined criteria) and the privacy rights of everyone is not obvious: different people will make different assessments. However, in order to have the discussion and debate, the technical possibilities for achieving useful surveillance of potential malefactors, while guarding privacy for political discussion, need to be presented to the public for examination. The best responses to this dilemma clearly still need to be worked out.

Having explored some of the current terrorist threats, some broad conclusions of policy and technology approaches should be considered:

- International cooperation to counter encrypted terrorist planning needs to be considered. Action by governments could include eliminating or closing down sites that use encrypted information to aid terrorist attacks that cause loss of life. In principle, laws could provide that operators or owners of such sites may face criminal penalties. If some jurisdictions provide safe havens for terrorist groups by furnishing communication technologies that are impervious to decryption, this serious vulnerability impedes the global struggle against terrorism.
- Near-real-time assessment of enormous amounts of data on the open internet, using analyses of vocabularies, semantics, images, etc. needs further development. This toolkit could be used to impede terrorists' logistical planning, and also to disrupt recruiting efforts by terrorist groups and attempts to be recruited by wannabe terrorists.
- The development of more sophisticated sting operations to disrupt terrorist recruitment and planning can contribute to countering terrorist efforts and may be able to disrupt terrorist plots at early stages.
- Denying territory to terrorist groups will not eliminate terrorism but will certainly render the terrorists' task more difficult by making it more complicated for them to recruit, train, plan, and provide logistic support for terrorist acts, especially those against remote targets. This argues, for example, for the importance of preventing significant pieces of territory (e.g., Raqqa in Syria and Mosul in Iraq) from serving as safe havens for terrorist planning. It is notable that recent efforts to commit terrorist acts in Europe have become more frequent as Daesh has lost territory but, on the other hand, many of their plots have become far less sophisticated as terrorist planning and training have been disrupted.[17]

Many problems connected with counterterrorism reside in the internet/data analytic/computer science fields. A newer generation of scientists has emerged that has, starting at a young age, acquired easy but also deep familiarity with current communications technologies. These scientists will be needed to address these problems successfully.

[17] A summary of recent terrorist attacks in Europe since 2015 may be found at http://www.express.co.uk/news/world/693421/Terror-attacks-timeline-France-Brussels-Europe-ISIS-killings-Germany-dates-terrorism, accessed on June 26, 2017.The first item recorded is an anti-Muslim terrorist attack on worshippers at a London mosque, and there was one other anti-Muslim event. The others are examples of jihadist attacks, most of which were generally accomplished with unsophisticated tools. The exceptions that used bombs rather than more primitive weapons were the Manchester attack and the Brussels airport and train station attacks, as noted earlier in the text. The civilian deaths at the Paris attacks of 2016 were nearly all due to automatic weapons probably obtained by Daesh agents in Belgium, which has looser gun control laws than most other European nations.

ACKNOWLEDGEMENTS

The author wishes to thank Dr. Fred Roder for his superb work, kindly reviewing this paper in great detail. He also wishes to thank Mr. Herbert Ferguson Tinsley of the National Consortium for the Study on Terrorism and Responses to Terrorism (START) at the University of Maryland, College Park, Maryland, who provided information and analysis regarding the projected future of terrorism and of countermeasures to terrorism.

REFERENCES

1. U.S. Congress, Office of Technology Assessment, *Technology Against Terrorism: The Federal Effort*, OTA-ISC-481, Washington, DC: U.S. Government Printing Office, July 1991. Also accessible online at http://ota.fas.org/reports/9139.pdf, last accessed June 26, 2017. This and the following reference provide a historical background of terrorism until that point and some significant technical information on counterterrorist technologies, much of which is still useful, particularly in the area of aviation security.
2. U.S. Congress, Office of Technology Assessment, *Technology Against Terrorism: Structuring Security,* OTA-ISC-511, Washington, DC: U.S. Government Printing Office, January 1992. Also accessible online at http://ota.fas.org/reports/9235.pdf, last accessed June 26, 2017.
3. The DHS Center of Excellence, START, has comprehensive archives on terrorist attacks and much related analytic work, which has been useful to the author of this paper. Online access is available at http://start.umd.edu, last accessed June 26, 2017.
4. The U.S. Department of State has published yearly reports on terrorism since 1995. Since 2004, these reports are titled *Country Reports on Terrorism.* Prior reports were called *Patterns of Global Terrorism.* These provide informative details on terrorist attacks and on summaries of trends from an official U.S. government source. The most recent one that is accessible is found at https://www.state.gov/j/ct/rls/crt/2015/index.htm, last accessed June 26, 2017.

The Impossibility of Probabilities

Peter D. Zimmerman

Emeritus Professor of Science and Security Kings College, London UK

Abstract. This paper discusses the problem of assigning probabilities to the likelihood of nuclear terrorism events, in particular examining the limitations of using Bayesian priors for this purpose. It suggests an alternate approach to analyzing the threat of nuclear terrorism.

INTRODUCTION

More than 40 years ago I picked up a slim paperback, *The Curve of Binding Energy*, by one of America's best prose stylists, John McPhee. It was a collection of articles originally published in *The New Yorker* magazine, recounting a long interview with one of McPhee's neighbors near Princeton, the physicist Theodore Taylor. Ted Taylor had been one of Los Alamos's finest fission weapon designers. He was skilled at visualizing the physical functioning of nuclear weapons and converting ideas to mathematics and then to explosive hardware. He had conceived the most efficient fission weapon ever built, and the most powerful pure fission atomic bomb in any nation's arsenal, one so large that few hydrogen bombs surpass its power.[i]

In retirement Taylor had a new worry, one which was first ridiculed. It was possible, he thought, that enough information about the design of nuclear weapons had seeped out around the borders of the classified world that any skilled and motivated physicist could replicate the designs of the bombs dropped in World War II, and very likely those that entered the U.S. stockpile in the subsequent decade.

The first generation weapons were designed with punch card tabulators and Marchand manual or electric-motor powered adding machines. In 1945, 'Computer' was a job title, jobs filled at Los Alamos mostly by the wives of the physicists working there. In 1974, computers were common in laboratories and universities. By 2017 far more powerful machines reside on single chips, need no more than the power of a flashlight battery, and are mere subcomponents of the ubiquitous smart phone in nearly everybody's pocket. Computational power was not a barrier to producing simple designs in 1974. It is of almost no importance as a barrier to far better weapons today.

Taylor noted that the real barrier to building a nuclear weapon was not the design and engineering, but the acquisition of rigorously pure ^{235}U or a supply of plutonium metal, also extremely pure. Indeed, roughly 90% of the entire Manhattan Project budget was spent on the reactors and reprocessing facilities at Hanford, Washington, and on the gaseous diffusion, magnetic separation, and gaseous thermal enrichment plants in Oak Ridge, Tennessee. Only about 8% went to Los Alamos.

By 1974 highly enriched uranium (HEU) was a common metal at university reactor laboratories, where it powered the early TRIGA[1] reactors. Some universities kept more than one complete core at the reactor site. HEU was used in many other reactors as well.

Although radioactive and toxic, plutonium existed in nuclear waste, at spent fuel reprocessing facilities, and elsewhere. Certainly it was dangerous, and likely lethal, to extract plutonium from spent fuel, but in recent years there has been no shortage of volunteers willing to sacrifice their lives for a cause. That was not so common in 1974 when McPhee and Taylor wrote, but the Japanese *kamikaze* pilots of World War II provided a proof of principle that such people could be recruited.

Nuclear Weapons and Related Security Issues
AIP Conf. Proc. 1898, 050004-1–050004-7; https://doi.org/10.1063/1.5009233
Published by AIP Publishing. 978-0-7354-1586-7/$30.00

Thus, Ted Taylor asserted with high confidence that it was possible for a non-state actor to fabricate a simple fission weapon, transport it to a target, and there destroy much of a city. Despite serious pushback from the Energy Research and Development Administration, successor to the Atomic Energy Commission and forerunner to the Department of Energy, and simultaneous discouraging statements from the then-new Nuclear Regulatory Commission, the concept of nuclear terror had arrived.

As if to prove the point, I decided to see if I could replicate the calculations behind Little Boy, the ^{235}U bomb dropped on Hiroshima in my spare time while teaching at Louisiana State University. At Princeton a young student, John Aristotle Phillips, accepted the challenge to design an implosion bomb using plutonium. Phillips became famous.[ii] I stored my designs in the only lockable filing cabinet I had, and then had the machine shop install a hasp and security padlock.

Novelizing the Idea

Nicolas Freeling, an English author of detective stories (and a personal friend) and I discussed the idea of nuclear terror, and decided it was worth a book, largely exploring the mind of a young physicist who was coerced into building the device. Taking its name from Los Alamos wartime slang, the book is called *Gadget.* It was first published in 1977 by Coward McCann in the United States. So far as I know, it only sold well at the University of California, Berkeley and Los Alamos, NM bookstores. Used copies are available on the Web if you care to read it.[iii] A terrorist nuclear explosive has acquired the general name of Improvised Nuclear Device (IND) in parallel with the acronym IED for Improvised Explosive Device, such as used against US troops in Iraq and Afghanistan.

Some parts of *Gadget* are based on my experiences; most is sheer fiction. The physics of the bomb is correct; much of the design is not, and deliberately so. That made certain that we would have no security clearance problems. The fact remains, however, that designing a uranium gun bomb using 80% or so ^{235}U is easy in principle. Much of the engineering is difficult, but none of it is beyond the capabilities of a group of physicists, electrical engineers, chemists, and demolition experts, who have the facilities of a good university physics or engineering laboratory and access to a few dozen acres where they can make loud noises.[iv]

After *Gadget* was published the Nuclear Regulatory Commission (NRC) stated that building any nuclear explosive device was impossible for anybody but a nation-state. The NRC's estimate was that it would require a cast of thousands and a budget of billions. The latter is true if you start with uranium ore; it is demonstrably false if the fuel is available.

Dormancy and Revival

From around 1977 until 11 September 2001 the matter of nuclear terror lay dormant. It figured in some analyses of the threat to the United States,[v] but in general, policy analysts were content to regard Brian M. Jenkins's work as correct. Jenkins wrote "Terrorists want a lot of people watching, not a lot of people dead."[vi] The United States' view of terrorism changed instantly when the World Trade Center towers crumbled. It became clear that terrorists might well not be content to have an audience, that terrorism was no longer the kind of individual or small group theater, embodied by attempts to assassinate the Russian tsar or the Austrian archduke (regardless of the consequences), but might well be members of large, well-supported and well-disciplined groups, capable of working in secret for long periods. Such groups might have the intent to kill tens of thousands, cause damage in the billions, and still commit an act that made the whole world watch.

The US Senate Committee on Foreign Relations held classified hearings on nuclear terror in 2002. At the request of then-Chairman and Senator Joseph R. Biden, Los Alamos and Livermore National Laboratories constructed complete (except for the substitution of inert nuclear and high explosive materials), full-scale models of INDs, which were exhibited during the hearings and made available to the full Senate to examine.[2]

Serious Analysis Begins

Nuclear terrorism suddenly shifted from an academic enterprise to an apparently real threat in a new paradigm. It was no longer a question of "can terrorists go nuclear?" but rather "*when will* terrorists go nuclear?"

Academics were eager to try to answer the question. These analysts split naturally into two groups:

1. Those who believed that the probability of a terrorist nuclear strike could be neglected because the likelihood of success was negligible.
2. Those who sought to quantify the probability that in any given year there would be a nuclear explosion detonated by a non-state actor and who believed such an enterprise was feasible.

The principal question asked of those analysts by policy makers boils down to estimating the annual risk of a terrorist nuclear detonation. Risk, of course, is the probability of a nuclear detonation during the year multiplied by the cost to remediate the damaged area. No remediation can replace the lives lost. Nevertheless, this annual risk, R_y, is approximately the insurance premium a rational nation would pay to prevent nuclear terror.

The cost of replacing a city, even a small one, is obviously very high. Even replacing a village is expensive, as demonstrated by damage amounts from tornadoes and floods in the United States. Unfortunately the equation

$$R_y = P_y * Cost \quad (1)$$

is inherently ill defined because the probability of occurrence is unknown and uncomputable, but I will try.

Probabilities

Generally, one thinks of probability in a frequentist setting. That is, one considers a large number of identical systems, identically prepared, and asks what fraction of those systems evolve to a given state. This works beautifully with roulette wheels and dice or coin flips. However a terrorist nuclear detonation is likely to be a single event, not repeated, and certainly not repeated indefinitely. At the best, any calculation of the probability of a terrorist group building or buying or stealing nuclear weapon is nothing but a guess, a hunch.

Skeptics

John Mueller of the Ohio State University and the Cato Institute delivered a paper at the American Political Science Association, Program on International Security Policy in Chicago, January 14, 2008, in which he derided those who believe that there is a real nuclear terror threat as "alarmists." He sought to demonstrate the proposition by hypothesizing that there were twenty steps which a nuclear terrorist gang must complete in order to succeed. The specific steps he postulates vary among his several papers on the subject. The steps in the APSA paper are summarized in Table 1:

Table (1). Mueller's Steps that Must be Completed for a Terrorist Group to Detonate a Nuclear Device

Step Number	Step Description
1	Find the HEU.
2	Have precise blueprints for device.
3	Avoid detection while stealing.
4	Nothing must go wrong.
5	Corrupt several insiders.
6	No leaks from team.
7	All gang members stay loyal.
8	Seize and transport adequate amount of adequately pure HEU.
9	Local and national police must see nothing despite desperate search.
10	Move HEU while being pursued and avoid capture.
11	Hire smugglers who stay bought.
12	No (other) criminal gangs sense something out of the ordinary.
13	The improvised nuclear device, with a mass around a ton, must be moved in secret.
14	Assemble a technical team.
15	Build a machine shop.
16	Assemble loyal team in target country.
17	Transport IND to target country.
18	Succeed in getting bomb into target.
19	Do it while nobody notices.
20	Set it off; not a fizzle

Mueller is a generous sort of analyst. That is, he suggests a 50% probability that each of his 20 steps can be accomplished. Although Mueller suggests that this will result in a probability of the enterprise succeeding, none of the steps can be estimated using frequentist statistics. That is, you don't get to make 50 tries at assembling a technical team, and count the number of successes. Nor is that possible for any of the other steps. At the very best, in another context they would be classed as Bayesian priors, with his best guesses (i.e. 50%) at the chances of success.

But note that Mueller's steps are not independent of one another. Neither are they all distinct. Nevertheless he calculates that

$$P = 1/2^{20}. \qquad (2)$$

That is a convenient number for him. It is precisely one in 1,048,576, or less than one chance in a million of a nuclear terror attack succeeding. Since it is so very low, it is utterly negligible. Since he always has 20 steps in his calculations and always uses 50% probabilities, it almost seems that he has deliberately sought that one in a million chance. It seems that an analyst can make the "probability" of a successful nuclear attack come out to be any desired value, just by manipulating the steps to a bomb.

Robin M. Frost, then working at the International Institute of Strategic Studies (IISS) in London, prepared an Adelphi Paper,[vii] which was a plea for calm in the face of predictions of nuclear terror. My response to Frost's technical arguments can be found in a comprehensive paper in *Survival.*[viii]

Friedrich Steinhausler[ix] compiled a list similar to that of Mueller:

- Access to a workshop with advanced equipment, such as precision calibrated, computer-guided machine tools with laser interferometer, airbearing lathe, and artificial room ventilation with built-in air cleaner;
- A supply of Krytron switches;
- Machining capabilities for the production of complex shapes (tolerance about 10^{-10} m).

This argument that the probability of success approaches zero is ludicrous; after all, the scientists at Los Alamos in 1944-45 had none of those capabilities and yet they built three nuclear devices between June and August 1945.

In short, all of the technical and semi-technical arguments that the probability of success of a nuclear terrorist approaches zero fail upon careful analysis, particularly when the experience of the Manhattan Project, with the technology available, then is factored in.

Optimists and a Different Sort of Probability Computation

However, many American analysts have taken the opposite approach. They have apparently looked at the consequences of the detonation of a crude nuclear weapon in a modern city and seem to have conflated consequences with probability.

The problem with any quantitative estimates of the probability of a terrorist nuclear detonation is that essentially all of the factors that would go into calculating a reasonable result are unknown. There are no unclassified, trustworthy data on the most important factor: how many groups are actively engaged in work that might lead to a weapon?

I do not believe the correct answer is "zero." I suspect that somewhere, a well-funded organization has begun thinking, working, organizing to gain a nuclear capability – but I do not *know.* I am morally certain that the number of groups is less than 1,000, likely very much less than 1,000. A range of one to one thousand is too large to approximate. All either could be, in any event, is a Bayesian prior, meaning my hunch.

Perhaps foremost among those who have argued that nuclear terrorism is either *the* greatest threat to the United States or at least *one of* the greatest threats are distinguished physicists Richard L. Garwin and William Perry. Others holding similar views include Harvard professor Graham Allison, whose book, *Nuclear Terrorism: The Ultimate Preventable Catastrophe,*[x] contains a wealth of references and citations to the literature and is fairly pessimistic about our ability to thwart a well financed and dedicate plot. Garwin and Perry estimated the probability of a nuclear terror attack at approximately 90 percent over the course of the next decade.[xi] That equates to about 20% each year, a frighteningly large risk. If policy were based on their calculation, the United States should embark on an immediate and intensive campaign to prevent such a holocaust.

It seems unlikely that any terrorist group is likely to be able to steal an intact and functioning nuclear weapon from the stockpiles of the five major nuclear weapons states, China, France, Russia, the United Kingdom, and the United States. All of these are considered to have reasonably robust security in place to thwart or deter the theft of a weapon.[3] However, perhaps another nuclear state would donate an intact device to the group or else has such lax safeguards that one could be stolen.

More Sophisticated Calculations

Matthew Bunn of Harvard developed a simple mathematical model to estimate the annual risk of a nuclear terror strike succeeding. The major driver is the number of groups which might attempt such an act. Bunn says that there have been at least two well-funded terror groups which gave serious consideration to the idea, two more than Mueller believes would be at all interested. Using two groups as his input, Bunn estimates a "significant acquisition attempt (to acquire fissile material) roughly once every other year."[xii] Bunn arrives at a probability of 29% over a ten year period, and at an annual probability of 5.6%.

One of my former graduate students in London, Michael Levi, has arrived at lower estimates, in part because he concentrated on ways to detect a project in progress and found them likely to succeed.[xiii] Another graduate student at Kings College London, the Norwegian physicist Simen Ellingsen, presented a thesis which I consider to be the best available work on determining the correct allocation of resources to thwart an IND attack.[xiv]

Presidents of the United States including George W. Bush and Barack H. Obama have each declared nuclear terrorism a threat. In a 2004 televised presidential candidates' debate Bush and his opponent Senator John Kerry agreed that it was the single greatest threat to U.S. national security.[xv] President Obama convened a nuclear security summit in 2016, in large measure to address the threat of nuclear terrorism.[xvi]

The next factor that should be understood is the length of time a group must work to produce its first nuclear explosive. National efforts to construct a fission explosive have taken at a minimum 3.5 years (United States, December 1941 to July 1945), although most programs have required four or more years, some close to a decade. The American program began before the first sample of plutonium had been produced and when even the chemical and metallurgical properties of the fissile fuels were essentially unknown.

So far as I am aware, no national program has ever been the beneficiary of a gift of fissile materials, although almost all analyses of nuclear terrorism assume that the uranium or plutonium is made available to the plotters in a convenient form. (Allison considers the difficulty of a plutonium reprocessing facility; Oak Ridge published a detailed study of the problem, and decided it was not insurmountable. But the papers have been withdrawn from public access).

An important question is, how much would it cost to make an improvised nuclear device? This has been carefully addressed, and the answer is not encouraging. In our article *The Bomb in the Back Yard*, Jeffrey Lewis and I thought that a group of about twenty properly chosen scientists, engineers, technicians and support personnel might be able to fabricate a workable, if unsophisticated and likely unpredictable, device in about a year starting from metallic uranium and assuming a gun-assembled device. That estimate resulted from careful calculation of the time to complete the appropriate processes and assumed that there were no budget constraints.[xvii]

The required fund raising, planning before beginning the engineering and construction work, acquisition of the special nuclear material, and the preparation of a site to do the work could well take at least another year.

We estimated the cost (in then-year money) at approximately $5,433,000, which is obviously too precisely specified. But $5,500,000 is certainly in the ball park, *including the cost of the fissile material.* Considering the resources that have been available to al Qaeda, Hezbollah, and the Islamic State, this is a rather paltry sum, providing almost no barrier to an attempt. Even were Lewis and I off by factors of two to ten low, the amount is still small.

CONCLUSION

It is in some sense gratifying to see that an idea which received government ridicule at its inception in 1976 has now been acknowledged as a serious threat to the country. It is even more gratifying that the possibility has not yet become reality, and it is my hope that it never does.

A great deal of effort has been devoted to analysis of nuclear terrorism in the open literature, much of it devoted to estimating the probability of occurrence. That is an incorrect procedure. Since we hope never to series of independent attempts to build an IND, it is impossible to use statistics, and the probability of a unique event occurring is inherently impossible to calculate in a useful sense. In any event, most of the inputs to a probabilistic calculation are unknown and generally unknowable.

The most useful approach seems to be to analyze the steps needed to go from the idea of a device to a workable, even if neither sophisticated nor reliable, device and to examine the signatures of each step on the road, how to detect it, and how to thwart the attackers.

REFERENCES

[i] John McPhee, "The Curve of Binding Energy," *The New Yorker*, December 3, 1973, available at http://www.newyorker.com/magazine/1973/12/03/i-the-curve-of-binding-energy. This refers to the original interviews published in *The New Yorker*, not to the complete book. Retrieved 2017/07/26.

[ii] https://en.wikipedia.org/wiki/John_Aristotle_Phillips, retrieved 2017-07-25. Phillips is now an "entrepreneur specializing in political campaigns."

[iii] Nicolas Freeling, *Gadget* (New York: Coward-McCann and Geoheghan, 1977), see particularly pp. 43–8, 63–7, 76–80, 90–93, 101–110, 226–31. Freeling (and Zimmerman, his physicist collaborator), have anticipated many of the problems facing nuclear terrorists, including criticality testing, criticality accidents, remote arming and detonation, and evasion of detection. The 'computations' to be found on pages 44, 45, 63, 64 and 65 must not be taken seriously.

[iv] "The Bomb in the Back Yard," Peter D. Zimmerman and Jeffrey G. Lewis, *Foreign Policy*, November-December 2006, pp 32-39 (cover article).

[v] *Nuclear Proliferation and Safeguards,* US Congress Office of Technology Assessment, Volume 1, p. 140 (Washington DC, 1977).

[vi] Brian M. Jenkins, *Will Terrorists go Nuclear?*, RAND Paper Series, no. P-5541, Santa Monica, Calif.: RAND, 1975, p. 5.

[vii] Robert M. Frost, *Nuclear Terrorism after 9/11,* Adelphi Paper 378 (Abingdon: Routledge for the IISS, 2005). [note: not available outside of a paywall on the Web]

[viii] Anna M. Pluta and Peter D. Zimmerman, "Nuclear Terrorism: A Disheartening Dissent," *Survival* **48**, 2 (Summer, 2006), pp 55-70.

[ix] Friedrich Steinhausler, "What it Takes To Become A Nuclear Terrorist," *American Behavioral Scientist,* **46,** No. 6, February 2003, p. 792.

[x] Graham Allison, *Nuclear Terrorism: The Ultimate Preventable Catastrophe*, Henry Holt and Company (New York: 2004).

[xi] Hearing of the Energy and Water Subcommittee of the U.S. House of Representatives Appropriations Committee: Subject: Weapons Activities Oversight, 29 March 2007.

[xii] Matthew Bunn, "A Mathematical Model of the Risk of Nuclear Terrorism," *The ANNALS of the American Academy of Political and Social Science*, **67**, 1 (2006), pp. 103-120. For the calculation see p. 106.

[xiii] Michael Levi, *On Nuclear Terrorism*, Harvard University Press, Cambridge, MA, 2007. Levi's dissertation done at Kings College London is more pessimistic and goes into much more detail on the nuclear weapons technology than seemed appropriate. It is no longer available.

[xiv] Simen Ellingsen, *Safeguards against nuclear terrorism: HEU vs Plutonium*, thesis presented for the Ph. D., 2007 Kings College London, Department of War Studies.

[xv] http://www.washingtonpost.com/wp-dyn/content/article/2007/09/24/AR2007092401154.html, accessed 29 July 2017.

[xvi] "Obama: How we can make our vision of a world without nuclear weapons a reality," *Washington Post,* March 30, 2016. https://www.washingtonpost.com/opinions/obama-how-we-can-make-our-vision-of-a-world-without-nuclear-weapons-a-reality/2016/03/30/3e156e2c-f693-11e5-9804-537defcc3cf6_story.html , accessed 29 July 2017.

[xvii] Peter D. Zimmerman and Jeffrey G. Lewis, *Foreign Policy,* November-December 2006, pp. 32-39 (note: cover article).

Drone Warfare

Hugh Gusterson

Professor of Anthropology and International Affairs
George Washington University
Anthropology Department
2110 G St NW
Washington DC 20052
Guster@gwu.edu

Abstract. Crude drones existed as early as World War I, but the technology matured in the Yugoslav Wars of the 1990s and the current wars around the Middle East. The U.S. first used a weaponized drone in late 2001, in Afghanistan. Drones may cause more or less civilian casualties depending on the targeting protocols employed by their operators. There is an inherent ambiguity in determining who is an insurgent from several thousand feet, but civilian casualties are likely to be higher if targeters emphasize "signature strikes" over "personality strikes," if they engage in "double-tap strikes," if they rely too much on local informants, and if they rely too heavily on cellphone identification in the absence of corroboration from other intelligence sources. The legality of drone warfare is fairly clear in established battle zones such as Afghanistan, but is more problematic in terms of both international and domestic law when it comes to drone strikes in countries such as Yemen, Pakistan and Somalia with which the U.S. is not at war. Looking to the future, the U.S. would be well advised to sponsor negotiations for an international drone convention that might establish clear international rules for the use of drones, ban autonomous smart drones, and establish adjudicatory procedures to handle allegations of war crimes.

INTRODUCTION[1]

Over the last two decades drones -- also known as Unmanned Aerial Vehicles (UAVs) and Remotely Piloted Aircraft (RPAs) -- have become a vital, and rapidly growing, component of U.S. airpower. The U.S. Air Force currently trains more people to operate drones than conventionally manned aircraft. (This necessitates drafting as drone operators some who would prefer to fly F-16s, B-1s, KC-135s and so on). The best known drones are the MQ-1 Predators (the last of which will be retired from active service in 2018[2]) and the MQ-9 Reaper drone, both of which are armed with air-to-ground missiles. Although this article will focus on the Predator and Reaper drones, the two weaponized drone platforms, it bears noting that the U.S. also deploys a number of other drones of varying sizes and capabilities, using them for reconnaissance rather than kinetic attacks. The largest and most expensive is the RQ-4 Global Hawk, which can operate at higher altitudes and with more expansive vision than the Predators and Reapers. At the other end of the spectrum, RQ-11 Ravens, WASP-IIIS, RQ-20 Pumas, RQ-16 T-Hawks, and RQ-7 Shadows are smaller systems, some hand-launched by forces in the field, that can be used for local tactical surveillance. Nautical drones are also under development.

Weaponized drones have been attractive to US military planners for three main reasons. First, they allow extended flexible aerial reconnaissance in a way that manned aircraft and satellites cannot. Manned aircraft can track developments on the ground, but can only stay in the air for a fraction of the 24 hours of which a Predator is

Nuclear Weapons and Related Security Issues
AIP Conf. Proc. 1898, 050005-1–050005-9; https://doi.org/10.1063/1.5009234
Published by AIP Publishing. 978-0-7354-1586-7/$30.00

capable; satellites are confined to programmed orbits that produce serial snapshots of developments on the ground rather than continuous tracking shots, and they cannot see through the clouds below which drones may fly. With their capacity to dwell flexibly in the air for long time periods, drones can track insurgents to see where they go, with whom they meet, and what they do, generating important intelligence about insurgent networks and patterns of life. Second, thanks to GPS technology and a drone's ability to linger in the air, moving slowly while it awaits the opportunity for a clear shot, drones potentially offer great precision in targeting. It is claimed that the Reaper drone, equipped with a wider range of firepower than the Predator, can destroy one room of a house without killing those in other rooms. And, third, drone warfare makes it possible to fight without losing pilots. In the words of General David Deptula, drones "allow you to project power without projecting vulnerability."[3] While one can debate whether war is still war if only one side dies (or has instead, as French theorist Gregoire Chamayou claims, become "hunting"), this radical asymmetry of vulnerability is deeply attractive to military planners.

Drones raise important questions about the relationship between embodied vulnerability, the battlefield, and martial bravery that have excited debate within the U.S. military, but lie beyond the scope of this article. This article discusses the operational protocols that guide drone strikes, debates about the legal status of drone operations, disagreements about civilian casualties due to drone attacks, and future horizons in drone warfare, including possible international frameworks to regulate drones. At present there are no international treaties that specifically regulate the use of armed drones.

DRONES 101

The first drones, radio-controlled bi-planes, were developed during World War I, but proved so hard to control that they were not much used. In World War II the U.S. used radio-controlled drones for target practice; it also experimented with kamikaze drones, loaded with explosives, from which the pilots could, in theory, parachute to safety after take-off, while pilots in other nearby planes guided the drones to their targets. (John F. Kennedy's older brother died when one of these drones exploded, as so many did, before he could parachute out). In 1946 the U.S. military sent remote-controlled B-17s to collect samples from mushroom clouds in the first nuclear tests after Hiroshima and Nagasaki.[4]

In the 1960s the U.S. developed the Lightning Bug drone, used during the Vietnam War for surveillance where manned flights would have been too dangerous, mainly over North Vietnam. Usually launched from under the wing of a larger plane, this small drone took pictures as it flew a programmed route, then deployed a parachute and was collected in the air or at sea.[5]

Israel has long been a pioneer of drone technology. During the Yom Kippur War of 1973 Israel lofted drones to draw Egyptian fire, then sent manned planes to destroy the Egyptian missile batteries thus identified. Israel used drones again, during its 1982 war in Lebanon, to scout for targets, and it sold its Pioneer drone to the U.S. for aerial surveillance during the 1991-2 Persian Gulf War.[6] It was one of Israel's leading drone designers, Abraham Karem, who designed the Predator drone after he emigrated to the United States.[7]

With the Predator, first deployed in 1995, we arrive at the workhorse of the U.S. drone fleet over the last decade. It has a wingspan of about fifty feet, an empty weight a little over 1,000 pounds, a single propeller at the back that can drive it at a maximum speed of 135 m.p.h., and the ability stay aloft for about 24 hours. Much of the $4 million cost of each Predator goes to daylight and infrared cameras, as well as the eavesdropping and communication equipment it carries. Until 2001 the Predator was used only for reconnaissance and to designate targets, via laser, for attack by manned planes. Since 2001 Predators have carried Hellfire missiles so that they can visually acquire and prosecute targets on the ground without the assistance of other aircraft.

Over time the Predator has been gradually supplanted by the second generation armed drone, the MQ-9 Reaper. First deployed in 2007, the Reaper has a longer wingspan than the Predator, a more powerful engine and, therefore, a higher cruising speed as well as the ability to carry more explosive ordinance. While the Predator conventionally carries just two Hellfire missiles, the Reaper can carry a range of armaments, offering drone operators a greater ability to customize an attack.

The Predator came into its own during the air war over Kosovo in 1998-9. The development of GPS technology and the availability of large quantities of satellite bandwidth made it possible to operate drones from considerable distances while extracting highly accurate information about the location of the drone and targets on the ground in real time. Unlike manned aircraft, drones could stay aloft for up to 24 hours, conducting extended surveillance; unlike most satellites, which pass over targets briefly and at predictable intervals that an adversary can determine, Predator drones could circle recurrently and redeploy, gathering information about patterns of movement and weapons deployments on the ground in flexible and unpredictable routines in response to real-time intelligence. The disadvantage of drones, on the other hand, is that, with their flimsy airframes and underpowered engines, they are vulnerable to bad weather; also, susceptible to dropped communications contact with their remote operators who, at the best of times, experience a two-to-four second delay in communicating with them, they have a high crash rate. Indeed, according to a *Washington Post* report, an average of over twenty drones per year crashed as recently as 2008-2015.[8]

Although there were occasional suggestions that the Predator could be weaponized to track and strike individual targets on the ground, it was official U.S. policy as recently as mid-2001, just a few months before the 9/11 attack on the United States, to oppose such "targeted killings," which were seen both as a violation of international law and of President Ford's ban on assassinations.[9] Thus, in July 2001, The U.S. ambassador to Israel, Martin Indyk, had condemned Israeli targeted killings of Palestinian operatives by saying, "the United States government is very clearly on record as against targeted assassinations. . . . They are extrajudicial killings, and we do not support that."[10] Meanwhile, according to the 9/11 Commission Report, CIA director George Tenet was "appalled" at the idea of assassinating people from the sky.[11]

This changed dramatically after the 9/11 attacks. Operating under the guidance of new legal opinions written after 9/11, U.S. leaders began to use Predator drones to hunt and kill al Qaeda and Taliban operatives in Afghanistan, hoping they might kill Osama bin Laden this way. Soon, drones were being employed by the U.S. in a number of countries, some operated by the Air Force, some by the CIA, and some under the control of the Joint Special Operations Command (JSOC). Many CIA officials, recalling public controversy in the 1970s about CIA assassination initiatives, were initially leery of targeted killing by drone, especially since capturing al Qaeda leaders and interrogating them offered the prospect of actionable intelligence. However, targeted killing by drone became more attractive to CIA leaders after CIA Inspector-General John Helgerson released a report in 2003 warning that senior CIA officials were vulnerable to criminal prosecution in connection with practices of extreme rendition and harsh interrogation practices employed in a network of prisons around the world, including Guantanamo. In Helgerson's opinion many of the CIA's interrogation practices violated the United Nations Convention Against Torture.[12]

At the time of writing, the U.S. has used drones to attack targets in Afghanistan, Iraq, Syria, Yemen, Somalia, Libya, the so-called "tribal areas" of Pakistan, and the Philippines.[13] As has been noted across the muslim world, in every case U.S. drones were used to attack muslims. While the public tends to have an image of drone strikes as attacks where a drone acts alone to hunt down insurgents on the ground, in fact drones more often act in concert with ground troops and other forms of air power. Thus the most heavily drone-bombed country in the world is Afghanistan, where U.S. drones have struck over 1,000 times as part of a wider war effort and about 25 percent of airstrikes in Afghanistan have been conducted by drones.[14]

Drone warfare can be broadly characterized as falling into two categories. In **mixed drone warfare** drones constitute part of a broader array of air power, they act in concert with U.S. ground forces, whom they may be protecting against ambush, and attacks are often directed by ground controllers in the thick of combat as much as by detached intelligence analysts on the other side of the world. This is the way drones have been deployed in Iraq and Afghanistan, where they have most commonly been under the control of the U.S. military operating under the international laws of war in what are broadly recognized as war zones. British drones have sometimes joined in these attacks, and other NATO allies have participated in the command structure that directs the strikes. On the other hand, drone strikes in Yemen, Somalia and Pakistan take the form of **pure drone warfare.** Here drones operate outside generally recognized warzones and hunt their prey alone, or in conjunction with small networks of spies or special forces on the ground who help develop targets. Although the leaders of Yemen and Pakistan gave quiet permission for many of these strikes, they denied that they had done so and even, on occasion, publicly condemned them.[15] Because these strikes are directed against countries with which the U.S. is not formally at war, their legal status is contested, and the United States' NATO allies have generally declined to get involved. These strikes have usually been under the control of the CIA, whose rules of engagement have not been publicly disclosed and are the subject of some public controversy.[16]

Regardless of whether they are used in mixed or pure drone warfare, drones are operated through a dual infrastructure of bases and pilots. For most of the time the planes are in the air, they are controlled by three-person teams of operators at U.S. military bases, especially at Creech Air Force based in Nevada. (These operators are, however, just one node in a network of intelligence analysts, military personnel and even military lawyers that may be distributed across the United States, even in different countries, constantly communicating in real time about the imagery and other information generated by an airborne drone). But the planes themselves are physically headquartered at bases near their area of aerial operations. Crews at these bases maintain and fuel the planes. And, since "the two-second delay between a pilot moving a joystick in Nevada and an aircraft responding in Afghanistan is enough to cause a crash during take-off and landing," takeoffs and landings are handled by operators at these local bases, who hand control of the aircraft over to U.S. based crews after half an hour, then resume control before landing.[17] Countries in which the U.S. is known to have based drones include Afghanistan, Iraq, Djibouti, Pakistan, Italy, Turkey, Mauritania, Qatar, the Seychelles, Tunisia, the United Arab Emirates, Burkina Faso, Chad, Ethiopia,Kuwait, Niger, Somalia, Saudi Arabia, and Uzbekistan.[18]

OPERATIONAL PROTOCOLS AND CIVILIAN CASUALTIES

Both critics and defenders of drones err in speaking of them as if they have an essential, inherent character when it comes to civilian casualties. Critics often represent drone strikes as inevitably producing large numbers of civilian deaths, while drone defenders often speak as if drones' potential to strike with high levels of precision and discrimination is invariably realized in practice. In fact, the degree to which drone strikes kill insurgents and spare civilians is highly dependent upon the targeting protocols employed, and these can be complex and variable. These operational protocols vary between the U.S. military and the CIA, and they have varied over time as well, becoming stricter toward the end of the second Obama Administration in response to a number of incidents in which several civilians were killed.

In 2013 CIA director John Brennan said, "We only authorize a particular operation against a specific individual if we have high confidence that the individual being targeted is indeed the terrorist we are pursuing… This is a very high bar. Of course, how we identify an individual naturally involves intelligence sources and methods, which I will not discuss."[19] This gives the impression that all U.S. drone strikes are what are known as **personality strikes:** strikes against individual leaders whose records have been vetted before they were placed on a death list, and who are then matched in real time with particular individuals on the ground in the cross-hairs of a drone. But, thanks to leaks to journalists, we now know that most drone strikes are not personality strikes, but **signature strikes** (a term that was classified until it was widely reported in the media). In signature strikes, people are targeted not because they are a confirmed individual on an official target list, but because they exhibit a behavioral signature that has become associated with insurgents or that looks suspicious to a drone operator watching through a camera several thousand feet in the air. Any "military aged male" can be an object of suspicion, especially if he appears armed (although civilians often carry guns in Afghanistan and Pakistan), if he appears to be digging, or if he is in a meeting with several other military aged males. According to one source men in Afghanistan or Pakistan observed to urinate while squatting rather than standing up may be killed because drone operators associate this behavior with Arab men, who are more likely to be "foreign fighters" for al Qaeda.[20]

Some behavioral signatures are quite simple to interpret: for example, someone observed firing on U.S. soldiers is clearly a hostile combatant, whether their identity is known or not. But the potential for misinterpretation, especially when discerning the intent of people from a different culture from the other side of the world, is demonstrated in a tragic story told, following a successful Freedom of Information Act request, by *Los Angeles Times* reporter David Cloud. In February 2010 a group of Afghan civilians had set out on a long trip before dawn, driving in convoy. They included shopkeepers buying supplies, students returning to school, and villagers visiting relatives or seeking medical treatment. Their convoy had the misfortune to be driving in the vicinity of a U.S. special forces team when radio intercepts suggested insurgents might be planning to attack them. A Predator drone was overhead, monitoring the area for insurgents. When a drone operator in Nevada first spotted the Afghans on the ground, they seemed to be kneeling on blankets. "Praying? I mean, seriously, that's what they [i.e. insurgents] do," said one drone operator. Another member of the team thought he caught the glint of a gun barrel in one of the pickup trucks, though the team was not completely sure about this. When one driver on the ground flashed his headlights at another, the pilot of an AC-130 above radioed that the two vehicles were, as one would expect of insurgents, "signaling." Assumptions (that turned out to be wrong) that the people were all "military age males" also excited suspicion. One person involved in the incident would later say, "we all had it in our head, 'Hey, why do you have 20 military age males at 5 a.m. collecting each other?' There can be only one reason." The drone struck and the U.S. now concedes that it killed at least fifteen civilians. According to Afghan villagers, it killed 23 civilians including three and four year-old children.[21]

The likelihood of civilian casualties is still higher in **double-tap strikes** – a controversial practice reportedly employed on occasion by the CIA in which drone operators strike those rescuing the victims of a drone strike, or attending their funeral, on the assumption that people rescuing or mourning insurgents are likely themselves to be insurgents.[22] Of course, they may also be good Samaritans or relatives of insurgents who are not themselves insurgents. In any case, the employment of "double-tap strikes" has now reportedly caused behavioral changes in Yemen and Waziristan, where bystanders often wait before pulling people out of the wreckage left by a drone strike.

In Waziristan, a "tribal area" of Pakistan on the border with Afghanistan that is the size of Rhode Island and has absorbed roughly 400 drone strikes, the CIA also relies on local paid informants to help identify and locate insurgent leaders according to investigative journalists.[23] Former senior State Department official Lawrence Wilkerson has claimed that bounties paid to such informants can go as high as $5,000 – several years' income in Waziristan.[24] While it is clear that CIA drone strikes in Waziristan have killed a large number of insurgents, there is also cause for concern here. The Pakistani-American anthropologist Akbar Ahmed points out that local informants tempted by large sums of money may finger innocents as a way to get the bounty, or use CIA drones as an opportunity to settle feuds, getting rich and eliminating the neighbor with whom they've been feuding for a decade in one blow.[25] This is not implausible since we now know that a number of detainees sent to Guantanamo on the basis of tips from local informants in Afghanistan turned out to be innocent victims of local disputes[26].

And, to return to where we began, what of "personality strikes" – those gold standard strikes on confirmed, identified insurgent leaders? One might assume that the error rate in such strikes – attacks on identified insurgent leaders on a named list of targets – would be low. However, the human rights organization Reprieve tracked official U.S. announcements of successful kills and found forty-one cases where the same "high-value" insurgent leader was declared dead in more than one drone strike. The report observes scathingly that these men "seemed to have achieved the impossible: to have 'died' in public reporting not just once, not just twice, but again and again. Reports indicate that each assassination target 'died' on average more than three times before their actual death."[27]

This begs the question of how drone operators, in a situation where their cameras lack the resolution to identify individual faces, come to believe that a person on the ground is a match for someone on their "kill list." When they do not rely on local informants, drone operators often identify insurgent leaders by tracking their cell phones. Sometimes this can be remarkably successful as in the first U.S. drone strike in Pakistan, in 2004, which killed the Pakistani Taliban leader Nek Mohammad, shortly after he gave away his location by giving an interview to the BBC on his cell phone.[28] But people may borrow one another's cell phones; moreover, insurgent leaders soon learned that they could be tracked through their cell phones, so many began to switch sim cards, even giving sim cards and cell phones they believed to have been compromised to children and other civilians in the expectation that their deaths would discredit the United States in the propaganda war. There have been disputes within the NATO ocommand structure over the strictness of protocols for confirming that a cell phone known to have been used by an insurgent in the past is at this moment in the hands of the same insurgent, with German commanders complaining that American protocols of verification are too permissive.[29]

It is hard to estimate accurately the number of civilians killed in drone strikes. Reports from journalists or human rights activists in Pakistan and Afghanistan may be exaggerated, but official U.S. assessments (done largely by the drone operators themselves) have their own liabilities: drone operators may not know how many civilians lie inside collapsed buildings and, in a war where insurgents do not wear uniforms, they have used a flawed methodology that "counts all military-aged males in a strike zone as combatants… unless there is explicit intelligence posthumously proving them innocent."[30] There is a further problem in that it may be wrong to speak of people as if they either are or are not insurgents, or as if their actions bespeak an ideological commitment to the insurgent cause. We know from journalists who have gained access to insurgents that people may work for the insurgents one month, but not the next, doing tasks for local insurgents if directed to do so by a clan leader with shifting allegiances, or if they need the money they will be paid for planting an IED. In other cases, insurgents may offer some local men a choice between joining them or being beheaded. As journalist Steve Coll says, "a young man of military age holding a gun outside a *hujra* might be a motivated Taliban volunteer, a reluctant conscript, or a victim of violent coercion."[31]

We should bear these caveats in mind as we assess varying estimates of civilian casualties from drone strikes. In 2016, the Obama Administration finally released an official estimate that drone strikes in Somalia, Yemen, Pakistan, and Libya had killed between 64 and 116 civilians, though this claim was met with widespread skepticism by journalists who had been covering the drone campaign.[32] (And it bears noting that the Obama estimate excludes Afghanistan, where over three quarters of U.S. drone strikes have taken place, as well as Iraq and Syria). Independent think tanks that track drone attacks put the numbers of civilian dead considerably higher. The New America Foundation (which only counts casualties confirmed by two independent and reputable media sources) estimated 376 to 474 civilian dead in Pakistan, Yemen and Somalia.[33] The Bureau of Investigative Journalism (which draws on a wider array of sources, including field investigations) estimates civilians dead at between 301 and 780 – a number over six times higher than the White House's.[34]

Whichever number one prefers, it seems clear that the proportion of civilian casualties declined in the last two years of the Obama Administration as targeting protocols were changed in response to drone strikes that had clearly killed significant numbers of civilians. Quite apart from humanitarian considerations, U.S. national security officials became concerned that popular anger arising from the killing of innocent civilians would make it easier for insurgents to enlist more recruits. The shift from Predators to Reapers helped here, since Reapers carry a wider array of firepower, including smaller ordinance that produces a smaller radius of destruction on the ground. Drone operators were also instructed to favor attacks on vehicles over attacks on houses, where unseen innocents were more likely to die. And stricter standards were adopted in confirming the identity of high-value targets.[35]

LEGAL QUESTIONS

The legal issues surrounding the use of drones on regular battlefields in what I am calling "mixed drone warfare" are relatively simple: here drone operators must follow the same general laws of war – not deliberately targeting civilians, making sure civilian deaths are proportionate to legitimate military objectives – that would guide the pilots of F-16s or Apache helicopters, for example. (In this regard, "double-tap" strikes, targeting with a follow-up missile those who rescue the victims of a drone strike, have proved particularly controversial. UN Special Rapporteur Christof Heyns has concluded that they constitute "a war crime in armed conflict and a violation of the right to life."[36]) The more complex issues arise in connection with "pure drone warfare" – blue-sky attacks outside a clearly defined theater of battle against the territory of countries with which the U.S. is not in a formally declared relation of war. These principal countries in question here are Yemen, Somalia, and the "tribal areas" of Pakistan. A number of international jurists, NGOs, and some American commentators (both Republican and Democrat) have claimed that drone strikes in Yemen, Somalia and Pakistan contravene international and domestic law, though these claims were vigorously disputed by the Obama Administration. The legal issues at stake can be grouped into two categories: those pertaining to international law and those pertaining to U.S. domestic law.

According to the UN charter, nations can legitimately use force on the territory of other nation-states only if they are authorized to do so by the United Nations, if they have been attacked, if they are acting to preempt an "imminent" attack on themselves, or if they have the consent of the other state to do so. Drone strikes in Yemen, Somalia and Pakistan have clearly not been authorized by the United Nations, and drone strikes in Somalia have been carried out in the absence of permission from any governmental entity there, but the Yemeni and Pakistani cases are more complicated. There is ample evidence that President Saleh of Yemen and Prime Minister Musharraf of Pakistan made deals in which they gave permission for U.S. drone strikes, at least in many cases, as long as this permission was not publicly advertised. (In Pakistan the U.S. was supposed to notify the Pakistani government of drone strikes ahead of time, but then started striking without notification as it grew more frustrated with Pakistan's intelligence services, which it believed to be collaborating with insurgents).[37] In the Pakistani case, at least, this raises the question of whether the Prime Minister had the legal ability to permit such strikes on his own authority since, in 2012, both Houses of Parliament in Pakistan voted to demand an immediate end to the drone strikes. There is also the question of whether Pakistan's government could authorize strikes on the "tribal areas," which were semi-autonomous regions. This led UN special rapporteur Ben Emmerson to conclude that, regardless of the continued cooperation of Pakistan's military and intelligence institutions, as a matter of law, "the continued use of remotely piloted aircraft in the Federally Administered Tribal Areas amounts to a violation of Pakistani sovereignty, unless justified under the international law principle of self-defence."[38]

International law also allows nations to use force to defend themselves against an imminent threat. In a much quoted phrase that has survived from a celebrated legal case in 1842, the threat of impending attack must be "instant, overwhelming, and leaving no choice of means, and no moment of deliberation."[39] For most lawyers, the paradigmatic example of this would be troops massing across the border, obviously preparing to attack. The U.S. has taken the position, unpersuasive to many international lawyers, that the existence of al Qaeda by definition constitutes an imminent threat and that there is therefore no need to demonstrate an imminent threat from the particular individuals or groups of individuals targeted by drones, even if their actions seem focused on local objectives.[40]

For the use of military force to be defensible under international law, not only must an attack be in progress or imminent, but it must be undertaken by an organization recognizable as a military force. (Thus an attack on a US embassy by a spontaneous mob would not constitute cause for war). UN special rapporteur Ben Emmerson defines an organized armed group as one that has "at least a common command structure, adequate communications, joint mission planning and execution, and cooperation in the acquisition and distribution of weaponry."[41] Emmerson, together with UN special rapporteurs Philip Alston and Christof Heynes, finds U.S. drone strikes illegal under international law on the basis of this criterion. This enunciation of international law is grounded in what we might call the "classic" vision of war as a contest between regular armies whose members are full-time soldiers. This definition of legitimate targets is harder to square with the practice of guerilla war where fighters do not wear uniforms, and they may drift in and out of an insurgency, taking up fighting as an occasional or part-time vocation. Guerilla war also is often practiced by decentralized networks of loosely affiliated groups rather than an organization coordinated by a military bureaucracy, leading some U.S. officials to claim that international law has been rendered obsolete in this regard and that it is unreasonable to hold the U.S. to an obsolete standard.

In terms of U.S. domestic law, the White House did not request Congressional approval under the War Powers Act to conduct military operations in Somalia, Pakistan or Yemen, but they claimed that such action was already authorized through the Congressional Resolution on the Authorization for the Use of Military Force (AUMF) issued immediately after the 9/11 attacks. The AUMF authorized the President "[t]o use all necessary and appropriate force against those nations, organizations, or persons he determines planned, authorized, committed, or aided the terrorist attacks that occurred on September 11, 2001, or harbored such organizations or persons, in order to prevent any future acts of international terrorism against the United States by such nations, organizations or persons."[42] While successive U.S. administrations have seen in this language a blank check to conduct military operations anywhere in the Middle East, others have questioned how this can be taken to allow drone attack on Colonel Ghadaffi's Libyan regime and on ISIS – both avowed enemies of al Qaeda.[43]

There has also been controversy over the use of drones to kill U.S. citizens without a judicial procedure in which they can question the evidence against them. To date the following U.S. citizens have been killed in drone strikes: Kemal Derwish, Anwar al-Awlaki, Samir Khan, Abdulrahman al-Awlaki (who was 16 at the time of his death), Jude Kenan Mohammad, and Warren Weinstein. While most of these deaths appear to have been accidental, a number of legal commentators in the U.S. were upset by the targeted killing of a U.S. citizen, Anwar al-Awlaki, on September 30, 2011.[44] Although there can be little doubt that al-Awlaki had made al Qaeda's cause his own, his government killed him without proving in court that his acts went beyond anti-American speech that might, despite its odious content, be protected by the First Amendment. Al-Awlaki preached vitriolic sermons against the U.S., helped produce the al Qaeda magazine *Inspire*, and was associated with a number of people who had carried out, or attempted, attacks on American targets: two of the 9/11 hijackers, the U.S. army psychiatrist (Nidal Malik Hasan) who shot thirteen people at the Fort Hood military base, and the Nigerian "underwear bomber" (Umar Farouk Abdulmutallab) who attempted to destroy a U.S. jetliner on Christmas day 2009. The U.S. claimed that al-Awlaki had become an "operational commander" of al Qaeda, though it has never made public its evidence for this claim, and that no judicial process is required to kill an American who is fighting with the enemy on the field of battle.

CONCLUSION: LOOKING TO THE FUTURE

On the one hand, drones offer the possibility of a more discriminate conduct of warfare, sparing the lives of innocent civilians, because of the deliberateness and precision with which they can strike. On the other hand, because they can be deployed semi-secretly and can kill people from the air in countries where the U.S. has no ground forces, they tempt presidents to use military force at or beyond the legal limits of international and domestic law, and to use force in situations where their predecessors would have been deterred by fear of casualties. (One thinks of President Reagan withdrawing U.S. troops from Beirut in 1983 and President Clinton withdrawing from Somalia after nineteen US servicemen died there in 1993). In the absence of any international norms or treaties regulating the use of drones, the George W. Bush and Obama Administrations have established a permissive template for their use that future administrations may regret if other countries similarly declare the right to attack countries with which they are not formally at war on the basis of rules of engagement that are closely held. In this context it bears mentioning that, according to the GAO, 76 other countries already have some sort of drone capability.[45] Meanwhile sub-state organizations such as Hezbollah and ISIS have already used crude tactical drones in their own military operations.

Looking to the future and extrapolating current political and technological trends, it is reasonable to expect (unless we take action to prevent it) that still more countries will acquire drones, and terrorists will start to use simple store-bought drones equipped with guns, explosives or chainsaws[46] for terrorist attacks within the United States and elsewhere while the U.S. military develops lethal micro drones disguised as birds or even insects whose use may plausibly be denied, as well as smart drones that can act in autonomous swarms without human direction. If autonomous smart drones are developed, they will raise the question of who or what is legally and morally responsible if a war crime is committed.[47]

We are only beginning to think about what drone arms control might look like, but here are some measures that would surely be worth debating: a ban on autonomous drones of the kind advocated by the International Committee on Robot Arms Control (ICRAC);[48] an explicit ban on double-tap strikes; a requirement that drone strikes be confined to recognized war zones or to countries that have given public, written permission for such strikes to go forward on their territory; international codification of targeting protocols in terms of best practices; a ban on police use of lethal (as opposed to surveillance) drones for law enforcement purposes; and a requirement that all video footage and operator communications leading up to a drone strike be archived and deposited with an international authority (modelled on the IAEA) that could adjudicate allegations of war crimes. Some will say that it is fanciful to think that any such measures could be feasible, but the alternative is an anarchic mess of practices that may act as a solvent on the international laws of war. In the cases of the Comprehensive Test Ban Treaty, the Nuclear Non-Proliferation Treaty, the Landmine Treaty, and the Convention on Cluster Munitions, bold multilateral arms control treaties have proven possible, even if they do not obtain completely universal adherence, and they have established new international norms in the process. What country would be in a better position to lead the way to an international convention on drone warfare than the one with the most knowledge and experience in this area, the United States?

---

[1] This article draws in parts on Hugh Gusterson, *Drone: Remote Control Warfare*, Cambridge, MA: MIT Press, 2016.

[2] https://www.defense.gov/News/Article/Article/1095612/air-force-to-retire-mq-1-predator-drone-transition-to-mq-9-reaper/

[3] Quoted in Gregoire Chamayou, *A Theory of the Drone*, New York: New Press, 2015, p.12

[4] Lone Wolf Media, *The Bomb*, PBS 2015. http://www.pbs.org/program/bomb/

[5] Adam Rothstein, *Drone,* London: Bloomsbury, 2015, pp.28-30.

[6] Gregoire Chamayou, *Theory of the Drone*, p.27; Rothstein, *Drone*, p.31.

[7] The Economist, "The Dronefather," *Economist*, December 1, 2012 http://www.economist.com/news/technology-quarterly/21567205-abe-karem-created-robotic-plane-transformed-way-modern-warfare

[8] Craig Whitlock, "More Air Force Drones are Crashing than ever as Mysterious New Problems Emerge." *Washington Post*, January 20, 2016. https://www.washingtonpost.com/news/checkpoint/wp/2016/01/19/more-u-s-military-drones-are-crashing-than-ever-as-new-problems-emerge/?utm_term=.14b1f565369c. For a histry of the Predator, see Richard Whittle, *Predator: The Secret Origins of the Drone Revolution*, New York: Henry Holt, 2014.

[9] Daniel Klaidman, *Kill or Capture: The War on Terror and the Soul of the Obama Presidency*, Boston: Houghton Mifflin Harcourt, 2012.

[10] Jane Mayer, "The Predator War," *New Yorker*, October 26, 2009, http://www.newyorker.com/magazine/2009/10/26/the-predator-war

[11] Brian Glynn Williams, *Predators: The CIA's Drone War on al Qaeda,* Potomac Books, 2013, pp.24-5; National Commission on Terrorist Attacks, *9/11 Commission Report*, 2004, pp.210-14. http://911.gnu-designs.com/

[12] Mark Mazzetti, "A Secret Deal on Drones, Sealed in Blood." *New York Times,* April 6, 2013, p.A1; Mark Mazzetti, *The Way of the Knife: The CIA, A Secret Army, and a War at the Ends of the Earth.* New York: Penguin, 2013.

[13] For discussion of drone strikes against insurgent training camps in the Philippines in 2006 and 2012, see Akbar Ahmed and Frankie Martin, "Deadly drone strikes on muslims in the Southern Philippines," Brookings Institute, March 5, 2012, http://www.brookings.edu/research/opinions/2012/03/05-drones-philippines-ahmed

[14] Alice Ross, "Who is dying in Afghanistan's 1,000-plus drone strikes?" July 24, 2014. https://www.thebureauinvestigates.com/2014/07/24/who-is-dying-in-afghanistans-1000-plus-drone-strikes/

[15] Klaidman, *Kill or Capture;* and Mazzetti, *The Way of the Knife.*

[16] The distinction between mixed and pure drone warfare is developed in greater detail in Hugh Gusterson, *Drone.*

[17] Bob Blackhurst, "The Air Force Men Who Fly Drones in Afghanistan by Remote Control," *Daily Telegraph*, September 24, 2012, http://www.telegraph.co.uk/news/uknews/defence/9552547/The-air-force-men-who-fly-drones-in-Afghanistan-by-remote-control.html

[18] Micah Zenko and Emma Jones, "Where the Drones Are," *Foreign Policy*, May 29, 2012, http://foreignpolicy.com/2012/05/29/where-the-drones-are/; Adam Entous and Missy Ryan, "U.S. Has Secretly Expanded its Network of Drone Bases to North Africa," *Washington Post*, October 26, 2016, https://www.washingtonpost.com/world/national-security/us-has-secretly-expanded-its-global-network-of-drone-bases-to-north-africa/2016/10/26/ff19633c-9b7d-11e6-9980-50913d68eacb_story.html?utm_term=.1b81fd7a90dd; https://en.wikipedia.org/wiki/List_of_United_States_drone_bases

[19] Charlie Savage, "Top U.S. Security Official Says 'Rigorous Standards' are Used for Drone Strikes," *New York Times*, April 30, 2012, http://www.nytimes.com/2012/05/01/world/obamas-counterterrorism-aide-defends-drone-strikes.html

[20] Andrew Cockburn, *Kill Chain: The Rise of the High-Tech Assassins*, New York: Henry Holt, 2015, p.224.

[21] David S. Cloud, "Anatomy of an Afghan War Tragedy," *Los Angeles Times*, April 10, 2011. This incident is also analyzed in detail in chapter 1 of Andrew Cockburn, *Kill Chain: The Rise of the High-tech Assassins,* NY: Henry Holt, 2015.

[22] Conor Friedersdorf, "drone attacks at funerals of people killed in drone attacks," *The Atlantic* October 24,2013, http://www.theatlantic.com/international/archive/2013/10/drone-attacks-at-funerals-of-people-killed-in-drone-strikes/280821/.

[23] Klaidman, *Kill or Capture;* Mazzetti, *The Way of the Knife.*

[24] AP. 2009. Ex-Bush official: Many at Guantanamo Bay are Innocent. *FoxNews.com*, March 19. http://www.foxnews.com/politics/2009/03/19/ex-bush-official-guantanamo-bay-innocent.html

[25] Akbar Ahmed, *The Thistle and the Drone: How America's War on Terror Became a Global War on Tribal Islam,* 2013, Washington DC: Brookings Institution Press.

[26] Sarah Holewinski, "Just Trust us: The Need to Know More About the Civilian Impact of Drone Strikes," in Peter Bergen and Daniel Rothenberg (eds.) *Drone Wars: Transforming Conflict, Law and Policy*, New York: Cambridge University Press, 2014, pp.42-70; Anand Gopal, 2014. *No Good Men Among the Living: America, The Taliban and the War Through Afghan Eyes.* New York: Metropolitan Books.

[27] Reprieve. 2014. You Never Die Twice: Multiple Kills in the US Drone Program. http://www.reprieve.org/wp-content/uploads/2014_11_24_PUB-You-Never-Die-Twice-Multiple-Kills-in-the-US-Drone-Program-1.pdf

[28] Mark Mazzetti, *The Way of the Knife,* pp.109-110.

[29] Der Spiegel. 2014. Obama's Lists: A Dubious History of Targeted Killings in Afghanistan. *Der Spiegel* December 28. http://www.spiegel.de/international/world/secret-docs-reveal-dubious-details-of-targeted-killings-in-afghanistan-a-1010358.html

[30] Jo Becker and Scott Shane, "Obama's Secret 'Kill List'." *New York Times* May 29, 2012.

[31] Steven Coll, "The Unblinking Stare," *The New Yorker*, November 24, 2014.

[32] See, for example, Karen DeYoung and Greg Miller. "White House Releases its Count of Civilian Deaths in Counterterrorism Operations." *Washington Post*, July 1, 2016.

[33] http://securitydata.newamerica.net/drones/methodology.html

[34] https://www.thebureauinvestigates.com/stories/2016-07-01/obama-drone-casualty-numbers-a-fraction-of-those-recorded-by-the-bureau This estimate, unlike the New America Foundation estimate, includes Libya. The Bureau's methodology is described in detail at https://www.thebureauinvestigates.com/explainers/our-methodology.

[35] Coll, "The Unblinking Stare"; Steven Coll, "Warren Weinstein and the long drone war," *New Yorker,* April 23, 2015, http://www.newyorker.com/news/news-desk/warren-weinstein-obama-drone-war

[36] Quoted in Chris Woods, *Sudden Justice: America's Secret Drone Wars*, New York: Oxford University Press, 2015, p.161. The report in question is Christof Heyns, *Report of the Special Rapporteur on Extrajudicial, Summary or Arbitrary Executions,* 2013.

[37] Mazzetti, *The Way of the Knife.*

[38] Ben Emmerson, *Report of the Special Rapporteur on the promotion and protection of human rights and fundamental freedoms while countering terrorism*, 2014 p.15 (accessed via http://www.ohchr.org/EN/Issues/Terrorism/Pages/SRTerrorismIndex.aspx)

[39] The case is discussed at http://avalon.law.yale.edu/19th_century/br-1842d.asp

[40] NBC News, "Justice Department memo reveals legal case for drone strikes on Americans," February 4, 2013, http://investigations.nbcnews.com/_news/2013/02/04/16843014-justice-department-memo-reveals-legal-case-for-drone-strikes-on-americans; Jeffrey Rosen, "Drone strike out," *New Republic* February 6, 2013.

[41] Ben Emmerson, *Report*, p.18, 20.

[42] Authorization for Use of Military Force, 115 Stat 224 (2001).

[43] Jack Goldsmith, "Obama's breathtaking expansion of a president's power to make war," *Time* 9/11/14 http://time.com/3326689/obama-isis-war-powers-bush/; Charlie Savage and Mark Landler, "War Powers Act Doesn't Apply for Libya, Obama says" *New York Times*, June 15, 2011

[44] See David Cole, "President Obama, did you or did you not kill Anwar al-Awlaki?" *Washington Post*, February 8, 2013.

[45] http://www.gao.gov/products/GAO-12-536

[46] See https://www.youtube.com/watch?v=6Viwwetf0gU

[47] Peter Singer, *Wired for War: The Robotics Revolution and Conflict in the 21st Century*, New York: Penguin Books, 2009.

[48] https://icrac.net/call/

Appendix A

WMD Chronology

David Hafemeister

Physics Department
California Polytechnic State University
San Luis Obispo, California

1763

—July 13: British give blankets with small-pox to Indians that attacked Fort Pitt.

1909

—Ernest Rutherford, Hans Geiger and Ernest Marsden discover the very small size of nuclei by scattering alphas particles from a gold foil.

1915

—April 22: Germans use chlorine gas weapons at Ypres.

1919

—June: Soon after World War I, Ernest Rutherford creates nuclei (^{4}He + ^{14}N => ^{1}H + ^{17}O). "Talk softly please. I have been engaged in experiments, which suggest that the atom can be artificially disintegrated. If it is true, it is of far greater importance than a war."

1920

—Rutherford speculates on existence of the neutron at the Royal Society of London.

1931

—November: Harold Urey discovers deuterium in hydrogen's optical spectra.

1932

—February 27: James Chadwick observes knock-on protons to discover the neutron. (^{4}He + ^{9}Be => ^{12}C + ^{1}n)

1933

—January 28: Adolph Hitler becomes Chancellor of Germany.

—March 4: Franklin Roosevelt becomes President of the United States.

—April: Eleven scientists who had won the Nobel Prize or would win it later were fired from German Universities because they were Jewish. Johannes Stark's and Philip Lenard's *Aryan physics* is not accepted by physicists.

—October: Leo Szilard recollects, "It occurred to me in October 1933, that a chain reaction might be set up if an element could be found that would emit two neutrons when it swallowed one neutron." This idea becomes a classified British patent in 1935 before uranium fission is discovered.

1934

—Artificial radioactivity discovered by Irene Curie and Frederic Joliet (alpha particles) and by Enrico Fermi (neutrons).

—June 30: Adolph Hitler leads assassination of storm-trooper rival, Ernst Rohm and others, army taken over.

Nuclear Weapons and Related Security Issues
AIP Conf. Proc. 1898, 060001-1–060001-14; https://doi.org/10.1063/1.5009235
Published by AIP Publishing. 978-0-7354-1586-7/$30.00

1938

—December: Fermi receives Nobel prize for physics for discovery of transuranic elements (actually fission of uranium) and departs for the "new world" of the United States.

—December 22: Otto Hahn and Fritz Strassman identify barium isotopes and a few days later, Lise Meitner and Otto Frisch conclude that uranium nuclei fission by neutron capture.

1939

—January to May: Many experiments on uranium fission.

—April 29: Conference in Berlin to consider a German bomb and nuclear reactor.

—August 2: Szilard and Edward Teller obtain a letter from Einstein on the possibility of uranium weapons, which Roosevelt receives from Alexander Sachs on October 11, 1939.

—September 1: Hitler invades Poland. US Ambassador to Germany William Dodd (1933-37) writes Roosevelt (Sept.18), "If the [democracies] had co-operated, they would have succeeded. Now it is too late."

1940

—June 3: Paul Hartek fails to produce neutron multiplication in Hamburg reactor (185 kg natural UO_2, 15 tonnes CO_2 ice).

1941

—January: Experiments at the Berlin "virus house," using a 300-kg uranium reactor with impure graphite, cause incorrect rejection of graphite as a moderator in favor of heavy water from Vermork, Norway.

—July: British "Maud" Committee reports that a weapon could be made with 10 kg ^{235}U; US National Academy Sciences endorses bomb program.

—August: Hamburg group begins construction of ultracentrifuges to obtain ^{235}U. Centrifuges explode in April 1942, but they attained 7% enrichment by March, 1943.

—October: Niels Bohr and Werner Heisenberg indirectly discuss nuclear weapons, described in Michael Frayn's play *Copenhagen*.

—December 6: Roosevelt directs substantial financial and technical resources to construct uranium bombs.

—December 7: Japan attacks Pearl Harbor.

1942

—May: Germans observe first neutron multiplication (13%) in Leipzig using 570 kg U and 140 kg heavy ice.

—October 2: First of 3700 German–V2 flights.

—December 2: First nuclear chain reaction at Chicago's Stagg Field by Fermi.

1943

—February 28: Vermork heavy water factory destroyed by Allies after a failed attack on November 19, 1942.

—March 15: J. Robert Oppenheimer moves to Los Alamos, New Mexico.

—March: Glenn Seaborg suggests Pu weapons might be jeopardized by ^{240}Pu as a spontaneous neutron emitter.

—Resumption of Soviet nuclear experiments.

1944

—Enigma Code decrypted at Bletchly Park, UK.

—August: Bohr presents memorandum on international control of nuclear weapons to British Prime Minister Winston Churchill and Roosevelt.

—November: First batch of spent fuel obtained from Hanford reactors.

—November: Samuel Goudsmit's Alsos mission obtains documents in Strasbourg, which imply that German weapons program made little progress.

1945

—January: First plutonium reprocessing at Hanford, Washington.

—January 20: First ^{235}U separated at the K25 gaseous diffusion plant, Oak Ridge, Tennessee.

—April 25: UN Charter signed by 50 nations in San Francisco.

—May 4: End of war in Europe.

—June 11: The Franck Report on the demonstration of the bomb and its international control sent to US Secretary of War.

—July 16: US explodes first atomic bomb, 19-kt *Trinity*, at Alamogordo, NM with electronics shielded to avoid EMP pulse, 300–m crater.
—August 6, 9: Atomic bombs destroy Hiroshima (Thin Man uranium, 13 kton, 3 m by 0.7 m diameter) and Nagasaki (Fat Man plutonium, 22 kton, 3.3 m by 1.5 m). Each weapon kills 70,000 prompt deaths and 140,000 within a year.
—August 15: End of war in the Pacific.

1946
—June 14: Bernard Baruch presents the Acheson–Lilienthal plan to the UN to internationalize the atom. "We are here to make a choice between the quick and the dead."
—June 30: First underwater detonation on ships by US at Bikini Atoll.
—July: Demonstrations in Times Square, New York City, against nuclear testing.
—December 31: Atomic Energy Commission takes over nuclear weapons program from US Army.

1948
—April, May: US atomic tests, Eniwetok Atoll in the Marshall Islands.
—November 4, 1948: UN General Assembly adopts US plan for international atomic control; USSR opposed.

1949
—April 4: NATO established.
—August 29: First Soviet atomic test, 22 kton of a US design at Semipalatinsk.
—October 30: AEC General Advisory Committee gives approval to build more powerful U bombs rather than the H bomb.

1950
—January 27: Klaus Fuchs confesses he transmitted atomic secrets to the Soviets.
—January 31: US President Harry Truman announces the decision to proceed with H bombs.
—March: Worldwide peace offensive to "ban the bomb" (the Stockholm Appeal) signed by more than 500 million people.
—June 25: North Korean army crosses 38th parallel, starting a war that lasted until July 1953.

1951
—January 21: First of 928 US tests at Nevada Test Site.
—May 9: *George,* DT–boosted atomic bomb, 225 kton.

1952
—January: UN Security Council establishes Disarmament Committee.
—June: Lawrence Livermore Laboratory established.
—October 3: First of 12 British nuclear tests at Monte Bello Islands, W. Australia.
—November 1: US explodes first fusion device *Mike*, 10.4 Mtons at Eniwetok (liquid deuterium, not deliverable).

1953
—April 18: *Badger* test in front of 2800 troops at Nevada Test Site.
—May: Utah officials ask AEC to stop Nevada tests because of fallout on St. George, Utah.
—August 12: First Soviet fusion device exploded on a tower in Siberia (probably not deliverable).
—Fall: India and Australia propose a total ban on nuclear weapons to the UN.
—December 8: "Atoms for Peace" speech by US President Dwight Eisenhower at UN.

1954
—January 21: First nuclear-powered submarine, USS Nautilus, launched.
—March 1: Castle–Bravo, 15–Mton H bomb test, affects Marshall Islanders and Lucky Dragon.
—April: British Parliament petitions Churchill, Eisenhower, and Malenkov to control nuclear weapons.
—April 12 to May 6: AEC hearings deny Oppenheimer his access to nuclear secrets.
—August 30: Atomic Energy Act of 1954 emphasizes peaceful uses of atomic energy.

1955

—January-December: Reports of increasing global nuclear fallout.
—May 6: West Germany joins NATO.
—May 14: Warsaw Pact Organization established.
—August 8–20: First International Conference on Peaceful Uses of Atomic Energy in Geneva.
—November 22: First deliverable, high yield, Soviet H bomb, 1.6 Mton, Semipalatinsk.

1956
—US National Academy of Sciences panel finds genetic effects from fallout of nuclear tests to be slight compared with effects from natural radiation background. Adlai Stevenson makes fallout an issue in US presidential campaign.

1957
—May 15: First British H bomb exploded of 0.3 Mton at Christmas Island in the Pacific Ocean.
—July 6–11: First Pugwash Conference advocates test ban; Soviet scientists attend.
—August: AEC inaugurates Plowshare Program for peaceful nuclear explosions.
—September 19: Rainier, the first underground test at 1.7 kilotons at Nevada Test Site.
—September 29: Soviets evacuate 10,000 from Mayak reprocessing plant explosion. US keeps the secret.
—October 1: International Atomic Energy Agency inaugurated in Vienna.
—October 4: First satellite Sputnik I put into orbit by USSR.
—December 2: Shippingport electric power reactor reaches full power of 60 MW.

1958
—January 1: European Atomic Energy Community (Euratom) established.
—January 31: Explorer I, first US satellite, put into orbit.
—November1958–September 1961: US, UK, USSR observe informal moratorium on nuclear tests.

1959
—September 2: Vela Uniform Seismic Project established by US Department of Defense.
—November 24: US/USSR sign a memorandum of cooperation for utilization of nuclear energy.

1960
—February 13: First nuclear test by France, 70 kton, in the Sahara Desert.
—May 1: U-2 spy plane shot down over USSR.
—July 26: USSR suggests 3 on-site inspections/year as part of a test ban.
—November 15: First Polaris missile launched from a US submarine.

1961
—January 17: Eisenhower's farewell address discourses on *the military-industrial complex*.
—February 1: US launches Minuteman-I missile.
—March 31: Pravda suggests the use of H–bombs to obtain fresh water from Soviet glaciers.
—April 12: Yuri Gagarin becomes the first cosmonaut in orbit.
—May 29: US and UK agree to draft Comprehensive Test Ban (CTB) Treaty with 12 on-site inspections/year.
—August: Installation of first US seismic stations in a network of 60 countries.
—September 1: USSR resumes nuclear tests.
—September 15: US resumes nuclear tests.
—October 30: Soviet *Tsar* bomb, 58-Mton partial test of 100–150 Mton, Novaya Zemlya.

1962
—February 20: John Glenn becomes first US astronaut in orbit.
—July 6: *Sedan* excavation experiment as a *Plowshare peaceful nuclear explosion*; 10–M tons of earth displaced.
—July 9: Electromagnetic pulse from *Fishbowl* test, 1.4 Mtons, 400 km above Johnston Atoll, turns off 300 streetlights on Oahu, Hawaii (1200 km away).
—September 3–7: Tenth Pugwash Conference proposes "black box" monitoring for CTBT verification.
—October: US reduces requirement to 8–10 inspections/year for a CTBT.
—October 15 to November 20: Cuban missile crisis, 162 warheads, and blockade.

—December 19: Communist Party Secretary Nikita Khrushchev accepts 2–3 inspections/year and 3 unmanned seismographic stations (black boxes) in USSR.

1963

—January: Secret test ban talks between US and USSR.
—February 19: US accepts 7 inspections/year, provided any mysterious event may be challenged.
—April: Khrushchev withdraws offer of 3 inspections/year.
—June 20: US and USSR sign "hot line" agreement.
—August 5: US/USSR/UK sign Limited Test Ban Treaty, banning nuclear explosions in atmosphere, space and underwater.

1964

—August 4: Gulf of Tonkin resolution sends US troops to Vietnam under President Lyndon Johnson.
—October 16: China explodes first nuclear bomb.
—October 22: Salmon, a US cavity-explosion, muffles seismic signal in a salt dome, near Hattiesburg, Mississippi.

1965

—June: Large Seismic Array (LASA, Montana) starts detecting nuclear explosions.

1966

—January 17: B-52 bomber crashes near Palomares, Spain with 4 unarmed H–bombs, 2 by sea and 2 by land.
—September 24: First French H–bomb, in Tuamotu Archipelago, Polynesia.

1967

—January 27: Outer Space and Celestial Bodies Treaty bans nuclear weapons in orbit and space.
—February 14: Treaty for Prohibition of Nuclear Weapons in Latin America signed in Mexico City (Tlatelolco); all nations must ratify for the treaty to enter into force.
—June 17: First Chinese H bomb, 3.3 Mton, Lop Nor.
—June 23–25: President Johnson and Soviet Premier Alexi Kosygin begin SALT and ABM talks in Glassboro, NJ,
—December 10: Gasbuggy Plowshare explosion in New Mexico to stimulate natural gas production.

1968

—July 1: Nuclear Non-Proliferation Treaty (NPT) opened for signature.
—August: Soviet GALOSH ABM system deployed around Moscow.
—August 24: French H bomb, 2.6 Mton, Fangataufa Atoll.

1969

—July 20: US Apollo 11 lands on moon.
—November 3: UN Committee on Disarmament proposes global exchange of seismic data to monitor a CTB treaty.
—November–December: Preliminary SALT talks in Helsinki.

1970

—January: Administration of President Richard Nixon cuts Plowshare budget.
—March 5: NPT enters into force with 46 nations in 1970 and 190 in 2013.
—November 30: Atlantic-Pacific Inter–oceanic Canal Commission rejects use of nuclear explosives.
—December 18: Largest US release of radioactivity from *Baneberry* test.

1971

—March 30: US deploys Poseidon SLBM.
—June 15: Three Russian nuclear explosions set off to increase oil productivity.

1972

—May 26: SALT–I and ABM treaties signed by Nixon and Soviet First Secretary Lenoid Brezhnev in Moscow.
—June: UN Conference votes 48 to 2 to halt nuclear testing; France/China oppose, US/UK abstain, USSR absent.
—October: US detects Soviet SS-18 flight test.
—November: SALT II negotiations begin.

1973

—May 30: Protocol establishes the ABM/SALT Standing Consultative Commission.
—June 22: World Court issues injunction to France to refrain from testing on Mururoa in Pacific Ocean.
—October 17, 1973 to March 17, 1974: OPEC embargoes oil to US/Netherlands.

1974

—May 18: India sets off a 12-kton, Pu–device under Rajasthan Desert, expands nuclear club beyond *WWII P-5*.
—November 24: US/USSR agree to limit strategic launchers (to 2400) and MIRV launchers (to 1320).

1975

—January 19: AEC reorganized into NRC (regulatory) and ERDA (development, became DOE on August 4, 1977 and weapons became NNSA in 1999)
—October 1: Grand Forks, North Dakota Safeguard ABM site is activated. It was closed 4 months later because of its ineffectiveness against Soviet MIRVs at a cost of $21 billion from 1968–78 (1998$).

1976

—Public key encryption invented.
—January 29: France cancels its reprocessing exports for South Korea and Pakistan (1978).
—October 28: US President Gerald Ford postpones reprocessing of spent fuel to obtain Pu.

1977

—April 7: US President Jimmy Carter postpones indefinitely commercial reprocessing, halts Clinch River Breeder Reactor and calls for an International Fuel Cycle Evaluation (INFCE) to make the nuclear cycle more resistant to proliferation.
—October 19, 1977, to February 26, 1980: 40 nations prepare INFCE report.

1978

—January 11: Nuclear Suppliers guidelines forwarded to IAEA.
—March 10: Nuclear Nonproliferation Act (NNPA) tightens nuclear export criteria.
—April 4: Anwar Sadat of Egypt and Menachem Begin of Israel sign Camp David Accord in Washington.
—July 13: Euratom ends US embargo of nuclear fuel to Europe (NNPA required renegotiation) with a brief meeting.

1979

—April, 6: US cuts economic and military aid to Pakistan because it is making weapons-grade uranium.
—June 8: First Trident with SLBM launched.
—June 18: Brezhnev and Carter sign SALT II Treaty in Vienna.
—September 22: Vela satellite bhangmeter picks up a signal over the South Atlantic, perhaps a 3 kton explosion.
—December 12: NATO agrees to negotiate INF Treaty, while building Pershing-II and ground launched cruise missiles.
—December 26: USSR invasion of Afghanistan removes SALT II from US Senate consideration.

1980

—July 15: Office of Science and Technology Policy concludes the light signals recorded by a Vela satellite over the South Atlantic on September 22, 1979 were probably not from a nuclear explosion. Others disagree.
—September 1980 to August 1988: Iran–Iraq war, with chemical weapons.
—December: Tomahawk submarine-launched cruise missile flies 600 miles to sea and returns to land with 5-m accuracy.

1981

—April 12–14: Space shuttle Columbia orbits Earth 36 times.
—June 7: Israel destroys Iraq's Osirak reactor, an act that was condemned but later appreciated after Gulf War.
—November 13: Senate ratifies Protocol I of Tlatelolco Treaty; US cannot deploy weapons in Puerto Rico, Virgin Islands, or Guantanamo Naval Base.
—November 30: US tables Zero-Zero INF proposal in Geneva.

1982

—June 14, Argentina surrenders to UK in the Falkland War.
—June 29: Strategic Arms Reduction Talks (START) begin in Geneva.
—July 16: Ambassadors Paul Nitze and Yuli Kvitsinsky walk in Geneva woods, trying to cut INF weapons by 50%.
—1982: USSR places 300 SS-20 missiles on its western border.
—November: US President Ronald Reagan adopts dense-pack basing for MX missiles.

1983

—March 23: Reagan calls for SDI "to make nuclear weapons impotent and obsolete."
—April 6: Scowcroft Commission recommends Midgetman (road-mobile, single-warhead) and 100 MX.
—Summer: Large phased-array radar construction discovered at Krasnoyarsk.
—November 23: Soviets leave Intermediate Nuclear Force (INF) negotiations as Pershing-II and GLCM are deployed.
—December 8: Soviets suspend START negotiations.

1984

—January: Strategic Defense Initiative Organization (SDIO) created.
—January 23: Reagan's first "Soviet Noncompliance" report released to Congress amid controversy.
—January 30: Soviets respond with an *aide memoire* to US State Department on US noncompliance.
—June 10: Homing Overlay Experiment (HOE) intercepts midcourse missile in limited test of kinetic kill vehicles.

1985

—January 8: Sec. State George Schultz and Ambassador Andrei Gromyko agree to negotiate on START/INF/ABM/SDI.
—February 20: Nitze–criteria for SDI deployment in terms of reliability, survivability and marginal cost.
—October 6: Reagan encourages broad interpretation of ABM Treaty to test other physical principles in space.
—October 14: Schultz states that a broad interpretation the ABM Treaty "is fully justified" to allow SDI testing.

1986

—July: Natural Resources Defense Council sets up seismographs in the Soviet Union.
—September: First B-1 bomber squadron becomes operational.
—October 7: Mordechai Vanunu publishes photos of the Israeli Dimona nuclear weapons facility in the London Times. He is drugged and taken to Israel for solitary confinement until April 22, 2004.
—October 11-12: Reagan and Soviet leader Mikhael Gorbachev agree to cut strategic weapons by 50% at Reykjavik.
—November 28: The 131st B-52 bomber with air launched cruise missiles is deployed, exceeding the 1320 warhead limit on MIRVed launchers, negating SALT II. US is first country to publicly renounce an arms control agreement.
—December: First 10 Peacekeeper missiles become operational in Minuteman-III silos in Wyoming.

1987

—February: Soviets end unilateral nuclear testing moratorium after 18 months.
—April 7: Missile Control Technology Regime signed to slow missile proliferation.
—April 23: APS Directed Energy Weapons Study (DEWS) Group "estimates that even in the best of circumstances, a decade or more of intensive research would be required [for an informed decision on DEWS]."
—August: First of six South African gun-type weapon is qualified.
—September 15: Schultz and Soviet Foreign Minister Eduard Shevardnadze sign Nuclear Risk Reduction Accord.
—December 8: Reagan and Gorbachev sign Intermediate Nuclear Force Treaty.

1988

—May 27: US Senate ratifies INF Treaty (93–5), the first ratification of an arms control treaty since 1972.
—June 28: Office of Technology Assessment testifies it disagrees with the charge of *likely violation of TTBT.*
—August–September: Joint Verification Experiment on nuclear tests carried out by US/USSR.

1989

—June 4: Chinese shoot 100's in Tiananmen Square. By 2012, many rebel students now national leaders.
—November 9: Fall of Berlin Wall.

1990

—October 24: Last Soviet nuclear test, as Gorbachev calls for a moratorium.
—November 19: NATO and Warsaw Pact sign Conventional Armed Forces in Europe (CFE) Treaty, limiting tanks/planes to same levels, a 60% reduction for WTO and no reduction for NATO. Enters into force in 1992, ending the Cold War.

1991

—January 17 to February 28, 1991: *Operation Desert Storm*, 34 nations defeat Iraq.
—April 3: UN Resolution 687 requires destruction of Iraq's nuclear/bio/chem weapons.
—July 10: South Africa signs NPT and later destroys its 6 nuclear weapons.
—July 31: US and USSR sign START I, each cutting from 12,000 to 7000 actual warheads.
—September 27/October 5: US President George H.W. Bush and Soviet leader Gorbachev reciprocal-unilateral measures place missiles to be cut by START I off alert, and remove surface-ship and ground-based tactical weapons.
—November 26: Last British test at NTS.
—November 27: Sam Nunn–Richard Lugar Act passes, giving Russia first $400 million to destroy weapons.
— December 1: Ukraine votes to become independent of USSR, followed by 13 other Soviet Republics and Russia.

1992

—March: US ends military plutonium reprocessing, HEU production stopped in 1964.
—March 9: China joins the NPT regime.
—March 24: Open Skies Treaty signed, allowing monitoring by over–flights of NATO and Warsaw Pact aircraft.
—May 23: START's Lisbon Protocol signed by US/Russia/Ukraine/Belarus/Kazakhstan, all agreeing to join NPT.
—August 3: France joins NPT regime.
—September 23: Last US nuclear test, *Divider*. Bush-I signs testing ban legislation.
—October 1: US Senate ratifies START I with the Biden condition calling for "the adoption of cooperative measures to enhance confidence in reciprocal declarations [for warheads and fissile materials]."

1993

—January 3: US/Russia sign START II, cutting deployed strategic warheads from 7000 to 3500.
—February 18: US/Russia agree on sale of 500 tonnes of HEU, blended to LEU reactor fuel.
—March 3: France joins the NPT.
—April 1: IAEA declares North Korea in violation of the NPT.
—September: US/Russia agree to start Materials Protection Control and Accounting program with lab-to-lab assistance.
—September 27: President Bill Clinton proposes Fissile Material Cut-off Treaty banning production of weapons Pu/HEU .
—December: UN adopts resolution calling for a Fissile Material Cut-off Treaty.

1994

—April 30: Nuclear Proliferation Prevention Act (Glenn–Pell–Helms Act) provides for sanctions on nuclear testing.
—May 30: US/Russia detarget strategic missiles, but they can be quickly reprogrammed.
—June 23: US Vice President Gore and Russian Prime Minister Viktor Chernomyrdin agree to close last 3 Russian plutonium-production reactors by 2000, with US aid to establish another power source.
—September 20: Clinton and Russian President Boris Yeltsin agree to exchange data on warheads and Pu/HEU.
—October 1: Russia stops plutonium production, except at 3 reactors where waste heat is used to heat cities.
—October 23: US/N–Korea sign "Agreed Framework" to eliminate N. Korean nuclear program and get S. Korean reactors.

1995

—March 1: Clinton announces permanent removal of 100 tonnes of Pu/HEU from US stockpile.
—May 10: Clinton and Yeltsin agree to declarations/inspections on excess Pu/HEU from weapons.

—May 12: NPT extended indefinitely after the 5 nuclear weapon states promise to join a CTBT.
—August 11: Clinton announces US will support a zero-yield CTBT.
—December 19: Wassenaar Arrangement on export controls replacing Cold War controls.

1996
—January 26: Senate ratifies START II by 87–4, cutting to 3500 total warheads each and bans land-based MIRV.
—January 27: France conducts last test and supports a zero-yield CTBT.
—March 25: US joins South Pacific Nuclear-Free Zone.
—April 11: Pelindaba Treaty for Nuclear-Free Africa signed by 43 African states.
—July 29: China's last nuclear test, 2 months before signing CTBT.
—September 10: UN opens CTBT for signature, banning nuclear tests everywhere, for all time, at any yield. Only India, Bhutan and Libya are opposed.
—September 17: US/Russia/IAEA agree to trilateral inspections on excess SNM.
—September 24: All five nuclear weapon states sign the CTBT at the UN.
—1996: Last nuclear warheads from former Soviet Republics returned to Russia.

1997
—April 29: Chemical Weapons Convention enters into force.
—March 21: Clinton and Yeltsin in Helsinki call for START III reductions to 2000–2500 deployed, strategic warheads and "measures relating to the transparency" for inventories and destruction of strategic warheads.
—May 15: IAEA adopts Additional Protocol to give IAEA more powers.
—May 27: Yeltsin signs NATO-Russian Founding Act.
—September 7: Russian General Alexander Lebed claimed "up to 100 suitcase bombs are not under the control of the Armed Forces of Russia."

1998
—May 11, 13, 28 and 30: India and Pakistan conduct multiple underground nuclear tests.
—August 31: North Korea launches a Dong-1 missile over Japan.
—September 22: US and Russia sign Nuclear Cities Agreement to help 10 Russian closed weapon cities.

1999
—October 13: US Senate rejects CTBT (51–48) in spite of 150 signatory nations.

2000
—April: Duma ratifies CTBT (April 21) and START II (April 14), contingent on constraint of ABM demarcation.
—May, 22: NPT review conference agrees to the 13 steps for progress on nuclear disarmament.
—September 1: Gore/Chernomyrdin agree to dispose of 34 tonnes Pu each, as MOX fuel or buried in a repository.

2001
—May 1: President George W. Bush states he will withdraw US from ABM Treaty to test/deploy ABM options.
—May 31: INF continuous Perimeter and Portal Monitoring ends as INF 13-year period ends.
—September 11: Osama bin Laden's al Qaeda uses US commercial planes to kill 3000 in destroying the World Trade Center and damaging the Pentagon. The US joins Afghan resistance forces to topple Taliban government, producing a new leader, Hamid Karzai. The 9-11 event creates a *major paradigm shift* to focus on terrorism over other matters.
—September-October: Letter-borne anthrax contamination in US postal system and buildings.
—December: G. W. Bush administration funds CTBT research, but not that which aids CTBT entry-into-force.
—December 4: US and Russia complete START I reductions to 6,000 warheads.
—December 7: US ends discussion on verification of Biological Weapons Convention.
—December 13: G.W. Bush formally starts process to withdraw from ABM Treaty 6 months hence.
—December 14: Navy Area Theater Ballistic Missile Defense system, covering small areas, is cancelled.

2002
—January 4: DoD Ballistic Missile Defense Office elevated to Missile Defense Agency.
—January 8: Nuclear Posture Review sets path to 1700–2200 deployed, operational US warheads by 2012, while retaining up to 5000 warheads on "active and inactive reserve" that can be deployed in days to years. The "new

triad" combines offense, defense, conventional arms and rapid information. Targeted countries are Russia, China, Iran, Iraq, Libya, N. Korea and Syria in possible conflicts between Israel/Iraq, N. Korea/S. Korea, China/Taiwan. The NPR suggests that new designs are needed to create a 5-kiloton earth-penetrating bomb to attack underground bunkers. Congress shortens preparation time for test-sites readiness for this weapon.

—January 23: Bush administration continues US/Russia Pu-MOX disposition but cancels geological immobilization.

—January 29: Bush labels North Korea, Iran and Iraq as an "axis of evil arming to threaten the peace of the world."

—March 14: India integrates the Agni-2 missile (2000 km, 1000 kg) into its armed forces.

— May 24: Presidents George W. Bush and Vladimir Putin sign the Strategic Offensive Reductions Treaty, which reduces *operational* warheads to 1700-2200. SORT allows response forces of some 2400 additional warheads, adds no new verification measures, and expires on December 31, 2012, the same date the limit of 1700-2200 is required. Russia retains 138 SS-18s and many SS-19s which would have been removed under START II..

—October 16: US announces North Korea admits having a clandestine centrifuge program to make HEU.

—November 4: Cuba accedes to the NPT, completing Western Hemisphere. UN conference lobbies nuclear weapon states to agree to a negative security assurance, so non-weapon states would not fear nuclear weapons.

— November 27: UN inspections begin in Iraq, pursuant to UN resolution 1444 that required Iraq to disarm its weapons of mass destruction, or be in "material breach." In 1997, Iraq declared 8 presidential palaces off limits, leading to an end of inspection.

— December 8: Iraq declares its lack of weapons of mass destruction as required by UN resolution.

2003

—January 10: North Korea withdraws from NPT after IAEA inspectors are removed (Dec. 27) and Governing Board "deplores in the strongest terms" (Jan. 6) the reprocessing restart, claimed two weapons and beginning centrifuges.

—March 2: Department of Homeland Security established with a budget of $30 billion.

—March 6: Senate ratifies Strategic Offensive Reduction Treaty. Iraq-war anger delays Russian Duma to May 14.

—March 19: US/UK invade Iraq for a 3-week war, without explicit support from the UN, which wanted more time for more inspections. Unrest in Iraq continues after the war, without evidence for weapons of mass destruction.

—March 31: US mildly sanctions Pakistan's Khan Research Institute for exporting centrifuge equipment to N. Korea in trade for missile technology.

—April 22: Los Alamos builds first US Pu pit in 15 years, a beginning for a plant to make 125 pits/year by 2018.

—April 25: North Korea tells US in Beijing that it has Nuclear weapons.

—June 6: IAEA Board of Governors charges Iran with a violation over clandestine centrifuges discovered on Feb. 20.

—August 28: North Korea informs China, Japan, Russia, South Korea and the US that it may test in the near future.

—December 14: Saddam Hussain captured.

—December 19: Libya announces it will dismantlement its nuclear weapons programs.

2004

—February 4: Pakistan President Pervez Musharraf pardoned Abdul Qadeer Khan for supplying centrifuge technology to Iran, Libya and North Korea.

—February 20: IAEA report states that Libya had ordered 10,000 centrifuge tubes, but was now cooperating.

—July 9: The Senate Intelligence Committee Report rebukes the October 2002 National Intelligence Estimate that "Baghdad has chemical and biological weapons by stating that this conclusion "overstated both what was known and what intelligence analysis judged abut Iraq's chemical and biological weapons holdings." The CIA did not invite DOE to test alleged centrifuge tubes. The administration's charge of Iraq's purchase of uranium was incorrect.

—July 22: The *9/11 Commission Report* states "we have seen no evidence that these or other earlier contacts ever developed into a collaborative operational relationship [Al Qaeda/Iraq]."

—September 30: Report on Iraq Weapons of Mass Destruction released by Charles Duelfer (Chief Iraq WMD Inspector). It agreed with the results of David Kay ("we were all wrong"). The report concluded Iraq's nuclear program ended in 1991, that Iraq unilaterally destroyed its chemical weapons in 1991 and that Iraq abandoned its ambitious biological weapon plans, but Saddam retained his desire for WMD.

—October 19: Brazil agrees to IAEA inspections of centrifuges, hoped as a precedent for Iran.

—December 7: US intelligence organized under centralized National Intelligence Director.

2005

—January 31: Elections in Iraq with high Shiite/Kurd turn-out but negligible Sunni participation.
—February: Iran nuclear program evident as FR/FRG/UK try diplomacy and US exerts pressure.
—February 10: North Korea claims it has nuclear weapons and refuses to join 6-party talks.
—March 6: China questions US data on DPRK nuclear weapons, suggests 2-party, US/NK talks.
—July 29: State department absorbs its arms control bureau into the non–proliferation bureau.
—September 19: North Korea commits to abandon its nuclear weapons program, returning to the NPT and safeguards.
—September 24: IAEA finds Iran in noncompliance to the NPT.

2006

—March 2: President George W. Bush and Prime Minister Manmohan Singh of India agree to cooperate on peaceful nuclear power, this requires amending the 1978 NNPA nonproliferation law and obtaining unanimous consent from the Nuclear Suppliers Group. India would place 14 of its 22 reactors under India-specific IAEA safeguards.
—April: Iran declares it has 164 centrifuges. US joins France, Russia and UK to dissuade Iran from enriching uranium on its soil. Brazil continues its work on its enrichment plant since 2004.
—October 9: DPRK test of 0.6 kton detected by 22 IMS seismic stations (with 30 certified). Radio–Xe detected by IMS and portable sensors.
—December 10: President Ehud Olmert lets slip that Iran is "aspiring to have nuclear weapons as America, France, Israel, Russia."

2007

—January 11: China uses a kinetic–kill, anti-satellite weapon to destroy a Chinese weather satellite. Arms race in space or a nudge to control ASAT weapons?
—February 13: Six party talks yields "initial actions" plan to implement September 2005 pledge by DPRK to abandon nuclear weapons and programs. This later fails.
—March 2: NNSA chooses the LLNL to design Reliable Replacement Warheads. Congress denies funding in 2008.
—July 14: Russia suspends CFE observance, followed by US in 2011.
—July: First W-88 produced in 18 years at Los Alamos, to build 20–50 per year.

2009

—April 5: President Obama states in Prague his intention to move towards abolition of nuclear weapons, with NEW START and CTBT ratifications. He stated his intention to remove troops from Iraq, but send more troops to Afghanistan, until perhaps 2013.
—May 25: DPRK 3-kt test detected by 61 IMS seismic stations, but not the radio–Xe signal.
—September 25: US, UK and France announce that Iran is building a covert enrichment facility inside a mountain near Qom. Previously Iran indicated it might restrain its enrichment program.
—November 29: Indian Prime Minister Singh states India is willing to join NPT as NWS.

2010

—April 5: Nuclear Posture Review removes 1st US response of nuclear attacks on NNWS bio–chem attacks, but retains this response for the 9 states, which have nuclear weapons. No new ICBMs proposed.
—April 8: New START signed in Prague with limits of 1550 operational strategic weapons and 700 deployed launchers. Tactical weapons not covered, as usual, but they consist of about 250 for the US and 2000 for RF, down from the Cold War numbers of 8000 and 23,000. Prompt Global Strike conventional weapon ICBMs to count as nuclear systems, not to be deployed until 2015-20.
—May: NPT Review Conference obtains consensus, with modest progress in a time of difficulty.
—June: Stuxnet cyberwar attack on Iran, destroying 1000 of 5000 centrifuges.
—June 9: UN Security Council sanctions cut military imports to Iran and restrict proliferating companies on a 12-3 vote. Brazil and Turkey offer a partial solution to allow reactor fuel, but not constrain higher enrichments.

2011

—February 5: New START enters into force, passing Senate Dec. 22, 2010 on 71–26 vote, 1550 operational warheads.
—November 22: US suspends implementation of CFE because Russian transfers troops to Georgia and Moldova.

2012

—NNSA and JASON declare Pu pit lifetimes is over 100 years, using ^{238}Pu fast decay.

—April 15, 2012: Saudi Arabian Armaco oil company under cyber attack.

—September 25: 100th Joint Actinide Shock Physics Experimental Research (JASPER) to obtain Pu equation of state from Pu projectiles.

—November 29: UN votes 138/9/41 to recognize Palestine as a nation.

—November: Oak Ridge's Titan computer does 17.6 petaflops, same as Earth's population each doing 3 M/sec.

2013

—January 23: Iran notified IAEA it is installing IR-2M centrifuges. Iran has 160 kg of 20%-HEU, needing 240 kg for a warhead's input steam.

—February: China's People's Liberation Army cyber-attacks over 60 US high tech companies.

—February 12: DPRK third nuclear test, 7 kton, under Kim Jong Un.

—May 13: CTBT ratified by 159 nations (not US, China) of 183 signatories (not India, N. Korea, Pakistan), with 36 of 44 nuclear capable nations ratified. Monitoring capability believed to be at 0.1–kt level for tamped explosions, and 1–2 kton for cavity shots. Certified IMS stations (275), testing (20), under construction (20), planned (22).

—March 11: N. Korea nullifies 1953 Korean armistice. Restarts Yongbyon reactor on April 2.

—March 13: 2 million small pox vaccines prepared, only US and RF have small-pox samples.

—April: 12 more ground-based interceptors sent to Alaska in response to N. Korea.

—April 2: UN votes 154/3/23 to regulate global conventional arms trade of $70 billion/year, U.S. uncertain.

—April 9: US ship-board lasers shoot down slow drones.

—April 15: Tsarnaev brothers attack Boston Marathon. US follows UK lead with 30 million cameras to observe.

—May 23: Command of US drones moved from CIA to DoD.

—May 23: Iran has 14,244 of 50,000 planed centrifuges, 700 are IR-2, more potent. Obama's red line of 240 kg of 20% medium enriched uranium may be crossed in the fall of 2013.

—May 31: Nuclear Regulatory Commission gives license for laser enrichment in North Carolina to GE–Hitachi.

—June 3: Supreme Court allows arrested suspects of serious crimes to be DNA tested for unknown other crimes.

—June 6: Edward Snowden leaks are published in the Washington Post and Manchester Guardian. NSA *PRISM* program tracks global email, web searches and other internet traffic. On June 9, *Boundless Informant* program tracks global phone calls. Snowden fights extradition from Hong Kong to Russia, but fails (July 2) to obtain passage to Equador or other states.

—June 6: Obama acknowledges phone-call and email lists of US citizens are kept, but he states they not accessed for US domestic citizens until evidence is presented to the FISA court.

—June 19: In Berlin, Obama states he is willing to reduce New START limit of 1550 by 1/3 in a reciprocal, unilateral agreement (not a treaty), monitored under New START. Putin wants ABM concessions to do this. The level of 1000 was last seen in 1954. Nunn Lugar renewed after 20 years with emphasis on securing nuclear materials, and not on missile and CW destruction.

—July 3: Prime Minister Mohamed Morsi of Egypt, representing the Muslim Brotherhood, is overthrown by Egyptian military with the support of secular parties. The Arab Spring oscillates away from democracy, without major regrets by the democracies.

—July: National Security Administration admits that domestic phone call, email and postal information is collected, but only read after approval by the Foreign Intelligence Surveillance Court (FISC). There are calls to expand oversight of FISC, which has been accused of leniency.

—August 9: UN panel of experts from 25 nations suggest ways to enhance cyber security with cooperative measures and with enhancements to international law.

—September 6: The National Security Agency foils encryption by hacking into computers via permitted backdoors. On September 10, Yahoo, Facebook, Google and Microsoft file suit against NSA.

—September 21: Old reports were released that describe a near catastrophe from the crash of a B52 bomber near Goldsboro, NC on January 23, 1961, 3 days after Kennedy's inauguration. Only a single switch prevented the 4–Mton bomb from exploding.

—September 26: Iraq ratifies the CTBT, after two US-Iraq wars and much chaos.

—September 27: Obama and Iranian President Hasssan Rouhani talk on the phone to attempt to start meaningful nuclear negotiations. This was the first leader-to-leader contact since 1979 when President Carter and the Shah of Iran talked, just before the 444-day hostage crisis began.

—October 2: UN dismantlement inspections begin in response to the death of 1400 from Sarin chemical weapons used by the Syrian government near Damascus on March 19 and August 21. Agreement between Russia, U.S.,

Syria and the UN Security Council (September 27) to end the Syrian CW program within a year came after threats of force by Obama and the lack of Congressional support to carry out such an attack. President Bashir al–Assad is helpful.

—November 24: Iran and *P5–plus–1 group* agree to a *Joint Plan of Action* to constrain Iran's nuclear production with instructive verification, in trade for a 10% reduction in economic sanctions. Inspections began on 20 January 2014, for six months, renewable for another six months. This could be followed with a more comprehensive agreement if the *Joint Plan of Action* is successful.

—December 18: Presidential review panel urges restraint on National Security Agency wiretaps.

2014

—January 17: Congress fully funds the B-61 bomb, with plans for force structures with interchangeable warheads.

—February 22: Ukrainian Parliament declares President Viktor Yanukovych unable to govern after riots, following his refusal to signing a European Union trade agreement. On March 1, Russian President Vladimer Putin deploys Russian troops into Crimea, which votes to join the Russian Federation on March 16, with accession on March 18.

—March 14: US mixed oxide fuel plant in South Carolina is mothballed.

—June 14: President Obama sends 5 nuclear bombers to EU to give assurance on the Russia/Ukraine issue.

—July: US State Department announces Russia violates INF Treaty by testing ground-based cruise missiles, that are not yet deployed. Russia claims US BMD deployments in Romania and Poland would be CFE violations.

—September 5: Ceasefire agreement in Ukraine's Donetsk and Luhansk regions. Putin makes German Chancellor Angela Merkel wait four hours in Milan after visit to Belgrade amid Western economic sanctions and falling oil prices.

—September 5: IAEA stated that Iran is complying with the interim agreement, but Iran failed to provide information about its past nuclear weapons activities. Iran had provided some information on exploding bridge detonators. Iran/P5+1 discussions continue.

—September 8: Syria exports its last CW stocks, but Syria has not destroyed its 12 remaining production facilities.

—September 11: US declares Russia in violation to INF treaty on its construction of a ground-based cruise missile with a range between 500 and 5,500 km.

—November 14: Russia terminates US–RF programs that prevent loss of nuclear materials.

—December 24: *Arms Trade Treaty* for transparent conventional weapon sales enters into force with 3 UN negative votes (Iran, Syria, DPRK), but US/RF/PRC and others have not ratified.

2015

—January 7: Charlie Hebdo magazine attacked in Paris by Al Qaeda.

—January 19: Google Glass camcorder no longer for sale to the general public.

—January 19: Cyber attack on Sony attributed to DPRK because of Kim Jong-un movie.

—January 26: Three–pound package delivered to White House by drone quadcopter.

—February 18: Federal Aviation Administration issues draft regulations for drones.

—March 3: Israeli Prime Minister Benjamin Netanyahu addresses Congress to undercut Iran–P5+1 agreement.

—May 9: Security Council sanctions DPRK for launching a submerged missile, violating UN Resolutions that ban ballistic missile tests.

—May 11: China deploys 3-warhead MIRV on 20 DF-5 ICBMs, capable of reaching US. All P5 nations now have MIRV.

—May 13: Saudi Arabian Prince Turki bin Faisal states, "Whatever the Iranians have, we will have, too."

—May 18: ISIS takes Ramadi, capital of Sunni Anbar Province in Iraq.

—May 26: NPT Review Conference fails to agree, as US, UK and Canada block a call for negotiations on Nuclear Free Zone in the Middle East.

—July 10: 20 million US security clearance forms for government jobs hacked, probably by China, includes all who applied over the past 15 years.

—July 14: Joint Comprehensive Plan of Action (JCPOA) between Iran and P5+1 signed, constraining Iran's enrichment, extending break-out time from a few months to a year, while reducing economic sanctions.

—September 11: House resolution of approval of the JCPOA fails 269 to 162, and Senate resolution of disapproval obtains 58 votes, shy of 60 needed to defeat, entering into force on October 18.

2016

—February 9: Obama Administration cancels MOX reprocessing plant, shifts to dilute and dispose of Pu.

—April 16: State Department claims 3rd year of Russian INF–violation of cruise missile possession.

—August 16: India deploys first SLBMs, with 12 warheads (750 km) or 4 warheads (3500 km).
—September 9: North Korea's fifth nuclear test, largest to date at 10 to 20 kt, with Pu for 6 to 8 weapons.
—October 16: Russia suspends the 2000 agreement with US to dispose of 68 tonnes Pu. Putin angry with sanctions over Crimea and Ukraine.
—October 27: General Assembly votes 123-38 (16 abstentions) to consider a total ban on nuclear weapons, with a vote during summer 2017. Sponsored by Austria, Brazil, Ireland, Mexico, Nigeria, S. Africa; with NATO and P5 against.
—November 30: UN Resolution caps DPRK coal exports to $700 million and other items.

2017
—March 23: SFRC considers sanctions on Iran because of Iran's missile testing (non nuclear).
—April 6: US launched 50 cruise missiles against Syria's Shayrat Air Base because of its April–4, sarin attack.
—April 18: Sec. State Rex Tillerson informed Congress that Iran is in compliance with JCPOA agreement.
—May 14: North Korea tests an intermediate–range ballistic missile with range of 3,000 to 5,500 km, capable of hitting Guam, but not the continental US. It can carry a large-size nuclear weapon.
—May 22: The first draft of the *Convention on the Prohibition of Nuclear Weapons* is released, before the July 7 deadline. It prohibits acquisition, possession, stockpiling, use, transfer, testing, and deployment. It is supported by most non-nuclear weapon states and opposed by all nuclear weapon states.
—June 18: US shoots down Syrian (Russia-aided) plane, the first air-to-air kill by US pilot in 18 years.
—July 25 and 27: Congress sanctions, almost unanimously, Russia, Iran and N. Korea [House 419–3, Senate 98–2].
—September 3: North Korean nuclear test of about 140 kT, possibly a boosted or thermonuclear weapon.

Appendix B: Author Bio-Briefs

Pierce Corden received his physics Ph.D. from the University of Pennsylvania in 1971. He worked at the Arms Control and Disarmament Agency (ACDA), the Department of State and as the senior American at the Provisional Technical Secretariat for the Comprehensive Test Ban Treaty Organization. In addition to his work on nuclear weapon development issues and test-ban verification, during his career he supported arms control for mass destruction and conventional weapons. He served as Deputy Executive Chairman of the UN Special Commission (for Iraq) in 1993. From 2008–2016 he was a Visiting Scholar at the American Association for the Advancement of Science. Currently he is an adviser on disarmament to the Holy See Mission to the UN.

Tony Fainberg was trained as an experimental particle physicist and worked as a researcher in this field for over 12 years in the United States and at CERN in Geneva, Switzerland. He then began another career, working at the interface of science and national security, which he has pursued for 35 years. He worked on nuclear safeguards issues at Brookhaven National Laboratory. He was a Congressional Science Fellow of the American Physical Society, during which time he worked in the US Senate. He then became a staff physicist at the Congressional Office of Technology Assessment, performing technical analyses of the "Star Wars" Strategic Defense Initiative of the Reagan Administration, and analyzing the potential uses of technologies in combating terrorism. He hen spent ten years in the US Government's Executive Branch on nonproliferation and counter terrorism issues. He ended his career as a technical analyst, pursuing these and other areas at the Institute for Defense Analyses, a not-for-profit corporation that performs studies for the US government.

Alexander Glaser is associate professor at the Woodrow Wilson School of Public and International Affairs and the Department of Mechanical and Aerospace Engineering at Princeton University, where he directs the Nuclearfutures Laboratory and co-directs the Program on Science and Global Security (SGS). His research interests include nuclear disarmament and nonproliferation, nuclear transparency and verification, the nuclear fuel cycle, and nuclear energy. Glaser is the co-editor of the journal Science & Global Security and a member of the International Panel on Fissile Materials, which publishes the Global Fissile Material Reports.

Charles L. Glaser is Professor of Political Science and International Affairs and Director of the Elliott School's Institute for Security and Conflict Studies at George Washington University. His research focuses on international relations theory, including the security dilemma, defensive realism, the offense-defense balance, and arms races; and on international security policy, including U.S. policy toward China, energy security, and U.S. nuclear weapons policy. Before joining George Washington University, Professor Glaser was the Emmett Dedmon Professor of Public Policy and Deputy Dean at the Harris School of Public Policy at the University of Chicago. He has also taught political science at the University of Michigan, was a visiting fellow at the Center for International Security and Cooperation at Stanford, and was a fellow in the Kissinger Institute at the Woodrow Wilson International Center for Scholars. He has served on the Joint Staff in the Pentagon and was a peace fellow at the United States Institute of Peace.

Hugh Gusterson is Professor of Anthropology and International Affairs at George Washington University. Gusterson is the author of Nuclear Rites (University of California Press, 1996), People of the Bomb (University of Minnesota Press, 2004), and Drone (MIT Press, 2016). He is co-editor of Cultures of Insecurity (University of Minnesota Press, 1999), The Insecure American (University of California Press, 2009), and Why America's Top Pundits Are Wrong (University of California Press, 2005). He has a regular column for the Bulletin of Atomic Scientists and for the new public anthropology website, Sapiens. He has also published in the Washington Post, the Los Angeles Times, the Boston Globe, the San Francisco Chronicle, the Philadelphia Inquirer, Science, Nature, New Scientist, American Scientist, and The Sciences.

David Hafemeister served on nuclear weapon and arms control issues in the U.S. Senate (Foreign Relations and Governmental Affairs Committees, Office of Senator John Glenn), the Office of Under Secretary of State and the PM and OES Bureaus, the Office of Strategic Negotiations of ACDA and as study director of Beyond START at the National Academy of Sciences and as technical-lead staff for CTBT-1 study at NAS. He also worked on these issues

Nuclear Weapons and Related Security Issues
AIP Conf. Proc. 1898, 060002-1–060002-5; https://doi.org/10.1063/1.5012598
Published by AIP Publishing. 978-0-7354-1586-7/$30.00

at Stanford, MIT, Princeton, Maryland universities, and was chair of the review committee on proliferation issues for Los Alamos National Laboratory. He is author of *Physics of Societal Issues* (Springer 2014) and *Nuclear Proliferation and Terrorism in the Post 9/11 World* (Springer 2016).

Siegfried S. Hecker initiated US-Russia Lab-to-Lab cooperative work to lessen nuclear dangers after the end of the Cold War as captured in Doomed to Cooperate, which is also the topic of his chapter. He is a senior fellow emeritus of the Freeman Spogli Institute for International Studies, and professor (research) emeritus in the Department of Management Science and Engineering at Stanford University. He served as the fifth director of the Los Alamos National Laboratory from 1986-1997 and as co-director of the Center for International Security and Cooperation at Stanford University from 2007-2012. His current professional interests include plutonium research, cooperative nuclear threat reduction, global nonproliferation, the expansion of nuclear energy, and threats of nuclear terrorism. He is a member of the National Academy of Engineering and a foreign member of the Russian Academy of Sciences. He is also a member of the Council on Foreign Relations, fellow of the American Physical Society, the American Academy of Arts and Sciences, and the American Association for the Advancement of Science. Among other awards, he received the Presidential Enrico Fermi Award, the Leo Szilard Prize, the Los Alamos National Laboratory Medal, the Department of Energy's E.O. Lawrence Award, and the American Nuclear Society Seaborg Award.

Theresa Hitchens is a Senior Research Scholar at the Center for International and Security Studies at Maryland (CISSM), University of Maryland, where she specializes in emerging technology and challenges to governance. She previously served as the Director of the United Nations Institute for Disarmament Research (UNIDIR) in Geneva, Switzerland. Ms. Hitchens has spent more than 15 years researching and writing on national military space issues and related technologies. She worked for more than a decade as a specialized journalist, writing on issues related to technology, arms control, the arms trade and international security affairs -- including as editor of *Defense News.* In 2009 she received the L.J. Horton Distinguished Alumna Award from Ohio University.

Edward Ifft, a retired member of the Senior Executive Service, has been involved in negotiating and implementing many of the key nuclear arms control agreements of the past 45 years. Dr. Ifft's career has been primarily in the State Department, with assignments in the U.S. Arms Control and Disarmament Agency, NASA and the Department of Defense. Early in his career, he served on the U.S. delegations to the negotiations on SALT and TTBT. He then served as Senior State Department Representative to both the START and CTBT negotiations in Geneva. After START concluded, he became Deputy Director of the On-Site Inspection Agency (OSIA). When OSIA was incorporated into the new Defense Threat Reduction Agency (DTRA) in 1998, he became Senior Advisor and State Department liaison to DTRA. As a U.S. START Inspector, he has participated in inspections of many sensitive military installations in the former Soviet Union. He also served as the last U.S. Commissioner (Acting) for the ABM Treaty until the U.S. withdrew from the Treaty in 2002. He participated in the Comprehensive Nuclear Test Ban Treaty Field Exercises in Kazakhstan in 2008 and Jordan in 2014.

Raymond Jeanloz, Professor of Earth and Planetary Science and of Astronomy and a Senior Fellow in the Miller Institute for Basic Research in Science at the University of California, Berkeley, chairs the US National Academy of Sciences Committee on International Security and Arms Control. He is also an Annenberg Fellow at Stanford's Hoover Institution, and serves on the Secretary of State's International Security Advisory Board. His scientific research uses lasers and diamond anvils to study materials at high pressures and temperatures, with applications to understanding the nature and evolution of planetary interiors. In addition to teaching and research, he is a longstanding advisor to the US Government and the University of California in areas ranging from Earth science and science education to national and international security. He is past chair of the National Research Council Board on Earth Sciences and Resources, and is a fellow of the American Academy of Arts and Sciences, American Association for the Advancement of Science, American Geophysical Union and American Physical Society, and a member of the National Academy of Sciences.

Stephen J. Karoly is the Acting Assistant Administrator within the Transportation Security Administration's (TSA) Office of Requirements and Capabilities Analysis (ORCA). In this role, Mr. Karoly serves as the principal advisor on matters relating to capabilities and requirements used in all modes of transportation ensuring that security technologies and processes are employed to meet the overall mission, vision, goals, and objectives of TSA and the Department of Homeland Security. Prior to joining TSA in 2014, Mr. Karoly served as Assistant Director for

Product Acquisition and Deployment at the Department of Homeland Security's (DHS) Domestic Nuclear Detection Office (DNDO). Prior to joining DNDO, Mr. Karoly served as the Director of Systems Engineering and Chief Systems Engineer at the Office of Technology Innovation and Acquisition, United States Customs and Border Protection (CBP).

Togzhan Kassenova is a fellow in the Nuclear Policy Program at the Carnegie Endowment. She currently works on issues related to the role of emerging powers in the global nuclear order, weapons of mass destruction nonproliferation issues, nuclear security, and strategic trade management. From 2011 to 2015 Kassenova served on the UN secretary general's Advisory Board on Disarmament Matters. She is the author of From Antagonism to Partnership: The Uneasy Path of the U.S.-Russian Cooperative Threat Reduction (2007) and Brazil's Nuclear Kaleidoscope: An Evolving Identity (2014) and a co-editor of Perspectives the Evolving Nuclear Order (2016).

Daryl G. Kimball became the Executive Director of the Arms Control Association in 2001. The Arms Control Association is an independent, non-profit membership organization based in Washington, D.C. that focuses on providing information, ideas, and solutions to address the dangers posed by the world's most dangerous weapons. He is the publisher of the Association's monthly journal, Arms Control Today. From 1997 to 2001, Kimball was the executive director of the Coalition to Reduce Nuclear Dangers, and from 1989-1997, Kimball was Director of Security Programs for Physicians for Social Responsibility.

Hans M. Kristensen is Director of the Nuclear Information Project at the Federation of American Scientists in Washington, D.C., where he is responsible for researching and documenting the status and operations of nuclear forces of the nine nuclear-armed states. Kristensen is co-author of the bi-monthly FAS Nuclear Notebook column in the Bulletin of the Atomic Scientists and the World Nuclear Forces overview in the SIPRI Yearbook. He is a frequent advisor to the news media on the status of nuclear forces and policy.

George Lewis George Lewis is currently a Visiting Scholar at the Judith Reppy Institute for Peace and Conflict Studies at Cornell University. He has a Ph.D. in experimental solid state physics from Cornell's Physics Department and is a Fellow of the American Physical Society and a recipient of its Joseph A. Burton Forum Award. Prior to returning to Cornell in 2005, he was in the Technology and Security Group of M.I.T.'s Security Studies Program for 16 years. His research over the last several years year has focused on the technology, capabilities, and implications of ballistic missile defenses systems, outer space surveillance, and future nuclear arms reductions. His blog is www.mostlymissiledefense.com.

Mark Lewis is the Director of Science and Technology Policy Institute at the Institute for Defense Analysis, which provides analysis of science and technology for the Executive Branch. Dr. Lewis formerly served as the Willis Young, Jr. Professor and Chair of the Department of Aerospace Engineering at the University of Maryland for 24 years. He was Chief Scientist for the U.S. Air Force from 2004 to 2008 and was President of the American Institute of Aeronautics and Astronautics (AIAA) from 2010 to 2011. He is the author of more than 300 technical publications and has been an advisor to more than 70 graduate students. He has served on advisory boards for NASA , the Air Force, DoD and the National Academy of Sciences. He was awarded the Meritorious Civilian Service Award and Exceptional Civilian Service Award and was the Aviation Week and Space Technology Laureate in 2007.

Allison Macfarlane is currently Professor of Science Policy and International Affairs at the George Washington University and Director of the Institute of Science and Technology Policy at the University's Elliott School of International Affairs. She served as Chairman of the U.S. Nuclear Regulatory Commission from July 2012 through December 2014. Dr. Macfarlane holds a PhD in geology from the Massachusetts Institute of Technology and a BSc degree in geology from the University of Rochester. She has been on the faculty at George Tech and George Mason University. From 2010 to 2012 she served on the Blue Ribbon Commission on American's Nuclear Future.

Robert Mikulak is currently an expert advisor on chemical and biological weapons issues at the Department of State. From November 2010 to July 2015, Ambassador Mikulak was the United States Permanent Representative to the Organization for the Prohibition of Chemical Weapons (OPCW), located in The Hague, The Netherlands. He previously served in the State Department and the U.S. Arms Control and Disarmament Agency in a series of increasingly responsible positions related to prohibition of chemical and biological weapons. During the decade-long negotiation of the 1992 Chemical Weapons Convention, he was the deputy U.S. negotiator and senior U.S. expert. He was awarded the Secretary of State's Award for Excellence in International Security Affairs in 2014, the Office of the Secretary of Defense Award for Exceptional Civilian Service in 2015, and the Hilliard Roderick Prize in Arms Control from the American Association for the Advancement of Science in 1992.

George Perkovich is vice president for studies at the Carnegie Endowment for International Peace. He works primarily on nuclear strategy and non-proliferation issues, cyberconflict, and on South Asian security. He is the author of the prize-winning book, India's Nuclear Bomb (University of California Press, 1999), and co-author of, Not War, Not Peace? Motivating Pakistan to Prevent Cross-Border Terrorism (Oxford University Press, 2016). Perkovich has advised many agencies of the U.S. government, and testified before both houses of Congress. He has been a member of the National Academy of Science's Committee on Arms Control and International Security, the Council on Foreign Relations Task Force on Nuclear Policy, and was a principal advisor to the International Commission on Nuclear Nonproliferation and Disarmament, a joint initiative of the governments of Japan and Australia. He served as a speechwriter and foreign policy advisor to Senator Joe Biden (D-Del.) from 1989-90.

Steven Pifer is a senior fellow and director of the Arms Control and Non-Proliferation Initiative at the Brookings Institution. A retired Foreign Service officer, his more than 25 years with the State Department included assignments as deputy assistant secretary of state with responsibilities for Russia and Ukraine, U.S. ambassador to Ukraine, and special assistant to the president and senior director for Russia, Ukraine and Eurasia on the National Security Council.

Miles Pomper is a Senior Fellow in the Washington DC office of the James Martin Center for Nonproliferation Studies (CNS) and chair of the Fissile Materials Working Group an international coalition of nearly 80 nongovernmental organizations and individuals worldwide working to prevent nuclear and radiological terrorism. His work focuses on nuclear security, nuclear nonproliferation, nuclear security, and nuclear arms control. He holds a master's degree in international affairs from Columbia University and a master's degree in journalism from Northwestern University. Before joining CNS he served as Editor-in-Chief of *Arms Control Today* from 2003-2009. Previously, he was the lead foreign policy reporter for *CQ Weekly* and *Legi-Slate News Service*, where he covered the full range of national security issues before Congress, and a Foreign Service Officer with the US Information Agency.

Joel Shurkin is a freelance writer in Baltimore, and the author of the recently published *True Genius: The Life and Work of Richard Garwin, The Most Influential Scientist You've Never Heard Of.* He is the author of 10 books on science and the history of science. He was science writer and instructor in journalism at Stanford University and founder of the science internship program. He was a national correspondent for Reuters, a bureau chief for United Press International, and science writer at the Philadelphia Inquirer and was a member of the Inquirer team that won the Pulitzer Prize for covering Three Mile Island. He has taught journalism at Stanford, the University of California Santa Cruz and the University of Alaska Fairbanks.

Leonard S. Spector is Deputy Director of the Middlebury Institute of International Studies' James Martin Center for Nonproliferation Studies and leads the Center's Washington D.C. office. Earlier in his career he served as Assistant Deputy Administrator for Arms Control and Nonproliferation at the National Nuclear Security Administration and founding director of the Nuclear Nonproliferation Project at the Carnegie Endowment for International Peace. His most recent publication proposes strategies for outlawing illicit nuclear procurement networks. Mr. Spector is the author or co-author of eight books and numerous articles on nonproliferation and comments frequently on this subject in the media.

Sharon Squassoni is a senior fellow at the Center for Strategic and International Studies (CSIS) where she has directed the Proliferation Prevention Program since 2010. Her work focuses on reducing nuclear risks, whether from nuclear weapons, nuclear terrorism or nuclear energy. Ms. Squassoni spent the first half of her career in the U.S. government, including as a senior specialist in weapons of mass destruction at the Congressional Research Service and in safeguards and policy planning positions in the State Department and the Arms Control and Disarmament Agency. She is on the Science and Security Board of the Bulletin of Atomic Scientists and on the Board of the Center for Arms Control and Nonproliferation. Her latest article on nuclear energy was "The Incredible Shrinking Nuclear Offset to Climate Change," *Bulletin of Atomic Scientists*, 2017.

Rachel Stohl is a Senior Associate with Stimson's Managing Across Boundaries Initiative and directs the Conventional Defense Program. Her areas of expertise focus on issues relating to the international arms trade, including drones, small arms and light weapons, and children in armed conflict. Prior to joining Stimson, Stohl was an Associate Fellow at Chatham House, the Royal Institute of International Affairs, from 2009-2011. She was a Senior Analyst at the Center for Defense Information in Washington, D.C. from 1998-2009. Stohl has been a consultant for a variety of international organizations, including Oxfam, Project Ploughshares, SIPRI, the Small Arms Survey, and World Vision. She served as a Scoville Fellow at the British American Security Information Council in D.C. and worked at the United Nations Center for Disarmament Affairs in New York and at the Program for Arms Control, Disarmament, and Conversion in Monterey, CA. Stohl is an adjunct professor in the Security Studies Program at Georgetown University. Stohl was the consultant to the U.N. ATT process from 2010-2013 and was previously the consultant to the U.N. Group of Governmental Experts (GGE) on the Arms Trade Treaty in 2008 and the U.N. Register for Conventional Arms in 2009.

Frank von Hippel is a Senior Research Physicist and Professor of Public and International Affairs emeritus at Princeton University where, in 1975, he co-founded what is now Princeton's Program on Science and Global Security and, in 1989, the journal *Science & Global Security* and, in 2006, the non-governmental International Panel on Fissile Materials. During 1993-4, he served as Assistant Director for National Security in the White House Office of Science and Technology Policy and played a major role in developing what was until recently called the International Nuclear Materials Protection and Cooperation Program. In 1991, the AIP published a collection of his policy articles in its "Masters of Modern Physics" series and, in 2010, he was awarded the APS Leo Szilard Award for his "outstanding work and leadership in using physics to illuminate public policy in the areas of nuclear arms control and nonproliferation, nuclear energy, and energy efficiency."

Amy F. Woolf is a Specialist in Nuclear Weapons Policy at the Congressional Research Service at the Library of Congress. She provides Congress with information and analysis on issues related to U.S. and Russian nuclear forces and arms control. She has addressed such topics as nuclear weapons strategy and doctrine and force structure, the U.S-Russian arms control agenda, ballistic missile defense policy, and threat reduction and nonproliferation programs in the former Soviet Union. Before joining CRS, Ms. Woolf was a member of the Research Staff at the Institute for Defense Analyses (IDA) in Alexandria, Virginia. She also spent a year at the Department of Defense, working on the 1994 Nuclear Posture Review.

Peter Zimmerman served as the last Chief Scientist of the Arms Control and Disarmament Agency before it was merged into the Department of State and as the first science adviser to the new arms control bureau. After leaving the Executive Branch, he was the Chief Scientist of the Senate Foreign Relations Committee, when former Vice President Joe Biden was the Committee Chairman. He then became founding director of the Kings College, London Centre for Science and Security Studies until his retirement as Professor Emeritus. He worked at several major international physics laboratories, and was a full professor of physics and astronomy at Louisiana State University. His interests turned towards nuclear arms control during a sabbatical at the University of California, San Diego, working the late Amb. Herbert F. York. In recent years, Dr. Zimmerman's primary interests have been in the study of mass casualty terrorism, both as a scholar and as a government adviser.

Appendix C: Conference Participants

Mark Acton
Andrei Afanasev
David Albright
Salman Al Saud
Bemnet Alemayehu
David Arceneaux
Daniel Bardayan
Michael Bennett
James Borgardt
Mark Bradisse
Jonathon Bush
Derek Boyd
Rachel Carr
Michael Carter
Veronica Cartier
Robert Cavallo
Sung Chang
Jared Chiu-Rourman
Sebastian Chmel
David Clemens
Isidor Coma
Pierce Corden
Arunjana Das
Bruno DeHarak
Samantha D'Dintrono
Alex Dunlop
Gregory Dutton
Sarah Eustace
Tony Fainberg
Jens Fiedler
M'Kai Folley
Bob Forcha
Emily Frame
Josh Frey
Edward Friedman
Elena Gramellini
Gillian Gayner
David Gerber
Graham Giovanetti
Alexander Glaser
Charles Glaser
Martin Glende
Henry Glyde
Bridget Golob
Katelyn Gough
John Gronager
Elena Gramellini
Matthias Grosse-Perdekamp
Hugh Gusterson
David Hafemeister
Theresa Hitchens
Matthias Hohenberger
Carlos Hojvat
Charles Horowitz
Eric Houser
Caroline Hughes
Vincent Ialenti
Edward Ifft
Raymond Jeanloz
Edward Jenner
Stephen Karoly
Togshan Kassenova
Hugh Kendrick
Stevie Kiesel
Daryl Kimball
Frank King
Marcus King
Eric Koboski
Michael Kordosky
Richard Kriske
Hans Kristensen
Joseph Lach
Ben Lamont
Anthony Lauren
Theodore Lavine
Robert Leachman
Alexander Lee
Christina Leehr
George Lewis
Lawrence Lewis
Mark Lewis
Evgenia Lieder
Tammy Ma
Meghan McCall
Miles McCord
Allison Macfarlane
Corey McRae
Laura McVey
Robert Mikulak
Michael Montemalo
Adam Morton
David Mosher
Richard Oakley
Lisa Oakley-Bogdewic
Kathryn Olesko
Joseph Paki
Diane Perlman
George Perkovich
Steven Pifer
Miles Pomper
Donald Raskin
Michael Reilly
August Ridenour
Charles Schlegel
Willi Schwarz
Joel Shurkin
Leonard S. Sector
Junaid Siddiqui
Mel Siegel
William Springer
Sharon Squassoni
Shirvel Stanislaus
Mallory Stewart
Rachel Stohl
Gian Surbella
Derek Thuecks
Eugene Tracy
Akriti Vasudeva
Matthew Walhout
Linton Wells III
Eleanor Williamson
Amy Woolf
Peter Zimmerman

Nuclear Weapons and Related Security Issues
AIP Conf. Proc. 1898, 060003-1–060003-1; https://doi.org/10.1063/1.5012603
Published by AIP Publishing. 978-0-7354-1586-7/$30.00